Ecosystem Functioning and Restoration

Ecosystem Functioning and Restoration

Editor: Francis Lawrence

Callisto Reference

www.callistoreference.com

Callisto Reference,
118-35 Queens Blvd., Suite 400,
Forest Hills, NY 11375, USA

Visit us on the World Wide Web at:
www.callistoreference.com

ISBN: 978-1-63239-831-4 (Hardback)

Printed in China.

Cataloging-in-publication Data

Ecosystem functioning and restoration / edited by Francis Lawrence.
p. cm.
Includes bibliographical references and index.
ISBN 978-1-63239-831-4
1. Biotic communities. 2. Biodiversity. 3. Ecology. I. Lawrence, Francis.
QH541 .E26 2017
577--dc23

Table of Contents

Preface

Ecosystem functioning is the process whereby ecosystems enable the interaction of various organisms in its environment. This book on ecosystem functioning and restoration provides in-depth analysis of various forms of ecosystem management and risk assessment. Timely intervention and analysis can provide adequate shelter and resources for a depleting ecosystem to restore itself. For all readers who are interested in ecosystem functioning and ecological risk assessment, the case studies included in this book will serve as excellent guide to develop a comprehensive understanding. It attempts to understand the multiple branches that fall under this discipline to help develop a better understanding of the latest advances within this area. The topics included in this book on ecosystem functioning and restoration are of utmost significance and bound to provide incredible insights to readers. It is appropriate for students seeking detailed information in this area as well as for experts.

It is often said that books are a boon to mankind. They document every progress and pass on the knowledge from one generation to the other. They play a crucial role in our lives. Thus I was both excited and nervous while editing this book. I was pleased by the thought of being able to make a mark but I was also nervous to do it right because the future of students depends upon it. Hence, I took a few months to research further into the discipline, revise my knowledge and also explore some more aspects. Post this process, I began with the editing of this book.

I thank my publisher with all my heart for considering me worthy of this unparalleled opportunity and for showing unwavering faith in my skills. I would also like to thank the editorial team who worked closely with me at every step and contributed immensely towards the successful completion of this book. Last but not the least, I wish to thank my friends and colleagues for their support.

Editor

1

An Ecosystem-Based Approach to Assess the Status of a Mediterranean Ecosystem, the *Posidonia oceanica* Seagrass Meadow

Sébastien Personnic[1]*, Charles F. Boudouresque[1], Patrick Astruch[2], Enric Ballesteros[3], Sylvain Blouet[4], Denise Bellan-Santini[5], Patrick Bonhomme[2], Delphine Thibault-Botha[1], Eric Feunteun[6], Mireille Harmelin-Vivien[1], Gérard Pergent[7], Christine Pergent-Martini[7], Jérémy Pastor[8], Jean-Christophe Poggiale[1], Florent Renaud[5], Thierry Thibaut[1], Sandrine Ruitton[1]

1 Aix-Marseille University, Mediterranean Institute of Oceanography (MIO), Université de Toulon, CNRS/INSU, IRD, UM 110, Marseille, France, **2** GIS Posidonie, Pytheas Institute, Aix-Marseille University, Marseille, France, **3** Centre d'Estudis Avançats de Blanes - CSIC, Blanes, Spain, **4** Aire marine protégée de la côte Agathoise, site natura 2000, Agde, France, **5** Aix-Marseille University, Institut Méditerranéen de Biodiversité et d'Ecologie (IMBE), UMR 7263, Station Marine d'Endoume, Marseille, France, **6** Museum National d'Histoire Naturelle, UMR 7208, Station Marine de Dinard, France, **7** Equipe Ecosystèmes Littoraux, FRES 3041, University of Corsica, Corte, France, **8** Université de Perpignan, Via Domitia, Centre de Formation et de Recherche sur les Environnements Méditerranéens, UMR 5110, Perpignan, France

Abstract

Biotic indices, which reflect the quality of the environment, are widely used in the marine realm. Sometimes, key species or ecosystem engineers are selected for this purpose. This is the case of the Mediterranean seagrass *Posidonia oceanica*, widely used as a biological quality element in the context of the European Union Water Framework Directive (WFD). The good quality of a water body and the apparent health of a species, whether or not an ecosystem engineer such as *P. oceanica*, is not always indicative of the good structure and functioning of the whole ecosystem. A key point of the recent Marine Strategy Framework Directive (MSFD) is the ecosystem-based approach. Here, on the basis of a simplified conceptual model of the *P. oceanica* ecosystem, we have proposed an ecosystem-based index of the quality of its functioning, compliant with the MSFD requirements. This index (EBQI) is based upon a set of representative functional compartments, the weighting of these compartments and the assessment of the quality of each compartment by comparison of a supposed baseline. The index well discriminated 17 sites in the north-western Mediterranean (French Riviera, Provence, Corsica, Catalonia and Balearic Islands) covering a wide range of human pressure levels. The strong points of the EBQI are that it is easy to implement, non-destructive, relatively robust, according to the selection of the compartments and to their weighting, and associated with confidence indices that indicate possible weakness and biases and therefore the need for further field data acquisition.

Editor: Arga Chandrashekar Anil, CSIR- National institute of oceanography, India

Funding: This study has been funded by the French Centre National de la Recherche Scientifique (CNRS), the French Réseau des Stations et Observatoires Marins (RESOMAR), the Muséum National d'Histoire Naturelle (MNHN, Paris), the French Agence de l'eau Rhône Méditerranée Corse through the French part of the European MSFD project (in French: Directive Cadre Stratégie pour le Milieu Marin, DCSMM). The funders had no role in study design, data collection and analysis, decision to publish, or preparation of the manuscript.

Competing Interests: The authors have declared that no competing interests exist.

* E-mail: sebastien.personnic@univ-amu.fr

Introduction

Human activities can deeply alter the environment, species composition and functioning of ecosystems. These alterations can be tracked by the use of biotic indices, i.e. species or groups of species whose function, population, or status reflect the environmental quality. Thus, biotic indices are monitored for changes in presence and abundance. The occurrence of an organism in a specific environment indicates that overall, its biological requirements are satisfied, whereas its disappearance suggests a change in the environment. Species are also monitored for changes in biochemistry, physiology or behaviour induced by environmental conditions. Biotic indices are used in terrestrial, freshwater and marine habitats, because they enable the quality of an environment to be characterized in an integrated way [1–7].

Biotic indices are widely used in the marine realm to **(i)** assess the quality of a water body, **(ii)** assess processes such as currents, sedimentation and climate under natural and anthropogenic forcing, and **(iii)** monitor the status of species or communities of interest, either emblematic species, indicators of some ecosystemic processes or indicators of pollution. Sometimes, key species and ecosystem engineers [8–10] are selected for this purpose. This is the case of the seagrass *Posidonia oceanica* [11–17].

In the European Union (EU), the so called Habitats Directive of 1992 (92/43/ECC) listed habitats and species that are used to designate areas ('Natura 2000 sites'), where they are strictly protected. While the Habitats Directive also considered the marine realm, more recently, the EU Marine Strategy Framework Directive (MSFD: 2008/56/EC) established a framework for conservation in the field of marine environmental policy. The

MSFD is considered to be the environmental pillar of the Integrated Maritime Policy adopted in 2010 by the European Commission (2010/477/EU). This directive established eleven criteria, based on the descriptors set out in the MSFD, to determine 'good environmental status': **(i)** Biological diversity is maintained, **(ii)** Introduced species are at levels that do not adversely alter the ecosystems, **(iii)** Populations of all exploited fish and shellfish are within safely biological limits, **(iv)** All elements of the food webs occur at levels capable of ensuring the long-term abundance of the species, **(v)** Human-induced eutrophication is at a minimum, **(vi)** Sea-floor integrity is at a level that ensures that the structure and functions of the ecosystems are safeguarded, **(vii)** Permanent alteration of hydrographical conditions does not adversely affect marine ecosystems, **(viii)** Concentration of contaminants are at levels not giving rise to pollution effects, **(ix)** Contaminants in fish and other seafood do not exceed levels established by Community legislation, **(x)** Properties and quantities of marine litter do not cause harm to the coastal and marine environment, and **(xi)** Introduction of energy is at levels that do not adversely affect the marine environment. The EU MSFD established a framework within which Member States agreed to take the appropriate measures to achieve or maintain good environmental status in the marine realm by the year 2020 at the latest.

A key point of the MSFD is the ecosystem-based approach. Most previous strategies only dealing with Biological Quality Elements (BQE: species and/or communities) used in the European Union Water Framework Directive (WFD) were not indicative of ecosystem status, but merely indicative of the environmental status of water bodies [18–22,17,23] for a review of the approaches for classifying and assessing quality of benthic habitats. As an example of the possible shortcomings that may arise from the WFD descriptors, a seagrass meadow characterized by normal leaf growth, shoot density and the absence of mechanical injuries would be ranked as 'good', even if deprived of some basic compartments of the ecosystem.

According to the MSFD, 'good environmental status' means that the marine environment is at a level that allows uses and activities by current and future generations, i.e. the structure, functions and processes of the constituent marine ecosystems, together with the associated physiographic, geographic, geological and climatic factors, allow those ecosystems to function fully and to maintain their resilience to human induced environmental change. Overall, marine species and marine habitats are protected, human-induced decline of biodiversity is prevented and diverse biological components function in balance.

While the ecosystem-based approach constitutes the background and the guidelines of the MSFD, the pristine state of an ecosystem is not clearly defined. Obviously, such a notion is very complex and prone to anthropocentrism; reference conditions, as observed in areas distant from known human impact, may constitute a more realistic notion. Here we try to define: **(i)** how the status of an ecosystem can be measured and how much its current status differs from 'reference conditions', and **(ii)** which parameters we need to monitor to ascertain these targets. The *Posidonia oceanica* seagrass meadow was chosen for this attempt because: **(i)** it is widely present in almost the whole of the European Mediterranean; **(ii)** it is the only marine ecosystem considered as 'priority habitat' by the EU Habitats Directive; **(iii)** its functioning is relatively well known [11,24–30]; and **(iv)** like many seagrass ecosystems in the world ocean, *P. oceanica* meadows have been impacted or lost under the influence of direct and indirect effects of human activities and are therefore regarded as threatened [31–33]. Similar ecosystem-based approaches have been attempted for fisheries (e.g. [34–38,40,41]) and for the management and conservation of ecosystem services (e.g. [34,41,42]).

Materials and Methods

The conceptual model

Posidonia oceanica (L.) Delile is a seagrass species (Magnoliophyta, kingdom Archaeplastida) endemic to the Mediterranean Sea [43], which dwells in the sublittoral zone, from mean sea level down to 30 to 40 m depth, depending upon water transparency [27,28,32]. Due to the length, up to 120 cm, and density of the leaves, the seagrass canopy decreases water movement and traps sediments [44–47]. Rhizomes resist burial by vertical growth, so that the sea bottom slowly rises. Within the sediment, the deeper parts of the rhizomes, attached leaf sheaths and roots die, but their decay is extremely slow, so that they can persist for millennia [24,27]. The terrace constituted by live and dead intertwined rhizomes, together with the sediment, which fills the interstices, is named 'matte' [27,48,49]. When *P. oceanica* dies, the matte persists (hereafter 'dead matte'), since the decay of the rhizomes proceeds extremely slowly [44,50]. *P. oceanica* is the engineer of an ecosystem of major importance in the Mediterranean Sea [27,28]. A conceptual model of the functioning of the *P. oceanica* ecosystem has been proposed [11,27,28]. Here, we use an updated version of this conceptual model (Fig. 1; C.F. Boudouresque, unpublished).

The simplified conceptual model of the functioning of the *P. oceanica* ecosystem in the north-western Mediterranean Sea used in the present study (Fig. 1) encompasses the following compartments (boxes); these compartments are listed hereafter and detailed later on:

- *Posidonia oceanica* roots and rhizomes (box 1).
- Multicellular Photosynthetic Organisms (MPOs) epibiotic on *P. oceanica* rhizomes.
- *P. oceanica* leaves (box 2).
- MPO leaf epibiota (box 3).
- Dissolved Organic Carbon (DOC).
- Pelagic microbial loop.
- Filter- and suspension-feeder leaf epibiota (box 4).
- Filter- and suspension-feeder benthic epibiota on *P. oceanica* rhizomes: the bivalve *Pinna nobilis* (box 5) and other species, e.g. bryozoans and ascidians (box 6).
- The litter detritus (essentially dead *P. oceanica* leaves and some broken rhizomes) (box 7).
- Detritus feeders 1, e.g. Amphipoda, Isopoda (crustaceans) and *Psammechinus microtuberculatus* (sea urchin).
- Detritus feeders 2 and 3, e.g. Amphipoda, Isopoda (crustaceans) and *Holothuria* spp. (sea cucumber) (box 8).
- Decomposers, namely *Bacteria*, *Archaea*, Fungi and heterotrophic stramenopiles (BAFHS) such as Labyrinthulomycota and Oomycota.
- The benthic microbial loop.
- The matte endofauna, e.g. annelids and mollusks.
- Herbivores 1, e.g. *Sarpa salpa* (teleost), *Paracentrotus lividus* (sea urchin), *Idotea* spp. and *Pisa* spp. (crustaceans) (box 9).
- Herbivores 2, e.g. Amphipoda (crustaceans), *Jujubinus* spp. and *Rissoa* spp. (gastropods).
- Predatory teleosts (e.g. *Diplodus* spp., *Sparus aurata*, *Labrus* spp. and *Symphodus* spp.), cephalopods and seastars (e.g. *Marthasterias glacialis*) (box 10).

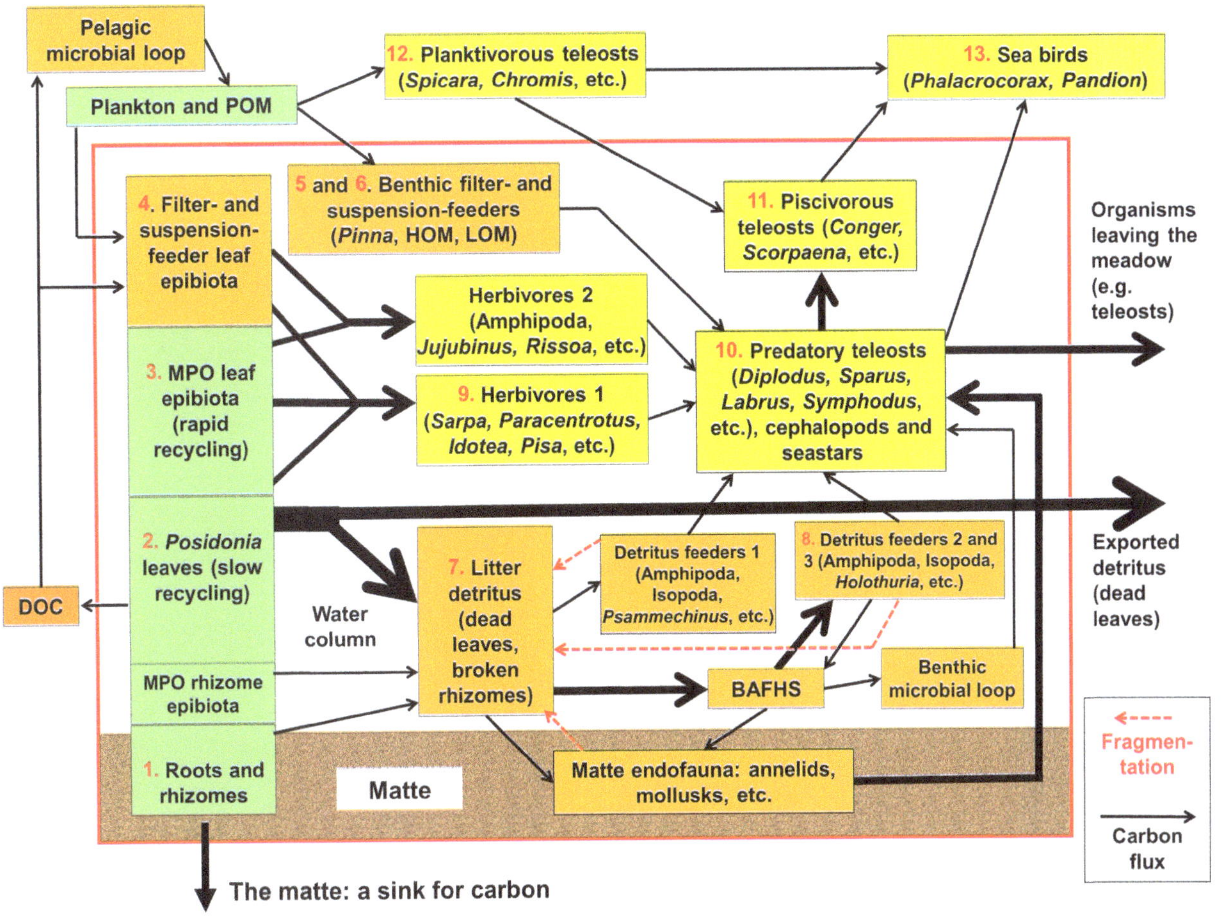

Figure 1. Conceptual model of the functioning of *Posidonia oceanica* seagrass ecosystem. For functional compartments and box numbers, see text. Primary producers are in green; filter-feeders, suspension-feeders, litter, detritus feeders, Dissolved Organic Carbon (DOC) and microbial loops are in orange; predators (including herbivores) are in yellow. The width of the arrows roughly represents the importance of the carbon flow. The proper *P. oceanica* ecosystem is included within the red rectangle. MPO: Multicellular Photosynthetic Organisms. POM: Particulate Organic Matter. From C.F. Boudouresque, unpublished.

- Piscivorous teleosts, e.g. *Conger conger*, *Scorpaena* spp. and *Serranus* spp. (box 11).
- Planktivorous teleosts of the water column, e.g. *Spicara* spp. and *Chromis chromis* (box 12).
- Sea birds, e.g. *Phalacrocorax aristotelis* ssp. *desmarestii* and *Pandion haliaetus* (box 13).
- Plankton (photosynthetic plankton and zooplankton) and non-living Particulate Organic Matter (POM).

Considered functional compartments (boxes)

Whenever possible, non-destructive methods were chosen to measure the parameters of the status of the functional compartments, as suggested by [51]. In the absence of further indications, measures are performed at a depth of ca. 15 meters (10 to 20 m), which is then considered as representative of the whole depth range [14,52]. The season of the sampling is indicated only for compartments which present seasonal variability. In the ranking of the status of the selected compartments (boxes), two cases were encountered; **(i)** a steady trend of the parameter, from very good to a low quality state; **(ii)** an upward/downward slope parameter, when the very good state corresponds to intermediate values.

***Posidonia* roots and rhizomes.** (box 1, Fig. 1). This compartment was estimated by the growth rate of vertical (orthotropic) rhizomes. The matte compartment results in carbon sequestration within the matte, which acts as a carbon sink [27,44,53,54] and was measured by means of lepidochronology [55,56]. Lepidochronology describes the annual cycle of leaf formation. A cycle includes a suite of dead leaf bases (called dead sheaths or scales) of increasing then decreasing thickness. An average of 7.5 leaves (generally between 6 and 9) are produced every year [57–60]. We considered that high and low rhizome growth rates are indicative of over sediment input or of deficit in sediment, respectively (Table 1). The highest growth rate of an orthotropic rhizome is 7 cm year^{-1}; higher sedimentation rates result in the death of the buried leaf bundle [44,61]. In contrast, deficit in sediment results in bare, non-sediment protected rhizomes, which are hence very vulnerable to storms, anchoring and trawling [27]. The nature of the substratum, namely meadows settled on rock, sand or matte, only slightly influences the growth rate of rhizomes [62] and no major differences were found at the community level [63] so that it was not necessary to adapt the scale. Thirty random *in situ* measures (growth of the rhizome corresponding to the last 8 dead leaf bases) are recommended. The obtained value was multiplied by 1.5 in order to take into account

Table 1. Relative weighting of each functional compartment (boxes; see Fig. 1 for box number) and ranges of each parameter defined for each grade of ecosystem status.

Box number	Functional compartment	Weighting (W)	Parameter	4	3	2	1	0
1	Roots and rhizomes ('rhizomes')	3	Growth of orthotropic rhizomes (mm a^{-1})	9 to 19	20 to 40	3 to 8	>40	<3
2	*Posidonia oceanica* leaves ('leaves')	5	- Density (shoots m^{-2})	≥490	489 to 370	369 to 250	249 to 130	<130
			- Cover (%)	>80	80 to 61	60 to 41	40 to 20	<20
3–4	MPOs, filter- and suspension-feeders leaf epibiota ('leaf epibiota')	4	Biomass (g DM $shoot^{-1}$) (only the 2 oldest - external - leaves)	0.3 to 0.7	0.1 to 0.2	0.8 to 1.5	<0.1	>1.5
5	Benthic filter-feeder: *Pinna nobilis* (bivalve) ('Pinna')	2	Density (individuals 100 m^{-2})	>3.0	3.0 to 1.1	1.0 to 0.6	0.5 to 0.1	<0.1
6	Other benthic filter- and suspension-feeders ('HOM LOM')	2						
	- HOM		- Density (m^{-2})	<0.1	0.1 to 0.9	1.0 to 1.9	2.0 to 5.0	>5.0
	- LOM		- Density (m^{-2})	>10.0	10.0 to 5.1	5.0 to 1.1	1.0 to 0.1	<0.1
7	Litter detritus: dead leaves and broken rhizomes ('litter')	2	g DM m^{-2}	>350	350-251	250-151	150-51	≤50
8	Detritus-feeders 2 and 3 (*Holothuria* spp.) ('Holothuria')	2	Individuals 10 m^{-2}	1.0 to 4.9	0.2 to 0.9	5.0 to 24.9	<0.2	≥25.0
9	Herbivores 1 ('herbivores')	5	- Density of *Paracentrotus lividus* (individuals m^{-2})	1.0 to 4.9	0.1 to 0.9	5.0 to 9.9	<0.1	≥10
			- Grazing index (% leaves)	30 to 59%	5 to 29%	60 to 95%	<5%	>95%
10	Predatory teleosts, cephalopods and seastars ('predators')	5	kg teleosts WM 100 m^{-2}	>1.5	1.5 to 1.1	1.0 to 0.6	0.5 to 0.3	<0.3
11	Piscivorous teleosts ('piscivores')	5	kg teleosts WM 100 m^{-2}	>1.0	1.0 to 0.6	0.5 to 0.3	0.2 to 0.1	<0.1
12	Planktivorous teleosts ('planktivores')	3						
	- Zooplankton feeders		- kg teleosts WM 100 m^{-2}	>3.0	3.0 to 1.6	1.5 to 0.8	0.7 to 0.3	<0.3
	- Omnivores		- kg teleosts WM 100 m^{-2}	>3.0	3.0 to 1.6	1.5 to 0.8	0.7 to 0.3	<0.3
9–12	All teleosts ('SRDI')	3	Specific Relative Diversity Index (SRDI)	>10	10 to 8	7 to 5	4 to 3	<3
13	Sea birds	1						
	- *Phalacrocorax* spp.		- Distance to the nearest nesting site (km)	<4	4 to 7	8 to 12	13 to 17	>17
	- *Pandion haliaetus*		- Distance to the nearest nesting site (km)	<4	4 to 7	8 to 12	13 to 17	>17

Between inverted commas: abbreviated name of the functional compartment in tables 4 and 7. DM: dry mass, including calcium carbonate. HOM: indicators of high level of organic matter. LOM: indicators of low level of organic matter. MPO: Multicellular Photosynthetic Organisms. POM: Particulate Organic Matter. WM: wet mass. Status: 4 (very good) through 0 (very low).

the fact that the rhizome continues to grow slowly during the following two years (Gérard Pergent, unpublished data).

***Posidonia* leaves.** (box 2, Fig. 1). A shoot is a rhizome tip with a bundle of living leaves. Shoot density is correlated with annual leaf primary production at local scale (patch) [64,65]. Primary production is a basic parameter for the functioning of the *P. oceanica* ecosystem. It was estimated by the number of shoots per square meter, measured within a small square frame (0.16 m^2 [66,67]) with at least 20 random replicates [5]. At a less local scale, the cover rate of the meadow is rarely 100%: it is broken by more or less extensive patches of either sand or dead matte, which reduce the overall cover. Cover rate was estimated by visual observation *via* a see-through plastic slide [68,69] or *via* vertical photographs [67,70]. Thirty random measures are recommended. Low cover rate is believed to characterize a poor condition of the meadow [5]. Some types of *P. oceanica* meadow, e.g. the hill meadow and the striped meadow, exhibit a low cover rate while in pristine state; these types of meadows do not, however, occur in the study area [27]. The final *P. oceanica* leaf index, for a given site, was the arithmetical mean between the density index, and the cover index.

MPOs, filter- and suspension-feeder leaf epibiota. (boxes 3 and 4, Fig. 1). *P. oceanica* leaf epibiota, both primary producers (MPOs and diatoms) and animals, share the same habitat (the leaf surface) and are usually co-consumed by the same species. In addition, some consumers eat simultaneously epibiota and the supporting leaf [68,71]. For these reasons, epibiota will be considered here as a single compartment. The colonisation of leaves by epibiota is a function of leaf age, the youngest leaves, in the center of the shoot, being less colonized, while the oldest, external leaves are the most colonized; in addition, leaf tips are more colonized than the lower parts of the leaves [72]. Leaf epibiota cover is believed to provide information on water quality, especially nutrient concentration in seawater [5,73,74]; however, it also reflects the herbivore pressure, epibiota biomass decreasing when macrograzer abundance increases [26]. By convention, the epibiota biomass was estimated on the two oldest (external) leaves, in July, on 30 randomly localized shoots (Table 1).

Filter- and suspension-feeder epibiota on *P. oceanica* rhizomes. (boxes 5 and 6, Fig. 1). A number of benthic filter- and suspension-feeders dwell on *P. oceanica* rhizomes, sometimes within the matte. They belong to bryozoans, hydroids, sponges, annelids (e.g. *Sabella spallanzani*), ascidians (e.g. *Halocynthia papillosa*, *Phallusia mammillata*, *P. fumigata*), gastropods and bivalves (such as the fan mussel *Pinna nobilis*) [75–79]. *Pinna nobilis* (box 5) density was estimated along 20 transects 10-m long and 1-m wide. Filter- and suspension-feeders other than *P. nobilis* (box 6) are indicators either of: **(i)** high level of organic matter in the water (hereafter HOM; e.g. *Sabella spallanzani*, *Phallusia mammillata*, *P. fumigata*, Didemnidae [80]); or **(ii)** low level of organic matter (hereafter LOM; e.g. bryozoans, sponges, *Halocynthia papillosa*, *Antedon mediterranea* [80]). HOM and LOM indicators are assessed within the same quadrats as the sea urchin *P. lividus* (see below, box 9). For colonial species, the number of colonies was taken into account, as suggested by [81]. For non-colonial species, the number of individuals was counted. Only individuals and colonies over 5 cm in diameter and/or height were considered. The final filter- and suspension-feeder epibiota (other than *P. nobilis*; box 6) on rhizomes index, for a given site, was estimated as the arithmetical mean between HOM and LOM indices.

Litter detritus. (box 7, Fig. 1). The litter detritus mass corresponds essentially to shed dead *P. oceanica* leaf blades and the epibiota they harbour, and some broken rhizomes. It therefore represents a kind of necromass [82]. Drift MPOs, exported from sublittoral reef habitats, can also occur within the litter. The litter detritus mass was estimated in July, within 5 randomly localized 0.1 m^2 quadrats; the litter was sucked up by an underwater vacuum cleaner. Litter detritus were dried at 50°C in an oven to constant weight.

Detritus-feeders 2 and 3. (box 8, Fig. 1). Detritus-feeders constitute a complex set of compartments. Here, the macro-detritus feeders *Holothuria* spp. were used as a proxy of detritus-feeders 2 and 3, as they are easy to count. Several species can be present, e.g. *H. polii* and *H. tubulosa* [83]. The abundance of *Holothuria* spp. was measured within the same quadrats as the sea urchin *P. lividus* (see below, box 9).

Herbivores 1. (box 9, Fig. 1). Macro-herbivores considered in this compartment were the sea urchin *Paracentrotus lividus* and the teleost *Sarpa salpa*, at 5 meters depth or, if the meadow is not present at this depth, at the upper limit of the meadow. They consume *P. oceanica* leaves, together with their epibiota, if present [68,71,84]. Other herbivores graze *P. oceanica* leaves, such as the spider crabs *Pisa* spp. and the isopod *Idotea hectica* [27,84], but their reduced size and habit make them more difficult to quantify and they were not considered. The abundance of *P. lividus* was assessed within 1-m^2 quadrats, with 30 replicates randomly localized. The census only considered individuals >3 cm (test diameter without spines), because small individuals can be hidden within the matte, generating bias in the census. A second proxy of the macro-herbivore pressure was the grazing index, i.e. the percentage of intermediate and adult leaves (*sensu* [85,86]) exhibiting bite scars due to *S. salpa* (all the intermediate and adult leaves of 30 shoots randomly localized). Bite scars were carefully distinguished from broken leaves, the latter being related to hydrodynamism. Most bite scars are due to the *S. salpa* browsing, and they are easy to distinguish from those due to other macro-herbivores [26,49]. The final macro-herbivore index was, for a given site, the arithmetical mean between the *Paracentrotus* index and the grazing index.

Predatory teleosts and cephalopods, piscivorous teleosts, planktivorous teleosts. (boxes 9 in part, 10, 11 and 12, Fig. 1). Teleost fishes associated with seagrass beds occupy different positions in and above the canopy during the day and at night, spend more or less time in this habitat depending on their life cycle, and naturally fluctuate in abundance and biomass with depth, season and years [87–90]. They participate actively in the functioning of the *P. oceanica* ecosystem (Fig. 1), but the perception of the composition and trophic structure of their assemblages largely depends on the methodology used [91]. Moreover, species richness, abundance and biomass of teleosts are favoured by the ecotones induced by the presence of rocky or sandy substrates in the middle of *Posidonia* beds. So these variations and methodological biases have to be taken into account to when measuring these compartments in order to assess the 'environmental status' of a site. These compartments were estimated *via* visual censuses at a standardized day time (10:00 to 16:00 UT) during the warm season (summer-autumn), preferentially in uniform beds (at least a long way from rocky substrates). All teleosts were counted within ten linear and 5-m wide transects, each census lasting 5 minutes. Around 50 m were covered so that each transect represents a surface area of nearly 250 m^2. Total length (at the nearest 2 cm) of individuals and the number of individual per species were noted. The Specific Relative Diversity Index (SRDI) is the mean number of species met with per transect. These data enable calculation of **(i)** predator biomass (predatory teleosts other than piscivorous and planktivorous; box 10), **(ii)** top predator biomass (piscivorous teleosts; box 11), **(iii)** planktivorous teleost biomass (box 12) and **(iv)** the Specific Relative Diversity Index (SRDI). The planktivorous teleost biomass is divided into 2 categories: the exclusive zooplankton feeders (*Chromis chromis*, *Spicara smaris*, *S. maena*, *Atherina* spp.) and the omnivorous feeders, which consume both zooplankton and POM (*Boops boops*, *Oblada melanura*). Some top predators (e.g. *Conger conger*, *Scorpaena* spp.) are active only during night time while *Serranus* spp. are active by day and thus more present in our visual counts. Some predators (e.g. *Symphodus rostratus*), though diurnal, are often hidden within the *P. oceanica* leaf canopy. As a result, their counts were underestimated to a greater or lesser extent. The parameter ranges within the status scale (Table 1) took into account these biases.

Sea birds. (box 13, Fig. 1). Most sea birds do not directly interact with the *P. oceanica* ecosystem, as they feed on offshore pelagic species, such as *Larus* spp. and *Puffinus* spp. The only exceptions are shags *Phalacrocorax* spp. and the osprey *Pandion haliaetus*. Shags can dive down to the benthic seagrass meadow; they mainly feed on pelagic planktivorous teleosts (e.g. *Spicara smaris*, *Chromis chromis*), but benthic teleosts (e.g. *Diplodus* spp., *Lithognathus mormyrus*, *Scorpaena notata*, *Serranus scriba* and *Symphodus mediterraneus*) account for 35% of the diet [92–94]. Osprey is an opportunistic fish-eating bird of prey. In Corsica, which harbours the only population within the study area, it mainly consumes mugilids (73% of the captures), together with *Diplodus* spp. (13%) and *Sarpa salpa* (11%) [95–97]. The sea bird compartment was estimated *via* the distance of the nearest *Phalacrocorax* spp. and *Pandion haliaetus*, respectively, nesting sites from the study site. As far as shags are concerned, the mean maximum foraging range is 16 km [95–97] (Table 1).

Some of the above-mentioned compartments correspond to inputs into the *P. oceanica* ecosystem from the pelagic habitat: plankton, POM and planktivorous teleosts. Outputs are also to be considered: **(i)** ca. 15–30% of the net primary production (NPP) corresponds to roots, rhizomes and dead sheaths buried and sequestrated within the matte [24,28,98,99]; **(ii)** ca. 6 to 50% of the NPP represents dead leaves that are exported as detritus towards beaches and adjacent habitats [28,31,98,100,101]; and **(iii)** a number of organisms (e.g. teleosts, crustaceans) leave the meadow, either temporarily, to feed in adjacent habitats, or permanently, to reach their adult quarters [87,102–104].

The Ecosystem-Based Quality Index (EBQI)

The rationale governing our ecosystem-based approach is trying to quantify and assess each compartment (box) of the conceptual model by means of a set of parameters, to balance their relative weighting and by using a simple algorithm to calculate a rank for the ecosystem status within a given area, matching the five Ecological Statuses of the Water Framework Directive (WFD), from bad to high. A great variety of parameters are available for the assessment of each compartment (box). Many of them are redundant. Others have been poorly used so that data are not available for most areas. For this reason a limited set of relevant parameters was selected for a restricted set of compartments.

Each parameter was assessed by means of a semi-quantitative scale (4 through 0), from very good (4) to very low (0). Calibration of the scale was based upon the available literature (e.g. [105]), including grey literature and expert judgment (the personal knowledge of the authors) based upon a Delphi process [106]. The highest grade (4) corresponds to the ecosystem status in the best-protected areas of well implemented MPAs, e.g. the Medes Islands reserve (Catalonia, Spain), the Port-Cros National Park (continental France), the Scandola reserve and the Bouches de Bonifacio reserve (Corsica, France).

Boxes were balanced, according to their relative weighting (W) in the ecosystem functioning, from 5 (highest weighting) to 1 (lowest weighting). The general principle in the ranking of the weighting of a box was that the boxes localized at the very base (bottom up control by primary producers), the herbivores (box 9) and the boxes localized at the very top (top down control via cascade effect) of the model were regarded as of major importance (with the exception of box 13), while intermediate ones were less weighted. Wasp-waist control has not been evidenced in that ecosystem [11,107]. The grade of each considered box was given by multiplying its status S (0 through 4) and weighting W (1 through 5), and therefore they are graded from 0 through 20. The grades of all considered boxes were added up, which gave the final grade of the ecosystem status (Ecosystem Based Quality Index, hereafter EBQI) at a given site. For practical purposes, the EBQI was converted to a scale from 0 to 10:

$$EBQI = \left[\sum_{i=1}^{13} (W_i \times S_i) / \sum_{i=1}^{13} (W_i \times S_{\max}) \right] \times 10$$

Where: W_i is the weighting of the box i, S_i the status of the box i, S_{max} the highest possible grade (= 4) for a box and i is the number of the box (1 through 13).

Five ecological status classes, from high to bad, according to the practice of the WFD, were delineated: bad (EBQI<3.5), poor (3.5≥EBQI<4.5), moderate (4.5≥EBQI<6), good (6.0≥EBQI<7.5) and high (EBQI≥7.5).

Since the box weightings were supported by partly subjective arguments (see above), we aimed to analyse the effect of the weighting choice on the EBQI and the ranking of the sites. In order to achieve our aim, we perturbed each weighting value and determined the new ranking obtained with the perturbed weightings. The perturbation on each weighting was obtained as follows. We first defined the maximum amplitude of the perturbation for all weightings and we then defined, for each weighting, a random perturbation, according to a uniform law between 0 and the maximum amplitude of perturbations. We then added or subtracted this perturbation term to or from the corresponding weighting. If the new weighting was less than 1, it was then set as equal to 1. If the new weighting was greater than 5, it was then set as equal to 5. We ended up with weightings between 1 and 5, close to the original ones if the maximum perturbation amplitude was low, and randomly selected otherwise. This has been calculated as follows: the perturbation method described previously was repeated 1,000 times and, for each site, we determined whether the rank of the site was the same as the initial one or if it had changed. An index of similarity was produced for each site, which was equal to 100% when the rank of the site was always unchanged after perturbation and 0% if it always changed.

For each compartment (box) status at each site, a Confidence Index (CI) was proposed (Table 2). The reason for the CI is **(i)** that data for one or several compartments may be missing or of poor quality in some sites, **(ii)** the reliability of available data may be different between compartments and sites, and **(iii)** it is worth drawing the attention of managers and scientists to those compartments that are poorly known and which merit further field studies. The grade of each considered compartment was given by its CI (0 through 4) and by its weighting (1 through 5), and therefore they are graded from 0 through 20. The grades of all considered compartments were added up, which gave the final mark of the CI at a given site. For practical purposes, the CI was converted to a scale from 0 to 4:

Table 2. Criteria to assess the Confidence Index (CI) of the status of a compartment (box).

CI	Criteria
4	Field data available, recent and suitable with the recommended methods
3	Field data recent, partially completed with expert judgment
2	No quantitative field data but recent expert judgment
1	No quantitative field data, but ancient expert judgment
0	No quantitative field data and no suitable expert judgment

$$CI_{EBQI} = \left[\sum_{i=1}^{13}(W_i \times CI_i) / \sum_{i=1}^{13}(W_i \times CI_{\max})\right] \times 4$$

Where W_i is the weighting of the box i, CI_i the Confidence Index of the box i, CI_{max} the highest possible Confidence Index (= 4) for a compartment (box) and i is the number of the box (1 through 13).

In order to test the efficiency of the proposed method, it was applied to seventeen sites (Table 3) with a variety of available data (published, unpublished, expert judgment). A site is defined as a *P. oceanica* meadow, from its upper limit down to the lower limit, covering in the order of a dozen to several hundred hectares. The sites are localized in the north-western Mediterranean Sea, including continental France (French Riviera, Provence, French Catalonia), Corsica, Balearic Islands and Spanish Catalonia. This area is considered as homogenous and is consistent with the marine subregions, as defined in the MSFD. These sites also cover of a wide range of human pressures, from lesser impact, within well implemented Marine Protected Areas (MPAs), such as the Port-Cros National Park, to highly impacted areas due to different disturbance and/or stress sources (pollution, overfishing, fish farms, port facilities, anchoring and mooring; see table 3).

Results

EBQI assessment

The results of the EBQI application for assessing the status of the *P. oceanica* ecosystem functioning in the sites across the NW Mediterranean area (Table 3) are presented in Table 4. An example of calculation of the EBQI is given (Table 5). The content of Table 4 is based upon unpublished observations from the authors of the present article, together with some published data [108–111] and whenever necessary (CI<4) on expert judgment; when several co-authors of the present work were involved in the assessment of a site, the Delphi method [106] was used for assessing the status of each compartment (box). According to the EBQI, the 17 study sites were placed within five Ecological Status classes, from Bad to High:

- Bad (EBQI<3.5): Sitges.
- Poor (3.5≥EBQI<4.5): Niolon, Gulf of Giens and Porquerolles Island northern coast.
- Moderate (4.5≥EBQI<6): Villefranche Bay, Saint-Cyr Bay, Plateau des Chèvres (Marseilles), Prado Bay (Marseilles), Valincu Gulf, Tossa de Mar, Scandola (Elbu Bay) and Peyrefite Bay.

Table 3. Sites used to test the proposed ecosystem-based approach to assess the status of the *P. oceanica* ecosystem. MPA: Marine Protected Area. NTZ: No Take Zone.

Site	Region	Protection status	Pressure
Espardell	Balearic Islands (Spain)	MPA, Natura 2000	Artisanal fishery
Sitges	Spanish Catalonia	Natura 2000	Pollution, artisanal and recreational fishery, sedimentation
Tossa de Mar	Spanish Catalonia	Natura 2000	Artisanal and recreational fishery
Medes Islands	Spanish Catalonia	MPA, NTZ, Natura 2000	River mouth
Peyrefite Bay	French Catalonia	MPA[a]	Artisanal fishery, anchoring[b]
Niolon (Côte Bleue)	Provence (France)	MPA, Natura 2000	River mouth, artisanal fishery, trawling
Prado Bay, Marseilles	Provence (France)		Coastal development, artisanal fishery, nutrients
Plateau des chèvres, Marseilles	Provence (France)	MPA[c], Natura 2000	Artisanal fishery, sewage outfall,
Saint Cyr Bay	Provence (France)		Coastal development, artisanal fishery, sewage outfall
Gulf of Giens	Provence (France)		Sewage outfall, artisanal fishery, trawling
Porquerolles Island, northern coast	Provence (France)	MPA, Natura 2000	Trawling, artisanal fishery, anchoring
Porquerolles Island, southern coast	Provence (France)	MPA, Natura 2000	Trawling, artisanal fishery
Bagaud Pass, Port-Cros Island	Provence (France)	National Park, MPA, Natura 2000	Artisanal fishery, anchoring
Port-Cros Island, southern coast	Provence (France)	National Park, MPA, Natura 2000	Artisanal fishery
Villefranche-sur-Mer Bay	French Riviera		Coastal development, sewage outfall artisanal fishery, anchoring
Scandola, Elbu Bay	West Corsica (France)	MPA, Natura 2000	Artisanal fishery
Valincu Gulf	West Corsica (France)		Artisanal fishery

[a]This area is close to the Natural Marine Reserve of Cerbère-Banyuls. Since October 2011, this area has been included within a natural marine park ('Parc naturel marin du golfe du Lion').
[b]Since 2010, anchoring is banned.
[c]Since May, 2012, this area has been included within a National Park ('Parc national des Calanques'). However, it is unlikely that this new status would have already resulted in a perceptible effect.

Table 4. Ecosystem status (EBQI) and Confidence Index (CI) for each sampling site, and scale value and CI of each parameter used to obtain them.

Box	1 Rhizomes	2 Leaves	3–4 Leaf epibiota	5 Pinna	6 HOM LOM	7 Litter	8 Holothuria	9 Herbivores	10 Predators	11 Piscivores	12 Planktivores	9–12 (SRDI)	13 Sea birds	EBQI and (CI)
Espardell	4(0)	4(4)	3(3)	3(4)	3(0)	3(0)	3(0)	3(3)	1(4)	1(4)	1(4)	3(4)	2(3)	6.4(2.9)
Sitges	2(0)	0(4)	0(3)	0(3)	2(0)	2(0)	2(3)	3.5 (3.5)	0(3)	0(3)	1(3)	0(3)	0(3)	2.3(2.7)
Tossa de Mar	2(3)	3(4)	4(4)	0(3)	2(0)	2(0)	4(2)	3(2.5)	2(4)	0(4)	2(4)	3(4)	1(2)	5.6(3.2)
Medes Islands	2(3)	3.5(4)	4(4)	3(3)	2(0)	2(0)	2(0)	2.5(3)	4(2)	4(2)	3(2)	4(2)	2(2)	7.9(2.4)
Peyrefite Bay	2(0)	3.5(4)	2(0)	4(4)	2(0)	2(0)	2(0)	2(4)	3(3)	1(3)	1.5(3)	4(3)	0(2)	5.8(2.3)
Niolon	2(2)	2.5(4)	2(2)	0(2)	1.5(2)	1(0)	3(0)	2(1.5)	1(2)	0(0)	2(0)	2(2)	1(4)	3.9(1.7)
Prado Bay	2(4)	2.5(3)	2(2)	0(2)	2.5(2)	2(2)	3(2)	2.5 (1.5)	3(2)	1(2)	1.5(2)	3(2)	2(4)	5.3(2.3)
Plateau des Chèvres	2(4)	2.5(4)	4(3)	0(2)	1.5(2)	2(2)	3(2)	2.5 (1.5)	2(4)	1(4)	0.5(2)	2(4)	2(2)	5.0(3.1)
Saint-Cyr Bay	1(2)	3(3)	2(0)	1(0)	2(0)	2(2)	2(0)	2(0)	2(0)	2(0)	2(0)	2(0)	0.5(4)	4.9(0.7)
Gulf of Giens	3(4)	4(4)	2(1)	2(0)	2(0)	1(0)	3(2)	1.5(2)	1(4)	0(2)	1(2)	1(3)	0.5(4)	4.3(2.4)
Porquerolles Island northern coast	3(3)	2(4)	3(3)	2(3)	2(3)	0(2)	1(2)	1.5(1)	1(2)	1(2)	2(2)	2(2)	1(4)	4.3(2.4)
Porquerolles Island southern coast	3(3)	4(5)	4(3)	3(3)	3(3)	3(2)	2(2)	2(1)	2(2)	2(2)	3(2)	3(2)	1(4)	6.9(2.6)
Bagaud Pass, Port-Cros Island	4(2)	3(4)	2(4)	4(4)	3(2)	4(2)	4(2)	3(2.5)	3(2)	2(2)	3(2)	4(2)	1(4)	7.6(2.6)
Port-Cros Island southern coast	4(0)	4(4)	4(3)	4(2)	3(2)	4(0)	3(0)	3.5(0)	4(2)	4(0)	3(2)	4(2)	1.5(4)	9.3(1.6)
Villefranche-sur-Mer Bay	2 (1)	1.5 (4)	2(1)	1(2)	3(2)	0(0)	0(2)	2(1)	3(0)	2(1)	1.5(1)	4(0)	0(4)	4.8(1.3)
Scandola, Elbu Bay	4(4)	3(4)	1(3)	4(4)	3(2)	2(2)	2(2)	2(1.5)	2(2)	1(2)	1.5(2)	3(3)	4(4)	5.7(2.6)
Valincu Gulf	(4.4)	3(4)	2(3)	2(2)	2(0)	2(2)	3(2)	2(0)	2(0)	2(0)	2(0)	2(0)	1(0)	5.4(1.4)

Boxes 1 through 13 and SRDI. The status and the Confidence Index (in brackets; CI) is indicated for each box. EBQI ranges from 0 to 10, while the CI is graded from 0 to 4).

Table 5. Example of calculation of the EBQI at the site Espardell (Balearic Islands).

Box number	Functional compartment	Weighting (W)	Parameter	Status and mean status when 2 parameters (S)	CI	Status grade: W×S	CI grade: W×CI
1	Roots and rhizomes ('rhizomes')	3	Growth of orthotropic rhizomes (mm a^{-1})	4	0	12	0
2	*Posidonia oceanica* leaves ('leaves')	5	- Density (shoots m^{-2})	(4)	(4)	(20)	(20)
			- Cover (%)	(4)	(4)	(20)	(20)
				4	4	20	20
3–4	MPOs, filter- and suspension-feeders leaf epibiota ('leaf epibiota')	4	Biomass (g DM $shoot^{-1}$)	3	3	12	12
5	Benthic filter-feeder: *Pinna nobilis* (bivalve) ('Pinna')	2	Density (individuals 100 m^{-2})	3	4	6	8
6	Other benthic filter- and suspension-feeders ('HOM LOM')	2					
	- HOM		- Density (m^{-2})	(3)	(0)	(6)	(0)
	- LOM		- Density (m^{-2})	(3)	(0)	(6)	(0)
				3	0	6	0
7	Litter detritus: dead leaves and broken rhizomes ('litter')	2	g DM m^{-2}	3	0	6	0
8	Detritus-feeders 2 and 3 (*Holothuria* spp.) ('Holothuria')	2	Individuals 10 m^{-2}	3	0	6	0
9	Herbivores 1 ('herbivores')	5	- Density of *Paracentrotus lividus* (individuals m^{-2})	(3)	(3)	(15)	(15)
			- Grazing index (% leaves)	(3)	(3)	(15)	(15)
				3	3	15	15
10	Predatory teleosts, cephalopods and seastars ('predators')	5	kg teleosts WM 100 m^{-2}	1	4	5	20
11	Piscivorous teleosts ('piscivores')	5	kg teleosts WM 100 m^{-2}	1	4	5	20
12	Planktivorous teleosts ('planktivores')	3					
	- Zooplankton feeders		- kg teleosts WM 100 m^{-2}	(1)	(4)	(3)	(12)
	- Omnivores		- kg teleosts WM 100 m^{-2}	(1)	(4)	(3)	(12)
				1	4	3	12
9–12	All teleosts ('SRDI')	3	Specific Relative Diversity Index (SRDI)	3	4	9	12
13	Sea birds	1					
	- *Phalacrocorax* spp.		- Distance to the nearest nesting site (km)	(4)	(3)	(4)	(3)
	- *Pandion haliaetus*		- Distance to the nearest nesting site (km)	0	(3)	0	(3)
				2	3	2	3
					Sum of the weighted status grades (left) and of the weighted CI grades (right) of the 13 boxes[a]	107	122
					EBQI[b] (left) and CI[c] (right) of the site	6.4	2.9

[a]The maximum value of the sum is 168 (see text).
[b]Maximum value: 10 (see text).
[c]Maximum value: 4 (see text).
CI: Confidence Index. DM: dry mass, including calcium carbonate. HOM: indicators of high level of organic matter. LOM: indicators of low level of organic matter. MPO: Multicellular Photosynthetic Organisms. POM: Particulate Organic Matter. WM: wet mass.

- Good (6.0≥EBQI<7.5): Espardell and Porquerolles Island southern coast.
- High (EBQI≥7.5): Bagaud Pass (Port-Cros Island), Medes Islands and Port-Cros Island (southern coast).

Redundancy of EBQI with already existing indices

Is the EBQI superfluous, i.e. redundant with already existing indices? A number of indices (EQR, Ecological Quality Ratio) based upon *P. oceanica* (the species itself, sometimes leaf epibiota,

Table 6. Comparison of EBQI with Ecological Quality Ratios (EQRs) based mainly upon *P. oceanica* (the organism itself) and aimed at establishing the ecological status of the seawater body.

Site	EBQI/10 (CI)	EQR/1	Type of index	Reference
Port-Cros Island, southern coast	9.3 (1.6)	0.802	PREI[a]	[14]
Medes Islands	7.9 (2.4)	0.752	POMI[b]	[12]
Scandola, Elbu Bay	5.7 (2.6)	0.802	BiPo[c]	[15]
Tossa de Mar	5.6 (3.2)	0.682	POMI	[12]
Valincu Gulf	5.4 (1.4)	0.386	PREI	[14]
		0.729	BiPo	[15]
Prado Bay, Marseilles	5.3 (2.3)	0.636	PREI	[14]
Plateau des chèvres, Marseilles	5.0 (3.1)	0.477	PREI	[14]
Saint Cyr Bay	4.9 (0.7)	0.682	PREI	[14]
Villefranche-sur-Mer Bay	4.8 (1.3)	0.280	PREI	[14]
Gulf of Giens	4.3 (2.4)	0.708	PREI	[14]
Porquerolles Island, northern coast	4.3 (2.4)	0.819	PREI	[14]
Niolon (Côte Bleue)	3.9 (1.7)	0.465	PREI	[14]
Sitges	2.3 (2.7)	0.238	POMI	[12]

[a]The metrics of PREI are: shoot density, shoot leaf surface area, ratio between epibiota biomass and leaf biomass, depth of the lower limit and type of this limit [14].
[b]The metrics of POMI are: shoot density, meadow cover, percentage of plagiotropic rhizomes, shoot leaf surface area, percentage of foliar necrosis, P, N and sucrose content in rhizomes, $\delta^{15}N$ and $\delta^{34}S$ isotopic ratio in rhizomes, N content in epiphytes, Cu, Pb and Zn content in rhizomes [12].
[c]The metrics of BiPo are: shoot density, shoot leaf surface area, lower depth limit and lower limit type [15].
PREI, POMI and BiPo indices are based upon distinct but similar metrics and range from 0 (lowest ecological status) to 1 (highest ecological status).

not the ecosystem) have been proposed and are currently routinely used, generally for the purpose of monitoring the ecological status of a water body, e.g. POMI [12], PREI [14] and BiPo [15]. They are based on different combinations of a number of parameters, e.g. shoot density, leaf surface area per shoot, depth of the lower limit of the meadow, percent cover of the dead matte and epibiota biomass. One or several of these indices are available from the literature for 13 out of the 17 study sites (Table 6). As these indices are more or less congruent [16], we have investigated the possible correlation between EBQI and the available EQR (POMI, BiPo or PREI) in these 13 sites, and between EBQI and PREI. There is no significant correlation (Spearman coefficient), either when all available EQRs are considered ($n = 13$, $r_s = 0.469$, p value $= 0.11$) or when only PREI data are concerned ($n = 9$, $r_s = -0.017$, p value $= 0.97$). This result is not unexpected: EQRs mainly assess the health of the seagrass, linked to e.g. the water quality, while EBQI measures the status of the ecosystem, linked not only to water quality but also to e.g. the overfishing. The contrasting ranks of Porquerolles Island (northern coast) through EBQI (poor) and EQR (high, first ranking, see Table 6), together with those of Valincu Gulf and Gulf of Giens, can be due to impacts other than the water quality, e.g. artisanal and recreational overfishing and anchoring of pleasure boats.

Redundancy of parameters to calculate the EBQI

Are some of the parameters used (functional compartments, boxes) redundant with other boxes or with the EBQI? Should such a redundancy exist, this could mean either that the box alone is sufficient to estimate the whole ecosystem's status, or that some boxes are useless. Obviously, the higher weighting is given to a box, the stronger is the probability of such a correlation. Some boxes (e.g. box 5 'Pinna' and box 10 'predators') are significantly correlated with others and/or with the EBQI (Table 7). However, the removal of the boxes correlated with others (boxes 5, 9, 11 and SRDI) results in changes (loss of accuracy?) in the EBQI: the five-class ranking (see below) of 53% of the study sites moves from one class to the next (result not presented). In addition, the removal of these possibly superfluous boxes would not save field time (data acquisition), since data for these boxes are acquired simultaneously with those of non-removed compartments.

Robustness of the EBQI

Is the EBQI robust, i.e. only slightly influenced by arbitrary choices in its conception and biases? Possible arbitrary choices are e.g. choice of the considered compartments (boxes) and the weighting (1 through 5) of the boxes. As far as the choice of the compartments taken into account is considered, it should be first emphasized that the choice is anything but arbitrary; boxes encompass the whole spectrum of an ecosystem functioning, from primary producers, herbivores, filter-feeders, suspension-feeders and detritus-feeders to top predators (Fig. 1). It is worth noting that the removal of up to 4 boxes (see above) only slightly changes the five-class ranking of the study sites: at the most, half of the sites move from one class to the next (e.g. good to moderate, poor to bad; results not presented). Finally, the changing of the weighting of the boxes (hereafter 'perturbation') alters the ranking of the sites; the changes increase with the importance of the perturbation (±1, ±2, fig. 2; ±3 and ±4, not presented). However, the changes due to weighting perturbation are relatively slight. In addition, an index of similarity was produced for each site, which is equal to 100% when the rank of the site is always unchanged after perturbation and 0% if it is always changed (Fig. 3). Some sites are very robust (e.g. Porquerolles Island northern coast and Bagaud Pass), other are more sensitive to the choice of weightings (e.g. Medes Islands and Tossa de Mar). The similarity ranges between 100 and 95% (perturbation ±1), 100 and 71% (±2), 100 and 57% (±3) and 100 and 12% (±4). The low values of the similarity index for high levels of perturbation of the weighting emphasize the fact that weighing the boxes is useful, despite the robustness of this parameter.

Table 7. Spearman's rank correlations between the status of each compartment (boxes 1 through 13 and SRDI), the other compartments (boxes) and the EBQI (see Table 5).

	1.Rhizomes	2. Leaves	3–4. Leaf epibiota	5. Pinna	6. HOM LOM	7. Litter	8. Holothuria	9. Herbivores	10. Predators	11. Piscivores	9–12 (SRDI)	12. Planktivores	13. Sea birds
1. Rhizomes	-												
2. Leaves	0.44	-											
3–4. Leaf epibiota	0.03	0.41	-										
5. Pinna	0.71	0.64	0.02	-									
6. HOM-LOM	0.57	0.33	−0.02	0.63	-								
7. Litter	0.44	0.55	0.29	0.48	0.51	-							
8. Holothuria	0.27	0.27	0.20	−0.09	−0.08	0.44	-						
9. Herbivores	0.07	0.03	0.20	−0.08	0.23	0.67	0.46	-					
10. Predators	−0.02	0.23	0.29	0.37	0.36	0.34	0.02	0.24	-				
11. Piscivores	0.22	0.30	0.33	0.51	0.46	0.42	−0.22	0.10	0.73	-			
12. Planktivores	0.18	0.28	0.38	0.34	0.24	0.34	0.05	0.06	0.48	0.62	-		
9–12 (SRDI)	0.16	0.32	0.30	0.56	0.59	0.41	−0.02	0.29	0.85	0.60	0.46	-	
13. Sea birds	0.37	0.22	0.37	0.12	0.19	0.36	0.30	0.26	0.20	0.18	0.08	0.19	-
EBQI	0.45	0.69	0.50	0.72	0.55	0.78	0.24	0.40	0.69	0.65	0.54	0.77	0.47

Bold italic characters: significant ($p<0.01$).

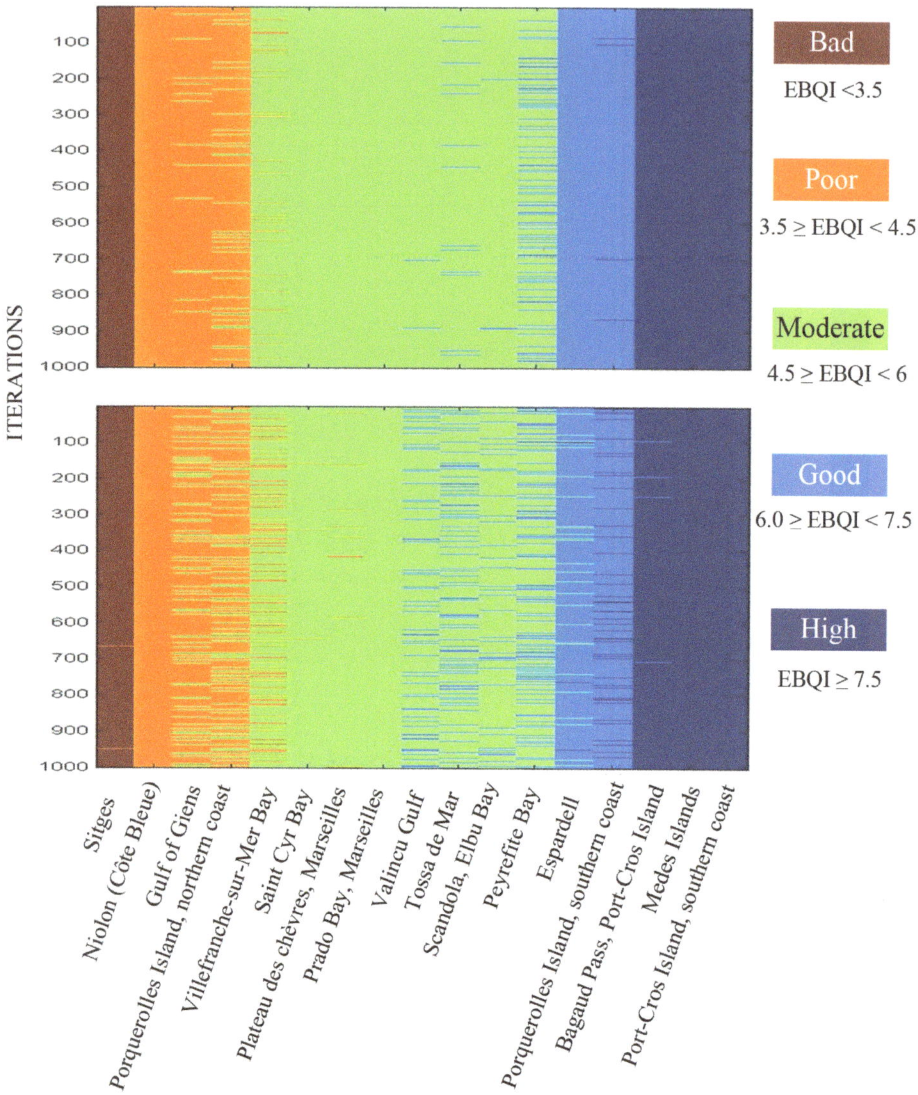

Figure 2. Robustness of the quality index (EBQI) with regard to the weighting of compartments (boxes). The 17 sites are arranged from left to right according to their growing EBQI (see Table 4) and ecological status (bad through high). Deep red = bad, orange = poor, green = moderate, light blue = good, deep blue = high. In order to test the effect of the weighting of the compartments (boxes) on the EBQI (robustness), weighting values have been randomly perturbed (above, ±1; below, ±2). 1000 iterations were performed. The change of the ecological status (bad through high) of a site, for a given iteration, is shown by the color of the new class within which it falls.

EBQI implementation

Is the EBQI excessively time-consuming or easy and rapid to implement? The acquisition of the parameters requires SCUBA diving field work. Considering that it is possible to work for around 1 hour at 15 m depth, we estimate that 6 dives involving 2 scientific divers are necessary to acquire all of the data for the assessment of one site (Table 8). For safety reasons, a diver can perform a maximum of two dives per day, so this will require 3 days of field work for a two man diving team. This field work requires expertise in seagrass and fish visual censuses. In addition, for a rapid and provisional assessment of the EBQI, already available data and expert judgment can be used (provided that the CI is specified).

Within the northern Mediterranean, the study sites are spread within three eco-regions, namely Corsica, Provence and French Riviera and Languedoc-Catalonia. Is the EBQI biased by regional environmental conditions, e.g. a higher mean temperature in Corsica and more water turbidity in Languedoc-Catalonia? The present study cannot answer this question as the sites were not randomly selected but chosen according to data availability. However, the EBQI range within each one of the three eco-regions does not exhibit obvious differences (Table 4), as far as the EBQI mean is concerned (~5.6 within each one of the three eco-regions).

Discussion

In the context of European Directives (Habitats Directive and WFD) and national regulations, a number of species and indices, based upon one or several species, have been designated to assess the quality of water bodies, the health status of emblematic species and habitats and to establish Marine Protected Areas (e.g. PMN [112], CI [113], EEI [2,3], SI [114], POMI [12], PREI [14], BiPo [15], PoSte [15], ZoNI [22] for European seagrasses

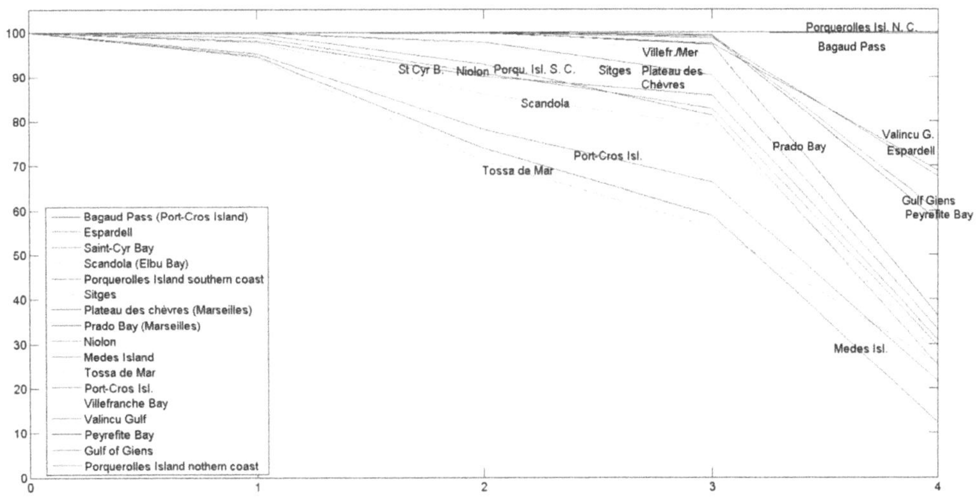

Figure 3. Robustness (index of similarity) of the quality index (EBQI) to the weighting of compartments (boxes). Percentage of times the ecological status of each site was unaltered by random perturbation of the weighting of the boxes (1,000 iterations) by ±1 through ±4. The similarity is equal to 100% when the class of the site is never changed after perturbation and 0% if it is always changed.

([17,51,115,116] for comprehensive reviews). These indices provide a valuable body of tools and information to coastal waters managers and make it possible to assess the status of a water body. However, the good quality of a water body and the apparent health of a species, whether emblematic or not, such as the seagrass *P. oceanica*, is not always indicative of the good structure and functioning of the whole ecosystem, a network of compartments, fluxes, functions, inputs and exportations. The most original contribution of the new MSFD European Directive is to provide an ecosystem-based approach to assess the ecological

Table 8. Estimation of time and diving effort needed for data acquisition within each box used in the EBQI.

Box	Proxy	Time	Dive organization
1 – *P. oceanica* roots and rhizomes	30 random *in situ* measures (growth of the rhizome corresponding to the last 8 leaf bases)	30 min	1 dive for 2 scientific divers
2 - *P. oceanica* leaves	20 random measures of shoot density in a square frame (0.16 m^2)	60 min	
	30 random measures of cover	10 min	
3 - MPO leaf epibiota	Sampling of the two oldest external leaves on 30 shoot randomly localized to estimate the epibiota biomass	10 min	
4 - Filter- and suspension-feeder leaf epibiota	Sampling of the two oldest external leaves on 30 shoot randomly localized to estimate the epibiota biomass	Same as box 3	
5 – Filter feeder benthic epibiota	Density of *Pinna nobilis* along 20 transects 10-m long and 1-m wide	120 min	1 dive for 2 scientists divers
7 - Litter detritus	Litter detritus mass collected in July, within 5 randomly localized 0.1-m^2 quadrats	90 min	1 dive for 2 scientist divers
6 – Other filter- and suspension-feeder benthic epibiota	Abundance of filter- and suspension-feeder benthic epibiota other than *Pinna nobilis* within 1-m^2 quadrats, with 30 replicates randomly localized	120 min	1 dive for 2 scientist divers
8 - Detritus feeders	Abundance of *Holothuria* spp. within 1-m^2 quadrats, with 30 replicates randomly localized	Same as box 6	
9 - Herbivores 1	Abundance of *Paracentrotus lividus* within 1-m^2 quadrats, with 30 replicates randomly localized at 5 m depth	60 min	1 dive for 2 scientist divers
	Percentage of intermediate and adult leaves exhibiting bite scars due to *S. salpa* (all the intermediate and adult leaves of 30 shoots randomly localized) at 5 m depth	30 min	
10 - Predatory teleosts	All teleosts counted within ten linear 50-m long and 5-m wide transects, each census lasting 5 minutes.	60 min	1 dive for 2 scientist divers
11 - Piscivorous teleosts	All teleosts counted within ten linear 50-m long and 5-m wide transects, each census lasting 5 minutes	Same as boxes 10 and 12	
12 - Planktivorous teleosts of the water column	All teleosts counted within ten linear 50-m long and 5-m wide transects, each census lasting 5 minutes	Same as boxes 10 and 11	
9–12 - SRDI	All teleosts counted within ten linear 50-m long and 5-m wide transects, each census lasting 5 minutes	Same as boxes 10, 11 and 12	

status of marine regions. The present approach constitutes a contribution towards this goal, and focuses on the *P. oceanica* seagrass meadow.

Here, we have developed, applied and tested an ecosystem-based index of the structure and functioning of the most emblematic and best-known Mediterranean coastal ecosystem, the *P. oceanica* meadow. This index (EBQI) is based upon a set of representative functional compartments, the weighting of these compartments and the assessment of each compartment quality by comparison of a supposed baseline based upon the available literature. The strong points of the EBQI are: **(i)** It is easy to implement, not too time-consuming and therefore relatively cheap; **(ii)** It is non-destructive, which is particularly important, dealing with protected species (e.g. *P. oceanica* and *Pinna nobilis*); sampling only concerns old leaves (not the shoot) and the litter, i.e. shed dead leaves; **(iii)** It is relatively robust, according to the selection of the compartments and to their weighting; **(iv)** It is associated with confidence indices CI (each compartment, and the overall mark) which indicate possible weakness and biases and therefore the need for further field data acquisition; **(v)** It can draw the attention of managers to sites whose CI is high and where routine monitoring can therefore be implemented, and conversely those with a low CI indicating a lack of knowledge.

The Integrated Coastal Zone Management (ICZM) contributes towards maximizing the benefits provided by the coastal zone and minimizing conflicts and the harmful effects of activities upon each other (e.g. [117]). ICZM, together with Ecosystem-Based Management (EBM; e.g. [118]) and, more generally, the implementation of European Directives, are in need of indices to assess ecological quality, either based upon a single or a few species, or ecosystem-based. The sites corresponding to well implemented MPAs, such as the Port-Cros National Park (Port-Cros Island southern coast, Bagaud Pass) and the Medes Islands marine reserve, not unexpectedly get the best marks, which is congruent with the overall excellent status of their habitats [108]. In contrast, sites localized in areas highly impacted by most human activities and waste, such as Sitges (South of Barcelona) and Niolon (close to the port of Marseilles and outfall of wastewater), obtain a very low mark. Though sites with obviously good or bad ecosystem status naturally obtain a good and bad, EBQI mark, respectively, some sites (e.g. Saint-Cyr Bay and Villefranche-sur-Mer Bay) do not obtain the mark they would seemingly have deserved through the apparent health of the seagrass itself. This confirms the usefulness of an index, EBQI, based upon the whole ecosystem rather than only upon the seagrass (sometimes also epibiota and other parameters). It is worth noting that the Confidence Index (CI) is overall relatively low. This is especially the case of Saint-Cyr Bay (CI = 0.7; table 4) and Villefranche-sur-Mer Bay (CI = 1.3), which could also account for the surprisingly low and not so low, EBQI, respectively, of these sites, according to literature data [119,120].

The weak points of the EBQI are: **(i)** The baselines used to assess the compartments, which may prove to be biased either by the poor knowledge or availability of totally non-human impacted areas, or by climate change which renders the supposed baseline obsolete; **(ii)** The probable need for testing and/or adapting the EBQI in areas distant from the study area -the north-western Mediterranean- and more generally to other seagrass ecosystems; (iii) The conceptual model which constitutes the basis for the present ecosystem-based approach is obviously oversimplified. For example, small-sized predatory 'invertebrates' were not taken into consideration. Some of them belong to the matte endofauna box, which encompasses detritus feeders, suspension-feeders and predators. Other are creeping or clinging organisms on the *P. oceanica* leaves. However, considering them would have made the model more complex and weighed down the assessment of ecosystem status. Moreover, other compartments are likely to provide redundant information, so that no improvement in accuracy is to be expected.

A number of attempts to review and compare the biotic indices using benthic (e.g. macrophytes) and pelagic marine, lagoon and estuarine organisms, have been performed [16,39,51,116,121]. For example, [116] analysed the strengths and weaknesses of 90 published indices. They identified several weaknesses: **(i)** problems of applicability due to practical and conceptual difficulties, affecting most of the indices related to ecosystem function, **(ii)** the failure of many indices using e.g. the taxonomic composition of the community to connect its condition with the stressors, and **(iii)**, as far as indices based upon the sub-individual level are concerned (e.g. multi-biomarkers), their poor strength to assess the ecological integrity of the habitat. They concluded that the most promising approach would be to aggregate indices with complementary strengths. In a sense, the EBQI could be considered as such an aggregative index. Although the goal of the EBQI is not to compete with the already available biotic indices utilizing the *P. oceanica* seagrass meadow for assessing the quality of a water body, but to assess the status of the proper seagrass ecosystem, the comparison of the EBQI with some other biotic indices will be of relevance. An attempt has been made here, using some published data; a more accurate comparison would require data acquisition at the same sites as those used for the EBQI, and is therefore beyond the scope of the present work. In addition, the correlation between the EBQI and anthropogenic gradients, including overfishing, will be useful, in order to assess which human impacts are of major relevance for the status of the ecosystem. This will constitute the next step.

Conclusions

Overall, the Ecosystem Based Quality Index (EBQI) is easy to implement, relatively robust and does not seem to be redundant with existing indices based upon the seagrass itself. In addition, the non-congruence of the EBQI, i.e. the quality of the ecosystem functioning, with the empirical idea one may have of the *P. oceanica* ecosystem at a given site, due e.g. to the health of the seagrass, the clearness of the water (and even the beauty of the landscape and the seascape) confirms the usefulness of the EBQI index, based upon the whole ecosystem rather than upon only the seagrass. Other anthropogenic impacts, e.g. overfishing, are putatively more important than the above mentioned parameters for the structure and functioning of the ecosystem. Finally, the EBQI that is proposed here for the *P. oceanica* ecosystem could constitute a model for similar indices designed for other ecosystems, such as coralligenous outcrops, underwater caves, soft bottoms and sublittoral reefs.

Acknowledgments

The authors are indebted to Fabien Morat for data concerning *Phalacrocorax aristotelis*, Philippe Lenfant, Boris Daniel (AAMP) and Pierre Boissery (AERMC) for contributions and valuable comments, Michael Paul for improvement of the English text and two anonymous reviewers for their highly valuable suggestions and improvement of the manuscript.

Author Contributions

Conceived and designed the experiments: CFB CPM DTB EB FR GP MHV SP SR. Performed the experiments: CPM EB GP JP PA PB SB SR TT. Analyzed the data: CFB JCP SP SR. Wrote the paper: DBS EB EF CFB JCP MHV SP SR.

References

1. Karr JR (1981) Assessment of biotic integrity using fish communities. Fisheries 6: 21–27.
2. Orfanidis S, Panayotidis P, Stamatis N (2001) Ecological evaluation of transitional and coastal waters: a marine benthic macrophytes-based model. Medit Mar Sci 2: 45–65.
3. Orfanidis S, Panayotidis P, Stamatis N (2003) An insight to the ecological evaluation index (EEI). Ecol Indic 3: 27–33.
4. White J, Irvine K (2003) The use of littoral mesohabitats and their macroinvertebrates in the assemblages of lakes. Aquat Conserv 13: 331–351.
5. Pergent-Martini C, Leoni V, Pasqualini V, Ardizzone GD, Balestri E, et al. (2005) Descriptors of *Posidonia oceanica* meadows: Use and application. Ecol Indic 5: 213–230.
6. Orfanidis S, Panayotidis P, Ugland KI (2011) Ecological Evaluation Index continuous formula (EEI-c) application: a step forward for functional groups, the formula and reference condition values. Medit Mar Sci 12: 199–231.
7. Hoare JM, Monks A, O'Donnel CFJ (2013) Do population indicators work? Investigating correlated responses of bird populations in relation to predator management. Ecol Indic 25: 23–34.
8. Jones CG, Lawton JH, Shachak M (1994) Organisms as ecosystem engineers. Oikos 69: 373–386.
9. Lawton JH (1994) What do species do in ecosystems? Oikos 71: 367–374.
10. Bond W (2001) Keystone species – Hunting the snark? Science 292: 63–64.
11. Boudouresque CF, Mayot N, Pergent G (2006) The outstanding traits of the functioning of the *Posidonia oceanica* seagrass ecosystem. Biol Mar Medit 13(4): 109–113.
12. Romero J, Martínez-Crego B, Alcoverro T, Pérez M (2007) A multivariate index based on the seagrass *Posidonia oceanica* (POMI) to assess ecological status of coastal waters under the water framework directive (WFD). Mar Poll Bull 55: 196–204.
13. Fernández-Torquemada Y, Díaz-Valdés M, Colilla F, Luna B, Sánchez-Lizaso JL, et al. (2008) Descriptors from *Posidonia oceanica* (L.) Delile meadows in coastal waters of Valencia, Spain, in the context of the EU Water Framework Directive. ICES J Mar Sci 65: 1492–1497.
14. Gobert S, Sartoretto S, Rico-Raimondino V, Andral B, Chery A, et al. (2009) Assessment of the ecological status of Mediterranean French coastal waters as required by the Water Framework Directive using the *Posidonia oceanica* Rapid Easy Index: PREI. Mar Poll Bull 58: 1727–1733.
15. Lopez y Royo C, Casazza G, Pergent-Martini C, Pergent G (2010) A biotic index using the seagrass *Posidonia oceanica* (BiPo), to evaluate ecological status of coastal waters. Ecol Indic 10: 380–389.
16. Lopez y Royo C, Pergent G, Alcoverro T, Buia MC, Casazza G, et al. (2011) The seagrass *Posidonia oceanica* as indicator of coastal water quality: experimental intercalibration of classification systems. Ecol Indic 11: 557–563.
17. Marbà N, Krause-Jensen D, Alcoverro T, Birk S, Pedersen A (2012) Diversity of European seagrass indicators: patterns within and accross regions. Hydrobiologia 704: 265–278.
18. Panayotidis P, Feretopoulos J, Montesanto B (1999) Benthic vegetation as an ecological quality descriptor in an Eastern Mediterranean coastal area (Kalloni Bay, Aegean Sea, Greece). Estuar Coast Shelf Sci 48: 205–214.
19. Ballesteros E, Torras X, Pinedo S, Garcia M, Mangialajo L, et al. (2007) A new methodology based on littoral community cartography by macroalgae for the implementation of the European Water Framework Directive. Mar Poll Bull 55: 172–180.
20. Guinda X, Juanes JA, Puente A, Revilla JA (2008) Comparison of two methods for quality assessment of macroalgae assemblages, under pollution types. Ecol Indic 8: 743–753.
21. Oliva S, Mascaró O, Llagostera I, Pérez M, Romero J (2012) Selection of metrics based on the seagrass Cymodocea nodosa and development of a biotic index (CYMOX) for assessing ecological status of coastal and transitional waters. Estuar Coast Shelf Sci 114: 7–17.
22. García-Marín P, Cabaço S, Hernández I, Vergara JJ, Silva J (2013) Multi-metric index based on the seagrass Zostera noltii (ZoNI) for ecological quality assessment of coastal and estuarine systems in SW Iberian Peninsula. Mar Poll Bull 68: 46–54.
23. Diaz R, Solan M, Valente R (2004) A review of approaches for classifying benthic habitats and evaluating habitat quality. J Environ Manage 73: 165–181.
24. Mateo MA, Romero J, Pérez M, Littler M, Littler D (1997) Dynamics of millenary organic deposits resulting from the growth of the Mediterranean seagrass *Posidonia oceanica*. Estuar Coast Shelf Sci 44: 103–110.
25. Alcoverro T, Manzanera M, Romero J (2001) Annual metabolic carbon balance of the seagrass *Posidonia oceanica*: the importance of carbohydrate reserves. Mar Ecol Prog Ser 211: 105–116.
26. Prado P, Tomas F, Alcoverro T, Romero J (2007) Extensive direct measurements of *Posidonia oceanica* defoliation confirm the importance of herbivory in temperate seagrass meadows. Mar Ecol Prog Ser 340: 63–71.
27. Boudouresque CF, Bernard G, Bonhomme P, Charbonnel E, Diviacco G, et al. (2012) Protection and conservation of *Posidonia oceanica* meadows. Marseilles: RAMOGE, RAC/SPA and Gis Posidonie publ. pp. 202 p.
28. Pergent G, Bazairi H, Bianchi CN, Boudouresque CF, Buia MC, et al. (2012) Mediterranean seagrass meadows: resilience and contribution to climate change mitigation. A short summary. Gland, Málaga: IUCN. 40 p.
29. Mazzuca S, Björk M, Beer S, Felisberto P, Gobert S, et al. (2013) Establishing research strategies, methodologies and technologies to link genomics and proteomics to seagrass productivity, community metabolism, and ecosystem carbon fluxes. Front Plant Sci 4(38): 1–19.
30. Sala E (2004) The past and present topology and structure of Mediterranean subtidal rocky-shore food webs. Ecosystems 7: 333–340.
31. Bellan-Santini D, Picard J (1984) Conséquences de la régression des herbiers de Posidonies pour le benthos animal des milieux méditerranéens. In: Boudouresque CF, Jeudy de Grissac A, Olivier J, editors. International Workshop on *Posidonia oceanica* beds. Marseilles: GIS Posidonie publ. pp. 423–429.
32. Boudouresque CF, Bernard G, Pergent G, Shili A, Verlaque M (2009) Regression of Mediterranean seagrasses caused by natural processes and anthropogenic disturbances and stress: a critical review. Botanica Marina 52: 395–418.
33. Coles R, Grech A, McKenzie L (2013) Seagrass under pressure. Seagrass-Watch 47: 2–6.
34. Ward T, Hegerl E (2003) Marine protected areas in ecosystem-based management of fisheries. Canberra: Commonweath of Australia publ. 66 p.
35. Pikitch EK, Santora C, Babcock EA, Bakun A, Bonfil R, et al. (2004) Ecosystem-based fishery management. Science 305: 346–347.
36. Turrell B (2004) The policy basis of the "Ecosystem Approach" to fisheries management. Norrköping, Sweden: EuroGOOS publ. 21: 1–28.
37. Rice JC (2005) Implementation of the Ecosystem Approach to Fishery Management – asynchronous co-evolution at the interface between science and policy. Mar Ecol Prog Ser 300: 265–270.
38. Tudela S, Short K (2005) Paradigm shifts, gaps, inertia, and political agendas in ecosystem-based fisheries management. Mar Ecol Prog Ser 300: 281–286.
39. Martínez-Crego B, Vergés A, Alcoverro T, Romero J (2008) Selection of multiple seagrass indicators for environmental biomonitoring. Mar Ecol Progr Ser 361: 93–109.
40. Christensen V, Maclean J (eds) (2011) Ecosystem approach to fisheries. A global perpective. Cambridge University Press. 342 p.
41. Kenchington E, Heino M, Nielsen EE (2003) Managing marine genetic diversity: time for action? ICES J Mar Sci 60: 1172–1176.
42. Rosenberg AA, McLeod KL (2005) Implementing ecosystem-based approaches to management for the conservation of ecosystem services. Mar Ecol Prog Ser 300: 270–274.
43. Ruíz JM, Boudouresque CF, Enríquez S (2009) Mediterranean seagrasses. Botanica Marina 52: 369–382.
44. Boudouresque CF, Jeudy de Grissac A (1983) L'herbier à *Posidonia oceanica* en Méditerranée: les interactions entre la plante et le sédiment. J Rech Océanogr 8: 99–122.
45. Gacia E, Granata TC, Duarte CM (1999) An approach to measurement of particule flux and sed-iment retention within seagrass (*Posidonia oceanica*) meadows. Aquat Bot 65: 255–268.
46. Sánchez-González JF, Sánchez-Rojas V, Memos CD (2011) Wave attenuation due to *Posidonia oceanica* meadows. J Hydraul Res 49: 503–514.
47. Infantes E, Orfila A, Simarro G, Terrados J, Luhar M, et al. (2012) Effect of a seagrass (*Posidonia oceanica*) meadow on wave propagation. Mar Ecol Prog Ser 456: 63–72.
48. Molinier R, Picard J (1952) Recherches sur les herbiers de Phanérogames marines du littoral méditerranéen français. Ann Inst Océanogr 27: 157–234.
49. Boudouresque CF, Meinesz A (1982) Découverte de l'herbier de posidonie. Hyères: Parc National de Port-Cros publ. iii + 79 p.
50. López-Sáez JA, López-Merino L, Mateo MÁ, Serrano Ó, Pérez-Díaz S, et al. (2009) Paleoecological potential of marine organic deposits of *Posidonia oceanica*: a case study in the NE Iberian Peninsula. Palaeogeogr Palaeoclimatol Palaeoecol 271: 215–224.
51. Montefalcone M (2009) Ecosystem health assessment using the seagrass *Posidonia oceanica*: a review. Ecological Indicators 9: 595–604.
52. Sartoretto S, Beaucour C (2011) Habitats particuliers de l'infralittoral : Herbier à *Posidonia oceanica* de la sous-région marine Méditerranée occidentale DCSMM/EI/MO. Ministère de l'Écologie, du Développement Durable, des Transports et du Logement DCSMM/EI/EE/MO/2.2.4.c. 14p.
53. Pergent-Martini C, Pergent G (1994) Lepidochronological analysis in the Mediterranean seagrass *Posidonia oceanica*: state of the art and future development. Oceanol Acta 17: 673–681.
54. Fourqurean JW, Duarte CM, Kennedy H, Marbà N, Holmer M, et al. (2012) Seagrass ecosystems as a globally significant carbon stock. Nat Geosci 5:505–509.
55. Boudouresque CF, Crouzet A, Pergent G (1983) Un nouvel outil au service de l'étude des herbiers à *Posidonia oceanica*: la lépidochronologie. Rapp PV Réun Commiss Internatl Explor Sci Médit 28(3): 111–112.
56. Pergent G, Boudouresque CF, Crouzet A, Meinesz A (1989) Cyclic changes along *Posidonia oceanica* rhizomes (lepidochronology): present state and perspectives. PSZN I Mar Ecol 10: 221–230.

57. Thélin I, Mossé A, Boudouresque CF, Lion R (1985) Le benthos littoral d'El Dabaa (Méditerranée, Égypte). II. L'herbier à *Posidonia oceanica*. Rapp PV Réun Commiss Internatl Explor Sci Médit 29(5): 247–248.
58. Pergent G, Pergent-Martini C (1991) Leaf renewal cycle and primary production of *Posidonia oceanica* in the bay of Lacco Ameno (Ischia, Italy) using lepidochronological analysis. Aquat Bot 42: 49–66.
59. Pergent G, Pergent-Martini C, Boudouresque CF (1995) Utilisation de l'herbier à *Posidonia oceanica* comme indicateur biologique de la qualité du milieu littoral en Méditerranée: état des connaissances. Mésogée 54: 3–27.
60. Martín MA, Sánchez Povedo M, Marhuenda E, Martínez JM, Sánchez Lizaso JL (1994) Estudio del estrato hipogeo de la pradera de *Posidonia oceanica* (L.) Delile de la reserva marina de Tabarca. In: Ramos Esplá A, editor. Trabajos de campo en la reserva marina de Tabarca (Alicante). Alicante: Universidad de Alicante publ. pp. 25–31.
61. Boudouresque CF, Jeudy de Grissac A, Meinesz A (1984) Relations entre la sédimentation et l'allongement des rhizomes orthotropes de *Posidonia oceanica* dans la baie d'Elbu (Corse). In: Boudouresque CF, Jeudy de Grissac A, Olivier J, editors. International Workshop on *Posidonia oceanica* Beds. Marseilles: GIS Posidonie publ. pp. 185–191.
62. Di Maida G, Tomasello A, Sciandra M, Pirrotta M, Milazzo M, et al. (2013) Effect of different substrata on rhizome growth, leaf biometry and shoot density of *Posidonia oceanica*. Mar Environ Res 87–88: 96–102.
63. Giovannetti E, Lasagna R, Montefalcone M, Bianchi CN, Albertelli G, et al. (2008) Inconsistent responses to substratum nature in *Posidonia oceanica* meadows: an integration through complexity levels? Chemistry and Ecology 24 (S1): 145–153.
64. Bay D (1984) A field study of the growth dynamics and productivity of *Posidonia oceanica* (L.) Delile in Calvi Bay, Corsica. Aquat Bot 20: 43–64.
65. Pergent-Martini C, Rico-Raimondino V, Pergent G (1994) Primary production of *Posidonia oceanica* in the Mediterranean basin. Mar Biol 120: 9–15.
66. Panayotidis P, Boudouresque CF, Marcot-Coqueugniot J (1981) Microstructure de l'herbier de *Posidonia oceanica* (Linnaeus) Delile. Botanica Marina: 115–124.
67. UNEP-MAP-RAC/SPA (2011) Draft guidelines for the standardization of mapping and monitoring methods of marine magnoliophyta in the Mediterranean. Working document, prepared by C Pergent-Martini for the tenth meeting of Focal Points for SPAs, Marseilles, France, 17–20 May 2011, UNEP(DEPI)/MED WG 359/9. Tunis: RAC/SPA Publ. 59 p.
68. Boudouresque CF, Verlaque M (2007) Ecology of *Paracentrotus lividus*. In: Lawrence J, editor. Edible sea urchins: biology and ecology, Second edition. Amsterdam: Elsevier. pp. 243–285.
69. Boudouresque CF, Bernard G, Bonhomme P, Charbonnel E, Le Diréach L, et al. (2007) Monitoring methods for *Posidonia oceanica* seagrass meadows in Provence and the French Riviera. Sci Rep Port-Cros Natl Park 22: 17–38.
70. Short FT, McKenzie LJ, Coles RG, Vidler KP (2002) SeagrassNet manual for scientific monitoring of seagrass habitat. Cairns: QDPI, QFS. 56 p.
71. Boudouresque CF, Verlaque M (2013) *Paracentrotus lividus*. In: Lawrence JM, editor. Sea urchins: Biology and ecology, Third Edition. Amsterdam: Elsevier. pp. 297–327.
72. van der Ben D (1971) Les épiphytes des feuilles de *Posidonia oceanica* Delile sur les côtes françaises de la Méditerranée. Mém Inst R Sci Nat Belg 168 : 1–101 + 6 tables.
73. Giovannetti E, Montefalcone M, Morri C, Bianchi CN, Albertelli G (2010) Early warning response of *Posidonia oceanica* epiphyte community to environmental alterations (Ligurian Sea, NW Mediterranean). Mar Poll Bull 60 (7): 1031–1039.
74. Mabrouk L, Hamza A, Ben Brahim M, Bradai MN (2013) Variability in the structure of epiphyte assemblages on leaves and rhizomes of *Posidonia oceanica* in relation to human disturbances in a seagrass meadow of Tunisia. Aquat Bot 108: 33–40.
75. Pérès JM, Picard J (1964) Nouveau manuel de bionomie benthique de la Mer Méditerranée. Rec Trav Stat Mar Endoume 31(47): 3–137.
76. Vicente N, De Gaulejac B (1993) *Pinna nobilis* L. (Mollusque Bivalve), indicateur biologique du littoral méditerranéen. In : Boudouresque CF, Avon M, Pergent-Martini C, editors. Qualité de milieu marin. Indicateurs biologiques et physico-chimiques. Marseilles: GIS Posidonie publ. pp. 141–150.
77. García-March JR, García-Carrascosa AM, Peña Cantero AL, Wang YG (2007) Population structure, mortality and growth of Pinna nobilis Linnaeus, 1758 (Mollusca, Bivalvia) at different depths in Moraira Bay (Alicante, Western Mediterranean). Mar Biol 150: 861–871.
78. García-March JR, Vicente N (2011) *Pinna nobilis* (Mollusca, Bivalvia) indicateur biologique de la qualité du littoral méditerranéen: densité des populations. In: Francour P, Bodilis P, Arceo H, editors. Actes du Congrès 2011 de la Société Zoologique de France. Nice: Université de Nice publ. pp. 80–81.
79. Urra J, Ramirez AM, Marina P, Salas C, Gofas S, et al. (2013) Highly diverse molluscan assemblages of *Posidonia oceanica* meadows in northwestern Alboran Sea (W Mediterranean): seasonal dynamics and environmental drivers. Estuar Coast Shelf Sci 117: 136–147.
80. Naranjo SA, Carball JL, Garcia-Gomez JC (1996) Effects of environmental stress on ascidian populations in Algeciras Bay (southern Spain). Possible marine bioindicators? Mar Ecol Progr Ser 144: 119–131.
81. Bianchi CN, Bedulli D, Morri C, Occhipinti Ambrogi A (1989) L'herbier de Posidonies: écosystème ou carrefour éco-éthologique? In: International Workshop on Posidonia Beds (CF . Boudouresque, A . Meinesz, E . Fresi & V . Gravez, eds). GIS Posidonie, Marseille 2: 257–272.
82. Romero J, Pergent G, Pergent-Martini C, Mateo M.A, Régnier C (1992). The detritic compartment in a *Posidonia oceanica* meadow: litter features, decomposition rates and mineral stocks. Mar Ecol PSZNI 13(1): 69–83.
83. Mezali K, Chekaba B, Zupo V, Asslah B (2003) Comportement alimentaire de cinq espèces d'holothuries aspidochirotes (*Holothuria*: Echinodermata) de la presqu'île de Sidi-Fredj (Algérie). Bull Soc Zool Fr 128: 49–62.
84. Verlaque M (1981) Preliminary data on some *Posidonia* feeders. Rapp PV Réun Commiss Internatl Explor Sci Méditerr 27(2): 201–202.
85. Giraud G (1979) Sur une méthode de mesure et de comptage des structures foliaires de *Posidonia oceanica* (Linnaeus) Delile. Bull Mus Hist Nat Marseille 39: 33–39.
86. Peirano A, Niccolai I, Mauro R, Bianchi CN (2001) Seasonal grazing and food preference of herbivores in a *Posidonia oceanica* meadow. Scientia Marina 65 (4): 367–374.
87. Francour P (1997) Fish assemblages of *Posidonia oceanica* beds at Port-Cros (France, NW Mediterranean): assessment of composition and long term fluctuations by visual census. PSZN I Mar Ecol 18: 157–173.
88. Bell JD, Harmelin-Vivien ML (1982) Fish fauna of French Mediterranean *Posidonia oceanica* seagrass meadows. 1. Community structure. Téthys 10: 337–347.
89. Harmelin-Vivien M (1982) Ichtyofaune des herbiers de posidonies du Parc national de Port-Cros: I. Composition et variations spatio-temporelles. Trav Sci Parc Natl Port-Cros 8: 69–92.
90. Francour P, Harmelin-Vivien M, Harmelin JG, Duclerc J (1995) Impact of *Caulerpa taxifolia* colonization on the littoral ichthyofauna of north-western Mediterranean Sea: preliminary results. Hydrobiologia 300/301: 345–353.
91. Harmelin-Vivien M, Francour P (1992) Trawling or visual censuses? Methodological bias in the assessment of fish populations in seagrass beds. PSZN I Mar Ecol 13: 41–51.
92. Guyot I (1990) Le cormoran huppé de Corse: biologie et interactions avec la pêche professionnelle. Trav Sci Parc Nat Rég Rés Nat Corse 28: 1–40.
93. Morat F (2007) Régime alimentaire de la population de cormoran huppé de Méditerranée (*Phalacrocorax aristotelis desmarestii*) de Riou. Marseilles: CEEP and Centre d'Océanologie de Marseille. 23 + 6 p.
94. Morat F, Mante A, Drunat E, Dabat J, Bonhomme P, et al. (2011) Diet of the Mediterranean European shag, *Phalacrocorax aristotelis desmaretii*, its ecological significance and interaction with local fisheries in the Riou Archipelago (Marseilles, France). Vie Milieu 61: 77–86.
95. Francour P, Thibault JC (1995) Le régime alimentaire du balbuzard pêcheur (*Pandion haliaetus*) dans la Réserve naturelle de Scandola. Trav Sci Parc Nat rég Rés Nat. Corse 53: 11–27.
96. Francour P, Thibault JC (1996) The diet of breeding osprey *Pandion haliaetus* on Corsica: exploitation of a coastal marine environment. Bird Study 43: 129–133.
97. Thibault JC, Bretagnolle V, Dominici JM (2001) Le balbuzard pêcheur en Corse. Ajaccio: Alain Piazzola. 184 p.
98. Pergent G, Romero J, Pergent-Martini C, Mateo MA, Boudouresque CF (1994) Primary production, stocks and fluxes in the Mediterranean seagrass *Posidonia oceanica*. Mar Ecol Prog Ser 106: 139–146.
99. Mateo MA, Romero J (1997) Detritus dynamics in the seagrass *Posidonia oceanica*: elements for an ecosystem carbon and nutrient budget. Mar Ecol Prog Ser 151: 43–53.
100. Pergent G, Rico-Raimondino V, Pergent-Martini C (1997) Fate of primary production in *Posidonia oceanica* meadows of the Mediterranean. Aquat Bot 59: 307–321.
101. Mateo MA, Sánchez-Lizaso JL, Romero J (2002) *Posidonia oceanica* 'banquettes': a preliminary assessment of the relevance for meadow carbon and nutrient budget. Estuar Coast Shelf Sci 56: 85–90.
102. Macquart-Moulin C (1968) Les amphipodes bentho-planctoniques du golfe de Marseille. Analyse des captures faites au cours de pêches planctoniques nocturnes régulières (années 1963–1964). Rec Trav Stat Mar Endoume 43(59): 311–332.
103. Harmelin-Vivien ML, Harmelin JG, Leboulleux V (1995) Microhabitat requirements for settlement of juvenile sparid fishes on Mediterranean rocky shores. Hydrobiologia 300–301: 309–320.
104. Jiménez S, Cano R, Bayle J, Ramos A, Sánchez Lizaso JL (1996) Las praderas de *Posidonia oceanica* (L.) Delile como zona de protección de juveniles de especies de interés comercial. Real Soc Esp Hist Nat, tomo extraordinario: 375–378.
105. UNEP-MAP-RAC/SPA (2011) Proposal of standard methods for inventorying and monitoring coralligenous and rhodoliths populations. Working document, prepared by G Pergent for the tenth meeting of Focal Points for SPAs, Marseilles, France, 17–20 May 2011, UNEP(DEPI)/MED WG 359/10. Tunis: RAC/SPA Publ. 23 p.
106. Dalkey N, Helmer O (1963) An experimental application of the Delphi method to the use of experts. Manage Sci 9: 458–467.
107. Ferrari B, Raventos N, Planes S (2008) Assessing effects of fishing prohibition on *Posidonia oceanica* seagrass meadows in the Marine Natural Reserve of Cerbère-Banyuls. Aquatic Botany 88: 295–302.
108. Astruch P, Boudouresque CF, Bonhomme D, Goujard A, Antonioli PA, et al. (2012) Mapping and state of conservation of benthic marine habitats and assemblages of Port-Cros national Park (Provence, France, northwestern Mediterranean Sea). Sci Rep Port-Cros Natl Park 26: 45–90.

109. Bazin N, Imbert M (2012) Le cormoran huppé de Méditerranée *Phalacrocorax aristotelis desmarestii*. Etat des connaissances et de la conservation actualisé des populations nicheuses des petites îles de Méditerranée. Marseilles: Initiative PIM. 18 p.
110. Romero J, Pérez M, Alcoverro T (2012) L'alguer de *Posidonia oceanica* de les illes Medes: més de trenta anys d'estudi. In: Quintana X, Hereu B, editors. El fons marí de les illes Medes i el Montgrí: quatre dècades de recerca per la conservatió. Toroella de Montrgrí: Càtedra d'Ecosistems Marins Mediterranis. pp. 81–100.
111. Rouanet E, Astruch P, Bonhomme D, Bonhomme P, Rogeau E, et al. (2012) Suivi de l'herbier de posidonie de la passe de Bagaud, impact de l'ancrage (Parc national de Port-Cros, Var, France). Marseilles: GIS Posidonie publ. 82 p.
112. Boudouresque CF, Charbonnel E, Meinesz A, Pergent G, Pergent-Martini C, et al. (2000) A monitoring network based on the seagrass *Posidonia oceanica* in the nortwestern Mediterranean Sea. Biol mar medit 7 (2): 328–331.
113. Moreno D, Aguilera PA, Castro H (2001) Assessment of the conservation status of seagrass (*Posidonia oceanica*) meadows: implications for monitoring strategy and the decision-making process. Biol Conservation 102: 325–332.
114. Montefalcone M, Albertelli G, Morri C, Bianchi CN (2007) Urban seagrass: status of *Posidonia oceanica* facing Genoa city waterfront (Italy) and implications for management. Mar Poll Bull 54: 206–213.
115. Van Hoey G, Borja A, Birchenough S, Buhl-Mortensen L, Degraer S, et al. (2010) The use of benthic indicators in Europe: from the Water Framework Directive to the Marine Strategy Framework Directive. Mar Poll Bull 60: 2187–2196.
116. Martínez-Crego B, Alcoverro T, Romero J (2010) Biotic indices for assessing the status of coastal waters: a review of strengths and weaknesses. J Environm Monit 12 (5): 1013–1028.
117. Pergent-Martini C, Pasqualini V, Ferrat L, Pergent G (2006) Ecological data in Integrated Coastal Zone Management: case study of *Posidonia oceanica* meadows along the Corsican coastline (Mediterranean Sea). Environ Manage 38: 889–895.
118. Parravicini V, Rovere A, Vassallo P, Micheli F, Montefalcone M, et al. (2012) Understanding relationships between conflicting human uses and coastal ecosystem status: a geospatial modelling approach. Ecol Indic 19: 253–263.
119. Meinesz A, Lefevre JR, Astier JM (1991) Impact of coastal development on the infralittoral zone along the southern Mediterranean shore of continental France. Mar Poll Bull 23: 343–347.
120. Leriche A, Pasqualini V, Boudouresque CF, Bernard G, Bonhomme P, et al. (2006) Spatial, temporal and structural variations of a *Posidonia oceanica* seagrass meadow facing human activities. Aquat Bot 84: 287–293.
121. Mascaró O, Alcoverro T, Dencheva K, Díez I, Gorostiaga JM, et al. (2013) Exploring the robustness of macrophyte-based classification methods to assess the ecological status of coastal and transitional ecosystems under the Water Framework Directive. Hydrobiologia 704: 279–291.

2

An Integrated Study to Analyze Soil Microbial Community Structure and Metabolic Potential in Two Forest Types

Yuguang Zhang[1]*, Jing Cong[1,2], Hui Lu[1], Caiyun Yang[3], Yunfeng Yang[4], Jizhong Zhou[3,4,5], Diqiang Li[1]*

1 Institute of Forestry Ecology, Environment and Protection, and the Key Laboratory of Forest Ecology and Environment of State Forestry Administration, Chinese Academy of Forestry, Beijing, China, **2** School of Mineral Processing and Bioengineering, Central South University, Changsha, China, **3** Institute for Environmental Genomics and Department of Microbiology and Plant Biology, University of Oklahoma, Norman, Oklahoma, United States of America, **4** State Key Joint Laboratory of Environment Simulation and Pollution Control, School of Environment, Tsinghua University, Beijing, China, **5** Earth Sciences Division, Lawrence Berkeley National Laboratory, Berkeley, California, United States of America

Abstract

Soil microbial metabolic potential and ecosystem function have received little attention owing to difficulties in methodology. In this study, we selected natural mature forest and natural secondary forest and analyzed the soil microbial community and metabolic potential combing the high-throughput sequencing and GeoChip technologies. Phylogenetic analysis based on 16S rRNA sequencing showed that one known archaeal phylum and 15 known bacterial phyla as well as unclassified phylotypes were presented in these forest soils, and *Acidobacteria*, *Protecobacteria*, and *Actinobacteria* were three of most abundant phyla. The detected microbial functional gene groups were related to different biogeochemical processes, including carbon degradation, carbon fixation, methane metabolism, nitrogen cycling, phosphorus utilization, sulfur cycling, etc. The Shannon index for detected functional gene probes was significantly higher ($P<0.05$) at natural secondary forest site. The regression analysis showed that a strong positive ($P<0.05$) correlation was existed between the soil microbial functional gene diversity and phylogenetic diversity. Mantel test showed that soil oxidizable organic carbon, soil total nitrogen and cellulose, glucanase, and amylase activities were significantly linked ($P<0.05$) to the relative abundance of corresponded functional gene groups. Variance partitioning analysis showed that a total of 81.58% of the variation in community structure was explained by soil chemical factors, soil temperature, and plant diversity. Therefore, the positive link of soil microbial structure and composition to functional activity related to ecosystem functioning was existed, and the natural secondary forest soil may occur the high microbial metabolic potential. Although the results can't directly reflect the actual microbial populations and functional activities, this study provides insight into the potential activity of the microbial community and associated feedback responses of the terrestrial ecosystem to environmental changes.

Editor: A. Mark Ibekwe, U. S. Salinity Lab, United States of America

Funding: This research was supported by the public welfare project of the national scientific research institution (CAFRIFEEP201101), China, and National Biological Specimens and Resources Sharing Platform in Nature Reserve (2005DKA21404).However, the funders had no role in study design, data collection and analysis, decision to publish, or preparation of the manuscript.

Competing Interests: The authors have declared that no competing interests exist.

* E-mail: yuguang@sina.com.cn (YZ); lidq@caf.ac.cn (DL)

Introduction

Microorganisms are the most abundant organisms on earth and play key roles in natural ecosystem, including the biogeochemical cycling of carbon, nitrogen, sulfur, phosphorus, and metals, and biodegradation or stabilization of environmental contaminants [1–3]. Because of their important roles, changes in soil microbial community may directly affect soil ecosystem function, particularly carbon and nitrogen cycling [4]. A deeply analysis of microbial community structure and their roles in ecological processes would improve our understanding of the biogeochemical elemental cycles affected by microbial communities in natural or man-made environments [5]. However, our understanding of soil microbial communities in terms of structure, composition, and functional activity are still limited, especially for soil microbial metabolic activity and ecosystem function have received little attention to date.

Numerous studies have investigated the structure and diversity of microbial communities and their relationships with the surrounding environments [4–8]. Previous studies have shown inconsistent or contradictory results, which might be caused by 1) high microbial community complexity and plasticity in natural environments, 2) different methods used to assess microbial diversity, 3) frequency and magnitude of environmental disturbance, and 4) limited knowledge on the temporal and spatial scales of microbial community in ecosystems [4–7,9]. To date, most studies have only described microbial community complexity at the phylogenetic level using 16S rRNA or single functional gene analysis [8]. Further studies of the metabolic potential and physiological traits of microorganisms are necessary to detect specific metabolic processes and functions [4]. Functional gene microarrays, with quantitative and high resolution capabilities, may be a useful tool for this purpose [10,11].

With the increasing efforts made towards implementing sustainable forest management and conservation strategies since last century [12], natural secondary forest, which form naturally after disturbance of the primary forest by human activities or by extreme natural events, are evident in temperate regions around the world [13]. Shennongjia Mountain is located in the northwestern region of Hubei Province, Central China, and this region is considered to have one of the highest levels of biodiversity in China owing to its unique geographic location and complex terrain [14]. In this study, we selected natural mature forest and a natural secondary forest, which formed in the 1970s following the fall of the primary deciduous broadleaf forest on Shennongjia Mountain and analyzed the soil microbial community and metabolic potential using both GeoChip and high-throughput 16S rRNA sequencing approaches. The aims of this study were to determine (i) the composition and structure of soil microbial community in two deciduous broadleaf forest types, (ii) the soil microbial functional gene diversity and the metabolic potential involved in carbon and nitrogen cycling, and (iii) the major environmental factors governing the soil microbial community structure.

Materials and Methods

Site and sampling

The study site was located on Shennongjia Mountain, Hubei Province, China, where deciduous broadleaf forest is a typical and important forest type. Two forest types were selected in this study: natural mature forest (MAT), and natural secondary forest (SEC), which was formed in the 1970s following the felling of the primary deciduous broadleaf forest. The two study sites had similar geography in terms of slope (30°), aspect (NE30°), and elevation (1715–1813 m), and were within 10 km of each other. The soil type was mountain yellow brown soil [15]. The plant survey and soil collected were permitted by the administrative bureau of Shennongjia National Nature Reserve. The detail location for the study site is N 31°29′, E110°21′ for MAT and N 31°25′, E11°20′ for SEC, respectively. At each study site, four plots (20 m×20 m) were selected with about 20 meters between adjacent plots. Ten to fifteen soil cores, at a depth of 0–10 cm, were taken from each plot and combined to obtain about 400 g of soil. Samples were sieved with 2 mm mesh to remove roots and stones, then mixed thoroughly. Soil samples were preserved at −80°C until DNA extraction.

Plant diversity, soil geochemical properties and microbial biomass analyses

Plant diversity was surveyed at each plot, including the plant species, number, height and canopy of each tree or shrub, and diameter at breast (1.3 m) height of trees (DBH>5 cm) and shrubs (DBH>1 cm). Average soil temperature at each plot was measured by placing the Long-Stem Thermomter (SPECTRUM, US) at a depth of 10 cm in relatively open patches. Soil moisture, soil pH, total soil organic carbon and nitrogen, available phosphorous, available water-dissolved organic carbon, and oxidizable organic carbon were measured as previously described [16]. Microbial biomass carbon was determined by the chloroform fumigation-extraction method [17].

DNA extraction, purification, and quantification

Soil microbial genomic DNA was extracted directly from each soil sample (5 g) using a protocol that included liquid nitrogen grinding, freezing and thawing, and treatment with sodium dodecyl sulfate for cell lysis, as previously described [18]. The freshly extracted DNA was purified twice using 0.5% low melting point agarose gel. Final DNA concentrations were quantified by the PicoGreen method [19] using a FLUOstar Optima microplate reader (BMG Labtech, Jena, Germany).

DNA amplification, sequencing, and data analysis

The V4 hypervariable region of the 16S rRNA gene was amplified in each soil sample using PCR primers 515F (5′-GTGCCAGCMGCCGCGGTAA-3′) and 806R (5′-GGACTACH VGGGTWTCTAAT-3′), using a sample tagging approach [20,21]. The amplification mix contained 10 units of AccuPrime High Fidelity *Taq* polymerase (Invitrogen, Grand Island, NY), 1× AccuPrime PCR buffer II (Invitrogen), 200 μM dNTPs (Amersham, Piscataway, NJ), and 10 μM of each primer in a volume of 25 μl. Genomic DNA (10 ng) was added to the PCR mix. Each sample was amplified under the following conditions: 94°C for 1 min, 30 cycles of 94°C for 20 s, 53°C for 25 s, and 68°C for 45 s, then 10 min at 68°C. The purified PCR amplicons were analyzed using a Miseq Benchtop Sequencer for 2×150 bp paired-end sequencing (Illumina, San Diego, CA).

To minimize the effect of random sequencing errors, only the first 250 bp after the proximal PCR primer of each sequence was analyzed. Sequence that did not perfectly match the PCR primer, had non-assigned tags, or had reads <250 bp was removed. All sequences were aligned using the Ribosomal Database Project Infernal Aligner [22], and the complete linkage clustering method was used to define operational taxonomic units (OTUs) using 97% identity as a cutoff [21]. The number of detected OTUs and sequences at different levels of classification were counted. Details of amplicon preparations, sequencing, and data analysis (e.g. classification, OTU identification) were described in previous reports [20,21]. To standardize samples, a sub-sample of 20,000 sequences per soil sample was used to compare relative difference among samples. Singletons were removed for downstream analyses. All the 16S rRNA sequences were deposited in GenBank database and the accession number is SRP035449.

GeoChip hybridization and data analysis

GeoChip 4.0 was used for DNA hybridization. This microarray contains >83,000 oligonucleotide probes targeting >150,000 genes in 410 gene categories involved in biogeochemical cycling of carbon, nitrogen, phosphorus, and sulfur, and bioremediation of metal and organic contaminants [24]. To produce consistent hybridizations from all samples, the amplification was used to generate approximately 3.0 μg of DNA from 50 ng of purified DNA using a TempliPhi Kit (GE Healthcare, Piscataway, NJ) following the manufacturer's instructions. Amplified DNA was labeled with Cy5 fluorescent dye (GE Healthcare) using a random priming method [23]. All hybridizations were carried out at 45°C for 10 h with 50% formamide using a TECAN HS4800 (TECAN, US) microarray hybridization station, and arrays were scanned using a ScanArray 5000 analysis system (Perkin-Elmer, Wellesley, MA).

Signal intensities of each spot were measured using ImaGene 6.0 (Biodiscovery, El Segundo, CA), and only the spots automatically scored as positive in the raw data output were used for further data analysis [25]. Spots with a signal-to-noise ratio (SNR = (signal intensity - background intensity)/standard deviation of the background) greater than 2.0 were used for further analysis [23]. The GeoChip data were further analyzed using the following steps: (i) genes detected in only two out of four samples from the same forest type were removed; (ii) the signal intensity of each spot was normalized by dividing by the mean value of eight samples of total signal intensity; and (iii) the data were transformed to the natural logarithmic form.

Statistical analyses

Relative abundance of the 16S rRNA gene was based on the proportional frequencies of those DNA sequences that could be classified at different OTUs. Diversity of the microbial community based on sequencing analysis and diversity of functional gene from GeoChip analysis were calculated using Simpson's reciprocal index (1/D) and the Shannon index (H′) using online software (http://ieg.ou.edu/). Detrended correspondence analysis (DCA) was used to determine the differences in overall microbial community structure and functional gene diversity between the two forest types. To examine the differences in soil microbial community structure and functional diversity between the two study sites, we used the Bray-Curtis similarity index to calculate distance matrices from OTUs and GeoChip hybridization data with the multi-response permutation procedure (MRPP) [26] and Adonis [27]. Mantel tests, canonical correspondence analysis (CCA), and variation partitioning analysis were used to evaluate the linkages between microbial community structure, metabolic potential related to carbon and nitrogen cycles, and environmental factors. The variance inflation factor was used for step-wise removal of redundant variables in CCA modeling [23]. All the analyses were performed by functions in the Vegan package (v. 1.15-1) in R (v. 2.9.1) (http://www.r-project.org/).

Results

Plant diversity and soil geochemical properties

Plant diversity and soil properties in the MAT and SEC forests were analyzed (Table 1). The number of plant species (including trees and shrubs) was 63 and 91 in MAT and SEC, respectively. The dominant trees were *Fagus engleriana*, *Quercus aliena var. acuteserrata*, and *Carpinus viminea* at the MAT forest site, and *Toxicodendron verniciflua*, *Tilia paucicostata*, *Juglans cathayensis*, and *Q. aliena var. acuteserrata* at the SEC site, respectively. The Shannon index of plant was significantly lower for MAT (1.18) than for SEC (2.12).

Soil organic carbon, oxidizable organic carbon, and microbial biomass carbon were significantly higher ($P<0.01$) in the SEC samples than in the MAT samples. Soil geochemical properties, such as total nitrogen, available nitrogen, available phosphorus, pH, and soil moisture, were also significantly different ($P<0.01$) between the two forest types (Table 1).

Soil microbial community composition and structure

To determine the overall composition of the soil microbial community in the two forest types, soil microbial communities were analyzed by 16S rRNA high throughput sequencing. After preprocessing of all reads, 65,666 and 65,584 high-quality sequences were obtained from the MAT and SEC sites, respectively (Table S1). The number of sequences ranged from 15,521 to 16,919 in MAT samples and from 16,002 to 16,970 in SEC samples. 10,160 OTUs were detected using 97% identity as a cutoff. Phylogenetic analysis showed that one known archaeal phylum and 15 known bacterial phyla as well as unclassified phylotypes were presented in these forest types (Table 2), and that all of the phylotypes were detected in both MAT and SEC, except for BRC1, which was only found in SEC. At the phylum level, the majority of OTUs were derived from *Proteobacteria* (4,048, 39.8%), followed by *Acidobacteria* (1,787, 17.6%), *Actinobacteria* (1,019, 10.0%), and *Verrucomicrobia* (896, 8.82%) (Table 2).

The Shannon index of 16S rRNA sequences was 8.05 and 8.28 for the MAT and SEC sites, respectively (Table S1). The number of OTUs and the Simpson index were significantly ($P<0.05$) different between MAT and SEC samples (Table S1). DCA was performed using the relative abundance values of sequencing data, resulting in two distinct clusters (Figure 1A). The results of nonparametric multivariate statistical tests, Adonis and MRPP, showed significant differences ($R^2=0.688$, $P=0.001$, and $\delta=0.387$, $P=0.026$, respectively) based on the abundance of all OTUs detected in MAT and SEC samples (Table S2). Significant ($P<0.05$) differences were also observed at the phylum level for *Acidobacteria, Actinobacteria, Armatimonadetes, Bacteroidetes, Proteobacteria, Planctomycetes*, and BRC1, except for WS3 (Table 2). The differences in microbial diversity at the phylum level in the two forest types varied. Ten known microbial phyla and unclassified taxa were significantly ($P<0.05$) higher in SEC samples, including the dominant phyla *Proteobacteria* and *Actinobacteria*. However, *Acidobacteria*, *Chlamydiae*, and *Planctomycetes* were significantly ($P<0.05$) lower in SEC soils. These results indicated that the

Table 1. Plant diversity and soil geochemical properties in two forest types.

Forest type	MAT	SEC	Forest type	MAT	SEC
non index of tree and shrubs**	1.18±0.27	2.12±0.17	Soil organic carbon (g/kg)**	36.40±3.41	86.07±9.70
Species number of tree*	6.00±0.41	12.50±2.10	Dissolved organic carbon (g/kg)	0.17±0.01	0.26±0.08
Height of tree**	14.14±0.54	8.73±0.34	Liable organic carbon (g/kg)**	2.00±0.48	12.37±1.58
Diameter at breast height of tree**	19.85±1.02	12.06±0.75	Microbial biomass carbon (g/kg)**	0.81±0.05	1.31±0.07
Soil Moisture (%) **	50.12±1.23	27.32±2.45	Total nitrogen (g/kg)**	2.33±0.45	5.83±0.45
Temperature at 10 cm depth (°C)	15.92±0.15	15.32±0.09	Available nitrogen (g/kg)**	0.22±0.02	0.41±0.03
Soil pH**	4.81±0.05	5.91±0.28	Available phosphorus (g/kg) *	0.004±0.00	0.015±0.00

Data present the mean value and standard error.
*, $P<0.05$,
**, $P<0.01$.

Table 2. Numbers of soil microbial OTUs and composition in two forest types.

Domain and phylum	Total[a] (%)	Average[b]		P (unpaired t test)
		MAT	SEC	
Archaea				
Crenarchaeota	19 (0.19)	4.50±0.96	10.00±0.82	0.005
Bacteria				
Acidobacteria	1787 (17.59)	777.75±22.21	685.50±14.06	0.017
Actinobacteria	1019 (10.03)	362.25±19.69	443.75±17.09	0.021
Armatimonadetes	22 (0.22)	5.25±0.63	8.50±1.19	0.065
Bacteroidetes	621 (6.11)	91.50±5.87	379.50±10.81	0.000
BRC1	3 (0.03)	0.00	2.25±0.25	0.000
Chlamydiae	17 (0.17)	8.00±1.47	1.50±0.28	0.019
Chloroflexi	109 (1.07)	26.25±2.32	42.25±3.71	0.014
Cyanobacteria	3 (0.03)	1.00±0.00	2.50±0.29	0.014
Firmicutes	129 (1.27)	45.50±4.17	60.75±5.07	0.061
Gemmatimonadetes	97 (0.95)	23.00±1.08	53.50±5.74	0.011
Nitrospirae	7 (0.07)	1.00±0.41	4.50±0.29	0.001
Planctomycetes	657 (6.47)	288.50±9.91	173.75±9.29	0.000
Proteobacteria	4048 (39.84)	1409.00±31.39	1826.25±52.01	0.001
Verrucomicrobia	896 (8.82)	459.00±41.12	349.75±3.07	0.076
WS3	12 (0.12)	0.50±0.28	6.50±0.87	0.004
Unclassified	714 (7.03)	212.25±5.96	326.75±11.06	0.000
Total	**10160**	**3623.00±77.66**	**4469.75±53.03**	**0.000**

[a]Data represent total numbers of OTUs detected by sequencing across all 8 samples.
[b]Data represent the mean value and standard error of OTUs detected using 4 samples in different forest types.

overall soil microbial taxonomic composition and structure were significantly different between the SEC and MAT samples.

Soil microbial functional gene diversity

A total of 38,925 and 50,068 gene probes were detected using GeoChip 4.0 in the MAT and SEC samples, respectively (Table S3). The detected gene groups were related to different biogeochemical processes, such as carbon degradation, carbon fixation, methane metabolism, nitrogen cycling, phosphorus utilization, stress responses, sulfur cycling, contaminant remediation, metal remediation, and energy processing (Table S4). The average number of gene probes detected was 43,427.3 for SEC

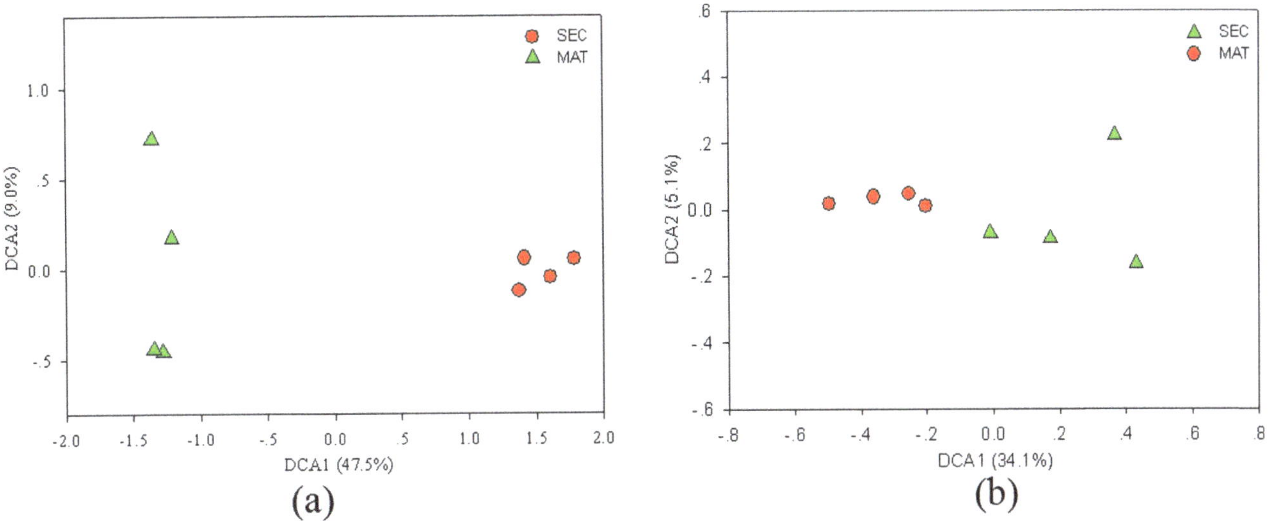

Figure 1. Detrended correspondence analysis (DCA) of soil microbial community structure. The DCA was analyzed based on the relative abundances of OTUs (A) and functional genes based on GeoChip 4.0 (B).

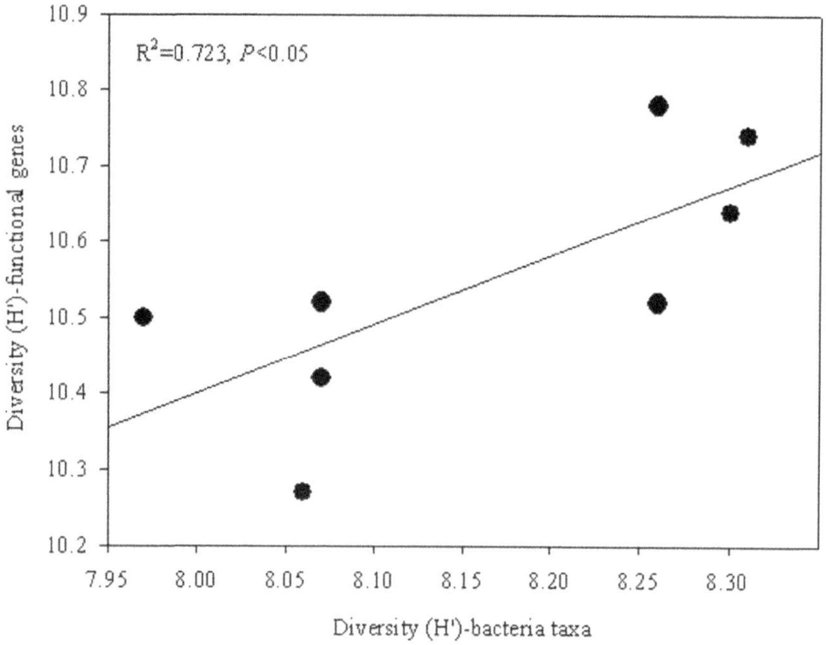

Figure 2. Relationship between soil microbial phylogenetic diversity index and functional gene diversity index.

samples and 33,982.8 for MAT samples. The Shannon index for detected functional microbial gene probes were significantly higher ($P<0.05$) at SEC sites (10.67) than at MAT sites (10.43) (Table S3). DCA showed that MAT samples were well separated from SEC samples (Figure 1B). The regression analysis showed that a strong positive ($P<0.05$) correlation was existed between the soil microbial functional gene diversity and phylogenetic diversity (Fig. 2).

Microbial functional genes and metabolic potential related to carbon cycling

To gain insight into the soil microbial metabolic potential related to carbon cycle, GeoChip 4.0 data for genes related to carbon cycling was analyzed further. Among the 5,792 gene probes related to carbon degradation, 4,276 (74%) were common between the two forest types, while 59 (1%) were presented only in MAT samples and 1,457 (25%) were presented only in SEC samples. The relative signal intensities of the different gene categories related to degradation of relatively active carbon (e.g. starch, hemicelluloses, pectin, and cellulose) and recalcitrant carbon (e.g. chitin and lignin) were significantly ($P<0.05$) higher in SEC than in MAT samples (Figure 1). This included genes encoding amylase and pullulanase (related to starch degradation), arabinofuranosidase and xylanase (related to hemicellulose degradation), cellobiase and exoglucanase (related to cellulose degradation), pectinase (related to pectin degradation), endochitinase and exochitinase (related to chitin degradation), and glyoxal oxidase and phenol oxidase (related to lignin degradation). These results indicated that the soil microbial metabolic capacity involved in carbon degradation were significantly different between the two different forest types.

Among the 1,085 genes involved in carbon fixation, the relative abundance of three important functional genes, ribulose-1,5-bisphosphate carboxylase/oxygenase (Rubisco), carbon monoxide dehydrogenase (CODH), and propionyl-coA (Pcc), was substantially different. A total of 306, 650 and 230 gene probes were detected for the Rubisco, Pcc, and CODH pathways, respectively. These genes had significantly higher signal intensity in SEC soils than in MAT soils (Figure S1). Geochip 4.0 also detected three enzymes involved in methane metabolism, namely the alpha-subunit of methyl coenzyme M reductase (mcrA), for methane production, and particulate methane monooxygenase (pmoA) and methane monooxygenase (mmoX) for methane consumption. Over 80% of the gene probes were associated with uncultured bacteria and archaea. The signal intensity of all the genes involved in methane metabolism was higher in SEC samples than in MAT samples (Figure 3). These results indicated that the microbial functional capacity related to carbon fixation and methane metabolism may be significantly affected by forest types. The relationships between soil microbial metabolic capacity related to carbon cycling and soil carbon components or enzyme activities were analyzed using the Mantel test (Table 3). The results showed that soil organic carbon and soil microbial biomass carbon were significantly linked ($P<0.05$) to the relative abundance of genes related to carbon cycling. Oxidizable organic carbon was significantly linked ($P<0.05$) to the relative abundance of genes involved in active organic carbon degradation (cellulase, hemicellulase, and starch). As expected, cellulase, glucanase, and amylase activities were significantly linked ($P<0.05$) to the relative abundance of genes encoding cellulase, endoglucanase and exoglucanase, and amylase, respectively.

Microbial functional genes and metabolic potential related to nitrogen cycling

Among the gene probes detected in the forest soil samples, 4,238 were involved in nitrogen cycling, including 767 related to nitrogen fixation, 513 related to ammonification (including 250 gene probes related to assimilatory N reduction), 1,734 related to denitrification, 351 related to dissimilatory N reduction, and 621 related to nitrification. These key functional gene categories were associated with different phylogenetic groups, for example, almost all gene probes related to the ammonification were derived from

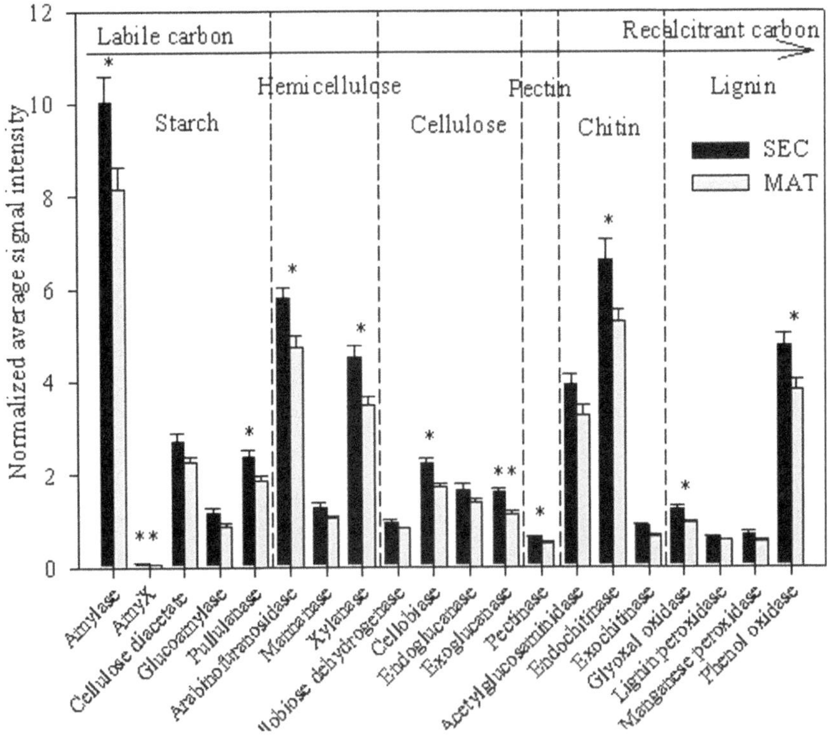

Figure 3. The normalized gene relative signal intensity of the detected key gene categories involved in carbon degradation. The signal intensities were the sum of detected individual gene for each functional gene, averaged among 4 soil samples. The complexity of carbon is presented in order from active to recalcitrant. All data are presented as mean ± standard error. Significant differences between two forest types are indicated above the bars. *, $P<0.05$, **, $P<0.01$.

cultured bacteria, including *Rhodococcus erythropolis*, *Pseudomonas entomophila*, and *Acinetobacter baumannii*. A total of 1,533 (88%) gene probes related to the denitrification were linked to unclassified bacteria, and 137 (22%) gene probes related to the nitrification were derived from archaea, which suggests that archaea maybe play an important role in nitrification process.

The abundance of several genes involved in nitrogen cycling were significantly ($P<0.05$) higher in SEC samples than in MAT samples (Figure 4), including *nifH* (related to nitrogen fixation), *amoA* (related to nitrification), *narG*, *nirS*, *nirK*, and *nosZ*, (related to denitrification), *napA* and *nrfA* (related to dissimilatory N reduction to ammonium), *gdh* and *ureC* (related to ammonification), and *niR* and *nirB* (related to assimilatory N reduction). The relationships between microbial functional genes and soil nitrogen components were analyzed using the Mantel test (Table 4). The results showed that soil total nitrogen was significantly linked ($P<0.05$) to the relative signal intensity of nitrogen cycling genes. The NH_4^+-N and NO_3-N concentrations were linked ($P<0.1$) to the signal

Table 3. The relationships between relative abundances of microbial functional genes involved in carbon and nitrogen cycling and soil carbon or nitrogen components and enzyme activity by Mantel-test.

Soil carbon parameters	Microbial relative abundance	R value	P
Soil organic carbon	Carbon cycling genes	0.389	**0.039**
Oxidizable organic carbon	Oxidizable organic carbon degradation genes(Cellulose, Hemicellulose, Starch)	0.403	**0.046**
Water dissolved organic carbon	Carbon cycling genes	0.114	0.216
Soil microbial biomass carbon	Carbon cycling genes	0.500	**0.020**
Total soil nitrogen	Nitrogen cycling genes	0.382	**0.050**
NH_4^+-N	Ammonification genes	0.296	**0.089**
No_3^-N	Nitrification genes	0.302	**0.079**
Cellulase activity	Cellulase genes	0.342	**0.046**
Glucanase activity	Endo- and exoglucanase genes	0.351	**0.022**
Polyphenol oxidase activity	Polyphenol oxidase genes	0.046	0.626
Amylase activity	Amylase genes	0.358	**0.027**

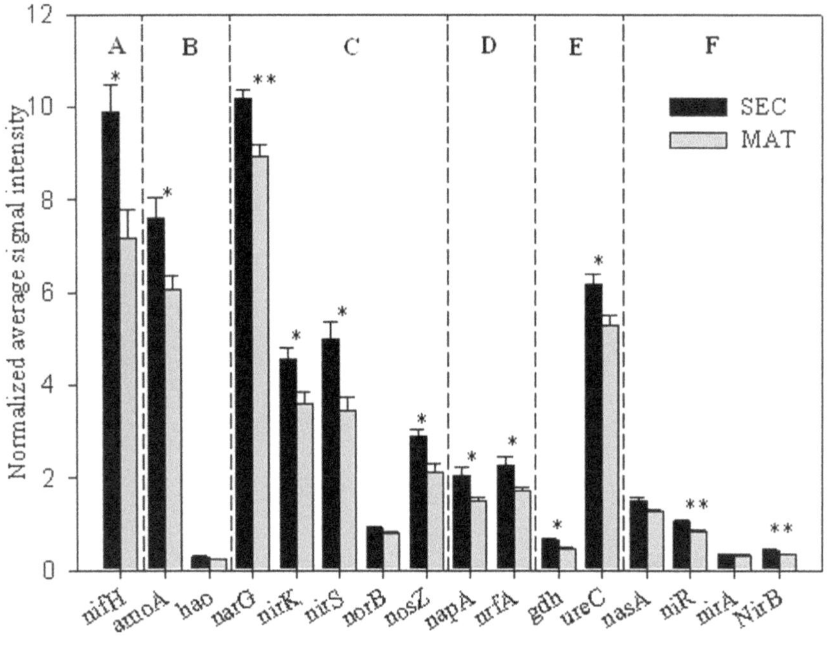

Figure 4. The normalized average signal intensity of detected key gene categories involved in the N cycling between SEC and MAT. The signal intensities were the sum of all detected individual gene for each gene category, and then averaged among 4 samples. (A). N_2 fixation, including nifH encoding nitrogenase; (B). Nitrification, including amoA encoding ammonia monooxygenase, hao for hydroxylamine oxidoreductase; (C). Denitrification, including narG for nitrate reductase, nirS for nitrite reductase, norB for nitric oxide reductase, and nosZ for nitrate reductase; (D). Dissimilatory N reduction to ammonium, including napA for nitrate reductase, and nrfA for c-type cytochrome nitrite reductase; (E). Ammonification, including gdh for glutamate dehydrogenase and ureC encoding urease; (F). Assimilatory N reduction, including nasA encoding nitrate reductase, niR, nirA and nirB encoding dissimilatory nitrite reductase. All data are presented as the mean±SE (error bars). **, $P<0.01$, *, $P<0.05$.

intensity of genes involved in the ammonification and nitrification processes, respectively.

Linking microbial community structure and environmental factors

Mantel tests and CCA were performed to analyze the major environmental factors responsible for shaping the microbial community structure and functional diversity. The Mantel test showed that microbial community structure and functional gene diversity were significantly ($P<0.05$) linked to soil and plant factors (Table 4). CCA was used to identify the major environmental variables controlling the soil microbial community structure, resulting in a model significant at a confidence level of $P=0.048$. The plant Shannon index and tree DBH appeared to be important environmental factors controlling the microbial community structure, as they had a long projection on Axis 1 ($P<0.01$), which represented the major variation among microbial communities (Figure 5A). Similarly, soil temperature and soil organic carbon were critical for shaping microbial communities.

Variance partitioning analysis was used to quantify the contributions of soil chemical factors (Soil), soil temperature (Temp), and plant diversity (Plant) to the variation in the microbial community. A total of 81.58% of the variation was significantly explained by these three environmental variables (Figure 5B). Soil, Temp, and Plant explained 26.02%, 8.58%, and 19.35% of variation, respectively, verifying that they were major factors in shaping the microbial community structure.

Table 4. Mantel test and partial Mantel test between 16S rRNA OTUs and Geochip data with environmental properties.

Environmental variable	16S OTUs		GeoChip 4.0	
	r value	P value	r value	P value
Whole variable	0.868	0.005	0.412	0.008
Soil organic carbon	0.770	0.003	0.392	0.013
Total nitrogen	0.832	0.003	0.402	0.019
pH	0.622	0.006	0.322	0.043
Temperature at the 10 cm depth	0.463	0.002	0.085	0.272
Shannon Index for trees and shrubs	0.779	0.006	0.337	0.028
Diameter at breast height of trees	0.532	0.008	0.125	0.158

The Mantel test was only performed for whole variable, which included all variables presented in this table.

Discussion

The development and increasing availably of high-throughput molecular techniques has helped improve our understanding of microbial composition, especially through the use of direct sequencing of metagenomic DNA. Fierer et al. indicated that all biomes were dominated by *Acidobacteria*, *Actinobacteria*, *Proteobacteria* and *Bacteroidetes*, and the bacterial community composition does not vary significantly across different biomes [1].Our results showed that microbial diversity at the phylum level differed between the two forest types, and *Acidobacteria*, *Protecobacteria*, and *Actinobacteria* were three of most abundant soil bacterial phyla. *Acidobacteria* is known to be dominant in other soil types, although they are often difficult to cultivate [8,28]. The relative abundance of *Acidobacteria* was also shown to be higher in forest, desert, and prairie soils compared with agricultural soils [28]. Recent studies

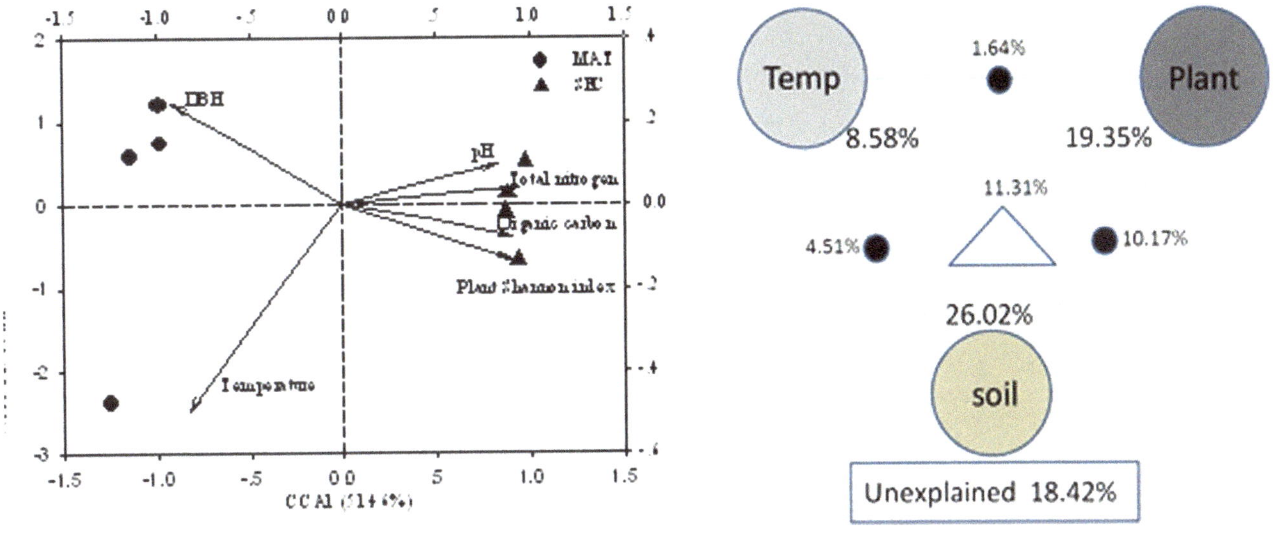

Figure 5. The Canonical correspondence analysis (CCA) and Variation partitioning analysis (VPA) between OTUs data and soil environmental variables. The left was CCA between microbial OTUs data and soil environmental factors, and the right was the VPA. DBH, diameter at breast height of tree. Soil, including soil organic carbon, total nitrogen and pH, soil temperature (Temp) and plant diversity (Plant, including the Shannon index of trees and shrubs, diameter at breast height of trees), and their relationships. Each diagram represents the biological variation partitioned into the relative effects of each factor or a combination of factors, in which geometric areas were proportional to the percentages of explained variation.

showed that *Acidobacteria* is in general oligotrophic, consistent with results showing that the proportion of *Acidobacteria* was significantly lower in nutrient-rich rhizosphere and agricultural soils than in bulk soil [4,29]. Some studies also found that the relative abundance of *Acidobacteria* was negatively linked to the level of nitrogen input [30]. In this study, the relative abundance of *Acidobacteria* was high in the natural forest, which had low soil organic carbon and nitrogen contents.

Due to microbial highly diversity and natural ecosystem complexity, it is a great challenging to set up the link of soil microbial structure and composition to functional activity related to ecosystem functioning in microbial ecology. Many studies have described microbial community structure in a range of different environments, but have not identified critical groups related to detailed functional processes [4–8,31]. Therefore, it remains unclear the relationship between microbial community structure and microbial functional gene diversity [32–34]. Functional redundancy is considerable debate in ecology as to how the relationship between the taxonomic diversity and observed functional diversity [35]. This study combines high-throughput sequencing to study the microbial community structure at the phylogenetic level and GeoChip to microbial potential metabolic activities at functional gene level. 16S rRNA gene-based high-throughput sequencing data provide information on microbial phylogenetic structure and composition, and GeoChip-based data provide quantitative information on various microbial functional groups [23]. These integrated technologies provide a reliable method for detection soil microbial structure, composition and potential functional activity [23]. In this study, a strong positive correlation were found between the soil microbial functional and phylogenetic diversity among sampling sites, which suggesting microbial phylogenetic and functional diversity aligned well and a low degree of functional redundancy were existed. Our result was consistent with a recent work on reconstructing the microbial diversity and function of pre-agricultural tallgrass prairie soils in the United States [36]. Furthermore, the higher phylogenetic diversity detected in SEC was consistent with the higher number of functional genes involved in the carbon and nitrogen cycles, unveiling the larger functional potential and capacity of the secondary forest soil.

Soil microbial metabolic potential and ecosystem function have received little attention owing to difficulties in methodology. As we all known, most of the biogeochemical processes were controlled and related to diverse and various functional genes, therefore, the conjunction between microbial community identified by using normal molecular methods (i.e. DGGE, T-RFLP) with soil enzyme activities or components is difficult. GeoChip-based data provides large-scale quantitative information on various biogeochemical functional groups, thus making it possible to link the functional structure of microbial communities with ecosystem processes [23]. Previous studies showed that change observed in the microbial community structure and functional groups has significant effects on soil enzyme activities or components related to carbon and nitrogen cycles. Yergeau et al. [37] and Reeve et al. [34] showed a significant correlation between cellulase enzyme activity and cellulose gene variants, detected using GeoChip, as well as significant correlations between dehydrogenase genes and enzyme activity, urease genes and soil ammonium, and sulfite reduction genes and soil sulfur level [34]. In this study, we also found the linked of soil microbial functional gene diversity to soil enzyme activities and components, such as the soil organic carbon and soil microbial biomass carbon were significantly linked ($P<0.05$) to the relative abundance of genes related to carbon cycling, oxidizable organic carbon was significantly linked ($P<0.05$) to the relative abundance of genes involved in active organic carbon degradation (cellulase, hemicellulase, and starch), respectively. Although the GeoChip data can't directly reflect the actual microbial populations and functional activities, it implied that the microbial populations carrying those genes could exist and have the functional capacity. Our GeoChip data showed that metabolic genes involved in the degradation of starch, cellulose, hemicellulose, chitin, lignin, and pectin, as well as those involved in nitrogen cycling, were significantly higher ($P<0.05$) in SEC than in MAT soils. Therefore, SEC soil had high microbial gene diversity and microbial metabolic potential. The use of GeoChip 4.0 may allow us to rapidly obtain community-wide information and observe differences functional capacity in different environments.

Only about 1% of the microorganisms in soil can be cultured, and researchers have increased our understanding of microbes in the natural environments by use of PCR amplification methods which reduce our dependence on the presence of culturable microbes. In our study, the majority of the genes related to carbon and nitrogen cycles were derived from the uncultured bacteria, suggesting that they are unique and may represent novel genes in the study area. For example, 88% gene probes related to the denitrification process were linked to unclassified bacteria. In previous studies, researchers received the same results, for example, Xie et al found 64.28% sequences were from uncultivated bacteria in the acid mine drainage [38], Zhang et al found the 320/372 genes were derived from the unidentified or uncultured organisms in the alpine meadows soils in the Qinghai-Tibetan plateau [39]. Therefore, a lot of the microbial functional gene resources were existed in natural environments.

Much recent work in soil microbial ecology has showed that soil microbial community structures are affected by specific environmental changes or disturbances [11,25]. Soil properties and plant diversity are key factors in shaping the microbial community composition and structure [40,41]. Organic carbon availability limited microbial communities in most soils [37], and addition of labile organic material rapidly alters microbial communities by selecting for populations that are most competitive in terms of growth rates and ability to absorb nutrients [41]. Thus, it was not considered to be the limiting factor for the microbial communities. Our data showed that plant diversity also significantly influenced the microbial community structure at the phylum level. Plant communities can influence the associated soil microbial communities through the types and amounts of organic carbon and nutrient inputs, and by altering the temperature and water content of the soil [42,43]. Therefore, changes in plant diversity, composition, and production affect the composition and diversity of soil microbial communities [9]. However, our understanding of differences in soil microbial diversity depending on various vegetation types is still poor.

In summary, this work studied the microbial community structure and metabolic potential by combining high-throughput sequencing and GeoChip technologies, and showed the soil microbial community structure and functional metabolic potential had a positive correlation. Although the microbial populations and metabolic activities could not be characterized comprehensively, the detected functional gene groups may represent the *in situ* microbial metabolic potential at the functional cluster level to some degree. Further analysis of the functional activity with different approaches such as mRNA-based analysis is needed to understand the biogeochemical processes related to microbial community.

Supporting Information

Figure S1 The normalized gene relative signal intensity of the detected key gene categories involved in carbon fixation and methane cycling. The signal intensities were the sum of detected individual gene for each functional gene, averaged among 4 soil samples. CODH: carbon monoxide dehydrogenase; Pcc: propionyl-coA; Rubisco: ribulose-1,5 -bisphosphate carboxylase/oxygenase; pmoA: methane monooxygenase; mcrA: methyl coenzyme M reductase. All data are presented as mean ± standard error. Significant differences between two forest types are indicated above the bars. *, $P<0.05$, **, $P<0.01$.

Table S1 16S rRNA diversity index. Summary of sequence numbers and OTUs, Shannon index and Simpson index for each sample at MAT and SEC based on sequencing data.

Table S2 The difference of soil microbial community composition and structure. Statistical analysis of differences in the microbial community composition and structure between SEC and MAT at the phylum level.

Table S3 The functional gene diversity index. Summary of numbers of detected gene probes, Shannon index and Simpson index for each sample at MAT and NAF based on GeoChip 4.0 data.

Table S4 The detected functional genes families. Summary of numbers of detected gene probes for different functional gene families at SEC and MAT based on GeoChip 4.0 data.

Acknowledgments

We gratefully acknowledge Dr. Ye Deng, Dr. Junjun Ding, Mr. Jingyuan Yang, Mr. Xiujiang Su, Dr. Guangliang Li and Dr. Yadong Xue for helping data analysis and field survey the plant plots.

Author Contributions

Conceived and designed the experiments: YZ JZ DL. Performed the experiments: YZ JC HL CY. Analyzed the data: YZ JC YY. Contributed reagents/materials/analysis tools: YZ DL. Wrote the paper: YZ.

References

1. Fierer N, Jackson RB (2006) The diversity and biogeography of soil bacterial communities. Proc Natl Acad Sci USA 103: 626–631.
2. Green JL, Bohannan BJM, Whitaker RJ (2008) Microbial biogeography: from taxonomy to traits, Science 320: 1039–1043.
3. He Z, Nostrnd JDV, Deng Y, Zhou J (2011) Development and applications of functional gene microarrays in the analysis of the functional diversity, composition, and structure of microbial communities. Front. Environ. Sci Engin China 5: 1–20.
4. López-Lozano NE, Heidelberg KB, Nelson WC, Garcia-Oliva F, Eguiarte LE, et al (2013) Microbial secondary succession in soil microcosms of a desert oasis in the Cuatro Cienegas Basin, Mexico. PeerJ 1: e47. Available: http://dx.doi.org/10.7717/peerj.47. Accessed 2013 March 5.
5. Fierer N, Nemergut D, Knight R, Craine JM (2010) Changes through time: integrating microorganisms into the study of succession. Res Microbiol 161: 635–642.
6. Kuramae EE, Gamper HA, Yergeau E, Piceno YM, Brodie EL, et al (2010) Microbial secondary succession in a chronosequence of chalk grasslands. ISME J 4: 711–715.
7. Banning NC, Gleeson DB, Grigg AH, Grant CD, Andersen GL, et al (2011) Soil microbial community successional patterns during forest ecosystem restoration. Appl Environ Microbiol 77: 6158–6164.
8. Ding G, Piceno YM, Heuer H, Weinert N (2013) Changes of soil bacterial diversity as a consequence of agricultural land use in a semi-arid ecosystem. PLOS one 8: e59497.
9. Zhu W, Cai X, Liu X, Wang J, Cheng S, et al (2010) Soil microbial population dynamics along a chronosequence of moist evergreen broad-leaved forest succession in southwestern China. J M Sci 7: 327–338.
10. He Z, Gentry TJ, Schadt CW, Wu L, Liebich J, et al (2007) Geochip: a comprehensive microarray for investigating biogeochemical, ecological and environmental processes. ISME J 1: 67–77.

11. Zhou J, Kang S, Schadt CD, Garten CTJ (2008) Spatial scaling of functional gene diversity across various microbial taxa. Proc Natl Acad Sci USA 105: 7768–7773.
12. Aerts R, Honnay O (2011) Forest restoration, biodiversity and ecosystem functioning. BMC Ecol 11: 29.
13. Wang CK, Yang JY (2007) Rhizospheric and heterotrophic components of soil respiration in six Chinese temperature forests. Global Change Biol 13: 123–131.
14. Ma C, Zhu C, Zheng C, Wu C, Guan Y, et al (2008) High-resolution geochemistry records of climate changes since late-glacial from Dajiuhu peat in Shennongjia Mountains, Central China. Chinese Sci Bull 53 (Supp.1): 28–41.
15. Zhao C, Chen W, Tian Z, Xie Z (2005) Altitudinal pattern of plant species diversity in Shennongjia Mountains, Central China. J Integr Plant Biol 47(12): 1431–1449.
16. Bao SD (1999) Soil and agricultural chemistry analysis. Beijing: China Agriculture Press, 25–150 p.
17. Chen CR, Xu ZH, Mathers NJ (2004) Soil carbon pools in adjacent natural and plantation forests of subtropical Australia. Soil Sci Soc Am J 68: 282–291.
18. Zhou JZ, Bruns MA, Tiedje JM (1996) DNA recovery from soils of diverse composition. Appl Environ Microbiol 62: 316–322.
19. Ahn S, Costa J, Emanuel J (1996) PicoGreen quantitation of DNA: effective evaluation of samples pre- or post-PCR. Nucleic Acids Res 24: 2623–2625.
20. Caporaso JG, Lauber CL, Walters WA, Berg-Lyons D, Lozupone CA, et al (2011) Global patterns of 16S rRNA diversity at a depth of millions of sequences per sample. Proc Natl Acad Sci USA 108: 4516–4522.
21. Caporaso JG, Lauber CL, Walters WA, Berg-Lyons D, Huntley J, et al (2012) Ultra-high-throughput microbial community analysis on the Illumina HiSeq and MiSeq platforms. ISME J 6: 1621–1624.
22. Cole JR, Wang Q, Cardenas E, Fish J, Chai B, et al (2009) The ribosomal database project: improved alignments and new tools for rRNA analysis. Nucleic Acids Res 37: 141–145.
23. He Z, Xu M, Deng Y, Kang S, Kellogg L, et al (2010) Metagenomic analysis reveals a marked divergence in the structure of belowground microbial communities at elevated CO_2. Ecol Lett 13: 564–575.
24. Lu Z, Deng Y, Van Nostrand JD, He Z, Voordeckers J, et al (2011) Microbial gene functions enriched in the deepwater horizon deep-sea oil plume. ISME J 5: 1–10.
25. Liang Y, Nostrand JD, Deng Y, He Z, Wu L, et al (2011) Functional gene diversity of soil microbial communities from five oil-contaminated fields in China. ISME J 5: 403–413.
26. McCune B, Grace JB (2002) Analysis of ecological communities. MJM Software Design, Gleneden Beach, OR.
27. Anderson MJ (2001) A new method for non-parametric multivariate analysis of variance. Aust J Ecol 26: 32–46.
28. Janssen PH (2006) Identifying the dominant soil bacterial taxa in libraries of 16S rRNA and 16S rRNA genes. Appl Environ Microbiol 72: 1719–1728.
29. Kielak A, Pijl AS, van Veen JA, Kowalchuk GA (2009) Phylogenetic diversity of Acidobacteria in a former agricultural soil. ISME J 3: 378–382.
30. Fierer N, Lauber CL, Ramirez KS, Zaneveld J, Bradford MA, et al (2012) Comparative metagenomic, phylogenetic and physiological analyses of soil microbial communities across nitrogen gradients. ISME J 6: 1007–1017.
31. Kirk JL, Beaudette LA, Hart M, Moutoglis P, Klironomos JN, et al (2004) Methods of studying soil microbial diversity. J Microbiol Meth 58: 169–188.
32. Kelly JJ (2003) Molecular techniques for the analysis of soil microbial processes: functional gene analysis and the utility of DNA microarrays. Soil Sci 168: 597–605.
33. Little AEF, Robinson CJ, Peterson SB, Raffa KE, Handelsman J (2008) Rules of engagement: inter-species interaction that regulate microbial communities. Annu Rev Microbiol 62: 375–401.
34. Reeve JR, Schadt CW, Carpenter-Boggs L, Kang S, Zhou J, et al (2010) Effects of soil type and farm management on soil ecological functional genes and microbial activities. ISME J 4: 1–9.
35. Hillebrand H, Matthiessen B (2009) Biodiversity in a complex world: consolidation and progress in functional biodiversity research. Ecol Lett 12: 1405–1419.
36. Fierer N, Ladau J, Clemente JC, Leff JW, Owens SM, et al (2013) Reconstructing the microbial diversity and function of pre-agricultural tallgrass prairie soils in the United States. Science 342: 621–624.
37. Yergeau E, Kang S, He Z, Zhou J, Kowalchuk GA (2007) Functional microarray analysis of nitrogen and carbon cycling genes across an Antarctic latitudinal transect. ISME J 1: 1–17.
38. Xie JP, He ZL, Liu XX, Liu XD, Van Nostrand JD, et al (2011) GeoChip-Based Analysis of the Functional Gene Diversity and Metabolic Potential of Microbial Communities in Acid Mine Drainage. Appl Environ Microbiol 77: 991–999.
39. Zhang Y, Lu Z, Liu S, Yang Y, He Z, et al (2013) Geochip-based analysis of microbial communities in alpine meadow soils in the Qinghai-Tibetan plateau. BMC Microbiol 13(1): 72.
40. Carney K, Maston P (2005) Plant communities, soil microorganism, and soil carbon cycling: does altering the world belowground matter to ecosystem functioning? Ecosystem 8: 928–940.
41. Drenovsky RE, Vo D, Graham KJ, Scow KM (2004) Soil water content and organic carbon availability are major determinants of soil microbial community composition. Microbial Ecol 48: 424–430.
42. Myers RT, Zak DR, White DC, Peacock A (2001) Landscape-level patterns of microbial community composition and structure in upland forest ecosystems. Soil Sci Soc Am J 65: 359–367.
43. Waldrop MP, Firestone MK (2004) Microbial community utilization of recalcitrant and simple carbon compounds: impact of oak-woodland plant communities. Oecologia 138: 275–284.

3

Contributions of Understory and/or Overstory Vegetations to Soil Microbial PLFA and Nematode Diversities in Eucalyptus Monocultures

Jie Zhao[1,3], Songze Wan[1], Chenlu Zhang[2], Zhanfeng Liu[1], Lixia Zhou[1], Shenglei Fu[1]*

1 Key Laboratory of Vegetation Restoration and Management of Degraded Ecosystems, South China Botanical Garden, Chinese Academy of Sciences, Guangzhou, China, **2** State Key Laboratory of Cotton Biology, Key Laboratory of Plant Stress Biology, College of Life Sciences, Henan University, Kaifeng, China, **3** Key Laboratory of Agro-Ecological Processes in Subtropical Region, Institute of Subtropical Agriculture, Chinese Academy of Sciences, Changsha, China

Abstract

Ecological interactions between aboveground and belowground biodiversity have received many attentions in the recent decades. Although soil biodiversity declined with the decrease of plant diversity, many previous studies found plant species identities were more important than plant diversity in controlling soil biodiversity. This study focused on the responses of soil biodiversity to the altering of plant functional groups, namely overstory and understory vegetations, rather than plant diversity gradient. We conducted an experiment by removing overstory and/or understory vegetation to compare their effects on soil microbial phospholipid fatty acid (PLFA) and nematode diversities in eucalyptus monocultures. Our results indicated that both overstory and understory vegetations could affect soil microbial PLFA and nematode diversities, which manifested as the decrease in Shannon–Wiener diversity index (H') and Pielou evenness index (J) and the increase in Simpson dominance index (λ) after vegetation removal. Soil microclimate change explained part of variance of soil biodiversity indices. Both overstory and understory vegetations positively correlated with soil microbial PLFA and nematode diversities. In addition, the alteration of soil biodiversity might be due to a mixing effect of bottom-up control and soil microclimate change after vegetation removal in the studied plantations. Given the studied ecosystem is common in humid subtropical and tropical region of the world, our findings might have great potential to extrapolate to large scales and could be conducive to ecosystem management and service.

Editor: Dafeng Hui, Tennessee State University, United States of America

Funding: This study was financially supported by the Knowledge Innovation Program of the Chinese Academy of Sciences (KSCZ2-EW-Z-6) and Natural Science Fund of China (31210103920). The funders had no role in study design, data collection and analysis, decision to publish, or preparation of the manuscript.

Competing Interests: The authors have declared that no competing interests exist.

* E-mail: sfu@scbg.ac.cn

These authors contributed equally to this work.

Introduction

Biodiversity plays a crucial role in ecosystem function and processes not only because of its importance in production of food, fiber and fuel, but also because of its roles in ground water replenishment, flooding controls, prevention of soil erosion and ecological invasion, influences on restoration succession, and so on [1,2]. Although the relationship between biodiversity and ecosystem functions and processes has been the focuses of quite a few studies over the last two decades, much controversy still exists [3]. The impact of plant diversity on ecosystem functions and processes has been explored mainly in grasslands (e.g. [4,5]); and primary productivity is frequently used to assess the ecosystem stability [4]. In general, ecosystems with high biodiversity level are more stable, sustainable, reliable and predictable [1,4,5,6]. By contrast, several studies reported no significant effect of plant diversity on plant productivity [7,8].

Soil contains the most diverse communities of organisms and soil biodiversity provides a number of ecosystem services [2,9]. Ecological interactions between aboveground and belowground biodiversity have received many attentions in the recent decades [10]. In one hand, plants play a key role in determining soil biota because plants are the main food sources (i.e. leaf litter, dead roots and root exudates) to soil biota, which may exert a bottom-up control on soil organisms. In another hand, soil biota may affect aboveground plant diversity and productivity through influencing nutrient cycling and distribution, and damaging plants by parasites, pathogens and herbivores [11,12].

Plant functional groups (PFG) or plant functional types (PFT) are assemblages of species with similar roles in ecosystem processes by responding in similar ways to multiple environmental factors [13]. They are necessary devices for reducing the complexity of plant community and often uncharted characteristics of species diversity in function and structure when attempting to project the nature and function of species assemblages into future environments [14]. However, due to the diverse aims of different researches, diverse plant functional groups are chosen to simplify the plant community, i.e. legumes, forbs/herbs, C3/C4-plants, overstory vegetation, and understory vegetation.

Vegetation removal has been frequently used to explore how specific plant species or functional groups affect ecosystem processes and functions. Overstory removal (timber harvest) has been a common forest management practice which also provides opportunities for soil ecologists to study relationships between overstory tree species and soil biota. In general, overstory removal suppresses soil biota such as bacteria, fungus, nematodes, and microarthropods in the long run [15,16,17]. However, there are studies which reported overstory removal had slight or non-significant impacts on soil biota, especially in the short term [18,19]. To date, less attention has been paid to understory vegetation in terms of their impacts on soil biota, probably because the biomass and economic value of the understory are less than overstory tree species. Understory vegetation is an important component in many forest ecosystems worldwide [20,21] and drives ecological processes and functions, such as biodiversity, stand productivity, tree-seedling regeneration, forest succession, litter decomposition, soil nutrient cycling and soil water conservation [19,22,23,24]. However, how understory vegetation affects the soil biodiversity has never been studied before.

In the present study, we monitored the soil microbial phospholipid fatty acid (PLFA) and nematode diversity after overstory and/or understory removal in eucalyptus monocultures in South China. The eucalyptus species is *Eucalyptus urophylla*, which is a fast-grow species and has been widely planted for reforestation in South China [19]. The understory vegetation was dominated by *Dicranopteris dichotoma*, a common fern species in humid subtropical and tropical region [20]. Other common understory species included *Miscanthus sinensis* and *Rhodomyrtus tomentosa*. Thus, the diversity of overstory and understory vegetations is not high in the studied ecosystems. Therefore, we did not calculate the plant diversity. In this experiment, overstory and understory vegetations were removed separately or in combination to assess their impacts on soil microbial PLFA profile and nematode community. Both *E. urophylla* and *D. dichotoma* are high in biomass and primary productivity [20,25]. Previous study, conducted at the same site as this study, reported that understory removal showed stronger effects on soil microbial community structure, and abundance of nematode and microarthropod than overstory removal [20]. Therefore, the hypothesis of this study was that the effects of understory removal on soil microbial and nematode diversity were greater than overstory removal. The primary goal of this study was to evaluate the contributions of the understory and/or overstory vegetation to soil microbial and nematode diversity. Our analyses also intend to address the question of which factors (i.e. resource and soil microclimate) affected soil biodiversity after altering of aboveground vegetations.

Materials and Methods

Site Description

This study was conducted at the Heshan Hilly Land Interdisciplinary Experimental Station (112°50′E, 22°34′N), property of the Chinese Academy of Sciences (CAS) in Guangdong Province of China. The climate is subtropical monsoon with a distinct wet and dry Season. The mean annual temperature and precipitation are 21.7°C and 1,700 mm, respectively. The soil is an acrisol [26]. Vegetation at the experimental site consisted of three (replications) 4-year-old *E. urophylla* plantations. The seedlings were planted with a spacing of 3×2 m.

Experimental Design

Our experiment was conducted in three replicated plots of 4-year-old *E. urophylla* monoculture plantations. In September 2009, we established an experiment with a split-plot design. Within each plantation, main plots (15×15-m) were overstory trees treatment (present or removed) and subplots were understory treatment (present or removed). In total, there were four treatments in each plantation, namely, control with intact overstory and understory (CK), understory removal (UR), overstory tree removal (TR) and all-plant removal (PR). The trees were cut with an electrical saw leaving 5 cm stumps aboveground. All woody debris was removed from the plots immediately after vegetation was removed. The shoots of all understory plants were removed manually with the aid of a machete. Trenches were created around each main-plot and subplot to prevent the interactions of the roots from other subplots and to prevent the exchange of nutrients among subplots. Germination of the stumps (TR plots) and understory vegetation (UR and PR subplots) were monthly removed.

Soil Sampling and Analysis

Soils were sampled on 20 September, 1 October and 6 November 2009, 11 March, 15 June and 14 October 2010, corresponding to 10 days before and 7, 37, 160, 256, 376 days after treatments were implemented. Soil cores (2.5 cm in diameter, 5 cm in length) were taken at 0–5 cm and 5–10 cm depths from eight locations selected randomly in each subplot within each plantation. Eight cores of the same depth from each subplot were combined to form one composite sample; there were three replicate samples for each treatment. The surface litter was removed carefully when soil sample was taken.

Soil water content (SWC %, g of water per 100 g dry soil) was measured by oven-drying for 48 h at 105°C. Soil temperature was recorded every 2 h with the DS1922L temperature logger iButtons (Dallas Semiconductor Corp., Dallas, TX) from September 2009 to October 2010. Soil microbial PLFAs were analyzed according to [27]. Briefly, the soil was extracted in a one-phase extraction mixture containing chloroform:methanol:phosphate buffer (1:2:0.8 v/v/v), with the amount of phosphate buffer corrected to account for existing soil water content mixture of chloroform:-methanol:citrate buffer (1:2:0.8 v/v/v). After extraction the phospholipids were separated from neutral lipids and glycolipids on solid phase extraction columns, 0.50 g Si (Supelco, Inc., Bellefonte, Penn). Polar lipids were eluted and were then subjected to mild alkaline methanolysis. Resulting fatty acid methyl esters (FAMEs) were separated, quantified, and identified using capillary gas chromatography (GC). Qualitative and quantitative fatty acid analyses were performed using an Agilent 6890 gas chromatograph (Agilent Technologies, Palo Alto, CA, USA) and the MIDI Sherlock Microbial Identification System (MIDI Inc., Newark, DE, USA). Concentrations of each PLFA were calculated relative to 19:0 internal standard concentrations. The following PLFAs were considered to be of soil microbial origin: i15:0, a15:0, 15:0, i16:0, 16:1ω7, i17:0, a17:0, 17:0, cy17:0, cy19:0 18:2ω6,9, 16:1ω5, 18:1ω7, and 18:1ω9 [28,29,30].

Nematodes were extracted from 50 g of fresh soil using the Baermann funnel method. After fixation in 4% formalin solution, nematodes were counted with a differential interference contrast (DIC) microscope (ECLIPSE 80 i, Nikon), and the first 100 individuals encountered were identified to genus level. All nematodes were identified to genus level when the sample contained fewer than 100 individuals.

Statistical Analysis

Shannon–Wiener diversity index, Pielou evenness index, Margalef richness index, and Simpson dominance index were employed to assess the soil microbial PLFA and nematode diversity.

$$\text{Shannon}-\text{Wiener diversity index}, H' = \sum_{i=1}^{S} Pi * \ln Pi \quad (1)$$

$$\text{Pielou evenness index}, \text{J} = \frac{H'}{H' \max} \quad (2)$$

$$H' \max = \ln S \quad (3)$$

$$\text{Margalef richness index}, \text{SR} = \frac{S-1}{\ln N} \quad (4)$$

$$\text{Simpson dominance index}, \lambda = \sum Pi^2 \quad (5)$$

Where '*Pi*' is the proportion of the individuals of "*i*th" group in the community; '*S*' is the total number of microbial PLFAs or nematode genera in the community; and '*N*' is the amount of microbial PLFAs or number of total nematodes in the community.

Repeated-measure ANOVA was employed to determine the time effect and treatment effect through the whole experimental period with TR and UR as between-subjects factors. To address the relationship between soil microclimates (i.e. soil water content and soil temperature) and soil biodiversity, linear and quadratic regression analysis were performed. Repeated-measure ANOVA and regression model were performed using SPSS software (SPSS Inc., Chicago, IL). Statistical significance was determined at $p<0.05$.

Results

Microbial PLFA Diversity

There was a significant time effect on all the microbial variables, namely, Shannon–Wiener diversity index (H'), Pielou evenness index (J), Margalef richness index (SR), and Simpson dominance index (λ) ($p<0.05$) (Fig. 1). There were trends that understory removal and overstory removal decreased soil microbial PLFA diversity at 0–5 cm soil depth by 0.024 ($p=0.08$) and 0.026 ($p=0.055$) units, respectively (Fig. 1a). Overstoty removal significantly reduced soil microbial PLFA diversity at 5–10 cm soil depth by 0.048 units ($p=0.034$) (Fig. 1b). Repeated measure ANOVA showed that understory removal and overstory removal apparently increased dominance index at 0–5 cm soil depth by 0.007 ($p=0.038$) and 0.008 ($p=0.035$) units, respectively (Fig. 1g); and overstory removal significantly increased dominance index at 5–10 cm soil depth by 0.011 units ($p=0.032$) (Fig. 1h). There was no remarkable effect of vegetation removal on evenness index and richness index of microbial PLFA profile at both 0–5 cm and 5–10 cm soil depths during the study (Fig. 1c, d, e, and f). No interaction effect of understory removal and overstory removal on soil microbial variables was found during the study.

Nematode Diversity

There was a significant time effect on all the nematode variables, namely, Shannon–Wiener diversity index (H'), Pielou evenness index (J), Margalef richness index (SR), and Simpson dominance index (λ) ($p<0.05$) (Fig. 2). There were trends that understory removal decreased soil nematode diversity index by 0.086 units ($p=0.096$) (Fig. 2a) and evenness index by 0.022 units ($p=0.058$) (Fig. 2c) at 0–5 cm soil depth, respectively. Overstory removal tended to increase the soil nematode evenness at 0–5 cm soil depth by 0.022 units ($p=0.062$) (Fig. 2c), and significantly increased the soil nematode evenness at 5–10 cm soil depth by 0.036 units ($p=0.004$) (Fig. 2d). There was a trend that interaction of understory removal and overstory removal tended to affect soil nematode diversity index at 0–5 cm soil depth during the study ($p=0.072$) (Fig. 2a). Neither soil nematode Margalef richness index (SR) nor Simpson dominance index (λ) were significantly affected by understory removal or overstory removal at either 0–5 or 5–10 cm depth (Fig. 2e, f, g, and h).

Correlations between Soil Biodiversity and Soil Microclimates

Regression analysis showed that soil water content and soil biodiversity were not well fitting in either linear or quadratic model (Appendix S1). Coefficients of determination (R^2) of both linear and quadratic regressions between soil water content and biodiversity indices (H', J, SR, and λ) of soil microbial PLFAs and nematodes were very low (Appendix S1). Shannon–Wiener diversity index (H') ($R^2=0.15$, $p=0.006$) and Margalef richness index (SR) ($R^2=0.02$, $p=0.001$) of microbial PLFA profile were quadratic correlated with soil temperature (Fig. 3a and c). Pielou evenness index (J) of nematode community was quadratic correlated with soil temperature ($R^2=0.111$, $p=0.024$) (Fig. 4b).

Discussion

Contributions of Understory or Overstory Vegetations to Soil Biodiversity

Apparently, all plant removal showed strongest suppression on soil microbial and nematode diversity in the present study. In addition, both understory removal and overstory removal reduced soil microbial PLFA and nematode diversities, which was inconsistent with our hypothesis that the effects of understory removal on soil biodiversity might be more powerful than overstory removal. Consistent with our finding, many studies reported that all plant removal had significant negative influence on soil biota in forest and grassland ecosystem [17,31,32]. Our previous study at the same site reported that understory removal had stronger negative effects on abundances and compositions of soil biota (microbes and nematodes) than overstory removal [20]. That was why our hypotheses stated as understory had greater contributions on soil biodiversity than overstory. Contrary to our hypotheses, our analyses clearly showed that the overstory vegetation impacted soil biodiversity too. In addition, overstory removal rather than understory removal significantly reduced amount of total PLFAs in 5–10 cm soil depth during the study (Appendix S2). There are many studies focused on the effects of overstory removal on soil biota because canopy tree harvest is a common forest management practice. For example, fifteen years after harvesting, diversity and structure of soil bacterial and fungal communities remained apparently altered by harvesting disturbances; and three fungi taxa (i.e. ectomycorrhizal fungi, saprobic taxa, and ascomycetes) and actinomycetes were most sensitive to harvesting disturbance in sub-boreal spruce forest in BC, Canada [15]. Both stem-only and whole-tree harvesting decreased microarthropod abundance in a temperate mixed conifer–hardwood stand [33] and in sub-boreal spruce stand [34]. However, these studies usually focused on size of populations and structure of communities of soil biota; there is no mention of how soil biodiversity responses to overstory removal in these studies. Furthermore, most of the studies did not clearly state

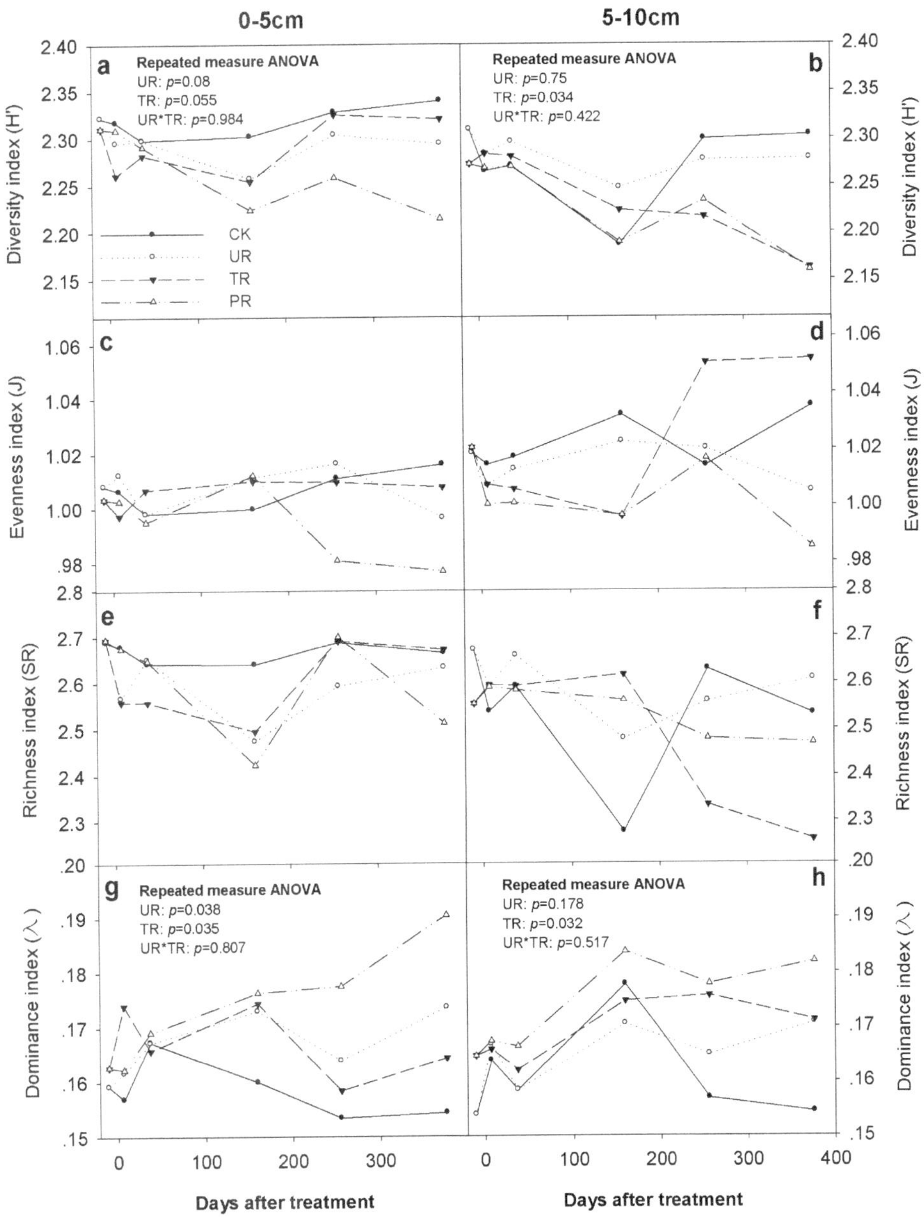

Figure 1. Microbial PLFA diversity as affected by vegetation removal during the study. Shannon–Wiener diversity index (H'), Pielou evenness index (J), Margalef richness index (SR) and Simpson dominance index (λ) under control (CK), understory removal (UR), tree removal (TR), and all-plant removal (PR) in each sampling event at 0–5 cm (a, c, e, g) and 5–10 cm (b, d, f, h) soil depths.

whether understory was remained un-harvesting or not when overstory was harvested. In addition, understory vegetation is important ecological component that plays important roles in driving forest ecosystem functions and processes [20,21]; however, it is usually overlooked in the previous studies. Several studies reported the relationship between understory vegetation and aboveground biodiversity (e.g., birds, reptiles, arthropods) [35,36,37]. Interactions between understory vegetation and soil biodiversity have not been documented yet. This is the first study reported the linkages between understory vegetation and soil microbial PLFA and nematode diversities.

Shannon–Wiener diversity index (H') gives weight to rarer taxa, whereas Simpson dominance index (λ) gives weight to more abundant taxa [38,39]. The suppressive effect of overstory removal on H' of soil microbial PLFA profile might indicate the linkage between overstory vegetation and soil rare microbial taxa. The promotive effect of overstory removal on λ of soil microbial PLFA profile might indicate the linkage between overstory vegetation and soil abundant microbial taxa. Therefore, overstory vegetation might link to both the rare and abundant taxa of soil microbial organisms. Although no significant effect of overstory removal on H' and λ of nematode community, overstory removal increased the Pielou evenness index (J) of nematode community, which implied that the abundance of nematode taxa did not differ from each other as much as their abundance before overstory removal. The likely reason might be that the rare species disappeared and were not collected after overstory removal, which eliminated the weight of the rare taxa to Pielou evenness

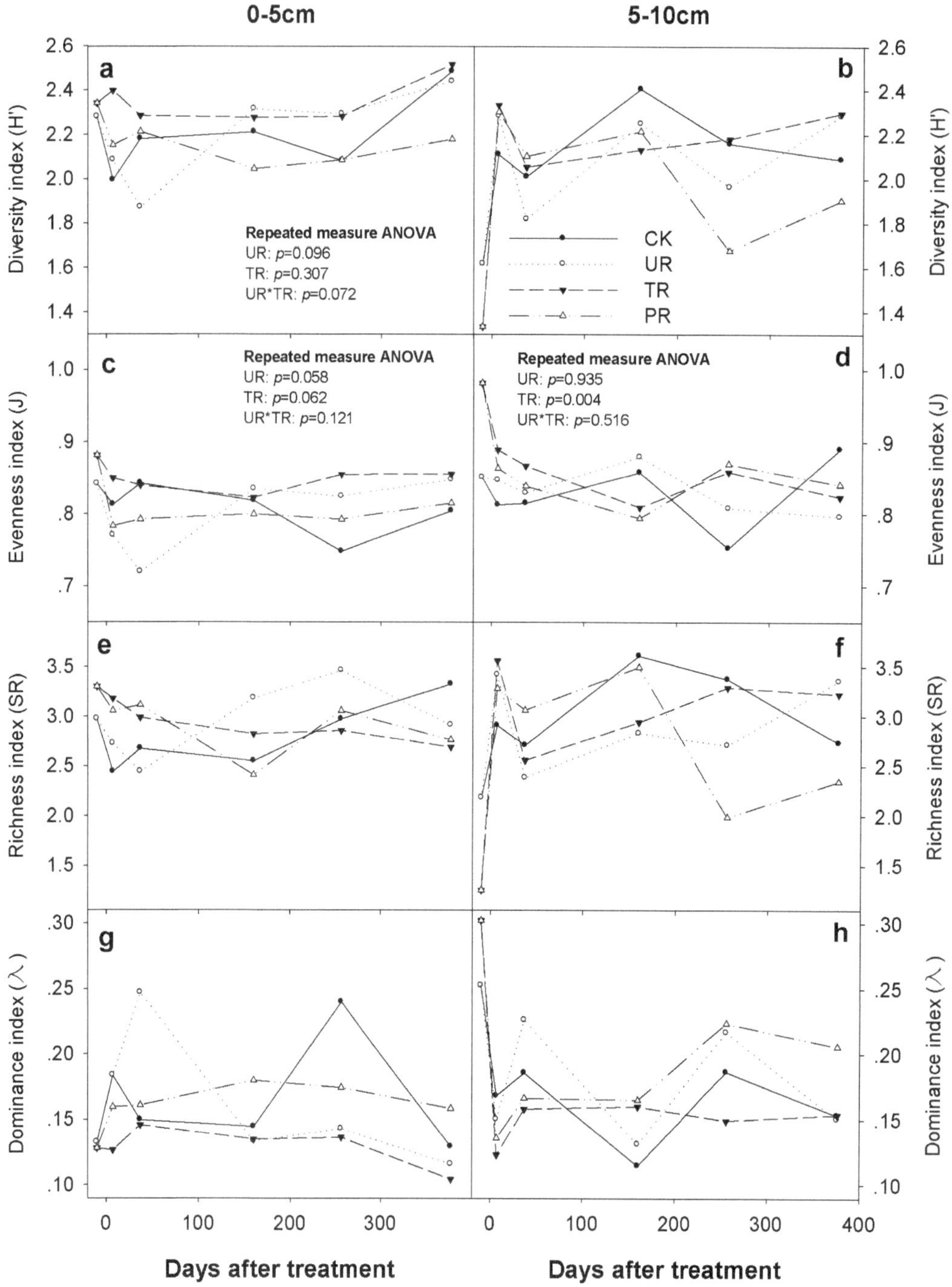

Figure 2. Nematode diversity as affected by vegetation removal during the study. Shannon–Wiener diversity index (*H'*), Pielou evenness index (J), Margalef richness index (SR) and Simpson dominance index (λ) under control (CK), understory removal (UR), tree removal (TR), and all-plant removal (PR) in each sampling event at 0–5 cm (a, c, e, g) and 5–10 cm (b, d, f, h) soil depths.

index. In the similar way, understory vegetation linked to both abundant and rare taxa at 0–5 cm soil depth.

Factor(s) that Affect Soil Biodiversity

Plant diversity may influence soil biodiversity via two main ways. High plant diversity may (1) enhance net primary productivity and (2) lead to high food resources (e.g. litter and root exudates) diversity, and subsequently supports high soil biodiversity [40]. However, many studies found plant species identities were more important than plant diversity in controlling soil biodiversity [41,42,43]; and sometimes the effect of plant identities can overrule the effect of plant diversity [41,44,45].

The bottom-up control might explain the alterations of soil biodiversity after vegetation removal [19,20,32]. *E. urophylla* is the only overstory species and *D. dichotoma* is the dominant understory species in the studied plantations. The productivity of both species are much high. The annual carbon increment of *E. urophylla* can reach 1960 gC m^{-2} yr^{-1} in the developmental stage in the studied

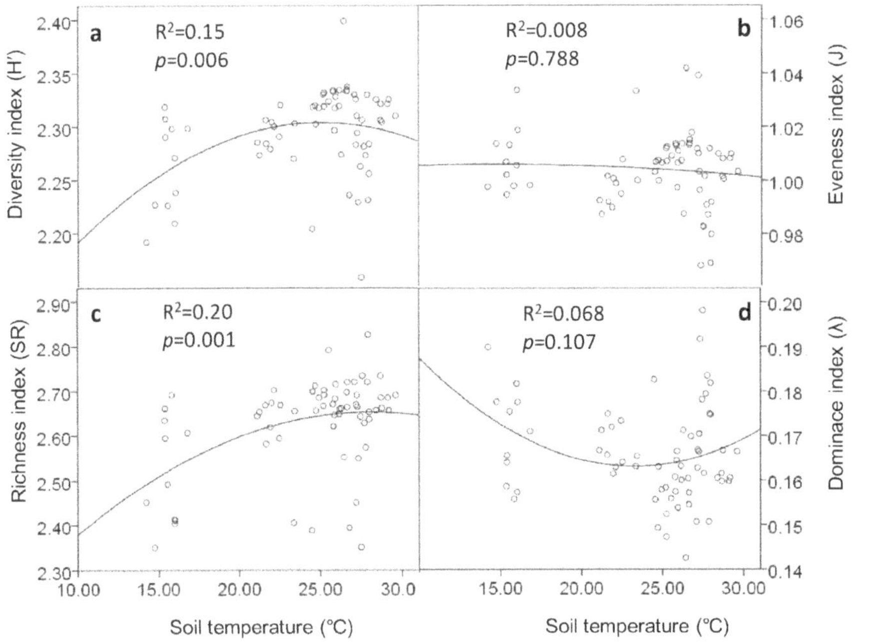

Figure 3. Relationships between indices of microbial PLFAs and soil temperature. Bivariate plots of the microbial PLFA diversity (a), evenness (b), richness (c) and dominance (d) indices versus soil temperature. Points represent results from all six sampling events at 0–5 cm soil depth. Soil temperatures used here were the mean daily temperatures of the six days that soil samplings were carried out.

area [25], while the biomass of *D. dichotoma* may contribute to 22%~48% of total live biomass in *E. urophylla* plantations in the studied area (unpublished data). Both *E. urophylla* and *D. dichotoma* removal resulted in the reduction of carbon input to soil food web [19], which in turn may change soil biodiversity. The bottom-up control opinion was also supported by the differential responses of soil biodiversity at different depths to overstory and understory vegetation. Overstory removal decreased soil biodiversity at both 0–5 cm and 5–10 cm soil depths while understory removal mainly suppressed soil biodiversity at 0–5 cm soil layer. The reason might be that the overstory *E. urophylla* is a deep-rooted species and the dominant understory *D. dichotoma* is a shallow-rooted fern species

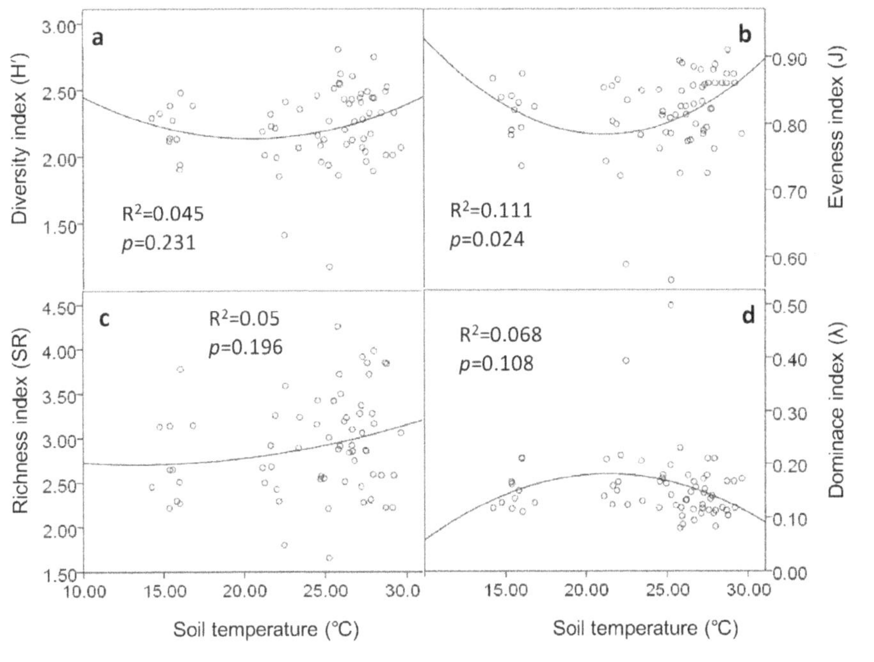

Figure 4. Relationships between indices of nematode diversity and soil temperature. Bivariate plots of the soil nematode diversity (a), evenness (b), richness (c) and dominance (d) indices versus soil temperature. Points represent results from all six sampling events at 0–5 cm soil depth. Soil temperatures used here were the mean daily temperatures of the six days that soil samplings were carried out.

with fibrous root on stolons. The deep-rooted species removal might reduce the availability of soil carbon and nutrients for soil biota at deeper soil depth. Previous studies reported that vegetation removal reduced fine root biomass in eucalyptus plantations in the studied area [19,25]. Unfortunately, none of these studies differentiated either overstory or understory roots from the total plant roots and they did not report the composition of root at different soil depths.

In the present study, soil microclimate (especially soil temperature) explained a certain proportion of variability of soil biodiversity indices. Therefore, soil microclimate change that induced by vegetation removal probably contributed to the alteration of soil biodiversity. Supporting this hypothesis, growth temperature affected the cellular fatty acids composition of sulphate-reducing bacteria that were of marine origin [46]. In addition, our own data revealed that vegetation removal increased soil temperature and decreased soil water content in the later stage of the study, and the effect of understory removal on soil microclimate was greater than overstory removal.

The negative effects of vegetation removal on soil biodiversity also gradually increased over time, which indicated that alterations in bottom-up control and soil microclimate induced by vegetation removal were time-dependent. Firstly, vegetation removal significantly reduced the food resource for soil biota. However, the decomposition of the remained soil residuals (e.g., roots and litter) is time consuming. Therefore, the soil biodiversity declined with the exhaustion of soil residuals. Secondly, the soil microclimate changes that induced by vegetation removal were from March to October (mainly in wet season) in 2010, which was overlapped with the significant declines of soil biodiversity in this study. Thus, the seasonal variation of soil microclimate might be another reason for the time-dependent response of soil biodiversity to vegetation removal.

Conclusions

Here we reported the first study designed to explore the relationship between overstory and understory vegetations and soil biodiversity. Our results showed that both overstory and understory vegetations correlated with soil biodiversity of Eucalyptus plantations in southern China. Interaction between overstory vegetation and soil biodiversity might be due to its high-biomass property that exerts a bottom-up control on soil biodiversity in the studied plantations. Interaction between understory vegetation and soil biodiversity might due to the mixing effect of vegetation identity and soil microclimate change after understory removal that drives soil biodiversity in the studied plantations. Eucalyptus has been extensively planted in many tropical and subtropical regions [25] and *Dicranopteris* is a common fern distributed through humid subtropical and tropical regions of the world [20]. Therefore, our findings might have great potential to extrapolate to large scales, especially to the humid subtropical and tropical region of the world.

Acknowledgments

This study was conducted at Heshan Hilly Land Interdisciplinary Experimental Station, CAS and Heshan National Field Research Station of Forest Ecosystem. We thank Dejun Li for refine the language of the manuscript. We also thank the editor Dafeng Hui and two anonymous reviewers for their comments.

Author Contributions

Conceived and designed the experiments: SF. Performed the experiments: JZ SW CZ ZL LZ. Analyzed the data: JZ SW CZ. Contributed reagents/materials/analysis tools: SF. Wrote the paper: JZ SF.

References

1. Naeem S, Li S (1997) Biodiversity enhances ecosystem reliability. Nature 390: 507–509.
2. Bardgett RD, Usher MB, Hopkins DW (2005) Biological diversity and function in soils. Cambridge, UK: Cambridge University Press.
3. Loreau M, Naeem S, Inchausti P, Bengtsson J, Grime JP, et al. (2001) Biodiversity and ecosystem functioning: current knowledge and future challenges. Science 294: 804–808.
4. Tilman D, Reich PB, Knops J, Wedin D, Mielke T, et al. (2001) Diversity and productivity in a long-term grassland experiment. Science 294: 843–845.
5. Hector A, Schmid B, Beierkuhnlein C, Caldeira MC, Diemer M, et al. (1999) Plant diversity and productivity experiments in European grasslands. Science 286: 1123–1127.
6. McGrady-Steed J, Harris PM, Morin PJ (1997) Biodiversity regulates ecosystem predictability. Nature 390: 162–165.
7. Huston MA (1997) Hidden treatments in ecological experiments: re-evaluating the ecosystem function of biodiversity. Oecologia 110: 449–460.
8. Huston MA, Aarssen LW, Austin MP, Cade BS, Fridley JD, et al. (2000) No consistent effect of plant diversity on productivity. Science 289: 1255.
9. Brussaard L (1997) Biodiversity and ecosystem functioning in soil. Ambio 26: 563–570.
10. Wardle DA, Bardgett RD, Klironomos JN, Setala H, van der Putten WH, et al. (2004) Ecological linkages between aboveground and belowground biota. Science 304: 1629–1633.
11. van der Heijden M, Klironomos J, Ursic M, Moutoglis P, Streitwolf-Engel R, et al. (1998) Mycorrhizal fungal diversity determines plant biodiversity, ecosystem variability and productivity. Nature 396: 69–72.
12. Van Der Heijden MGA, Bardgett RD, Van Straalen NM (2008) The unseen majority: soil microbes as drivers of plant diversity and productivity in terrestrial ecosystems. Ecology Letters 11: 296–310.
13. Lavorel S, McIntyre S, Landsberg J, Forbes TDA (1997) Plant functional classifications: from general groups to specific groups based on response to disturbance. Trends in Ecology & Evolution 12: 474–478.
14. Woodward FI, Cramer W (1996) Plant functional types and climatic change: Introduction. Journal of Vegetation Science 7: 306–308.
15. Hartmann M, Howes CG, VanInsberghe D, Yu H, Bachar D, et al. (2012) Significant and persistent impact of timber harvesting on soil microbial communities in Northern coniferous forests. The ISME Journal.
16. Seastedt T, Crossley Jr D (1981) Microarthropod response following cable logging and clear-cutting in the southern Appalachians. Ecology: 126–135.
17. Forge T, Simard S (2001) Structure of nematode communities in forest soils of southern British Columbia: relationships to nitrogen mineralization and effects of clearcut harvesting and fertilization. Biology and Fertility of Soils 34: 170–178.
18. Panesar TS, Marshall VG, Barclay HJ (2000) The impact of clearcutting and partial harvesting systems on population dynamics of soil nematodes in coastal Douglas-fir forests. Pedobiologia 44: 641–665.
19. Wu J, Liu Z, Wang X, Sun Y, Zhou L, et al. (2011) Effects of understory removal and tree girdling on soil microbial community composition and litter decomposition in two Eucalyptus plantations in South China. Functional Ecology 25: 921–931.
20. Zhao J, Wan S, Li Z, Shao Y, Xu G, et al. (2012) *Dicranopteris*-dominated understory as major driver of intensive forest ecosystem in humid subtropical and tropical region. Soil Biology and Biochemistry 49: 78–87.
21. Nilsson MC, Wardle DA (2005) Understory vegetation as a forest ecosystem driver: evidence from the northern Swedish boreal forest. Frontiers in Ecology and the Environment 3: 421–428.
22. Barrett J, Cochran P, Busse M (1996) Changes in ponderosa pine site productivity following removal of understory vegetation. Soil Science Society of America Journal 60: 1614–1621.

23. Zhao J, Shao Y, Wang X, Neher DA, Xu G, et al. (2013) Sentinel soil invertebrate taxa as bioindicators for forest management practices. Ecological Indicators 24: 236–239.
24. Zhao J, Wan S, Fu S, Wang X, Wang M, et al. (2013) Effects of understory removal and nitrogen fertilization on soil microbial communities in *Eucalyptus* plantations. Forest Ecology and Management 310: 80–86.
25. Chen D, Zhang C, Wu J, Zhou L, Lin Y, et al. (2011) Subtropical plantations are large carbon sinks: Evidence from two monoculture plantations in South China. Agricultural and Forest Meteorology 151: 1214–1225.
26. FAO (2006) World reference base for soil resources 2006. World Soil Resources Report 103 FAO, Rome.
27. Bossio DA, Scow KM (1998) Impacts of carbon and flooding on soil microbial communities: Phospholipid fatty acid profiles and substrate utilization patterns. Microbial Ecology 35: 265–278.
28. Frostegård A, Bååth E (1996) The use of phospholipid fatty acid analysis to estimate bacterial and fungal biomass in soil. Biology and Fertility of Soils 22: 59–65.
29. Joergensen RG, Wichern F (2008) Quantitative assessment of the fungal contribution to microbial tissue in soil. Soil Biology & Biochemistry 40: 2977–2991.
30. Frostegård Å, Tunlid A, Bååth E (2011) Use and misuse of PLFA measurements in soils. Soil Biology and Biochemistry 43: 1621–1625.
31. Wardle DA, Bonner KI, Barker GM, Yeates GW, Nicholson KS, et al. (1999) Plant removals in perennial grassland: vegetation dynamics, decomposers, soil biodiversity, and ecosystem properties. Ecological Monographs 69: 535–568.
32. Zhao J, Wang X, Shao Y, Xu G, Fu S (2011) Effects of vegetation removal on soil properties and decomposer organisms. Soil Biology and Biochemistry 43: 954–960.
33. Bird G, Chatarpaul L (1988) Effect of forest harvest on decomposition and colonization of maple leaf litter by soil microarthropods. Canadian Journal of Soil Science 68: 29–40.
34. Battigelli JP, Spence JR, Langor DW, Berch SM (2004) Short-term impact of forest soil compaction and organic matter removal on soil mesofauna density and oribatid mite diversity. Canadian Journal of Forest Research 34: 1136–1149.
35. Atwell RC, Schulte LA, Palik BJ (2008) Songbird response to experimental retention harvesting in red pine (Pinus resinosa) forests. Forest Ecology and Management 255: 3621–3631.
36. López G, Moro M (1997) Birds of Aleppo pine plantations in south-east Spain in relation to vegetation composition and structure. Journal of Applied Ecology: 1257–1272.
37. Humphrey J, Hawes C, Peace A, Ferris-Kaan R, Jukes M (1999) Relationships between insect diversity and habitat characteristics in plantation forests. Forest Ecology and Management 113: 11–21.
38. Neher DA, Lee Campbell C (1994) Nematode communities and microbial biomass in soils with annual and perennial crops. Applied Soil Ecology 1: 17–28.
39. Zhao J, Zhang W, Wang K, Song T, Du H (2013) Responses of the soil nematode community to management of hybrid napiergrass: The trade-off between positive and negative effects. Applied Soil Ecology. DOI: 10.1016/j.apsoil.2013.10.011.
40. Wardle DA (2006) The influence of biotic interactions on soil biodiversity. Ecology Letters 9: 870–886.
41. De Deyn GB, Raaijmakers CE, Van Ruijven J, Berendse F, Van Der Putten WH (2004) Plant species identity and diversity effects on different trophic levels of nematodes in the soil food web. Oikos 106: 576–586.
42. Viketoft M, Palmborg C, Sohlenius B, Huss-Danell K, Bengtsson J (2005) Plant species effects on soil nematode communities in experimental grasslands. Applied Soil Ecology 30: 90–103.
43. De Deyn GB, Quirk H, Bardgett RD (2011) Plant species richness, identity and productivity differentially influence key groups of microbes in grassland soils of contrasting fertility. Biology Letters 7: 75–78.
44. Wardle DA, Yeates GW, Williamson W, Bonner KI (2003) The response of a three trophic level soil food web to the identity and diversity of plant species and functional groups. Oikos 102: 45–56.
45. Wardle DA, Yeates GW, Barker GM, Bonner KI (2006) The influence of plant litter diversity on decomposer abundance and diversity. Soil Biology and Biochemistry 38: 1052–1062.
46. Könneke M, Widdel F (2003) Effect of growth temperature on cellular fatty acids in sulphate-reducing bacteria. Environmental Microbiology 5: 1064–1070.

4

Resource Availability and Spatial Heterogeneity Control Bacterial Community Response to Nutrient Enrichment in Lakes

KathiJo Jankowski*, Daniel E. Schindler, M. Claire Horner-Devine

University of Washington, School of Aquatic and Fisheries Sciences, Seattle, Washington, United States of America

Abstract

The diversity and composition of ecological communities often co-vary with ecosystem productivity. However, the relative importance of productivity, or resource abundance, versus the spatial distribution of resources in shaping those ecological patterns is not well understood, particularly for the bacterial communities that underlie most important ecosystem functions. Increasing ecosystem productivity in lakes has been shown to influence the composition and ecology of bacterial communities, but existing work has only evaluated the effect of increasing resource supply and not heterogeneity in how those resources are distributed. We quantified how bacterial communities varied with the trophic status of lakes and whether community responses differed in surface and deep habitats in response to heterogeneity in nutrient resources. Using ARISA fingerprinting, we found that bacterial communities were more abundant, richer, and more distinct among habitats as lake trophic state and vertical heterogeneity in nutrients increased, and that spatial resource variation produced habitat specific responses of bacteria in response to increased productivity. Furthermore, changes in communities in high nutrient lakes were not produced by turnover in community composition but from additional taxa augmenting core bacterial communities found in lower productivity lakes. These data suggests that bacterial community responses to nutrient enrichment in lakes vary spatially and are likely influenced disproportionately by rare taxa.

Editor: Martin Heil, Centro de Investigación y de Estudios Avanzados, Mexico

Funding: University of Washington Keeler Professorship to Daniel Schindler, Seattle ARCS Foundation and EPA STAR Fellowship to KathiJo Jankowski. The funders had no role in study design, data collection and analysis, decision to publish, or preparation of the manuscript.

Competing Interests: The authors have declared that no competing interests exist.

* E-mail: kathijo@u.washington.edu

Introduction

Ecosystem productivity is an important driver of the diversity and composition of ecological communities. Much attention has been given to understanding how communities change with increased productivity, due to the desire to understand how species and their threats are distributed globally [1] and the widespread increase in nutrient enrichment and primary productivity of many ecosystems [2]. Productive ecosystems often support high species richness [3], as evidenced by diversity hotspots in ecosystems such as marine upwelling zones [4] and tend to host distinct communities from low productivity ecosystems. Productivity is thought to promote changes in species richness and composition due to the increased energy available to support the coexistence of multiple species and trophic levels [5,6], as well as by promoting shifts to species that dominate in productive environments. However, productivity is not always a good predictor of species richness [7], and the mechanisms behind observed richness and compositional changes in response to increased ecosystem productivity remain obscure [8].

Spatial or temporal heterogeneity in resource availability can also facilitate the coexistence of species in many environments [9,10], and is commonly used to explain why species richness varies with ecosystem productivity [11–13]. Yet, the relative importance of resource availability and heterogeneity in influencing patterns of species richness and composition in productive ecosystems remains unclear for many ecological communities [8,14–16], especially for prokaryotes. Bacteria are a fundamental component of food webs and provide the foundation for overall ecosystem functioning, yet we know relatively little about how bacterial communities respond to increases in productivity in most ecosystems [17–19]. In addition, bacteria have unique characteristics, such as metabolic flexibility and dormancy that might make their response to productivity and resource heterogeneity unique. In addition, bacteria can acquire new functional capacities through the exchange of genetic material [20], thus, taxonomic richness may be unresponsive to changes in productivity [21].

Lakes vary widely in productivity and the heterogeneity of resource distribution in response to variation in nutrient loading from human and watershed sources [15,22]. Increased primary production, or trophic status, in lakes is associated with changes in species richness and composition of many ecological communities, including bacteria [23,24]. The richness of macroorganisms often declines at richer trophic state, due to the dominance of phytoplankton that are less palatable or toxic to consumers [25], declines in littoral productivity [26], and changes to the physical and chemical characteristics of the lake environment [22]. Therefore, changes associated with increased lake trophic status often negatively impact diversity of lake communities, change their composition, and lead to the dominance of a few species through homogenization of food resources and reduction in habitat availability [15,27]. Bacterial communities are known to shift in response to increased lake trophic status [28], but the fundamental mechanisms and importance of resource distribution in mediating those changes have not been fully explored.

Nutrient enrichment in lakes tends to magnify the vertical differences in physical and chemical characteristics such as nitrogen (N), phosphorus (P) and dissolved oxygen (DO) among lake strata [22]. However, existing studies of the response of bacterial communities to eutrophication have only evaluated the responses of surface communities or the integrated water column rather than habitat-specific responses [28,29]. In stratified lakes, the surface layer (epilimnion) is typically warm, nutrient-poor, and productive, whereas the deep layer (hypolimnion) is cooler, richer in nutrients, and often low in dissolved oxygen (DO). These differences may be especially important when considering how the response of lake bacteria may differ from eukaryotic communities since vertical differences in physical and chemical conditions are known to structure bacterial communities in stratified lakes [30–32]. For example, low DO in the hypolimnion promotes the use of diverse energy pathways by bacteria such as denitrification and sulfate reduction that are not energetically advantageous in the oxic epilimnion, and therefore, could promote higher diversity of bacterial communities in the entire water column in response to increased trophic status. Therefore, bacterial communities are likely less similar among lake strata in high productivity (eutrophic) than in low productivity (oligotrophic) lakes.

In addition, although several studies have observed changes in bacterial communities with increased lake trophic status [24,29,33], few have identified which type of bacterial taxa are responsible for driving shifts in overall composition [34,35]. For example, while there is increasing evidence that some taxa flourish in high productivity lakes [36], it is unclear whether taxonomic changes result from a complete turnover in the community [37], an increase in the relative abundance of a few key taxa [38], or the increased presence of previously rare or novel taxa that augment a core community of taxa present in low nutrient lakes. For example, a study that evaluated how dominant, common and rare taxa responded to another important disturbance in lakes, lake mixing, found that shifts in the bacterial community were driven by the increased dominance of a few taxa [38].

We evaluated how bacterial abundance, taxonomic richness, and composition changed among and within lakes along a gradient of increasing trophic state. In particular, we quantified the amount of variation in the bacterial response that was explained by trophic state, resource heterogeneity, and their combination. Second, we evaluated whether communities associated with different lake habitats (specifically the epi-, meta- and hypolimnion) responded differently to increased trophic status than communities assessed in the surface layer or integrated at the whole-lake scale. Finally, we evaluated which taxa were responsible for changes in community characteristics; specifically, we asked whether patterns were driven by turnover in the community or by additional taxa augmenting a core community present across all lakes. Thus, in this study we were able to address whether changes in the observed number of taxa and composition of bacterial communities followed the same patterns as eukaryotic communities in response to productivity in lakes and whether the distribution or abundance of resources was more important in shaping those patterns.

Methods

We sampled 21 lakes in the Puget Sound region of western Washington (USA) and southern British Columbia (Canada, Figure 1) that spanned a large gradient of anthropogenic nutrient loading and productivity [39]. We sampled during the summer-stratified period of July and August 2008. Therefore, our samples reflected the communities that had developed following two to three months of stratified conditions within the water column [31]. As previously described [39], the lakes included in this study were physically similar. Twenty of the 21 lakes were monomictic, and one lake was too shallow to develop thermal stratification. No permissions were required to access 17 of these lakes since they were accessible via a public boat launch. We obtained permission from the University of British Columbia to access the remaining four lakes, which were on the property of their Malcolm Knapp Research Forest. No endangered or protected species were involved in this research.

All bacterial community samples and measurements of lake environmental characteristics were collected over the deepest point in each lake. Water samples for nutrient and chlorophyll *a* analyses were collected from the epilimnion (surface), metalimnion (thermocline depth), and hypolimnion (within 3 m of the lake bottom) with a van Dorn bottle. Total N (TN) was determined using the perchloric acid digestion method [40] followed by analysis with automated colorimetry on a Lachat autoanalyzer (Lachat Instruments, Loveland, CO, USA). Total P (TP) concentration was determined colorimetrically after persulfate digestion and reaction with molybdate and stannous chloride [40]. Water samples for inorganic N and P determination were pre-filtered through a 0.2 mm Supor filters (Supor-200, Pall Gelman, East Hills, NY) and then analyzed colorimetrically using the same methods as above without a pre-digestion step. Chlorophyll *a* concentration was determined fluorometrically (Turner Designs, Sunnyvale, California) and used as a surrogate for algal community biomass. Temperature, dissolved oxygen (DO), and pH measurements were taken at 1-m depth intervals with a YSI sonde 6600 (YSI Integrated Systems & Services, Yellow Springs, OH, USA). Other physical lake data such as mean and maximum depth, lake area, and drainage area were obtained from the King County Water and Land Resources Division and the Washington Department of Ecology.

Two water samples for bacterial community analysis were collected from the epilimnion, metalimnion and hypolimnion of each lake with a Van Dorn Bottle. Two 300-mL samples were pooled and bacteria collected on 0.2-μm filters (Supor-200, Pall Gelman, East Hills, NY). Filters were frozen immediately and stored at −80°C until further processing. DNA was extracted from replicate filters using the Qiagen DNEasy Blood and Tissue Mini-kit (Qiagen, Valencia, CA). Samples for bacterial cell enumeration were preserved with 2% formalin, filtered onto a 0.2 μm black polycarbonate filter, stained with 4′, 6-diamidino-2-phenylindole (DAPI), and viewed with a Nikon Eclipse 80i digital microscope at 1000× magnification.

Bacterial community composition and observed richness were assessed using automated ribosomal intergenic spacer analysis [41]. ARISA generates fingerprints of the microbial community based on the length heterogeneity in the intergenic spacer region between the 16S and 23S rRNA genes, which varies among organisms. ARISA has similar limitations as other PCR-based fingerprinting approaches [41] and tends to only survey dominant taxa in a community, thus our assessment of bacterial community composition is really a comparison of the community of dominant taxa among lakes. However, ARISA has been shown to give a robust, high-resolution view of bacterial assemblages in aquatic ecosystems [42,43], to generate results that are consistent with more high resolution techniques [42,44], and can represent species-level taxonomic resolution (98–99% sequence similarity;

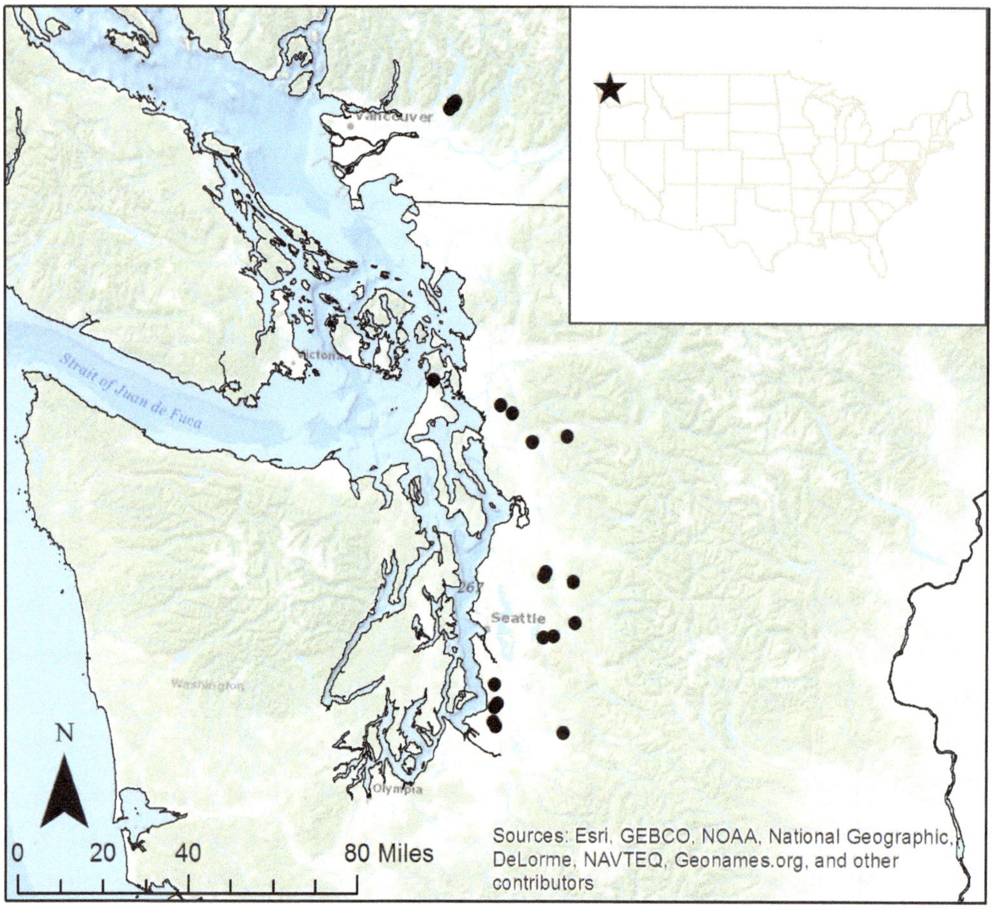

Figure 1. Map of study sites. Lakes included in the study were located in the Puget South Basin in Washington state, USA and British Columbia, Canada. Lakes are indicated by black points.

[42]). The 16S-23S intergenic region was amplified using the polymerase chain reaction (PCR) from the total extracted DNA using 6-FAM-labelled universal 1406-F primer (5′ TGYACA-CACCGCCCGT-3′) and bacterial specific primer 23S-R(5′-GGGTTBCCCCATTCRG-3′) [41,45]. PCRs were conducted on a Mastercycler gradient thermocycler (Eppendorf, New York). PCR products were pooled, quantified, and analyzed on a MegaBACE 1000 automated capillary sequencer (GE Healthcare Corporation, New Jersey). Operational taxonomic units (OTUs) were generated by binning ARISA fragments into successively larger length bins based on their size and eliminating fragments that were <150 and >1300 bp [42]. We used peak area to estimate relative abundance of OTUs in our samples [43], which we considered to be the ratio of the peak area of an OTU in a sample to the total peak area of the sample. We also converted the peak area matrix to presence-absence to assess the composition of bacterial communities in the ARISA profiles by occurrence patterns. We calculated observed richness from ARISA profiles by summing the number of OTUs observed in each sample, hereafter referred to as profile richness. We found no differences in bacterial community patterns using peak height vs. peak area.

Statistical Analyses

We used a principal components analysis (PCA) to summarize physical and chemical variation related to trophic status among lakes. We found that lakes varied little in relevant physical characteristics (lake area and mean depth), and thus we used the first axis of the resulting PCA as a multivariate proxy for increasing lake trophic state (Figure 2A). Although we did not measure primary productivity directly, other studies have found good agreement between primary productivity and chlorophyll *a* and nutrient concentrations in lakes with similar concentrations as lakes in this study [46].

To quantify vertical heterogeneity in chemical and physical variables within each lake (e.g., TN, TP, temperature, and DO), we used the standard deviation of measurements among lake strata (Table S1). We then performed a PCA that included only the standard deviations of these physical and chemical variables to establish a gradient of resource heterogeneity among lakes.

To compare the influence of increasing trophic state ("trophic status"), depth variation in resource availability ("resource heterogeneity"), and their combination (trophic status and resource heterogeneity) on the bacterial community, we then took the scores from the first principal component (PC 1) of each PCA and regressed them against metrics of bacterial abundance, ARISA profile richness, and an index of community similarity among ARISA profiles (see below for description). The combined trophic status and resource heterogeneity model contained two predictor variables: PC 1 of the trophic status PCA, and PC1 of the resource heterogeneity PCA. All variables were transformed to meet the assumptions of normality prior to the PCA. We evaluated the support for each of the three candidate models describing the

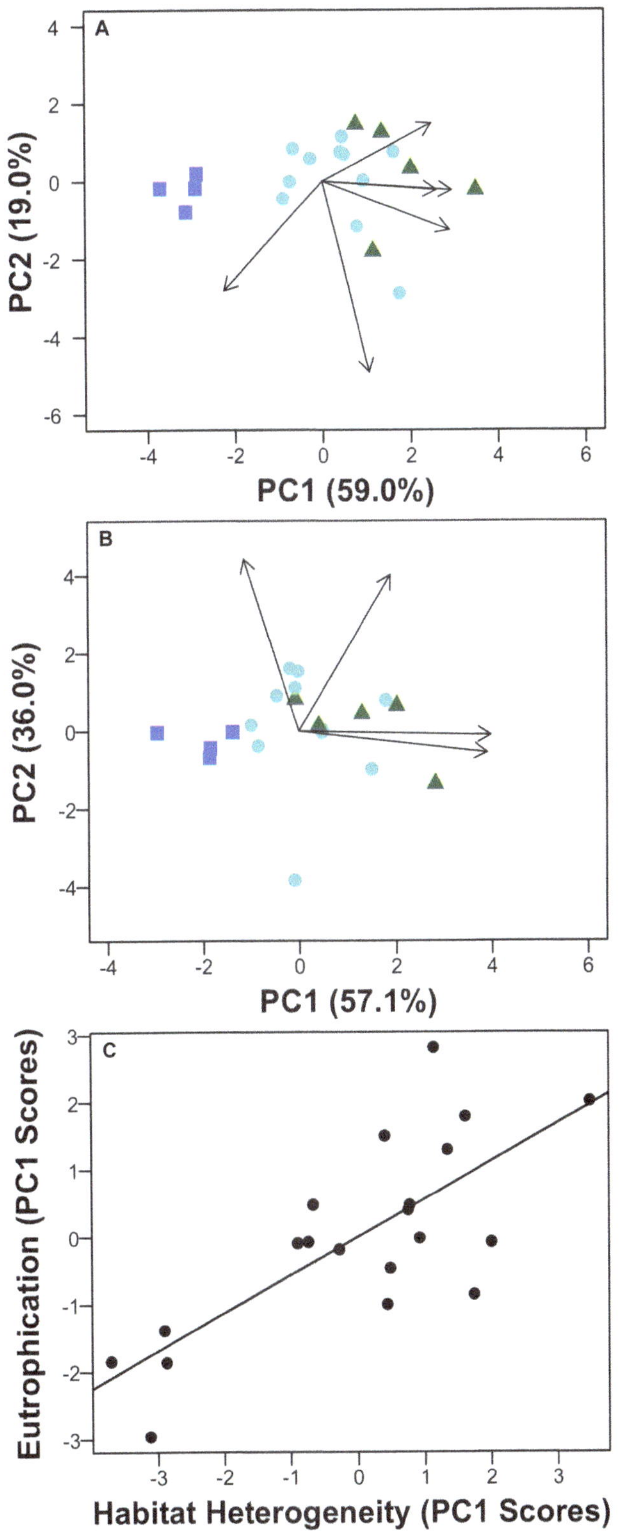

Figure 2. Principal component analyses (PCAs) showing environmental variation across lakes in this study. Panel A shows PCA results based on trophic state variables, panel B shows PCA results based on Heterogeneity variables (standard deviation among depths in variables) and panel C shows the correlation between the trophic state PC 1 scores and heterogeneity PC 1 scores ($R^2 = 0.57$). Arrows show significant variables ($p<0.05$) and values in parentheses show percentages of total environmental variation among lakes explained by each axis. 'EpiTN' = epilimnetic TN, 'EpiTP' = epilimnetic TP, 'HypoTP' = Hypolimnetic TP, 'HypoDO' = hypolimnetic DO, and DO = epilimnetic DO. '.std' indicates that the standard deviation of measurements of the specified variable among layers. Triangles = eutrophic, stars = mesotrophic, and squares = oligotrophic lakes.

relationships between environmental conditions and the bacteria community attributes using Akaike's Information Criteria adjusted for small sample sizes [47]. The model with the lowest AIC_c was considered the best model, and models within 2 AIC_c units of one another were considered to be equally good [47]. In addition, we calculated AIC weights (w_i) for each individual model, which estimates a probability that model *i* is the best model given the set of models we considered. Finally, to evaluate overall importance of the individual variables, trophic status and resource heterogeneity, we calculated w_i for each term across the three models we compared (Table 1).

Bacterial community similarity among samples was assessed using Sorensen's coefficient for occurrence data [48] and the Chao-Sorensen abundance estimator for relative abundance data [49]. We assessed the overall similarity of communities within a lake by using an average dissimilarity value among ARISA profiles from all two-layer comparisons. We used a constrained analysis of principal coordinates (CAP) to evaluate if changes in community composition were associated with increasing trophic state [50] since it allowed us to use the Chao-Sorensen similarity index.

Finally, we investigated whether changes in the bacterial community with increased lake trophic status were realized by shifts in "widespread" or "narrowly distributed" taxa. We assessed how the relative contribution of widespread OTUs (taxa observed in the majority of lakes) changed with trophic state and habitat heterogeneity. We considered widespread taxa to be those that were observed in 90% of lakes in our study (but see Table S2 for evaluation of different thresholds). We then assessed whether the occurrence and relative abundance of these taxa changed across the lake trophic gradient and with increasing heterogeneity (i.e., PC1 of trophic status and resource heterogeneity PCAs). All analyses were done in R Version 2.14.0 [51] using the vegan package [52].

Results

Lake Characteristics

Productivity-related variables such as TP, TN, and chlorophyll a explained a substantial portion of the environmental variation among lakes in this study (59%; Figure 2, Table S1). Epilimnetic TP concentrations ranged from 4.6 $\mu g\ L^{-1}$ to over 30 $\mu g\ L^{-1}$, and chlorophyll *a* ranged from 0.23 to 10.2 $\mu g\ L^{-1}$, thus the lakes ranged from oligotrophic to eutrophic [22]. Environmental conditions did not change similarly in each layer with increased trophic state; for example, the epilimnion was less variable among lakes than either the metalimnion or hypolimnion in most environmental characteristics (Table S3). As a result, conditions within the water column were more heterogeneous as trophic state increased ($R^2 = 0.57$, Figure 2C). We observed the most significant differences in TN and TP concentrations among layers as trophic state increased (Figures 2B and C). TN and TP were correlated with the availability of NH_4 ($r = 0.71$) and PO_4 ($r = 0.86$), respectively. Thus, nutrient availability was variable within the water column. Finally, the percent change in nutrient concentrations was greater in the hypolimnion than in either the epi- or metalimnion (Table S3).

Table 1. Comparison of models evaluating the effects of trophic state, heterogeneity, and their combination on bacterial communities.

	Model	n	k	R^2	AIC_c	ΔAIC_c	w_i*	w_i of T^+	w_i of H^{++}
ABUNDANCE									
	Trophic State	21	3	0.46	612.2	9.1	0.01	0.34	0.83
	Heterogeneity	21	3	0.66	603.1	0.0	0.50		
	PC1 T+PC1 H	21	4	0.69	603.3	0.0	0.33		
RICHNESS									
	Trophic State	17	3	0.30	129.3	2.9	0.16	0.33	0.84
	Heterogeneity	17	3	0.41	126.4	0.0	0.67		
	PC1 T+PC1 H	17	4	0.40	129.2	2.8	0.17		
DISSIMILARITY									
	Trophic State	17	3	0.12	−23.8	0.9	0.35	0.46	0.65
	Heterogeneity	17	3	0.17	−24.7	0.0	0.54		
	PC1 T+PC1 H	17	4	0.12	−21.5	3.1	0.11		
WIDESPREAD TAXA	Trophic State	18	3	0.29	−36.9	1.5	0.26	0.44	0.74
	Heterogeneity	18	3	0.35	−38.4	0.0	0.56		
	PC1 T+PC1 H	18	4	0.35	−36.1	2.3	0.18		

*AIC_c weight,
$^+$Trophic State,
$^{++}$Heterogeneity.
Relationship of trophic state, heterogeneity, and their combination with abundance, richness, dissimilarity of bacterial communities among lake habitats ("Dissimilarity"), and the proportional abundance of common taxa. 'n' = sample size and 'k' = number of parameters in each model. 'w_i' refers to the AIC_c weight calculated for each model and the weights for the individual terms 'T' and 'H' across all models.

Did Lake Bacterial Communities Shift in Response to Increasing Trophic State?

Bacterial communities at the whole-lake scale shifted significantly in association with increasing lake trophic state (Figure 3, Figure S1). Average bacterial abundance ($R^2 = 0.46$, Table 1, Figure 2A) and ARISA profile richness increased linearly with our proxy for lake eutrophication ($R^2 = 0.30$, Table 1, Figures 3A & E) and ranged from 66 to 106 OTUs per lake. Our CAP model showed that bacterial community composition shifted with increased lake trophic status and shifts were strongly associated with increasing chlorophyll *a* ($r = 0.99$) and epilimnetic TN ($r = 0.76$) concentration (Figure S1). The CAP model explained 29% of the total variation in community composition among lakes. The first CAP axis captured the majority of that explained variation (71.4%), indicating that community composition changed in response to increased lake trophic state.

Did the Responses of the Bacterial Community Vary among Habitats in Lakes?

Bacterial communities associated with surface and deep habitats displayed different patterns of abundance, and the richness and composition of ARISA profiles changed significantly as lake trophic state increased (Figure 3, Figure S1). Average bacterial abundance increased with trophic state ($R^2 = 0.46$), was highest in the metalimnion (ANOVA; $F = 5.6$, $p = 0.006$), but increased significantly in all layers across the trophic gradient (Figure 3). The richness of ARISA profiles also varied significantly among layers (ANOVA, $F = 10.6$, $p<0.001$), but only increased notably in the hypolimnion in response to trophic state ($R^2 = 0.15$, Figure 3H). Profile richness was highest on average in the hypolimnion (55±7 SD), which also had the largest range of observed richness, ranging from 37 OTUs in Gwendoline Lake, an oligotrophic lake, to 71 OTUs found in more nutrient-rich Geneva Lake. Therefore, increases in the profile richness in the hypolimnion accounted for the increases we observed in overall lake richness ($R^2 = 0.39$).

When all lake communities were considered together, there were significant, but small, compositional differences among epi-, meta- and hypolimnetic communities (ANOSIM, $R = 0.16$, $p = 0.001$), and surface and deep communities shared the fewest taxa (data not shown). Furthermore, surface communities were significantly less variable than deep communities across the trophic gradient (Homogeneity of dispersion, $p<0.001$), and surface and deep communities within a given lake tended to become less similar to one another as trophic state increased ($R^2 = 0.12$). However, heterogeneity in nutrient concentrations among strata explained slightly more of that variation than trophic state alone ($R^2 = 0.17$, Figure 4, Table 1).

In all cases, the heterogeneity model or the trophic status plus heterogeneity model explained more variation in bacterial communities among lakes than the trophic status model alone (Table 1). We found that greater vertical heterogeneity of nutrient availability (Figure 2) was strongly related to increased abundance, ARISA profile richness, and decreasing similarity of communities among lake strata (Table 1, Figure 4). Total abundance and observed richness were both more strongly related to increases in the heterogeneity of P and N than increases in their concentrations alone or to differences in temperature and DO among strata, which other studies have shown to be associated with heterogeneity in bacterial community composition (Table S1; [31,32]). Therefore, although we saw an increase in observed richness with increased trophic state ($R^2 = 0.30$), observed richness was more closely linked to greater heterogeneity of nutrients within the water column ($R^2 = 0.41$). Overall, the AIC_c shows that the heteroge-

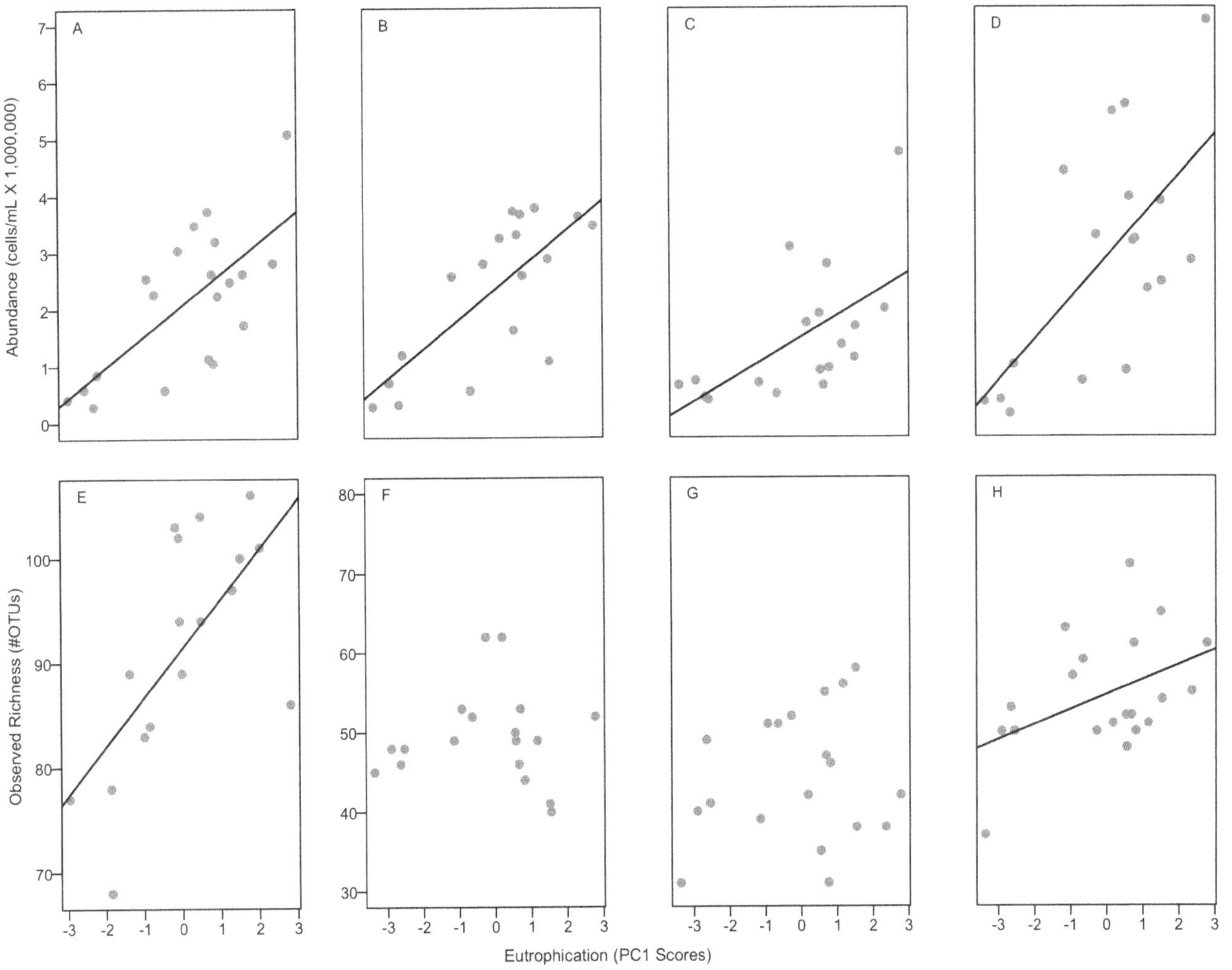

Figure 3. Relationships of increasing trophic state (PC 1 scores) with abundance and richness. A) Whole-lake average abundance ($R^2 = 0.46$, $p < 0.001$), B) Epilimnetic abundance ($R^2 = 0.52$, $p < 0.001$), C) Metalimnetic abundance ($R^2 = 0.33$, $p = 0.007$), D) Hypolimnetic abundance ($R^2 = 0.40$, $p = 0.003$), E) Whole-lake average richness ($R^2 = 0.30$, $p = 0.003$), F) Epilimnetic richness ($R^2 = -0.06$, $p = 0.98$), G) Metalimnetic richness ($R^2 = -0.03$, $p = 0.49$), and H) Hypolimnetic richness ($R^2 = 0.15$, $p = 0.05$).

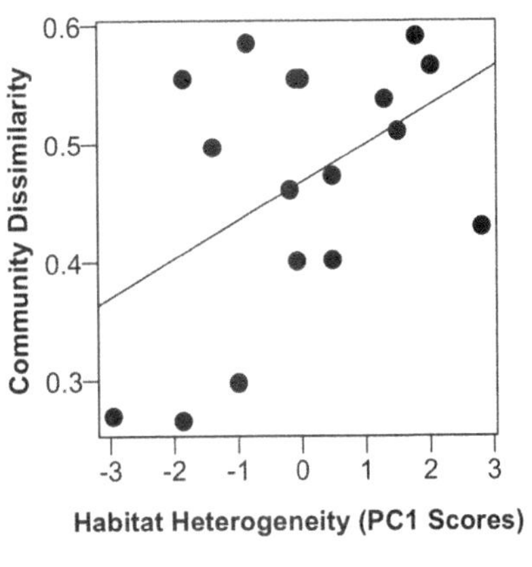

Figure 4. Bacterial communities were less similar among lake strata as chemical heterogeneity increased ($R^2 = 0.17$, $p = 0.05$). Similarity of communities was based on the average Chao Sorensen Dissimilarity from comparisons of relative abundance of taxa among three depth strata.

neity model had the most support, and that, in fact, the heterogeneity term had the most weight across models (Table 1). Thus, as measured here, bacterial communities exhibited habitat-specific responses to lake eutrophication, and spatial variation in resource availability often influenced bacterial community composition more than simple increases in nutrient concentration and productivity (Figures 4 and 5).

Which Taxa Accounted for changes in Community Composition with Eutrophication?

We observed a total of 221 OTUs across all lakes and found that some of those taxa were widespread among lakes. We observed that while a core community of 11 OTUs was present and detected using ARISA in ~90% of lakes in this study ("widespread taxa"), and while still present across lakes, made up a decreasing proportion of both the number ($R^2 = 0.23$) and relative abundance of taxa in lakes as lake trophic state increased ($R^2 = 0.35$, Figure 5). Specifically, while these widespread taxa comprised 35–40% of the relative abundance of the community in the more homogenous oligotrophic lakes, they were less than 15% in the more heterogeneous eutrophic lakes. These trends were

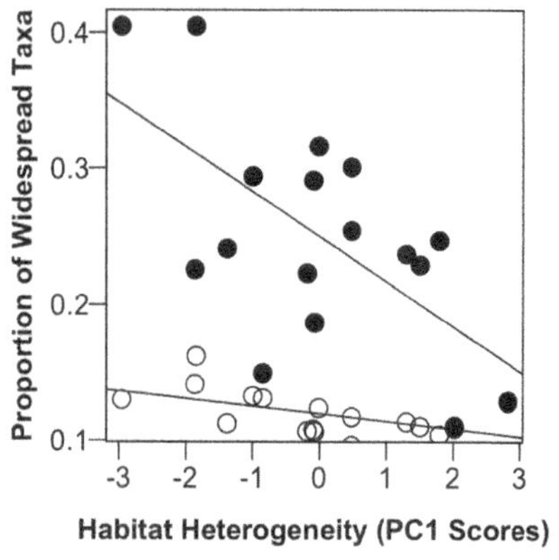

Figure 5. Relationship of habitat heterogeneity with the proportional representation of widespread taxa in each lake. Heterogeneity was measured as the PC 1 scores of 'heterogeneity' model. Filled circles show the total proportion of relative abundance made up by widespread taxa ($R^2 = 0.35$, $p = 0.007$) and open circles show widespread taxa as a proportion of the total number of taxa observed ($R^2 = 0.23$, $p = 0.03$).

robust to using different thresholds to define "widespread" (e.g., present in 70–90% of lakes), and strengthened as we considered thresholds up to 90% (Table S2). We only observed four OTUs that were present in >90% of lakes, which likely were ubiquitous taxa that would be present regardless of lake trophic state [36]. In addition, we found that the increasing representation of previously low abundance or new taxa in response to trophic status was explained most by increasing heterogeneity among habitats (Table 1). Although the AIC_c values for all three models were very close, the AIC value and variable w_i's suggest that heterogeneity explained the declining contribution of the widespread OTUs we observed to the overall lake community (Table 1). This suggests that either novel, previously low abundance, (i.e., below the detection limit of ARISA), or dormant taxa [35] increased in their relative importance as nutrient status of lakes increased, and that chemical heterogeneity in the lake environment most likely facilitated the increased prevalence of these taxa in the lake community.

Discussion

We found that the abundance and richness of bacterial communities increased as lake nutrient status increased, in parallel to what has been reported in other studies [28,33,45]. However, we found even stronger relationships between the overall abundance and number of OTUs observed and the spatial heterogeneity in nutrient conditions among lake layers as trophic status increased. Additionally, because environmental conditions among lake strata diverged and communities associated with these habitats responded differently to lake trophic status (Figure 3), bacterial communities as measured by ARISA fingerprinting were much less similar among lake strata as lake trophic state increased (Figure 4, Table S3). Furthermore, our analyses demonstrated that a core bacterial community of dominant bacteria did not change systematically in high nutrient lakes but rather that increasing habitat heterogeneity, specifically due to large changes in conditions in the deepest layer of lakes, provided additional habitat for previously low abundance or absent taxa that became detectable in more eutrophic conditions. Our study shows that the response of bacterial communities to increased productivity in lakes may differ from that of other lake organisms as a result of spatial heterogeneity in resources specifically affecting the ecology of bacteria and supports the idea that the recruitment of rare or dormant bacterial taxa in lake communities may facilitate much of this response (e.g., [35]).

Effects of Trophic State on Lake Bacterial Communities

We observed significant increases in bacterial abundance and changes in the richness and composition of ARISA profiles at the whole-lake scale in response to increases in lake trophic status. As expected, abundance was positively related to increases in nutrient availability (TN and TP) and chlorophyll *a* (Table 1) across the entire range of ecosystem productivity we observed. High abundances of bacteria are often correlated with high rates of bacterial productivity [46], which can be nutrient-limited in oligotrophic lakes. We also observed increased ARISA profile richness as trophic state increased (Figure 3E), suggesting that more taxa coexisted as dominant types (and thus able to be detected by ARISA) as resource availability increased. As has been observed for macroorganisms in lakes [23], other studies have shown lake bacterial richness to exhibit a range of linear and unimodal responses to productivity depending on the spatial and taxonomic resolution of the study [17,18,29]. We likely observed a linear increase in observed richness since our lakes represented a relatively modest gradient in nutrient loading and chlorophyll *a* compared to the global distribution of lake productivity [22], thus, we may have only captured the initial upward slope of a unimodal response across the global range of lake trophic states. Alternatively, is also possible that we observed a consistently increasing number of taxa rather than a unimodal trend due an effect of increased sampling; in other words, because we took discrete samples from each lake habitat and ran a separate ARISA analysis on each sample rather than on one integrated sample from each lake, we were able to detect more taxa than previous studies [23,29].

Bacterial Community Response to Resource Heterogeneity

Bacterial community shifts were most strongly associated with changes in the heterogeneity of the lake environment in all cases (Table 1). Specifically, we observed more OTUs and decreased similarity among bacterial communities in surface and deep habitats in response to heterogeneity in N and P concentrations (Figures 2B and 3). Environmental conditions within lakes changed to different degrees among the epi-, meta- and hypolimnion (Table S3), which translated to habitat specific responses of bacterial communities in each layer across lakes (Figure 3). For instance, although bacterial abundance increased in all layers, we only observed increases in richness in the hypolimnion with increased trophic status (Figure 3H), which also had the largest range in nutrient concentrations across the trophic gradient (Figure S1). Interestingly, communities in the surface layer varied less across the trophic gradient than communities in the hypolimnion, suggesting that studies that evaluate trends in bacterial communities in response to eutrophication miss an important aspect of the bacterial response.

While our study design does not allow us to definitely conclude whether patterns in bacterial communities were related more to increased trophic state or to habitat heterogeneity because the two were correlated (Figure 2C), we found strong support for our heterogeneity model (Table 1) and patterns suggesting that both were important for bacterial communities. For instance, we observed lower richness (66 OTUs) and higher similarity of communities among lake strata (Sorensen similarity = 0.86) in the only lake in our study that did not stratify into discrete habitats.

Although this lake had similar nutrient concentrations as eutrophic lakes, environmental conditions were more homogeneous within the water column in this lake suggesting that heterogeneity in resource availability influenced bacterial richness more than nutrient concentration alone in all lakes in our study.

While changes in bacterial communities in response to increasing lake trophic status have been widely observed [28,29,45] as have differences in communities among lake strata [30,31,53] previous studies have not linked these observations to evaluate the combined effects of lake resource availability and resource heterogeneity on the richness and composition of lake bacterial communities. Thus, while similar communities may inhabit eutrophic lakes [36], communities become increasingly distinct from one another as surface and deep habitat conditions diverge and biotic interactions change [54]. Furthermore, these findings contrast with studies of other lake organisms, such as zooplankton and fish, which find increased productivity reduces diversity as a result of homogenization of food resources and loss of habitat [15,26,27]. Thus, our study demonstrates that the type of heterogeneity that influences communities varies among macro- and microorganisms and that the response to a large-scale environmental change is expressed differently among habitats within ecosystems.

In addition, our results suggest that changes in bacterial communities with increased trophic status may be more strongly related to vertical differences in nutrient concentrations than dissolved oxygen, temperature, and light, which are typically thought to structure differences among communities in different strata within lakes [24,30,31,53]. We tested this by regressing axis 1 and 2 of our environmental PCA (Figure S1), which represented differences among layers in nutrient concentrations and DO/temperature, respectively. We found that in all cases axis 2 (variation in DO and temperature) did not have a strong relationship with any bacterial response measure. Few studies have tested for this and those that have, have often focused on a single lake or small set of lakes [30,53]. Thus, these results suggest that increased vertical differences in nutrient concentrations may be more important in structuring the bacterial community than vertical differences in DO and temperature as the trophic state of lakes increases.

Were Widespread or Narrowly Distributed Taxa Responsible for Shifts in Community Composition?

We found that habitat heterogeneity played an important role in shaping the bacterial response to increasing lake trophic status through enabling the increased contribution of new or previously low abundance taxa in lake communities. Furthermore, we found that increased resource heterogeneity simultaneously allowed for the retention of a core group of taxa that were widespread among lakes while also providing new habitat and resources for these previously unobserved or low abundance taxa (Figure 5). Thus, our results support the general notion of the importance of rare taxa in microbial communities [35,55] and suggest that bacterially-relevant habitat heterogeneity may be an important mechanism driving the bacterial response to increased lake ecosystem productivity. Additionally, this pattern may be common to many types of planktonic communities [34]. For instance, a study of the response of a lake phytoplankton community to eutrophication and recovery found that there were phytoplankton that were consistently present through time, but that temporally rare taxa were most responsive to changes in lake nutrient status and drove changes in community composition [56].

While the use of ARISA allowed us to screen the bacterial community in a large number of lakes and has been shown to have species-level taxonomic resolution, as with so many methods used to sample bacterial communities, there are limitations associated with using this approach [57]. For example, ARISA underestimates the total richness of the bacterial community and has biases such as preferential amplification of abundant organisms [41]. For example, many of the additional taxa that we observed in more eutrophic lakes could have been present at low abundances in oligotrophic lakes, and therefore below the detection limit of ARISA. However, increased detection of these taxa in eutrophic lakes suggests that they are at higher abundances and thus may be more functionally important in those communities. In addition, other studies have shown that ARISA captures similar patterns in diversity among communities as more high-resolution techniques such as clone libraries [58]. Further, although sequencing and clone library techniques would have allowed us to identify specific taxa, the higher costs associated with those techniques would have limited our ability to sample the entire trophic gradient in our study. Finally, our results are comparable to other studies using similar techniques [24,33,35], and recent studies of bacterial responses to nutrient additions that have used high throughput sequencing techniques suggest that using a more thorough sampling approach likely would not likely reveal a different trend in how richness and composition respond to increased ecosystem productivity (e.g., [19,44]). Thus, the fact that we observed such striking trends using this approach suggests that higher resolution sampling would have only strengthened observed patterns.

In summary, we showed that bacterial community composition changed and was richer and more heterogeneous within lakes as trophic status increased. In contrast to trends in macroorganisms whose diversity is often negatively associated with increases in lake productivity [15,23,27], we showed that the high degree of heterogeneity in bacterial resources in eutrophic lakes promoted higher richness as a result of differentiation of bacterial taxa among lake habitats. We found that eutrophication alters the drivers of bacterial community differences within lakes from physical and redox related variables to changes in nutrient availability. Furthermore, our results suggest that rare or dormant taxa may be most responsible for changes in bacterial communities with increased lake trophic state [35,59]. This "seed bank" of taxa has increasingly been recognized to be important in responding to changes in many types of ecosystems [60], and understanding the role of rare and dormant taxa an important frontier for understanding the processes that regulate how microbial communities respond to ecosystem change in general.

Supporting Information

Figure S1 Constrained analysis of principal coordinates (CAP) of bacterial community composition with environmental variables.

Table S1 Results of eutrophication and heterogeneity Principal Components Analyses (PCAs).

Table S2 Model results from comparing effects of trophic state, heterogeneity, and the combination of the two (T+H) on the relative abundance of widespread taxa in lakes.

Table S3 Means and variation of key environmental variables among and within lakes.

Acknowledgments

The authors would like to acknowledge M. Dyen, H. Bekris, and A. Coogan for extensive help with field data collection and laboratory analyses. We are grateful to R. Lange, J. Griffiths, J. Armstrong and two anonymous reviewers for helpful reviews on the manuscript.

Author Contributions

Conceived and designed the experiments: KJJ DES MCHD. Analyzed the data: KJJ. Contributed reagents/materials/analysis tools: DES MCHD. Wrote the paper: KJJ DES MCHD.

References

1. Reid WV, Mooney HA, Cropper A, Capistrano D, Carpenter SR, et al. (2005) Ecosystems and human well-being: synthesis. Washington, D.C.: Island Press.
2. Carpenter SR, Caraco NF, Correll DL, Howarth RW, Sharpley AN, et al. (1998) Nonpoint pollution of surface waters with phosphorus and nitrogen. Ecological Applications 8: 559–568.
3. Rozenzweig ML (1995) Species diversity in space and time. Cambridge, U.K.: Cambridge University Press.
4. Ribalet F, Marchetti A, Hubbard KA, Brown K, Durkin CA, et al. (2010) Unveiling a phytoplankton hotspot at a narrow boundary between coastal and offshore waters. Proceedings of the National Academy of Sciences of the United States of America 107: 16571–16576.
5. Wright DH (1983) Species-energy theory - An extension of species-area theory. Oikos 41: 496–506.
6. Mittelbach GG, Steiner CF, Scheiner SM, Gross KL, Reynolds HL, et al. (2001) What is the observed relationship between species richness and productivity? Ecology 82: 2381–2396.
7. Adler PB, Seabloom EW, Borer ET, Hillebrand H, Hautier Y, et al. (2011) Productivity Is a Poor Predictor of Plant Species Richness. Science 333: 1750–1753.
8. Cardinale BJ, Hillebrand H, Harpole WS, Gross K, Ptacnik R (2009) Separating the influence of resource 'availability' from resource 'imbalance' on productivity-diversity relationships. Ecology Letters 12: 475–487.
9. Hutchinson GE (1961) The paradox of the plankton. American Naturalist 95: 137–145.
10. Tilman D (1982) Resource Competition and Community Structure. Princeton, New Jersey: Princeton University Press.
11. Chase JM, Leibold MA (2002) Spatial scale dictates the productivity-biodiversity relationship. Nature 416: 427–430.
12. Shurin JB, Winder M, Adrian R, Keller W, Matthews B, et al. (2010) Environmental stability and lake zooplankton diversity - contrasting effects of chemical and thermal variability. Ecology Letters 13: 453–463.
13. Reich PB, Frelich LE, Voldseth RA, Bakken P, Adair EC (2012) Understorey diversity in southern boreal forests is regulated by productivity and its indirect impacts on resource availability and heterogeneity. Journal of Ecology 100: 539–545.
14. Stevens MHH, Carson WP (2002) Resource quantity, not resource heterogeneity, maintains plant diversity. Ecology Letters 5: 420–426.
15. Barnett A, Beisner BE (2007) Zooplankton biodiversity and lake trophic state: Explanations invoking resource abundance and distribution. Ecology 88: 1675–1686.
16. Eilts JA, Mittelbach GG, Reynolds HL, Gross KL (2011) Resource Heterogeneity, Soil Fertility, and Species Diversity: Effects of Clonal Species on Plant Communities. American Naturalist 177: 574–588.
17. Horner-Devine MC, Leibold MA, Smith VH, Bohannan BJM (2003) Bacterial diversity patterns along a gradient of primary productivity. Ecology Letters 6: 613–622.
18. Smith VH (2007) Microbial diversity-productivity relationships in aquatic ecosystems. Fems Microbiology Ecology 62: 181–186.
19. Bowen JL, Ward BB, Morrison HG, Hobbie JE, Valiela I, et al. (2011) Microbial community composition in sediments resists perturbation by nutrient enrichment. Isme Journal 5: 1540–1548.
20. Madigan MT, Martinko JM, Stahl DA, Clark DP (2010) Brock biology of microorganisms. San Francisco: Benjamin Cummings.
21. Green JL, Bohannan BJM, Whitaker RJ (2008) Microbial biogeography: From taxonomy to traits. Science 320: 1039–1043.
22. Wetzel RC (2001) Limnology: lake and river ecosystems. San Diego: Academic Press.
23. Dodson SI, Arnott SE, Cottingham KL (2000) The relationship in lake communities between primary productivity and species richness. Ecology 81: 2662–2679.
24. Longmuir A, Shurin JB, Clasen JL (2007) Independent gradients of producer, consumer, and microbial diversity in Lake Plankton. Ecology 88: 1663–1674.
25. Smith VH (2003) Eutrophication of freshwater and coastal marine ecosystems - A global problem. Environmental Science and Pollution Research 10: 126–139.
26. Vadeboncoeur Y, Jeppesen E, Vander Zanden MJ, Schierup HH, Christoffersen K, et al. (2003) From Greenland to green lakes: Cultural eutrophication and the loss of benthic pathways in lakes. Limnology and Oceanography 48: 1408–1418.
27. Vonlanthen P, Bittner D, Hudson AG, Young KA, Muller R, et al. (2012) Eutrophication causes speciation reversal in whitefish adaptive radiations. Nature 482: 357–U1500.
28. Kolmonen E, Haukka K, Rantala-Ylinen A, Rajaniemi-Wacklin P, Lepisto L, et al. (2011) Bacterioplankton community composition in 67 Finnish lakes differs according to trophic status. Aquatic Microbial Ecology 62: 241–U249.
29. Korhonen JJ, Wang JJ, Soininen J (2011) Productivity-Diversity Relationships in Lake Plankton Communities. Plos One 6.
30. De Wever A, Muylaert K, Van der Gucht K, Pirlot S, Cocquyt C, et al. (2005) Bacterial community composition in Lake Tanganyika: Vertical and horizontal heterogeneity. Applied and Environmental Microbiology 71: 5029–5037.
31. Shade A, Jones SE, McMahon KD (2008) The influence of habitat heterogeneity on freshwater bacterial community composition and dynamics. Environmental Microbiology 10: 1057–1067.
32. Garcia SL, Salka I, Grossart HP, Warnecke F (2013) Depth-discrete profiles of bacterial communities reveal pronounced spatio-temporal dynamics related to lake stratification. Environ Microbiol Rep 5: 549–555.
33. Lindstrom ES (2000) Bacterioplankton community composition in five lakes differing in trophic status and humic content. Microbial Ecology 40: 104–113.
34. Galand PE, Casamayor EO, Kirchman DL, Lovejoy C, Karl DM (2009) Ecology of the rare microbial biosphere of the Arctic Ocean.. Proceedings of the National Academy of Sciences of the United States of America 106: 22427–22432.
35. Jones SE, Lennon JT (2010) Dormancy contributes to the maintenance of microbial diversity. Proceedings of the National Academy of Sciences of the United States of America 107: 5881–5886.
36. Newton RJ, Jones SE, Eiler A, McMahon KD, Bertilsson S (2011) A Guide to the Natural History of Freshwater Lake Bacteria. Microbiology and Molecular Biology Reviews 75: 14–49.
37. Bell T, Bonsall MB, Buckling A, Whiteley AS, Goodall T, et al. (2010) Protists have divergent effects on bacterial diversity along a productivity gradient. Biology Letters 6: 639–642.
38. Shade A, Chiu CY, McMahon KD (2010) Seasonal and Episodic Lake Mixing Stimulate Differential Planktonic Bacterial Dynamics. Microbial Ecology 59: 546–554.
39. Jankowski KJ, Schindler DE, Holtgrieve GW (2012) Assessing non-point source nitrogen loading and nitrogen fixation in lakes using δ15N and nutrient stoichiometry. Limnology and Oceanography 57: 671–683.
40. Eaton A, Greenberg AE, Rice EW (2005) Standard Methods for the Examination of Water and Wastewater: American Public Health Association.
41. Fisher MM, Triplett EW (1999) Automated approach for ribosomal intergenic spacer analysis of microbial diversity and its application to freshwater bacterial communities. Applied and Environmental Microbiology 65: 4630–4636.
42. Brown MV, Schwalbach MS, Hewson I, Fuhrman JA (2005) Coupling 16S-ITS rDNA clone libraries and automated ribosomal intergenic spacer analysis to show marine microbial diversity: development and application to a time series. Environmental Microbiology 7: 1466–1479.
43. Yannarell AC, Triplett EW (2005) Geographic and environmental sources of variation in lake bacterial community composition. Applied and Environmental Microbiology 71: 227–239.
44. Logue JB, Langenheder S, Andersson AF, Bertilsson S, Drakare S, et al. (2012) Freshwater bacterioplankton richness in oligotrophic lakes depends on nutrient availability rather than on species-area relationships. ISME J 6: 1127–1136.
45. Yannarell AC, Kent AD, Lauster GH, Kratz TK, Triplett EW (2003) Temporal patterns in bacterial communities in three temperate lakes of different trophic status. Microbial Ecology 46: 391–405.
46. Pace ML, Cole JJ (1994) Comparative and experimental approaches to top-down and bottom-up regulation of bacteria. Microbial Ecology 28: 181–193.
47. Burnham KP, Anderson DR (2002) Model selection and multi-model inference: A practical information-theoretic approach. New York: Springer-Verlag.
48. Legendre P, Legendre L (1998) Numerical Ecology. Amsterdam: Elsevier Science B.V.
49. Chao A, Chazdon RL, Colwell RK, Shen TJ (2005) A new statistical approach for assessing similarity of species composition with incidence and abundance data. Ecology Letters 8: 148–159.
50. Legendre P, Anderson MJ (1999) Distance-based redundancy analysis: Testing multispecies responses in multifactorial ecological experiments. Ecological Monographs 69: 1–24.
51. R Core Team (2012) R: A language and environment for statistical computing. Vienna, Austria: R Foundation for Statistical Computing.
52. Oksanen J, Guillaume Blanchet F, Kindt R, Legendre P, Minchin PR, et al. (2012) vegan: Community Ecology Package. R package version 2.0–5 ed.
53. Dorigo U, Fontvieille D, Humbert JF (2006) Spatial variability in the abundance and composition of the free-living bacterioplankton community in the pelagic zone of Lake Bourget (France). FEMS Microbiol Ecol 58: 109–119.
54. Kent AD, Jones SE, Yannarell AC, Graham JM, Lauster GH, et al. (2004) Annual patterns in bacterioplankton community variability in a humic lake. Microbial Ecology 48: 550–560.
55. Sogin ML, Morrison HG, Huber JA, Mark Welch D, Huse SM, et al. (2006) Microbial diversity in the deep sea and the underexplored "rare biosphere".

Proceedings of the National Academy of Sciences of the United States of America 103: 12115–12120.
56. Schindler DE, Chang GC, Lubetkin S, Abella SEB, Edmonson WT (2003) Rarity and functional importance in a phytoplankton community.. In: Kareiva P, Levin SA, editors. The importance of species: perspectives on expendability and triage. Princeton, New Jersey: Princeton University Press.
57. Bent SJ, Forney LJ (2008) The tragedy of the uncommon: understanding limitations in the analysis of microbial diversity. ISME J 2: 689–695.
58. Kovacs A, Yacoby K, Gophna U (2010) A systematic assessment of automated ribosomal intergenic spacer analysis (ARISA) as a tool for estimating bacterial richness. Res Microbiol 161: 192–197.
59. Lyons KG, Brigham CA, Traut BH, Schwartz MW (2005) Rare species and ecosystem functioning. Conservation Biology 19: 1019–1024.
60. Lennon JT, Jones SE (2011) Microbial seed banks: the ecological and evolutionary implications of dormancy. Nature Reviews Microbiology 9: 119–130.

5

Ammonium as a Driving Force of Plant Diversity and Ecosystem Functioning: Observations Based on 5 Years' Manipulation of N Dose and Form in a Mediterranean Ecosystem

Teresa Dias[1]*, Adelaide Clemente[1,2], Maria Amélia Martins-Loução[1], Lucy Sheppard[3], Roland Bobbink[4], Cristina Cruz[1]

1 Centro de Biologia Ambiental, Faculdade de Ciências da Universidade de Lisboa, Lisboa, Portugal, **2** Museu Nacional de História Natural e da Ciência, Jardim Botânico, Lisboa, Portugal, **3** Centre for Ecology and Hydrology – Edinburgh, Bush Estate, United Kingdom, **4** B-Ware Research Centre, Radboud University Nijmegen, Nijmegen, The Netherlands

Abstract

Enhanced nitrogen (N) availability is one of the main drivers of biodiversity loss and degradation of ecosystem functions. However, in very nutrient-poor ecosystems, enhanced N input can, in the short-term, promote diversity. Mediterranean Basin ecosystems are nutrient-limited biodiversity hotspots, but no information is available on their medium- or long-term responses to enhanced N input. Since 2007, we have been manipulating the form and dose of available N in a Mediterranean Basin maquis in south-western Europe that has low ambient N deposition (<4 kg N ha^{-1} yr^{-1}) and low soil N content (0.1%). N availability was modified by the addition of 40 kg N ha^{-1} yr^{-1} as a 1:1 NH_4Cl to $(NH_4)_2SO_4$ mixture, and 40 and 80 kg N ha^{-1} yr^{-1} as NH_4NO_3. Over the following 5 years, the impacts on plant composition and diversity (richness and evenness) and some ecosystem characteristics (soil extractable N and organic matter, aboveground biomass and % of bare soil) were assessed. Plant species richness increased with enhanced N input and was more related to ammonium than to nitrate. Exposure to 40 kg NH_4^+-N ha^{-1} yr^{-1} (alone and with nitrate) enhanced plant richness, but did not increase aboveground biomass; soil extractable N even increased under 80 kg NH_4NO_3-N ha^{-1} yr^{-1} and the % of bare soil increased under 40 kg NH_4^+-N ha^{-1} yr^{-1}. The treatment containing less ammonium, 40 kg NH_4NO_3-N ha^{-1} yr^{-1}, did not enhance plant diversity but promoted aboveground biomass and reduced the % of bare soil. Data suggest that enhanced NH_y availability affects the structure of the maquis, which may promote soil erosion and N leakage, whereas enhanced NO_x availability leads to biomass accumulation which may increase the fire risk. These observations are relevant for land use management in biodiverse and fragmented ecosystems such as the maquis, especially in conservation areas.

Editor: Shuijin Hu, North Carolina State University, United States of America

Funding: This study was supported by the Fundação para a Ciência e Tecnologia (FCT) through the projects PTDC/BIA-BEC/099323/2008 and PTDC/BIA-ECS/122214/2010. The funders had no role in study design, data collection and analysis, decision to publish, or preparation of the manuscript.

Competing Interests: The authors have declared that no competing interests exist.

* E-mail: mtdias@fc.ul.pt

Introduction

High biodiversity can stabilize ecosystems through functional complementarities, which can buffer the impacts of environmental change [1,2]. Biodiversity and ecosystem functions are influenced by several drivers (*e.g.* land use change, increased nutrient availability), but ecosystem responses (and mechanisms) to those drivers remain unclear [3], especially responses of ecosystems under persistent anthropogenic influence such as enhanced nitrogen (N) deposition.

Enhanced N availability has been acknowledged as a global and increasing threat to biodiversity [4–7] and ecosystem function [8,9]. However, most of our knowledge of the impacts of increased N availability on ecosystems comes from northern Europe and America [6]. Mediterranean-type ecosystems appear on the 'neglected ecosystems list' [6,10] despite being a global conservation priority [11,12], rivalling tropical rainforests [12,13]. In fact, not much is known about the impacts of increased N availability on Mediterranean-type ecosystems, other than those found in California [10,14]. Ecosystems in the Mediterranean Basin that have experienced intensive human development and impact for millennia [15], and where N deposition is expected to increase threefold by 2050 [5,16] are particularly deserving of study.

The most distinctive features of Mediterranean-type ecosystems likely to influence responses to increased N availability are: climate (highly seasonal, with warm dry summers that contrast with cool rainy winters); soils (low nutrient levels and organic matter and high contents of bases such as carbonates); dominance of dry N deposition; asynchrony between N availability and biological activity [10]; and spatial and temporal heterogeneity [17]. On the basis of these distinguishing features, it seems likely that Mediterranean-type ecosystems could respond rather differently from north temperate ones. Even within Mediterranean-type ecosystems, differences in soil fertility [13] and phosphorus (P) availability in particular can undermine extrapolations from Californian ecosystems to those in the Mediterranean Basin [10]. For these reasons, in 2007 an N-manipulation (dose and form) field experiment was established in a severely nutrient-limited Mediterranean Basin maquis. Contrasting with most studies (north temperate systems [6–9] and Mediterranean Basin [18,19]), within one year plant richness increased and changes in plant community composition were observed [20]. Recently, Forest et al. [3] provided data supporting the hypothesis that long-term impacts of N enrichment on ecosystem functioning depend strongly on biodiversity changes, especially non-random changes in species composition. As a result, our observation raises key questions:

(i) Is this N-driven increase in plant richness transient? This is especially relevant since N deposition in most European ecosystems has already reached a threshold, beyond which diversity has declined [21]. Accordingly, given our N doses, we expected that after 5 years, plant diversity would have stabilized or be in decline.

(ii) Are there consistent plant responders to increased N availability in Mediterranean maquis? We hypothesized that the changes in plant cover and species composition along the ecological succession [22] may confound the identification of consistent plant responders to enhanced N availability.

(iii) Are the N-driven changes in diversity linked with plant productivity? Experiments that directly manipulate species diversity often report a positive impact on productivity, whereas observations of natural communities reveal more complex relationships [2], and nutrient additions generally increase productivity but decrease diversity [23]. In our case, because we expected that the N-benefited species would be small, short-lived plants [24,25] that contribute little to ecosystem functioning, even if plant richness continued to increase, we did not expect an increase in productivity.

Inherent to all these questions is the need to understand whether the form of N matters more than the dose. This is especially relevant since the co-existing plant species occupy distinct N niches, with ammonium (NH_4^+) availability as a determinant [17,26], so that the plant's response to enhanced N availability may vary according to the N form irrespective of the dose.

We report 5 years of results from an N-manipulation field experiment, focusing on N-driven changes among the plant community and soil characteristics in a Mediterranean Basin maquis. Specifically, we focused on the following ecosystem features: vascular plant community (composition, richness and evenness), aboveground biomass (standing biomass and litter production), soil extractable N (NH_4^+ and nitrate – NO_3^-) and organic matter, and percentage of bare soil.

Materials and Methods

Study site

We are grateful to Arrábida Natural Park for making the experimental site available and allowing the N manipulation experiment to which this paper refers. The study site (38°29′N - 9°1′W) is located in Serra da Arrábida in the Arrábida Natural Park, Portugal (a Natura 2000 site - PTCON0010 Arrábida/Espichel), which is within the sub-humid thermomediterranean bioclimatic domain [27]. According to records (1971–2000 - Instituto Nacional de Meteorologia e Geofisica), mean annual precipitation is 730 mm; mean maximum temperature, 27.8°C (August); and mean minimum temperature, 8.1°C (January). Over the experimental period (2007–2011), mean annual precipitation was 870 mm (±231 mm, SE); mean maximum temperature, 21.6°C (±0.7°C, SE); and mean minimum temperature, 12.9°C (±0.5°C, SE). Total monthly precipitation and mean monthly temperatures over the experimental period are shown in Figure 1.

The site is located on a southeast-facing slope (5%) at 130 m altitude, which is protected from public access and has not been managed in recent decades. The soils of Serra da Arrábida are classified, according to the FAO system, as calcic rhodo-chromic luvisols and calcareous chromic cambisols [28]. The soil is skeletal (topsoil layer of approximately 15 cm) with a bulk density of 1.3 g cm^{-3}. Silt predominates (50%), while sand and clay contents are 32% and 18% (silt-sand-loam texture).

Mediterranean maquis vegetation dominates the site, comprising closed vegetation: mainly shrubs with few annuals and some geophytes and normally with some trees, some of which may be in shrub form (Eunis class F5.2 – Mediterranean maquis). The standing community developed after a fire event in summer 2003, four years before the first N addition of this experiment. The dominant plant species was a Cistaceae, *Cistus ladanifer* L. [20,29]. Other abundant plant species were *Erica scoparia* L. (Ericaceae), *Calluna vulgaris* (L.) Hull (Ericaceae), *Genista triacanthos* Brot. (Fabaceae), *Ulex densus* Welw. ex Webb (Fabaceae), *Dittrichia viscosa* L. (Asteraceae) and *Myrtus communis* L. (Myrtaceae). Herbaceous species, of which many were annual plants, comprised ≈10% of the total plant cover [20].

Experimental design and fertilization schedule

During the experimental period (2007–2011), estimated background N deposition ranged between 2.9 kg ha^{-1} yr^{-1} (1.8 kg NO_x+1.1 kg NH_y) in 2008 and 3.8 kg ha^{-1} yr^{-1} (2.2 kg NO_x+1.6 kg NH_y) in 2010. These values were estimated based on the model used by the European Monitoring and Evaluation Programme (grid location: x = 53 and y = 4 - http://www.emep.int/mscw/index_mscw.html). The chosen N doses were high enough to simulate 'worst case' scenarios of N enrichment in this type of habitat, but lower than the N deposition reported for highly N polluted areas in Mediterranean-type ecosystems [30,31]. The N forms mimicked the most likely N pollution scenarios within the Mediterranean Basin (*e.g.* agricultural sources alone or combined with urban/industrial sources). Control plots received no added N, while there were three N treatments: **40A** received 40 kg NH_4^+-N ha^{-1} yr^{-1} as a 1:1 mixture of NH_4Cl and $(NH_4)_2SO_4$; **40AN** received 20 kg NH_4^+-N ha^{-1} yr^{-1} and 20 kg NO_3^--N ha^{-1} yr^{-1} as NH_4NO_3; and **80AN** received 40 kg NH_4^+-N ha^{-1} yr^{-1} and 40 kg NO_3^--N ha^{-1} yr^{-1} as NH_4NO_3. Thus the 40A and 40AN treatments provided the same N dose, while 40A and 80AN provided the same NH_4^+ dose. To prevent N 'contamination' through runoff from the N-plots, the experimental plots were randomly distributed in three rows across the slope, with the controls being located in the top row.

Beginning in January 2007, the dry N salts were homogenously added, by hand, sprinkled over the soil surface, in three equal applications over the year: mid-autumn/winter, spring and summer (Fig. 1). Each treatment was replicated three times (3 plots of 400 m^2 each). To restrict boundary effects and dilution processes, all measurements, analyses and sample collection were performed within the central 100 m^2 square.

Plant diversity and % of bare soil

The composition of the vascular plant community was assessed in June 2007, May 2008 [20] and June 2011 within one 5×5 m

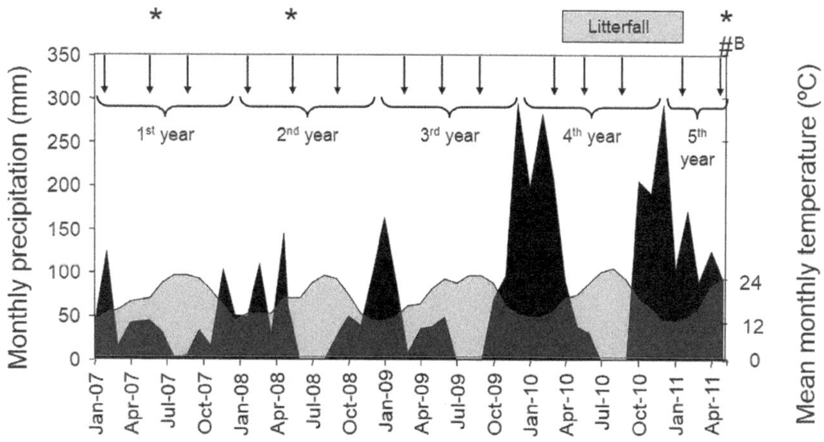

Figure 1. Weather conditions and main experimental events over the experimental period. Mean monthly temperature (light grey) and total monthly precipitation (black). Arrows represent the time of N additions from January 2007 to June 2011. Asterisks refers to the occasions of plant community assessments (2007, 2008 and 2011). Litterfall was collected from April to December 2010. "#B" refers to the time of aboveground biomass sampling (June 2011).

square per experimental plot (within the internal 100 m^2). Percentages of species cover calculated from the total projected crown area and of bare soil (as a measure of erosion potential) were recorded. Plant species observed in the three community assessments are grouped by life form in Table S1 using the Flora Digital de Portugal database (http://jb.utad.pt/pesquisa). This classification was considered more informative than others (*e.g.* by forb, grass, shrub and tree) because it provides some clues to plant size and life cycle. From the vascular plant community assessments in 2007 and 2011, it was possible to calculate plant richness and evenness [32] for the first and fifth springs of the experiment.

Soil extractable N and organic matter

Soil inorganic N pools were measured to estimate soil N retention (the lower the soil inorganic N pools, the higher the N retention and vice versa), and organic matter was measured as a proxy for belowground C sequestration. Soil was sampled from the four corners and the centre of the central 100 m^2 square of each plot. Soil samples (2 cm diameter and 15 cm depth) were removed, sieved (2 mm) and stored at 4°C until analysis. Sampling took place in May 2007 (first spring) and 2011 (fifth spring of the N additions). Individual soil samples (five per plot) were analysed for nitrate (NO_3^--N [33]), ammonium (NH_4^+-N [34]), pH (H_2O extract) and organic matter [20]. Soil inorganic N (N_{in}) was the sum of the water-extracted NH_4^+ and NO_3^-. NO_3^-, NH_4^+ and N_{in} were expressed as μg N per gram of dry soil. Bulk soil samples (equal mixtures of the five soil samples from each experimental plot) were dried (at 60°C until constant mass), ground (MM 2000) and used for quantifying total soil N and C, and hence the C/N ratio, by dry combustion using an elemental analyser (EuroVector, Italy).

Aboveground biomass

Aboveground biomass included standing biomass and leaf litter production. The standing plant biomass was determined in June 2011 (the fifth spring of N additions – shown in Fig. 1) by removing all aboveground plant biomass within three randomly located 1 m^2 squares per plot. Plant biomass was dried to constant mass at 60°C. Litterfall was assessed using litter traps (1.5 mm mesh screen with 0.04 m^2 collecting surface; at ca. 15 cm above the ground) placed under the canopy of five *C. ladanifer* shrubs (located close to the four corners and at the centre of the internal 100 m^2 square) in each plot. Litter was collected fortnightly from April to December 2010 (shown in Fig. 1) and weighed after drying to constant mass at 60°C. Given that the vegetation is dominated by summer semi-deciduous species that shed most of their leaves and twigs in the summer [17,35], the litterfall collected between April and December 2010 was considered to represent the annual fall (Fig. 4-b). Since *C. ladanifer*'s aboveground biomass did not respond to the N addition treatments (data not shown) and litterfall in Californian ecosystems was not significantly affected by five years of 50 kg NH_4NO_3-N ha^{-1} y^{-1} additions [36], no changes in litterfall were also assumed in our study. The amount of biomass that was produced and 'lost' every year through litterfall was estimated by multiplying this annual value by four, representing years 2007 to 2010 (2011 was not considered since most leaf shedding occurs in summer/autumn). The sum of this value and the standing aboveground plant biomass was considered to correspond to the total aboveground biomass produced by the plant community during the experiment.

Derived variables and statistics

The cumulative N, NH_4^+ and NO_3^- load (estimated N deposition +N additions) at each plant assessment were calculated using the above-mentioned EMEP estimates and assuming that N, NH_y and NO_x were deposited homogenously over the year (Fig. 3).

Summary statistics of soil properties, plant species and community responses (richness, evenness, cover, biomass and litterfall) of the various N additions were compared. Two-way ANOVA was applied to determine if there were significant interactions between time and treatment for soil and plant variables. Differences per treatment in biomass and litterfall were analysed by a one-way ANOVA. Both types of ANOVA were followed by a Bonferroni test ($p<0.05$ or $p<0.1$), or by a Games-Howell test whenever homogeneity of variances was not confirmed by the Levene's test. Between treatment differences for change in plant cover were analysed by a one-way ANOVA (followed by a Bonferroni test $p<0.1$) and a Kruskal-Wallis test ($p<0.1$) for normal and non-normal samples respectively (Table S1). Linear correlations between plant richness and cumulative N, NH_4^+ and NO_3^- were also studied (Pearson's correlations). Correlation between plant richness and cumulative NH_4^+ and that between plant richness and cumulative NO_3^- were compared using the

Steiger's Z test ($p<0.05$). In all cases, analyses were performed to ensure that the assumptions regarding the tests' application were not violated. SPSS software, version 20.0, was used for all tests.

Results

Impacts of the N treatments on plant diversity

As usual in the Mediterranean region, the annual rainfall was highly variable (ranging from 513 mm in 2007 and 1541 mm in 2010), with rainfall being greatest during the last two rainy seasons of the experiment (2009/2010 and 2010/2011) (Fig. 1). Initially (spring 2007), richness and evenness were similar in all treatments (Fig. 2). Between treatment differences were, however, evident after one year [20], and remained throughout (Table S1 and Fig. 2). Richness and composition changed the most (≈60% in relation to the control) in treatments receiving 40 kg NH_4^+-N ha^{-1} yr^{-1} (40A and 80AN – Fig. 2 and Table S1). In contrast, between 2007 and 2011, plant species richness (Fig. 2-a) did not change in the control or 40AN plots. Linear, positive and significant correlations were found between plant richness and cumulative N, NH_4^+ and NO_3^- (Fig. 3). However, the correlation between plant richness and cumulative NH_4^+ addition was higher than that between plant richness and cumulative NO_3^-. During the same period, plant evenness decreased with time, decreasing least in the 40AN plots (Fig. 2-b).

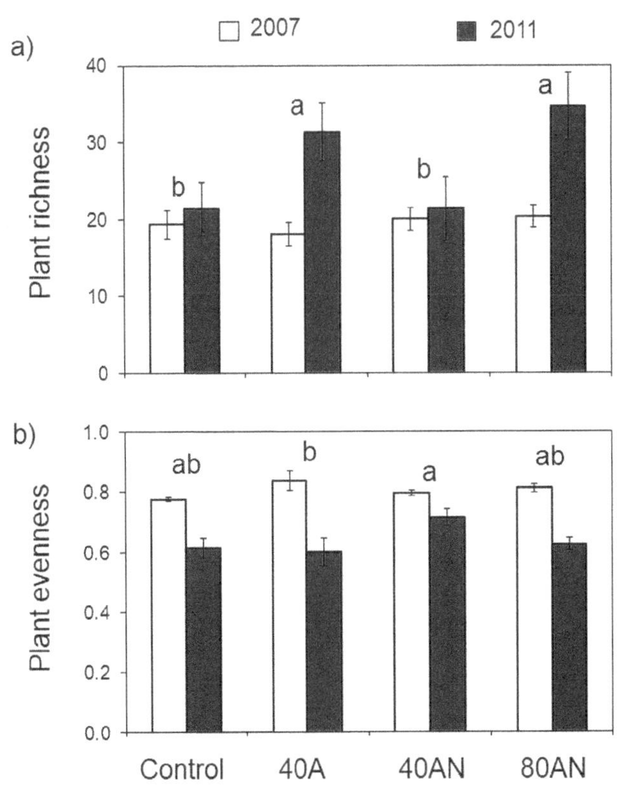

Figure 2. Impacts of the N treatments on vascular plant diversity. Response of the vascular plant community to the N treatments (Control, 40A, 40AN and 80AN), in terms of plant richness (a) and evenness (b). Community assessments were performed in the first and fifth springs of N additions: 2007 and 2011 respectively. Different letters refer to statistically significant differences between treatments (two-way ANOVA $p<0.05$ followed by a Bonferroni test; there were no significant interactions between treatment and time). Bars represent the mean (n = 3 experimental plots per treatment) ±SE.

Plant community composition changed (between 2007 and 2011) as a result of both the ongoing post-fire ecological succession, indicated by the changes in the controls, and the N additions (Table S1). The life forms that were most responsive to N, and to NH_4^+ in particular, were small and ephemeral plants (geophytes, hemicryptophytes and therophytes, Table S1). Initially (2007–2008), *Dittrichia viscosa* benefited most from the increased N (irrespective of dose and form), while *Cistus ladanifer* was affected (negatively impacted) by 80AN. In the longer-term (2007–2011), *C. ladanifer* continued to respond negatively to 80AN, while *Ulex densus* was affected by 40A. These N-affected (negatively impacted by N) species were perennial shrubs. Small species with short life cycles benefited most from N (Table S1): *Hypericum* sp (hemicryptophytes), *Gastridium ventricosum* (annual) and *Sonchus* sp (annual).

Based on a consistent response over 5 years of N addition (Table S1), groups of species that responded similarly to the N dose or form were identified (Tables 1 and S2). No species showed a preference for low N availability, *i.e.* none significantly reduced its cover or disappeared from all the N treatments while remaining in the control plots. The covers of *D. viscosa* and *Sonchus* sp. decreased in all plots, although N appeared to offset some of the cover loss that was observed in the controls. The presence of *Carlina corymbosa*, *Gladiolus illyricus* ssp. *reuteri* and *Galium* sp. may indicate increased N availability, while that of *Salvia sclareoides*, *Asphodelus ramosus*, *Blackstonia perfoliata* and *Dactylis glomerata* may indicate high N availability (associated with the 80AN treatment). In terms of response to the form of N, *Pulicaria odora* appears to prefer NO_3^- (only present in 40AN and 80AN plots), while *Sanguisorba hybrida* appears to avoid NO_3^- (decreased cover in 40AN and 80AN), consistent with being characteristic of later phases of succession. Finally, *Rubia peregrina* and *Brachypodium phoenicoides* appear to prefer high NH_4^+ availability (associated with 40A and 80AN), whereas *Anemone palmate* appears to be rather sensitive to NH_4^+ (found in neither 40A nor 80AN).

Impacts of N on ecosystem processes

N additions did not affect total N or C concentrations after one and five years (Tables 2 and S3). Soils contained very little N (0.1%) or C (~2%), resulting in a C/N ratio <20. N treatments did increase soil available N and organic matter. In relation to the N dose, control plots contained significantly lower concentrations than the 80AN treatment plots, and plots receiving 40 kg N ha^{-1} yr^{-1} (40A and 40AN) showed intermediate levels. Soil NO_3^- and soil inorganic N (these two variables were highly correlated, $r = 0.98$, $p<0.001$) increased over time. The soil also acidified over the course of the experiment, independent of treatment (Tables 2 and S3).

The addition of 40AN and 80AN increased aboveground standing biomass over the control (Fig. 4-a). When estimated leaf litter (Fig. 4-b) produced between 2007 and 2011 is taken into account, aboveground biomass produced by the 40AN plants also exceeded controls (Fig. 4-c).

Initially control plots had most bare ground (Fig. 5), but by the fifth spring of the experiment, differences between treatments were visible: addition of 40AN reduced the area of bare ground whereas 40A increased it.

Discussion

N-driven enhanced plant richness: is the impact transient?

To our knowledge, this is the first integrated study of the impacts of different N doses and forms on a Mediterranean Basin

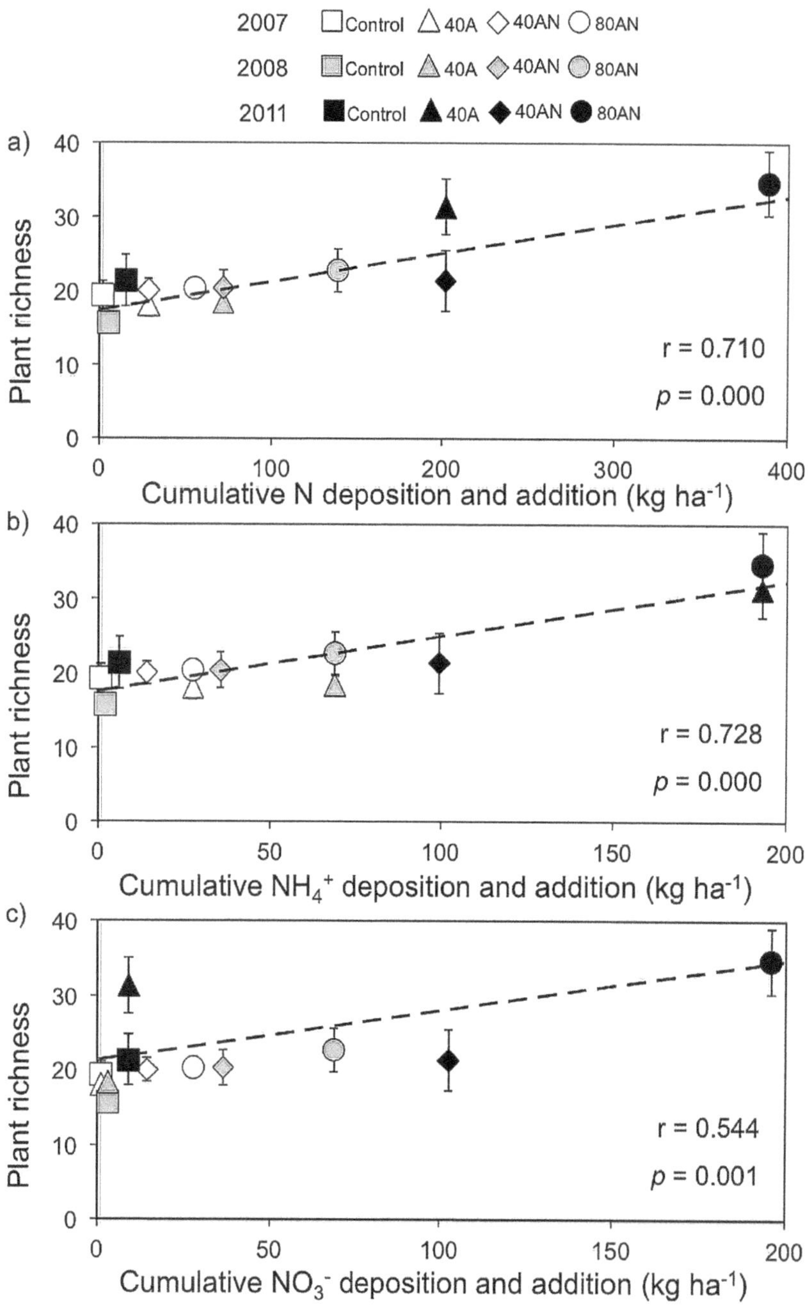

Figure 3. Relation between cumulative N and plant richness. Pearson's correlations between plant richness (number of vascular plant species per 25 m^2) and cumulative N (a), NH_4^+ (b) and NO_3^- (c) over the experiment, accounting for estimated background deposition (EMEP- see material and methods). Correlation between plant richness and cumulative NH_4^+ differed significantly from that between plant richness and cumulative NO_3^- [Steiger's Z test ($p<0.05$)]. Symbols represent the mean (n = 3 experimental plots per treatment and per year) ±SE, but correlations were based on the individual values (n = 36).

ecosystem. In this post-fire successional ecosystem, addition of N continued to promote vascular plant richness, even after 5 years. Assuming that, in this ecosystem, N availability is a severe environmental limitation, the N-driven increase in richness may be explained by the revised Grime's humped-back model [37,38]. The richness increment would reflect the initial alleviation of the stress condition (N limitation), allowing the coexistence of the characteristic site species [35,39] with the incoming exploitative ones [40,41]. The sustained increase in plant richness during the 5 years of N additions may be due to four non-exclusive causes:

(i) The difference in life span between the N-benefited (with short life cycles, including annuals) and the N-affected species (perennials), which would allow the detection of the incomers but only drastic impacts on the losers;

Table 1. Plant species potentially indicative of the N dose and form.

		Benefited	Affected
N dose	**40 kg N ha^{-1} yr^{-1}**	- *Carlina corymbosa*	
		- *Gladiolus illyricus ssp reuteri*	
		- *Galium sp*	
	80 kg N ha^{-1} yr^{-1}	- *Salvia sclareoides*	- *Cistus ladanifer*
		- *Asphodelus ramosus*	
		- *Blackstonia perfoliata*	
		- *Dactylis glomerata*	
N form	**NO_3^-**	- *Pulicaria odora*	- *Sanguisorba hybrida*
	NH_4^+	- *Rubia peregrina*	- *Anemone palmata*
		- *Brachypodium phoenicoides*	

Plant species that responded consistently (over 5 years of N addition treatments) to the N dose and/or form (Tables S1 and S2).

(ii) The ecology of the N-benefited species would allow them to colonize the increased bare soil left by the death of some perennial N-affected plants (*e.g. C. ladanifer* [29]);

(iii) The ecosystem is also limited by lack of water and phosphorus (P), so that the most aggressive/competitive species do not have the best conditions for their development. Not surprisingly, most N-benefited species were annuals that avoid summer drought as seeds but are highly dependent on adequate supplies of N and P [42], which can become more available with an improved N supply [20]. Even though it rained most in years 4 and 5 of the experiment, water alone cannot explain the enhanced plant richness as it did not increase in the control or 40AN plots;

(iv) A large part of the added N is being lost from the ecosystem. This is an obvious hypothesis but one for which there is no supporting data (*i.e.* runoff, nitrate leaching, ammonia volatilization and denitrification were not measured). Although NO_3^- leaching is considered to be an early indicator of N 'saturation' [8] in Mediterranean ecosystems, it has only been measured in Californian chaparral [43] and even then the N loss was less than expected due to immobilisation by the rapidly growing vegetation [44]. N accumulation in vegetation in that system was estimated to be 73 kg N ha^{-1} yr^{-1} [45], which is similar to our high N dose (80 kg N ha^{-1} yr^{-1}). To quantify how much N the system could potentially 'process' before entering the descending phase of the humped model, longer-term studies are necessary.

The observed reduction in *C. ladanifer*'s cover [29] in this experiment probably resulted from its sensitivity to increased NH_4^+ availability [17,26], even in the presence of high NO_3^- availability [26]. The reduction of its canopy will have greatly increased light availability near the soil surface, allowing 'exploitative' species to establish themselves, increasing plant richness. In the presence of NO_3^- but not too much NH_4^+ (40AN treatment), the overshading by *C. ladanifer* is likely to have excluded these small exploitative species. Keeping in mind that the range in

Table 2. Impact of the N treatments on soil properties.

Soil properties		Control	40A	40AN	80AN
N (%)	2007	0.1±0.0	0.1±0.0	0.1±0.0	0.1±0.0
	2011	0.1±0.0	0.1±0.0	0.1±0.0	0.1±0.0
C (%)	2007	1.6±0.2	1.6±0.3	1.8±0.1	2.0±0.6
	2011	1.8±0.1	1.9±0.2	2.1±0.1	2.2±0.3
C/N ratio	2007	18.5±0.8	16.6±0.1	17.4±1.9	16.6±1.0
	2011	18.6±0.4	17.4±0.7	18.1±0.5	17.6±0.8
	*	b	ab	ab	a
N_{in}	2007	5.9±0.9	9.4±2.0	9.6±2.1	10.5±1.9
(μg g^{-1})	2011	7.9±1.3	12.2±1.9	12.9±1.6	19.2±2.5
	*	b	ab	ab	a
NO_3^--N	2007	5.6±0.8	8.3±1.9	8.9±2.2	8.1±1.7
(μg g^{-1})	2011	7.3±1.3	9.8±1.1	12.0±1.5	17.4±2.3
	*	b	ab	ab	a
NH_4^+-N	2007	0.3±0.1	1.1±0.3	0.7±0.1	2.4±0.3
(μg g^{-1})	2011	0.6±0.1	2.3±1.0	0.9±0.1	1.8±0.5
	*	b	ab	ab	a
OM (%)	2007	5.7±0.7	5.8±0.3	6.4±0.3	7.0±1.2
	2011	4.9±0.5	6.5±0.3	7.0±0.5	7.8±0.5
	*				
pH (H_2O)	2007	5.8±0.2	6.5±0.3	6.4±0.1	6.6±0.4
	2011	5.1±0.2	4.9±0.3	5.1±0.1	5.6±0.3

Soil surface (0–15 cm) properties [total N and C, C/N ratio, concentrations of N_{in} – extractable inorganic N, NO_3^- – nitrate, and NH_4^+ – ammonium, OM – organic matter, and pH (H_2O)], in the first (May 2007) and fifth (May 2011) springs of the experiment, according to the N addition treatment (Control, 40A, 40AN and 80AN). Different letters refer to significant differences between treatments (ANOVA $p<0.05$ followed by a Bonferroni test), while * refers to significant differences between 2007 and 2011 (there were no significant interactions between time and treatment, $p<0.05$ – Table S3). Values represent the mean (n = 3 experimental plots per treatment) ±SE.

NO_3^- in our experiment is smaller than that of NH_4^+, the sensitivity of *C. ladanifer* to increased NH_4^+ availability would explain why plant richness was more related to the cumulative ammonium inputs than to the cumulative nitrate inputs.

This NH_4^+-driven increase in plant richness is unlikely to be observed when ecosystems are, or become, dominated by late successional species as these are more NH_4^+ tolerant [17,26]. However, it is possible that climbing plants already present in the community (*e.g. Lonicera implexa, Rubia peregrina, Rubus ulmifolius*, etc.) may become dominant in later stages of succession, as in tropical ecosystems [6]. In conclusion, even though there are no data on the impacts of long-term enhanced N availability on Mediterranean Basin ecosystems, we consider that the N-driven increase in plant richness is transient.

Are there consistent plant responders to increased N availability in Mediterranean maquis?

The composition of the plant community found at the experimental site was consistent with that expected in a Mediterranean maquis regenerating after a fire event [46]. After a fire, there is a 'window of opportunity', *i.e.*, a period of reduced competition for light, nutrients and water [40,41], which favours the establishment of many annuals, hemicryptophytes, geophytes and subshrubs as well as ruderal species colonizing from the surroundings [46]. In a post-fire succession, as the 'window of

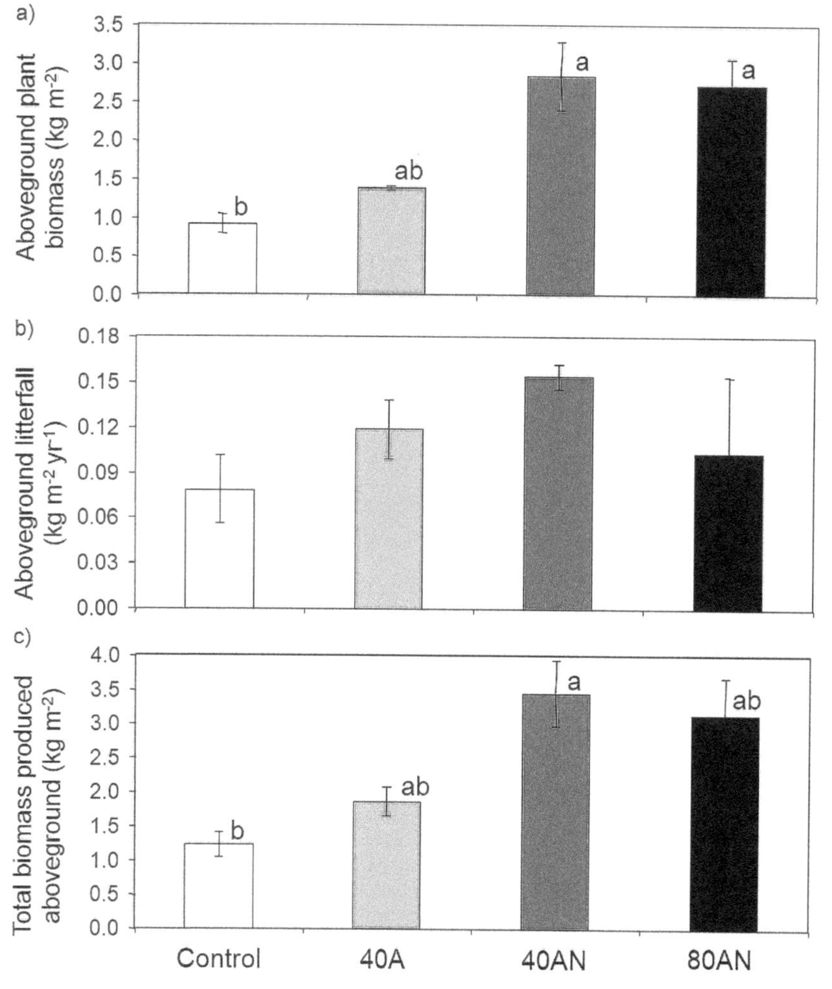

Figure 4. Impacts of the N treatments on plant biomass. Aboveground plant biomass (a), estimated litterfall production per year (b) and total aboveground biomass (sum of the standing biomass and the cumulative litterfall produced on the previous four years - c) according to the N treatments (Control, 40A, 40AN and 80AN). Aboveground plant biomass was harvested in June 2011 (the fifth spring of the experiment) from three 1-m^2 squares per experimental plot. Litterfall was collected from April to December 2010 using litter traps (five 0.04 m^2 per experimental plot). Different letters refer to statistically significant differences between treatments (ANOVA $p<0.05$ followed by a Bonferroni test). Bars represent the mean (n = 3 experimental plots per treatment) ±SE.

opportunity' passes (~5 years after the fire), some plant species tend to disappear from the community [47]. This natural dynamic of the plant community [22,46] has to be taken into account as the background over which the population dynamics associated with the N additions have to be superimposed.

After one year, *D. viscosa* was the only species which had benefited from N addition (irrespective of dose and form). In subsequent years, its cover, together with that of *Sonchus* sp., began to decrease, which is consistent with their disappearance in late succession [40,41,48]. However, N additions appeared to offset some of the cover loss observed in the control, suggesting that increasing N availability extends the 'window of opportunity' for more nutrient-demanding plants. Other plant species responded more consistently to the N dose and/or form over the 5-year study period suggesting that they could be used as indicators of the N status of the ecosystem for at least a period of 5 years.

Are the N-driven changes in diversity linked with functional alterations?

N-addition treatments failed to change soil C and N concentrations, or the C/N ratio, which remained within the range reported for the Mediterranean Basin [49]. In contrast to most studies [6], soil acidity was unaffected by the N additions, possibly due to the high soil calcium carbonate content [50]. However, after 5 years, application of the highest N dose (80AN) had increased soil extractable N, possibly indicating the transition from a closed N cycle to an open and leaky one where N may be lost through NO_3^- leaching [51]. However, NO_3^- leaching can occur much later than many diversity and internal N cycle changes [8]. On the other hand, and in contrast to observations of several studies [52], high N (80AN) addition does appear to have increased soil organic matter, most likely reflecting a decrease in decomposition [53]. This is of particular importance for Mediterranean Basin soils due to their naturally low organic matter concentration and hence higher susceptibility to erosion and desertification [54]. Altogether, it can be concluded that the

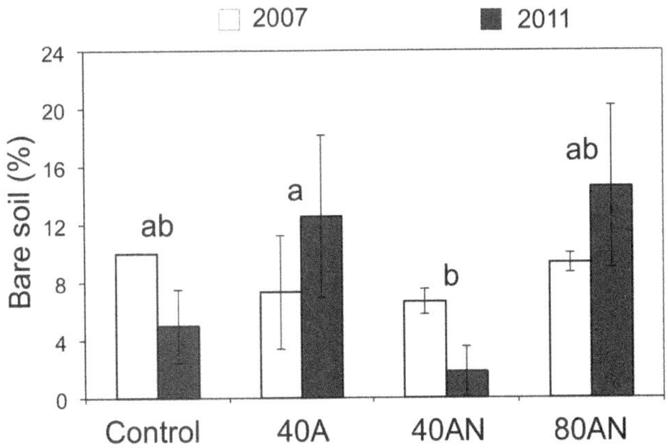

Figure 5. Impact of the N treatments on the % of bare soil. Response of the % of bare soil to the N treatments (Control, 40A, 40AN and 80AN) on the first and fifth springs of N additions (2007 and 2011). Different letters refer to statistically significant differences between treatments (two-way ANOVA $p<0.1$ followed by a Bonferroni test; there were no significant interactions between treatment and time). Bars represent the mean (n = 3 experimental plots per treatment) ±SE.

cumulative N inputs have increased the 'N status' of the ecosystem [7,55].

The aboveground biomass accumulated under control conditions was within the range of the equivalent Californian habitat – coastal sage scrub [36]. Under increased N availability, and in agreement with reports of other temperate [6] and Mediterranean ecosystems [18,30,56], aboveground plant biomass increased. Positive relationships between plant richness and productivity have been reported for Mediterranean [57] and several other ecosystems [58,59]. However, in this study, plant richness and plant productivity were not related, as these were promoted by different N sources. The treatments with more NH_4^+ (40A and 80AN) promoted species richness, but the incomers were small plants that contributed very little to productivity or to protect the soil from erosion that may result from the increased % of bare soil. Given that the non-random N-driven loss of the dominant plant species has been shown to reduce productivity [3], the NH_4^+-driven reduction in the cover of the most abundant plant species (*e.g. Cistus ladanifer, Ulex densus*) would explain the absence of a productivity increase.

In general, N-enhanced plant richness was not linked with functional alterations, mainly because the N-benefited species were small short-lived plants, providing a minor contribution to ecosystem functions while the N-affected plants were perennial shrubs, which underpin ecosystem functions.

Is the form of N more important than the dose?

The data suggest that plant community composition is driven by the amount of NH_4^+-N, highlighting the importance of NH_4^+ as a driving force in Mediterranean ecosystems [17,26]. Given the characteristics of the Mediterranean climate (long dry periods interspersed with torrential rain events), which facilitates both rainfall and wind erosion [60], plant cover and the spatial structure of vegetation play a significant role in preventing soil loss and erosion in Mediterranean areas [61]. Due to the likely loss of cover with increasing NH_4^+ deposition, agricultural emissions are likely to increase erosion in the maquis. This suggests that enhanced NH_y availability (resulting mainly from agriculture) may affect the structure of the surrounding maquis, decrease soil protection (as a consequence of an increase in bare soil) and promote N leakage (as a consequence of an increase in soil extractable N).

In contrast, the lower NH_4^+-N dose applied to the combined N treatment (40AN) plots, with fewer plant species, improved aboveground biomass and soil protection most effectively by almost fully covering the soil with plants. Erosion can also be increased indirectly as a result of NO_3^- deposition increasing aboveground biomass and the risk of wildfires, indicating that increasing urbanisation also threatens these ecosystems. Land managers and planners need to be aware of these effects before they authorise expansion of such N sources close to these ecosystems.

Overall, these results highlight the importance of cumulative N and enhanced N availability, particularly of NH_4^+, as a driving force behind the dynamics and stability in plant community structure and composition of Mediterranean ecosystems [17,26]. The high resilience to disturbance of plant communities in the Mediterranean Basin has been related to their evolutionary history of disturbance, namely anthropogenic pressure [62]. Also, biodiversity may have its greatest impact on the functioning of diverse, naturally heterogeneous ecosystems [63], so it is possible that patchiness and heterogeneity may confer higher resilience to increased N availability (and hence higher reliability on ecosystem functions) than homogeneous un-fragmented ecosystems. Finally, the effectiveness of the critical N loads [64] established for this European habitat, at between 20 and 30 kg N ha^{-1} yr^{-1} [65] would be improved by inclusion of the impact of the N form.

Supporting Information

Table S1 Effect of the N treatments on plant community composition and cover. List of the vascular plant species observed in the three assessments, and their respective changes in cover (2008–2007/2011–2007) according to the N additions.

Table S2 List of the plant species that responded consistently (after one and 5 years of N addition treatments) to the N dose and/or form. The species' most common habitats and, when available, their responses to N enrichment in other studies are shown in the right column.

Table S3 Statistical analyses of soil surface properties.

Acknowledgments

We are grateful to Steve Houghton for help with the manuscript's preparation and to the two reviewers for the comments and suggestions, which greatly improved the present paper.

Author Contributions

Conceived and designed the experiments: TD MAML LS CC. Performed the experiments: TD AC. Analyzed the data: TD AC MAML LS RB CC. Contributed reagents/materials/analysis tools: TD AC MAML CC. Wrote the paper: TD AC MAML LS RB CC. Obtained permission for use of field site: TD CC.

References

1. Hooper DU, Adair EC, Cardinale BJ, Byrnes JEK, Hungate BA, et al. (2012) A global synthesis reveals biodiversity loss as a major driver of ecosystem change. Nature 486: 105–108.
2. MacDougall AS, McCann KS, Gellner G, Turkington R (2013) Diversity loss with persistent human disturbance increases vulnerability to ecosystem collapse. Nature 494: 86–89.
3. Forest I, Reich PB, Tilman D, Hobbie SE, Polasky S, et al. (2013) Nutrient enrichment, biodiversity loss, and consequent declines in ecosystem productivity. Proceedings of the National Academy of Sciences of the United States of America 110: 11911–11916.
4. Sala OE, Chapin FS, Armesto JJ, Berlow E, Bloomfield J, et al. (2000) Biodiversity - Global biodiversity scenarios for the year 2100. Science 287: 1770–1774.
5. Phoenix GK, Hicks WK, Cinderby S, Kuylenstierna JCI, Stock WD, et al. (2006) Atmospheric nitrogen deposition in world biodiversity hotspots: the need for a greater global perspective in assessing N deposition impacts. Global Change Biology 12: 470–476.
6. Bobbink R, Hicks K, Galloway J, Spranger T, Alkemade R, et al. (2010) Global assessment of nitrogen deposition effects on terrestrial plant diversity: a synthesis. Ecological Applications 20: 30–59.
7. De Schrijver A, De Frenne P, Ampoorter E, Van Nevel L, Demey A, et al. (2011) Cumulative nitrogen input drives species loss in terrestrial ecosystems. Global Ecology and Biogeography 20: 803–816.
8. Phoenix GK, Emmett BA, Britton AJ, Caporn SJM, Dise NB, et al. (2012) Impacts of atmospheric nitrogen deposition: responses of multiple plant and soil parameters across contrasting ecosystems in long-term field experiments. Global Change Biology 18: 1197–1215.
9. Southon GE, Field C, Caporn SJM, Britton AJ, Power SA (2013) Nitrogen deposition reduces plant diversity and alters ecosystem functioning: field-scale evidence from a nationwide survey of UK heathlands. PLoS one 8: e59031.
10. Ochoa-Hueso R, Allen EB, Branquinho C, Cruz C, Dias T, et al. (2011) Nitrogen deposition effects on Mediterranean-type ecosystems: An ecological assessment. Environmental Pollution 159: 2265–2279.
11. Myers N, Mittermeier RA, Mittermeier CG, da Fonseca GAB, Kent J (2000) Biodiversity hotspots for conservation priorities. Nature 403: 853–858.
12. Klausmeyer KR, Shaw MR (2009) Climate change, habitat loss, protected areas and the climate adaptation potential of species in Mediterranean ecosystems worldwide. PLoS One 4: e6392.
13. Cowling RM, Rundel PW, Lamont BB, Arroyo MK, Arianoutsou M (1996) Plant diversity in Mediterranean-climate regions. Trends in Ecology & Evolution 11: 362–366.
14. Fenn ME, Allen EB, Weiss SB, Jovan S, Geiser LH, et al. (2010) Nitrogen critical loads and management alternatives for N-impacted ecosystems in California. Journal of Environmental Management 91: 2404–2423.
15. Blondel J, Aronson J (1999) Biology and wildlife of the Mediterranean region. Biology and wildlife of the Mediterranean region: i–xxii, 1–328.
16. Galloway JN, Dentener FJ, Capone DG, Boyer EW, Howarth RW, et al. (2004) Nitrogen cycles: past, present, and future. Biogeochemistry 70: 153–226.
17. Cruz C, Bio AMF, Jullioti A, Tavares A, Dias T, et al. (2008) Heterogeneity of soil surface ammonium concentration and other characteristics, related to plant specific variability in a Mediterranean-type ecosystem. Environmental Pollution 154: 414–423.
18. Bonanomi G, Caporaso S, Allegrezza M (2006) Short-term effects of nitrogen enrichment, litter removal and cutting on a Mediterranean grassland. Acta Oecologica-International Journal of Ecology 30: 419–425.
19. Ochoa-Hueso R (2011) Efectos de la deposición atmosférica de nitrógeno en ecosistemas Mediterráneos. Madrid: Universidad Autonoma de Madrid. 412 p.
20. Dias T, Malveiro S, Martins-Loucao MA, Sheppard LJ, Cruz C (2011) Linking N-driven biodiversity changes with soil N availability in a Mediterranean ecosystem. Plant and Soil 341: 125–136.
21. Stevens CJ, Gowing DJG, Wotherspoon KA, Alard D, Aarrestad PA, et al. (2011) Addressing the impact of atmospheric Nitrogen deposition on western European grasslands. Environmental Management 48: 885–894.
22. Trabaud L (1994) Postfire plant community dynamics in the Mediterranean basin. Ecological Studies; The role of fire in Mediterranean-type ecosystems 107: 1–15.
23. Jiang L, Wan S, Li L (2009) Species diversity and productivity: why do results of diversity-manipulation experiments differ from natural patterns? Journal of Ecology 97: 603–608.
24. Suding KN, Collins SL, Gough L, Clark C, Cleland EE, et al. (2005) Functional- and abundance-based mechanisms explain diversity loss due to N fertilization. Proceedings of the National Academy of Sciences of the United States of America 102: 4387–4392.
25. Allen EB, Temple PJ, Bytnerowicz A, Arbaugh MJ, Sirulnik AG, et al. (2007) Patterns of understory diversity in mixed coniferous forests of southern California impacted by air pollution. Thescientificworldjournal 7: 247–263.
26. Dias T, Neto D, Martins-Loucao MA, Sheppard L, Cruz C (2011) Patterns of nitrate reductase activity vary according to the plant functional group in a Mediterranean maquis. Plant and Soil 347: 363–376.
27. Rivas-Martínez S, Penas A, Díaz TE (2004) Bioclimatic map of Europe-Thermoclimatic belts.
28. Specht RL, Rundel PW, Westman WE, Catling PC, Majer JD, et al. (1988) Mediterranean type ecosystems. A data source book. Dordrecht: Kluwer Academic Publishers.
29. Dias T, Martins-Loucao MA, Sheppard L, Cruz C (2012) The strength of the biotic compartment in retaining nitrogen additions prevents nitrogen losses from a Mediterranean maquis. Biogeosciences 9: 193–201.
30. Fenn ME, Baron JS, Allen EB, Rueth HM, Nydick KR, et al. (2003) Ecological effects of nitrogen deposition in the western United States. Bioscience 53: 404–420.
31. Meixner T, Fenn M (2004) Biogeochemical budgets in a Mediterranean catchment with high rates of atmospheric N deposition - importance of scale and temporal asynchrony. Biogeochemistry 70: 331–356.
32. Krebs CJ (1989) Ecological methodology. Ecological methodology: i–xii, 1–654.
33. Matsumura S, Witjaksono G (1999) Modification of the Cataldo method for the determination of nitrate in soil extracts by potassium chloride. Soil Science and Plant Nutrition 45: 231–235.
34. Cruz C, Martins-Loução MA (2000) Determination of ammonium concentrations in soils and plant extracts. In: Martins-Loução MA, Lips SH, editors. Nitrogen in a Sustainable Ecosystem. The Netherlands: Leiden: Backhuys Publishers. pp. 291–297.
35. Clemente AS, Rego FC, Correia OA (2005) Growth, water relations and photosynthesis of seedlings and resprouts after fire. Acta Oecologica-International Journal of Ecology 27: 233–243.
36. Vourlitis GL, Pasquini SC, Mustard R (2009) Effects of dry-season N input on the productivity and N storage of Mediterranean-type shrublands. Ecosystems 12: 473–488.
37. Grime JP (1973) Competitive exclusion in herbaceous vegetation. Nature 242: 344–347.
38. Michalet R, Brooker RW, Cavieres LA, Kikvidze Z, Lortie CJ, et al. (2006) Do biotic interactions shape both sides of the humped-back model of species richness in plant communities? Ecology Letters 9: 767–773.
39. Casals P, Romanya J, Vallejo VR (2005) Short-term nitrogen fixation by legume seedlings and resprouts after fire in Mediterranean old-fields. Biogeochemistry 76: 477–501.
40. Lavorel S, Lepart J, Debussche M, Lebreton JD, Beffy JL (1994) Small-scale disturbances and the maintenance of species-diversity in Mediterranean old fields. Oikos 70: 455–473.
41. Lavorel S, Oneill RV, Gardner RH (1994) Spatio temporal dispersal strategies and annual plant-species coexistence in a structured landscape. Oikos 71: 75–88.
42. Fanelli G, Lestini M, Sauli AS (2008) Floristic gradients of herbaceous vegetation and P/N ratio in soil in a Mediterranean area. Plant Ecology 194: 231–242.
43. Fenn ME, Allen EB, Geiser LH (2011) Mediterranean California. In: Pardo LH, Robin-Abbott MJ, Driscoll CT, editors. Assessment of N deposition effects and empirical critical loads of N for ecoregions of the United States: General Technical Report, USDA Forest Service, Northern Research Station, Newtown Square, PA. pp. 143–170.
44. Meixner T, Fenn ME, Wohlgemuth P, Oxford M, Riggan P (2006) N saturation symptoms in chaparral catchments are not reversed by prescribed fire. Environmental Science & Technology 40: 2887–2894.
45. Riggan PJ, Goode S, Jacks PM, Lockwood RN (1988) Interaction of fire and community-development in chaparral of southern-California. Ecological Monographs 58: 155–176.
46. Thompson JD (2005) Plant evolution in the Mediterranean. New York: Oxford University Press.
47. Trabaud L (1987) Dynamics after fire of sclerophyllous plant communities in the Mediterranean Basin. Ecologia Mediterranea 13: 25–37.
48. Bonet A, Pausas JG (2004) Species richness and cover along a 60-year chronosequence in old-fields of southeastern Spain. Plant Ecology 174: 257–270.
49. Sardans J, Penuelas J, Estiarte M, Prieto P (2008) Warming and drought alter C and N concentration, allocation and accumulation in a Mediterranean shrubland. Global Change Biology 14: 2304–2316.
50. Schultz J (1995) The ecozones of the world: the ecological divisions of the geosphere. The ecozones of the world: the ecological divisions of the geosphere: i–x, 1–449.
51. De Schrijver A, Verheyen K, Mertens J, Staelens J, Wuyts K, et al. (2008) Nitrogen saturation and net ecosystem production. Nature 451: E1.
52. Knorr M, Frey SD, Curtis PS (2005) Nitrogen additions and litter decomposition: A meta-analysis. Ecology 86: 3252–3257.
53. Dias T, Oakley S, Alarcon-Gutierrez E, Ziarelli F, Trindade H, et al. (2013) N-driven changes in a plant community affect leaf-litter traits and may delay organic matter decomposition in a Mediterranean maquis. Soil Biology & Biochemistry 58: 163–171.
54. Jones A, Panagos P, Barcelo S, Bouraqui F, Bosco C, et al. (2012) The state of soil in Europe.
55. Padgett PE, Allen EB, Bytnerowicz A, Minich RA (1999) Changes in soil inorganic nitrogen as related to atmospheric nitrogenous pollutants in southern California. Atmospheric Environment 33: 769–781.
56. Allen EB, Sirulnik AG, Egerton-Warburton L, Kee SN, Bytnerowicz A, et al. (2005) Air pollution and vegetation change in southern California coastal sage scrub: a comparison with chaparral and coniferous forest. Albany: U.S.

Department of Agriculture, Forest Service, Pacific Southwest Research Station. 79–96 p.
57. Caldeira MC, Hector A, Loreau M, Pereira JS (2005) Species richness, temporal variability and resistance of biomass production in a Mediterranean grassland. Oikos 110: 115–123.
58. Tilman D, Wedin D, Knops J (1996) Productivity and sustainability influenced by biodiversity in grassland ecosystems. Nature 379: 718–720.
59. Cardinale BJ, Wright JP, Cadotte MW, Carroll IT, Hector A, et al. (2007) Impacts of plant diversity on biomass production increase through time because of species complementarity. Proceedings of the National Academy of Sciences of the United States of America 104: 18123–18128.
60. Pimentel D, Harvey C, Resosudarmo P, Sinclair K, Kurz D, et al. (1995) Environmental and economic costs of soil erosion and conservation benefits. Science 267: 1117–1123.
61. López-Bermúdez F, García-Gómez J (2006) Desertification in the arid and semiarid Mediterranean regions. A food security issue. Desertification in the Mediterranean Region A Security Issue 3: 401–428.
62. Lavorel S (1999) Ecological diversity and resilience of Mediterranean vegetation to disturbance. Diversity and Distributions 5: 3–13.
63. Tylianakis JM, Rand TA, Kahmen A, Klein A-M, Buchmann N, et al. (2008) Resource heterogeneity moderates the biodiversity-function relationship in real world ecosystems. Plos Biology 6: 947–956.
64. Nilsson J, Grennfelt P (1988) Critical loads for sulphur and nitrogen. Sweden.
65. Bobbink R, Hettelingh J-P (2011) Review and revision of empirical Critical Loads and dose-response relationships. In: Effects CCf, editor. Netherlands: National Institute for Public Health and the Environment (RIVM).

6

Leaf and Life History Traits Predict Plant Growth in a Green Roof Ecosystem

Jeremy Lundholm[1]*, Amy Heim[1], Stephanie Tran[1], Tyler Smith[2]

1 Departments of Biology and Environmental Science, Saint Mary's University, Halifax, Nova Scotia, Canada, **2** Eastern Cereal and Oilseen Resarch Centre, Agriculture and Agri-Food Canada, Ottawa, Ontario, Canada

Abstract

Green roof ecosystems are constructed to provide services such as stormwater retention and urban temperature reductions. Green roofs with shallow growing media represent stressful conditions for plant survival, thus plants that survive and grow are important for maximizing economic and ecological benefits. While field trials are essential for selecting appropriate green roof plants, we wanted to determine whether plant leaf traits could predict changes in abundance (growth) to provide a more general framework for plant selection. We quantified leaf traits and derived life-history traits (Grime's C-S-R strategies) for 13 species used in a four-year green roof experiment involving five plant life forms. Changes in canopy density in monocultures and mixtures containing one to five life forms were determined and related to plant traits using multiple regression. We expected traits related to stress-tolerance would characterize the species that best grew in this relatively harsh setting. While all species survived to the end of the experiment, canopy species diversity in mixture treatments was usually much lower than originally planted. Most species grew slower in mixture compared to monoculture, suggesting that interspecific competition reduced canopy diversity. Species dominant in mixture treatments tended to be fast-growing ruderals and included both native and non-native species. Specific leaf area was a consistently strong predictor of final biomass and the change in abundance in both monoculture and mixture treatments. Some species in contrasting life-form groups showed compensatory dynamics, suggesting that life-form mixtures can maximize resilience of cover and biomass in the face of environmental fluctuations. This study confirms that plant traits can be used to predict growth performance in green roof ecosystems. While rapid canopy growth is desirable for green roofs, maintenance of species diversity may require engineering of conditions that favor less aggressive species.

Editor: Andrew Hector, University of Oxford, United Kingdom

Funding: This work was supported by an NSERC Discovery grant to J. Lundholm. The funders had no role in study design, data collection and analysis, decision to publish, or preparation of the manuscript.

Competing Interests: The authors have declared that no competing interests exist.

* Email: jlundholm@smu.ca

Introduction

Green roofs consist of vegetation and growing media, deployed on building rooftops. They are constructed to provide a range ecosystem services including reduction of stormwater runoff [1], reduction of energy fluxes through the building envelope [2], and provision of habitat for urban biodiversity [3,4]. Green roofs with shallow growing media (usually less than 200 mm) are termed "extensive" green roofs, and are increasingly popular as they provide key services, while minimizing weight loading on buildings. Shallow growing media and exposed rooftop conditions are challenging for plant growth and survival, due to drought stress [5] and winterkill [6], thus the green roof industry relies on drought-tolerant plant species for these extensive green roofs.

Green roofs are commonly planted with succulents such as *Sedum* spp. However, there is a movement to incorporate other plant life-forms, such as grasses, to increase functional diversity [7], provide visual interest [5], and to provide better habitat for invertebrates [3]. Other life-form groups may not survive well, depending on roof conditions [8,9]. Poor plant survival leads to lower vegetation coverage which tends to reduce the provision of key ecosystem services from the green roof [7,10]. Studies of the dynamics of green roof plant species composition show that original species composition shifts over time, with some species becoming extirpated from the system reflecting unsuitability of species to the green roof environment and/or negative interactions with other plant species [11,12].

Ecosystem services from green roofs depend in part on vegetation composition [13]. Recent work has identified the potential for more diverse plant assemblages to provide enhanced functioning [14,15], but green roof service provisioning may change as species abundances shift and high species diversity may not persist over time. While there are thousands of plant species that could be used on extensive green roofs, and even more possible combinations of species, horticultural trials to test so many species under controlled conditions would be logistically impossible. To streamline the process of plant selection for green roofs, ecologists have turned to plant traits to determine which general plant characteristics are related to survival, growth and performance of key ecosystem services [16]. Plant leaf traits are commonly used as predictors of ecosystem functioning [17–22] and community assembly processes [23,24]. Leaf and other plant characteristics such as height, flowering phenology, and method of vegetative reproduction have been combined as indicators of general life history strategies, most notably Grime's Competitive-Stress-tolerant-Ruderal strategies [25,26]. Such strategies are

Table 1. A list of all the species used in this experiment including their growth form, the treatments they were used in and their origin.

Species	Code	Growth Form	Treatments	Origin
Sagina procumbens L.	Sag. p	Creeping Forb (C)	Mono., C, C/T/S, G/C/S, G/C/T, D/C/S, D/C/T, D/G/S, All	Native
Danthonia spicata (L.) Beauv.	Dan. s	Graminoid (G)	Mono., G, G/C/S, G/C/T, G/T/S, D/G/S, D/G/T, D/G/C, All	Native
Deschampsia flexuosa (L.) Trin.	Des. f	Graminoid (G)	Mono., G, G/C/S, G/C/T, G/T/S, D/G/S, D/G/T, D/G/C, All	Native
Poa compressa L.	Poa. C	Graminoid (G)	Mono., G, G/C/S, G/C/T, G/T/S, D/G/S, D/G/T, D/G/C, All	Introduced
Empetrum nigrum L.	Emp. n	Creeping Shrub (D)	Mono., D, D/C/S, D/C/T, D/G/C, D/G/S, D/G/T, D/T/S, All	Native
Gaultheria procumbens L.	Gau. p	Creeping Shrub (D)	Mono., D, D/C/S, D/C/T, D/G/C, D/G/S, D/G/T, D/T/S, All	Native
Vaccinium vitis-idaea L.	Vac. v	Creeping Shrub (D)	Mono., D, D/C/S, D/C/T, D/G/C, D/G/S, D/G/T, D/T/S, All	Native
Sedum acre L.	Sed. a	Succulents (S)	Mono., S, C/T/S, G/C/S, G/T/S, D/C/S, D/G/S, D/T/S, All	Introduced
Sedum rosea (L.) Scop.	Sed. r	Succulents (S)	Mono., S, C/T/S, G/C/S, G/T/S, D/C/S, D/G/S, D/T/S, All	Native
Sedum spurium M. Beib.	Sed. s	Succulents (S)	Mono., S, C/T/S, G/C/S, G/T/S, D/C/S, D/G/S, D/T/S, All	Introduced
Campanula rotundifolia L.	Cam. r	Tall Forbs (T)	Mono., T, G/T/S, D/G/T, D/T/S, G/C/T, D/C/T, C/T/S, All	Native
Plantago maritima L.	Pla. m	Tall Forbs (T)	Mono., T, G/T/S, D/G/T, D/T/S, G/C/T, D/C/T, C/T/S, All	Native
Solidago bicolor L.	Sol. b	Tall Forbs (T)	Mono., T, G/T/S, D/G/T, D/T/S, G/C/T, D/C/T, C/T/S, All	Native

Mono. refers to the treatment planted with the species in monoculture; Nomenclature follows [Zinck 1998].

meant to integrate various components of plant anatomy, morphology, physiology and habitat preferences into general syndromes representing divergent evolutionary responses to the abiotic and biotic environment. In this study, we examined plant species abundance and diversity as it changed over four years in a green roof experiment, and identified leaf traits and life history characteristics that predicted population growth rates (changes in abundance) and final biomass in monocultures and mixed-species plantings. We hypothesized that traits associated with stress-tolerance, such as low stature and low specific leaf area would be the best predictors of growth and biomass in this system.

Methods

Ethics Statement

Seeds and leaves were collected from a field site (Duncan's Cove Nature Reserve) under permit from Nova Scotia Environment; live plants were collected from public but not protected land and no species under protected status was collected. The experiment was carried out on the campus of Saint Mary's University with permission.

This study examined the changes in species abundance and survival of thirteen perennial plant species in a biodiversity-ecosystem functioning experiment, in a replicated green roof system over four growing seasons (2007–2010) (described previously, see [7]). Each replicate consisted of a black plastic module (36 cm×36 cm×12 cm), with a free-draining base, lined with a composite non-woven water-retention layer (Huesker Inc., Charlotte, NC, USA), followed by an Enkamat (Colbond Inc., Enka, NC, USA) above to act as a drainage/filter layer which was topped with growing medium. We used a commercially available green roof growing medium (Sopraflor X, Soprema Inc., Drummondville, QC, Canada) to a depth of approximately 6 cm (above the Enkamat).

Three species in each of five life-form groups (Table 1) were included in this experiment. Two of the creeping forbs originally included were extirpated by the end of the first year and were not included in any subsequent analyses (these turned out to be annuals that did not re-seed). Total initial density of plants (21) was controlled in each replicate, while composition and/or diversity varied as follows: three replicates of each species in monoculture; five replicates containing each of the three species belonging to a single life-form group; five replicates of each three life-form group combination (ten combinations total), with all three species of each life-form included every time its particular group was included and 20 replicates of the mixture of all five life form groups (15 species initially; 13 species from years 2–4) (Table 1). We planted 21 individual plants in each module (regardless of the number of life-form groups present) in four rows of four plants (on 9 cm centers) and a center row of five plants (on 7 cm centers). For mixtures, we alternated life form groups so that species from a group were well dispersed throughout the module. The planting sequence involved randomly choosing the life-form and species pattern (without replacement) until all species to be included had been selected once, after which the same pattern was repeated throughout the module until a total of 21 plants had been included. Modules were installed on a roof approximately 5 m above the ground on Saint Mary's University campus in Halifax, Nova Scotia (44° 39′ N, 63° 35′ W). For more details on roof setting and climate see [7]. Modules were placed randomly into five blocks representing a shading gradient based on proximity of surrounding buildings; block 1 had the least insolation, resulting in the lowest substrate temperatures [27], and average substrate temperatures increase linearly from block 1 to 5. The monocultures only had three replicates each, which were randomly placed into blocks 1, 3, and 5. The complete design consists of 11 levels of the planting treatment factor, which can be grouped into 1-, 3- or 5-life-form treatments, with a random block factor. Response variables (described below) are continuous variables: final above-ground biomass and change in abundance, separately for each species in each replicate.

Plant communities were maintained by weeding out any species not originally present in the mixture, but new seedlings or vegetative growth from species originally planted were left undisturbed.

Species chosen for this experiment were selected either from naturally occurring habitats in the region (coastal barrens e.g. [28]) or from urban habitats within Halifax, Nova Scotia for the three

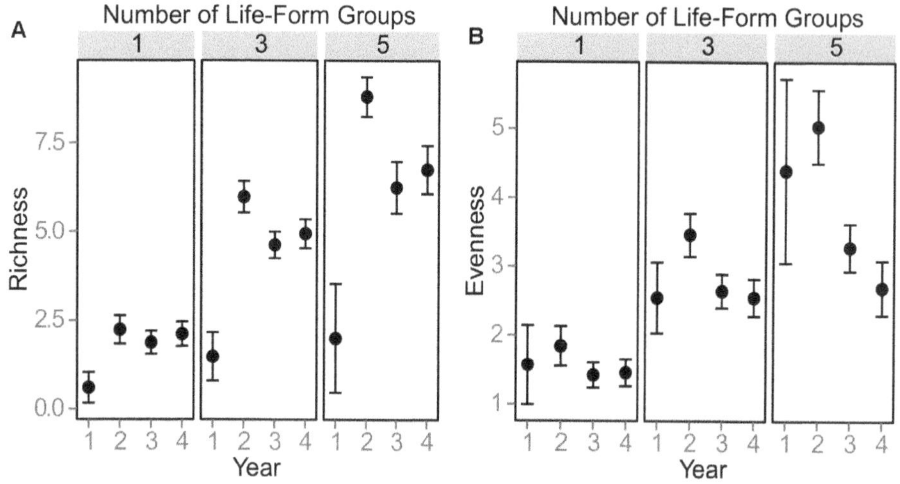

Figure 1. Canopy species richness (A) and evenness (B) over four years (means±95% CI). Treatments grouped by number of life-forms planted; original planted species richness was 3, 9, 15 for 1, 3, and 5 life-form groups respectively. Species evenness is 1/Simpson's index. Sample sizes: 1 life-form treatements: n = 25; 3 life-form treatments: n = 50; 5 life-form treatment: n = 20.

non-native species: *Poa compressa*, *Sedum acre*, and *Sedum spurium*. These non-natives were included to provide comparisons with species commonly used by industry (the *Sedum* spp.) or with a species commonly occurring spontaneously on European green roofs (*Poa compressa* [29]).

Species abundances were quantified in each year during the biomass peak (late July-early August) using a point intercept

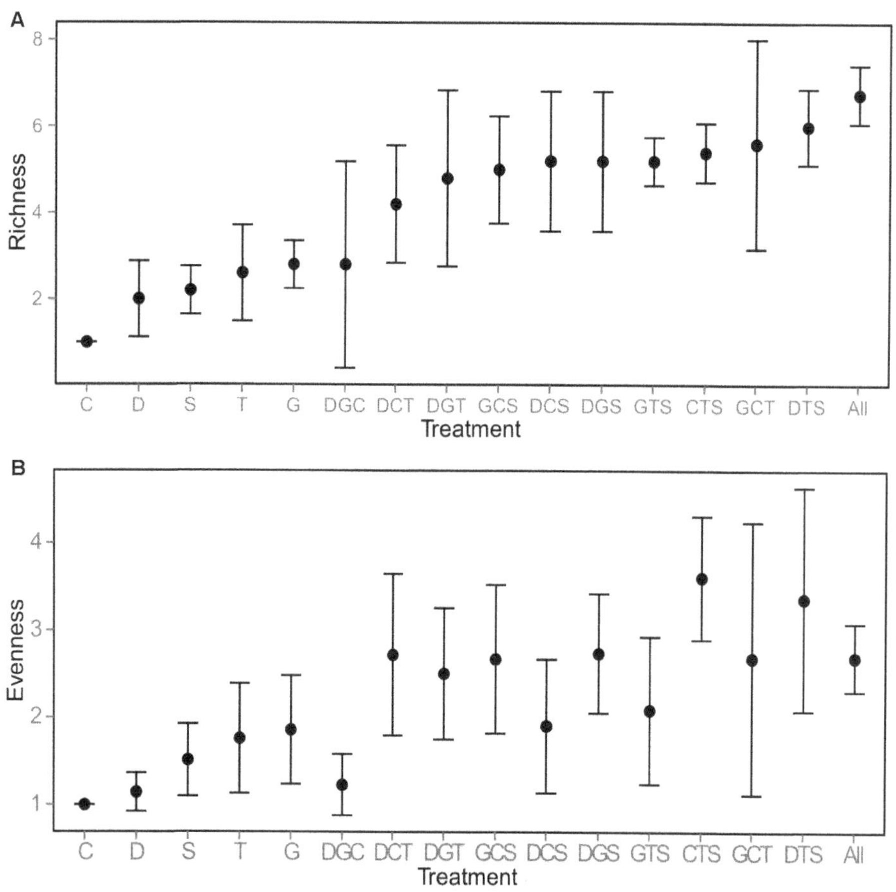

Figure 2. Canopy species richness (A) and evenness (B) in year four, by specific life-form combination (means±95% CI). Sample sizes: 1 life-form treatments: n = 5 each; 3 life-form treatments: n = 5 each; 5 life-form treatment: n = 20.

method [30], using a pin frame (Domenico Ranalli, Regina, Saskatchewan, Canada). The frame was 30 cm high with a length of 36 cm and a width of 36 cm, contained 16 equally spaced rods (6 mm diameter), and rested on the edge of the module such that the base of the pins was approximately 2 cm above the substrate surface. During sampling we recorded the number of contacts with the rods by parts belonging to each plant species. The sum of contacts was termed the "canopy density" for each species. We calculated the change in abundance (population growth) as (ln(canopy density in year 4) - ln(canopy density in year 1))/# days [31]. All aboveground plant material was clipped at the end of the growing season in year 4, sorted to species, dried and weighed to determine final aboveground biomass for each species in each replicate module. Because several species grew slowly, year 1 canopy density was sometimes zero even though the species was present in the module (it was not large enough to register a contact with the pin frame). To determine change in abundance, these zero values were recoded as 1 contact so that the species could be included in the ln-transformed data. Canopy density for species extirpated from a module by 2010 (not detected in pin frame sampling) was recorded as 0.5 instead of 0, so that we could register a negative change in abundance for such species instead of an undefined value. For calculations of canopy species richness and evenness, these species were left as zeroes. This study addresses only changes in abundance of species, quantified using canopy densities, and final biomass, not other ecosystem functions or services, which are the subject of detailed analyses elsewhere [7].

We determined traits for the 13 species using field sampled material and values from the literature. Five plants per species were randomly selected at a natural coastal barrens site and one leaf was harvested from each plant to determine fresh and dry leaf weights, and leaf area (see Table S1). Non-native plants were sampled from urban populations in Halifax, Nova Scotia. Single-sided leaf area was obtained using ImageJ software (Image Processing and Analysis in Java, http://rsbweb.nih.gov/ij/). Leaves were pressed and dried at 55°C for two days before recording leaf dry weight. Canopy height [26] and plant height were measured in the field. Month of flower initiation, flowering period, maximum height, and lateral spread, following [26] were determined from the literature [32,33] and corrected based on local field observation if literature values did not correspond with local plant communities. Raw C-S-R (C: competitive; S: stress-tolerant; R: ruderal) scores were derived from the basic leaf and other traits according to [26]. The traits used as predictors in subsequent analyses were leaf area, specific leaf area (SLA: leaf area/leaf dry weight), leaf dry matter content (dry weight/fresh weight), average height and the C, S and R scores. SLA is negatively related to stress-tolerant strategies and positively related to competitive strategies [18,34]. Stress-tolerant plants are typically short, competitive usually tall while ruderals can vary. Leaf dry matter content is usually negatively associated with growth rates [35].

We compared species richness and evenness (1/Simpson index) separately across all years (combining all planting treatments within a number of life forms (1, 3 or 5)), and for year four species richness and evenness by plotting means and 95% confidence intervals (CI). We used principal components analysis (PCA) to reduce the number of variables and produce composite axes using LA, SLA, LDMC and C, S, and R scores, with variables ln transformed to improve linearity and homogeneity of variance (transformations: ln(LA, SLA, height); square root(LDMC); ln (x+10) used for C,S,R scores). To determine predictors of final biomass and change in abundance in monoculture and the five life-form treatment, we used all-subsets selection using the AICc criterion. Since many models had Δ_i values of below 7 for both response variables, this suggested that there was no best model, so model averaging was used to create a predictive model (using the subset of models with Δ_i values <7)[36,37]. We separately performed the same analyses using the first three principal components of the trait dataset, but the raw trait values always yielded lower AICc values, so we only used the raw traits in the model fitting. Statistical analyses were completed using the R-package, v. 3.0.2 [38], with model averaging completed using the MuMIn package.

Table 2. Rainfall recorded at green roof site.

Date Range	Total Rainfall* (mm)
June 12-August 31, 2008	298.2
June 12-August 31, 2009	349.9
June 12-August 31, 2010	236.1
April 1-August 31, 2009	639.7
April 1-August 31, 2010	357.6

*Tipping bucket (TE525M, Campbell Scientific, Edmonton, AB) mounted 4 m above roof surface (installed June 2008).

Results and Discussion

Changes in Species Richness, Evenness and Abundance

Canopy species richness and evenness peaked in 2008 (Figure 1). Richness of species detected in vegetation canopies never reached the number of species planted originally in any of the mixture treatments (Figure 1), with the five life-form group (originally planted with 15 species) ending up with an average richness of seven species in the canopy. Single life-form treatments, which started with three species, had equivalent richness (ranging from two to three species) and evenness in year four (Figure 2). The highest richness levels in the three life-form treatments were detected in GCT (grasses, creeping forbs, tall forbs) and DTS (dwarf shrubs, tall forms, succulents) treatments with between five and six species on average. Evenness varied considerably between the three life-form treatments, with CTS (creeping forbs, tall forbs, succulents) having the highest and DGC (dwarf shrubs, creeping forbs, grasses) the lowest. While this method of determining canopy density only samples plants large enough to be detected in the pin frame, the biomass harvest revealed that all species survived in the five life-form treatment (see Table S1) and in most of the other mixture treatments (unpublished data), thus species richness and evenness calculated from canopy density alone are not incorporating all of the species in the community. While canopy biomass and diversity are important predictors of ecosystem functions in this system [7], it is still possible that these smaller individuals with little presence in the canopy and low aboveground biomass could make contributions to overall functioning of the green roof system. At low substrate depths comparable to those in our study, loss of species over time is a common result in green roof studies [11,12], and competition is frequently cited as the main reason for such declines in richness.

The dwarf shrub treatment ended up dominated by *E. nigrum* by year four, with very little biomass in the canopy contributed from the other two dwarf shrub species (Table S2). All three grass species made up substantial portions of the canopy in the grass single life-form treatment, with *P. compressa* at about double the

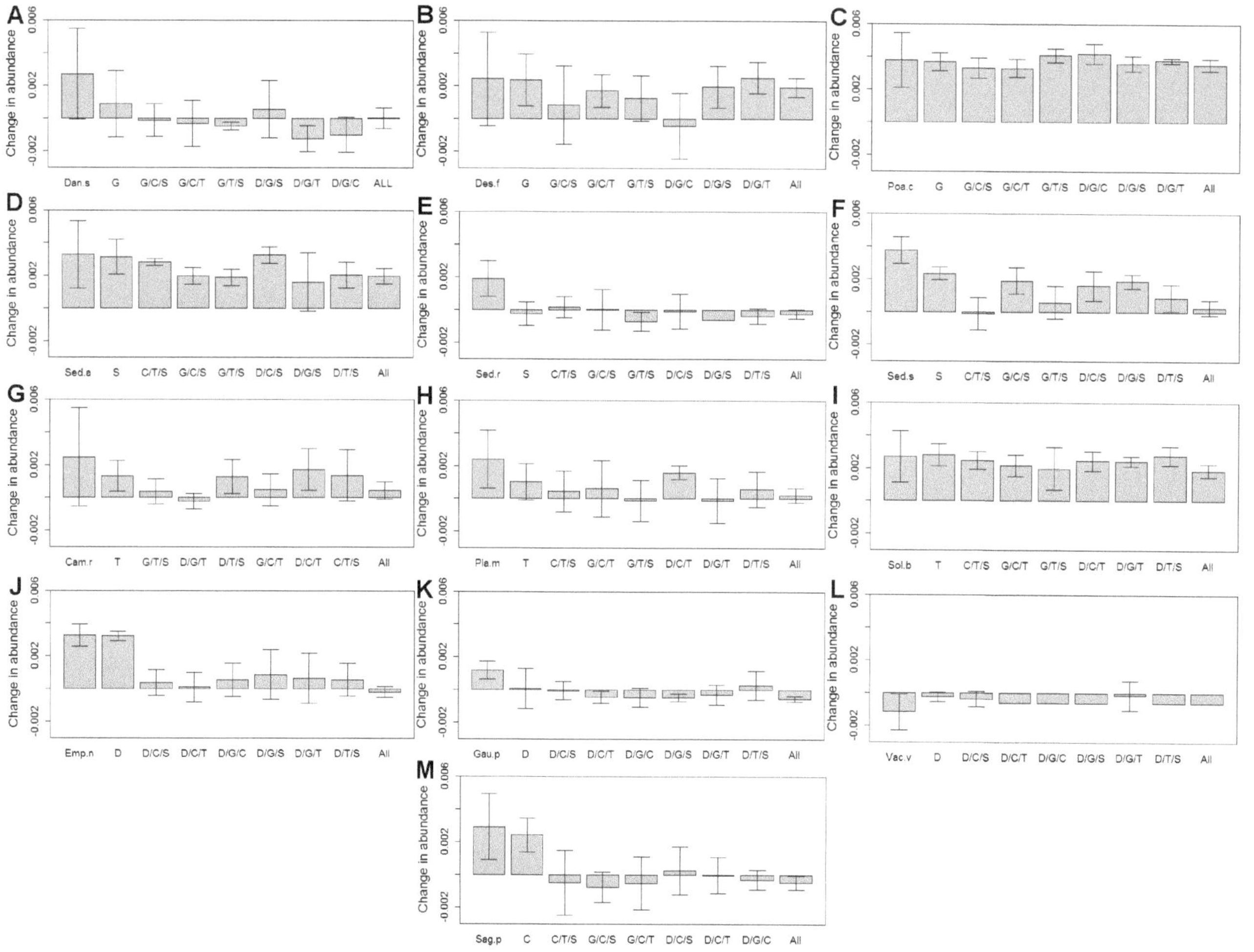

Figure 3. Change in abundance of 13 species in monoculture and mixtures over four years (mean±95% CI). Change in abundance calculated using canopy density: (ln(canopy density in year 4) - ln(canopy density in year 1))/# days. A: *D. spicata*; B: *S. acre*; C: *C. rotundifolia*; D: *E. nigrum*; E: *D. flexuosa*; F: *S. rosea*; G: *P. maritima*; H: *G. procumbens*; I: *S. procumbens*; J: *P. compressa*; K: *S. spurium*; L: *S. bicolor*; M: *V. vitis-idaea*. In each panel, the farthest left bar is the monoculture; mixture life form codes: C = creeping forb; D = dwarf shrub; G = grass; S = succulent; T = tall forb. Sample sizes: 1 life-form treatments: n = 5; 3 life-form treatments: n = 5; 5 life-form treatment: n = 20.

canopy density as *D. flexuosa*, and 10 x as much as *D. spicata*. In the succulent treatment, final canopy composition was heavily weighted toward *S. acre*, although this species had much reduced canopy density in 2009 (Table S2). *S. rosea* had very low density in the canopy through the entire experiment. The tall forb treatment was dominated by *S. bicolor* throughout, with approximately five and three times the canopy density of *P. maritima* and *C. rotundifolia*, respectively. *S. bicolor* canopy density in the tall forb treatment peaked in 2009. It is possible that weather conditions favoured *S. bicolor* in 2009, while being disadvantageous for *S. acre*. This is borne out in other mixture treatments. *S. acre* was denser in 2008 and 2010 compared with 2009 for the DGS, DTS, GCS, GTS, and five life-form mixture treatments; *S. bicolor* peaked in 2009 in all of the three and five life-form treatments in which it was planted. Weather station data from the site (Table 2) shows that 2009 was substantially wetter than 2008 or 2010. Powdery mildew (Erysiphaceae) was commonly observed on all three succulent species, and it is possible that fungal pathogens associated with wetter conditions decreased growth of *S. acre* in 2009. This finding indicates that fluctuations in species abundance may have been caused by environmental variables, in addition to the successional changes resulting from competition among species. In 2009, when *S. acre* had low abundance in the canopy, the other species in the succulent treatment did not make up for the decline in *S. acre* (see Table S2), but in the treatments with three or five life-forms containing both *S. acre* and *S. bicolor* (CTS, DTS, GTS, CDGST, see Table S2), there was evidence of compensatory dynamics, such that the increase in *S. bicolor* in 2009, the wet year, made up for the decrease in *S. acre*. This suggests another value of planting green roof ecosystems with multiple life-forms: canopy biomass may be more resilient in the face of climatic variation or other disturbances when species or life-forms with contrasting responses to environmental conditions are included.

The three life-form mixtures tended to be dominated by *P. compressa*, *D. flexuosa*, *S. acre*, and *S. bicolor* in treatments where these species were planted originally. The treatments including tall forbs with no grasses showed relatively high canopy densities of all three tall forbs (e.g. CTS, Table S2), but when grasses were present, *P. maritima* and *C. rotundifolia* had much lower abundances compared with *S. bicolor* (e.g. GTS, Table S2). *S. procumbens* had low abundances throughout the mixed life-form group treatments. The five life-form treatment was dominated by *P. compressa* by year 4,

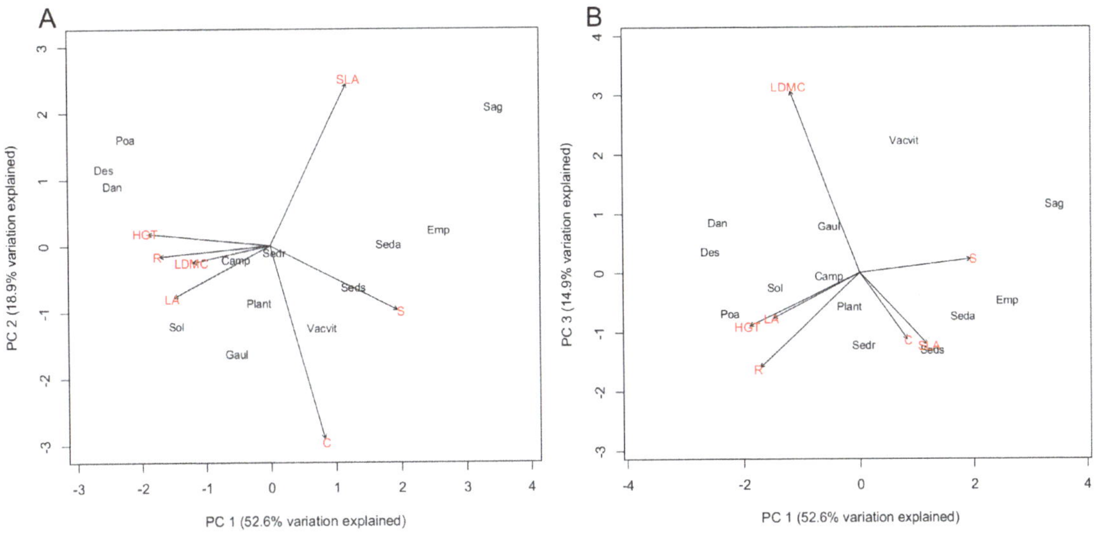

Figure 4. Principal components analysis of species, by leaf traits and life history variables. Species abbreviations: Poa = P. compressa; Des = D. flexuosa; Dan = D. spicata; Traits in red: SLA = specific leaf area; LA = leaf area; LDMC = leaf dry matter content; HGT = plant height; R = ruderal score; S = stress-tolerance score; C = competitive score.

with over five times the canopy density of the next most abundant species (*D. flexuosa*, *S. acre* and *S. bicolor*). All the other species had low abundances in this treatment, with most showing low value over all four years; *D. spicata* and *P. maritima* had relatively high canopy densities in 2008 but then declined.

The changes in canopy density in monoculture varied over 20-fold, between *P. compressa* and *S. spurium* canopies growing the fastest, and *V. vitis-idaea*, which had a net negative average change in abundance (Figure 3). Most species had equivalent increases in abundance in monoculture and the corresponding mixture of three species from the same life-form group (Figure 3), suggesting that the net effect of growing with neighbours of the same species is equivalent to that of growing with other species of the same life form group. *S. rosea* was an exception, suggesting a net negative effect of growing with the two other succulents. Most species had a relatively lower change in abundance in the three- and five-group treatments than in monoculture, suggesting a net negative effect of growing with neighbours from other life form groups. In contrast, while *V. vitis-idaea* had negative changes in abundance in all treatments, changes in abundance were less negative in some mixed species treatments compared to the monoculture (Figure 3), suggesting net positive effects of growing with other groups. In this species, individual plants were very small and it seems likely that intraspecific competition was low in the monocultures, and the positive effects of growing with heterospecific neighbours resulted from amelioration of harsh physical conditions (exposure, bare soil) or rather, that lack of plant biomass in the monoculture resulted in conditions that were too exposed, compared with the shelter provided by larger plants in the mixed group treatments. This was the only species that showed a positive response to interspecific neighbours. Given the harsh conditions on extensive green roofs, some authors have suggested that stress-tolerant plants might facilitate other species [39], but only one species out of our 13 showed a positive response to interspecific neighbors.

Five species showed no evident changes in abundance between monoculture and mixture treatments: *P. compressa*, *S. bicolor*, *S. acre*, *C. rotundifolia*, *P. maritima*. Of these species, *P. compressa*, *S. bicolor* and *S. acre* had some of the highest positive changes in abundance in monoculture and no apparent net negative effect of growing with other species and groups. *C. rotundifolia* and *P. maritima* showed markedly lower changes in abundance in some mixture treatments, but there was high variability within a treatment; shortages of plant material restricted monoculture treatments to three replicates each. Other species only showed lower changes in abundance in particular group combinations; *D. flexuosa* had low rates of abundance change in the dwarf shrub, graminoid, and creeping shrub mixture but elsewhere exhibited similar changes in abundance to the monoculture. Two tall forbs, *C. rotundifolia* and *P. maritima*, had the lowest abundance growing in a mixture with the dwarf shrubs and graminoids (Figure 3).

Plant Traits

Principal components analysis indicated three meaningful components (eigenvalues >1) (Table S3). The three grass species are very close in trait space, scoring high on the negative end of PC 1, indicating tall plants with relatively large leaf area and ruderal tendencies (Figure 4; Table S1). While these species diverge strongly in their functioning in this system [7], similar heights, leaf traits and other features characterize this group. In contrast, the dwarf shrub group (*E. nigrum*, *V. vitis-idaea* and *G. procumbens*) show greater spread along the first two PC axes, with *E. nigrum* having the highest scores on Axis 1, associated with high stress-tolerance scores, low ruderal scores and low heights. *V. vitis-idaea* and *G. procumbens* have lower axis two scores, indicating lower SLA and greater competitiveness, although these species typically inhabit infertile environments. The other life-form groups show some spread in this trait space. The creeping forb *S. procumbens*, *E. nigrum* and two of the succulents *S. spurium* and *S. acre* have relatively high values of SLA and stress-tolerance. The three grasses have low values of stress-tolerance and high values of SLA. The third principal component is positively correlated with LMDC and negatively correlated with R score. The three grasses show more differentiation along PC 3 with *D. spicata* having the highest LMDC. The dwarf shrub group is also spread out along this axis, with *V. vitis-idaea* loading positively (high LDMC, low C and SLA), followed by *G. procumbens* and *E. nigrum*. Of the species with large positive changes in abundance in both monoculture and mixture *P. compressa* and *D. flexuosa* showed similar aggregate trait

Table 3. Multiple linear regression showing standardized coefficients from model averaging for final biomass and change in abundance in monoculture and five life-form treatments.

Dependent Variable	Predictors	Model Averaged β Coefficient	95% Confidence Interval Lower bound	95% Confidence Interval Upper bound
Monoculture biomass	**SLA**	0.73	0.30	1.16
	LA	−0.11	−0.57	0.35
	LDMC	0.09	−0.36	0.54
	Height	0.41	−0.22	1.03
	R score	0.63	0.16	1.04
	S score	−0.43	−0.94	0.07
	C score	−0.19	−0.68	0.30
Five life-form treatment biomass	**SLA**	0.39	−0.03	0.81
	LA	0.02	−0.49	0.53
	LDMC	−0.25	−0.76	0.25
	Height	0.24	−0.54	1.03
	R score	0.68	0.18	1.18
	S score	−0.61	−1.15	−0.07
	C score	0.04	−0.49	0.58
Monoculture change in abundance	**SLA**	0.76	0.29	1.23
	LA	0.10	−0.43	0.64
	LDMC	−0.60	−1.10	−0.09
	Height	0.53	−0.07	1.14
	R score	0.53	0.11	0.94
	S score	−0.61	−1.14	−0.07
	C score	0.21	−0.31	0.73
Five life-form treatment change in abundance	**SLA**	0.48	0.02	0.94
	LA	0.37	−0.19	0.94
	LDMC	−0.14	−0.75	0.48
	Height	0.30	−0.57	1.18
	R score	0.63	0.09	1.16
	S score	−0.62	−1.18	−0.06
	C score	−0.07	−0.69	0.55

values (Figure 3); the grasses, *S. acre* and *S. bicolor* were highly divergent in terms of their trait values, suggesting that high positive rates of change in abundance in this system can be attained by different trait combinations. The native species in this category are typically found in the periphery of coastal barrens heathland habitat where disturbances open up the matrix of dominant ericaceous plants, whereas *P. compressa* and *S. acre* are commonly found in disturbed, hard-surfaced environments in urban areas. The dwarf shrubs, dominant on organic soils on coastal barrens [28], showed lower changes in abundance when combined with the other species suggesting lower competitive ability in the green roof system. The growing medium used in this experiment was designed for extensive green roofs planted with *Sedum* spp., and had a relatively high pH, low organic matter content and high soluble nutrient concentrations, in contrast to the native soils where the dwarf shrubs dominate [7,28]. With a more suitable substrate, we might expect these native shrubs to grow faster, and perhaps to out-compete ruderal species that are adapted to more fertile conditions. Nevertheless, high changes in abundance in *S. bicolor* and *D. flexuosa* suggest that native species can have high positive changes in canopy abundance in a green roof system, if suited to substrate properties.

SLA was a strong positive predictor of final biomass in monoculture (Table 3); R and S scores were positive and negative predictors, respectively, in both monoculture and the five life-form treatments. In general, the traits predicted biomass in both monoculture and five life-form treatments in the same directions but with variation in the size of the coefficients (Table 3). Coefficients for LA, LDMC and C scores in the biomass models were weakly negative or close to zero, those for height were positive.

The models for change in abundance had relatively high positive coefficients for SLA and R score in both monoculture and the five life-form mixture (Table 3). LDMC and S score were strong negative predictors of change in abundance in monoculture, and height was a relatively strong positive predictor. C score and LA were weaker positive predictors. Change in canopy abundance in the five life-form treatment had positive coefficients for LA and height, weak negative coefficients for LDMC and C score, and a strong negative S score.

While we expected more stress-tolerant species to perform the best in this shallow-substrate environment, we suspect that the high nutrient levels and pH of the green roof substrate gave the more ruderal species an advantage, leading to shifts in species composition over time. While this was not quantified, many of the species that we weeded out of the modules over the experiment were non-native, ruderal species common in urban areas, further supporting the idea that this particular green roof system in this climate promotes the growth of ruderals over stress-tolerant species or "faster" over "slower" species (*sensu* [40]), at least in the first four years post-establishment. It should also be pointed out that we only examined aggregate changes in abundance over the entire four years. Early establishment growth during year 1 may reflect a different environment, dominated by abiotic factors such as high soil temperatures and moisture limitations, whereas growth in later years may reflect plant responses to more intense competition with intra- or interspecific neighbours. These possible effects cannot be sorted out with these data as initial post-planting sampling resulted in many zeroes in the canopy density measurements, so initial vs. end of year 1 comparisons would result in unreliable change in abundance calculations, and were not included here. Longer-term studies of green roofs could be expected to show greater dominance of stress-tolerant species over time, as nutrients become more limiting, but empirical studies in Europe show that a suite of ruderals dominates after 20–100 years [29,41], including some dominant in our study (*S. acre* and *P. compressa*). While the regressions suggest that species with ruderal traits have the highest increases in abundance here, *S. acre* was associated with higher S scores, but also had high change in abundance. Overall, SLA predicted changes in canopy abundance and biomass in monocultures and mixtures and could be used to select species for extensive green roofs in this climate, in fertile substrates. Reich [40] posits an overall differentiation of plant species along an access of "slow" to "fast", referring to the rates at which they can acquire resources, with slow plants dominating in low resource areas, which have been characterized by others as stressful [25], and faster plants dominant in more resource-rich environments. The traits that predicted successful growth and biomass accumulation were similar in the monocultures and the mixture combining all the species, and were more strongly related to "fast" traits than those possessed by species typically growing in low-resource environments. Increases in abundance in the mixtures occurred primarily between years 1 and 2 for most species.

All mixture treatments ended up with fewer species in the canopy than originally planted, and the change in abundance results indicate that the majority of species performed poorly in multi-species mixtures. While environmental fluctuations can account for some of the year-to-year variation, we suggest that most of these species experienced net negative effects of interspecific competition. The species that performed well in both monocultures and mixtures were predicted by leaf and aggregate life history traits, suggesting that this method can be used to select green roof plant species. However, if high diversity in green roof systems is to be maintained, selection of species that can coexist, alterations to substrate conditions to favor more stress-tolerant natives, or incorporation of spatial heterogeneity into the system might all help promote longer-term species coexistence.

Acknowledgments

Melissa Ranalli contributed to the original experimental design and data collection; J. Scott MacIvor, Zac MacDougall, Cory Blanchard and André Thibault also assisted with data collection.

Author Contributions

Conceived and designed the experiments: JL. Performed the experiments: JL ST TS. Analyzed the data: JL AH ST TS. Contributed reagents/materials/analysis tools: JL. Wrote the paper: JL AH ST TS.

References

1. VanWoert ND, Rowe DB, Andresen JA, Rugh CL, Fernandez RT, et al. (1995) Green roof stormwater retention: effects of roof surface, slope, and media depth. J Environ Qual 34: 1036–1044.
2. Eumorfopoulou E, Aravantinos D (1998) The contribution of a planted roof to the thermal protection of buildings in Greece. Energ Buildings 27: 29–36.
3. Kadas G (2006) Rare invertebrates colonizing green roofs in London. Urban Habitats 4: 66–86.
4. McGuire KL, Payne SG, Palmer MI, Gillikin CM, Keefe D, et al. (2013) Digging the New York City skyline: soil fungal communities in green roofs and city parks. PLoS ONE 8(3): e58020. doi:10.1371/journal.pone.0058020.
5. Dunnett NP, Kingsbury N. (2004) Planting Green Roofs and Living Walls. Portland (OR): Timber Press. 254 p.
6. Boivin M, Lamy M, Gosselin A, Dansereau B (2001) Effect of artificial substrate depth on freezing injury of six herbaceous perennials grown in a green roof system. HortTechnology 11: 409–412.
7. Lundholm J, MacIvor JS, MacDougall Z, Ranalli M (2010) Plant Species and Functional Group Combinations Affect Green Roof Ecosystem Functions. PLoS ONE 5(3): e9677. doi:10.1371/journal.pone.0009677.
8. Monterusso MA, Rowe DB, Ruby CL (2005) Establishment and persistence of Sedum spp. and native taxa for green roof applications. Hortscience 40: 391–396.
9. Butler C, Butler E, Orians CM (2012) Native plant enthusiasm reaches new heights: perceptions, evidence, and the future of green roofs. Urban Forest Urban Green 11: 1–10.
10. Speak AF, Rothwell JJ, Lindley SJ, Smith CL (2013) Reduction of the urban cooling effects of an intensive green roof due to vegetation damage. Urban Climate 3: 40–55.
11. Dunnett N, Nagase A, Hallam A (2008) The dynamics of planted and colonising species on a green roof over six growing seasons 2001–2006: influence of substrate depth. Urban Ecosyst 11: 373–384.
12. Rowe DB, Getter KL, Durhman AK (2012) Effect of green roof media depth on Crassulacean plant succession over seven years. Landscape Urban Plan 104: 310–319.
13. MacIvor JS, Lundholm JT (2011) Performance evaluation of native plants suited to extensive green roof conditions in a maritime climate. Ecol Eng 37: 407–417.
14. Ranalli MA, Lundholm JT (2008) Biodiversity and ecosystem function in constructed ecosystems. CAB Reviews: PAVSNNR 3: #034; doi:10.1079/PAVSNNR20083034.
15. Cook-Patton SC, Bauerle TL (2012) Potential benefits of plant diversity on vegetated roofs: a literature review. J Environ Manage 106: 85–92.
16. Farrell C, Szota C, Williams NGS, Arndt SK (2013) High water users can be drought tolerant: using physiological traits for green roof plant selection. Plant Soil 372: 177–193.
17. Díaz S, Cabido M (1997) Plant functional types and ecosystem function in relation to global change. J Veg Sci 8: 463–474.
18. Grime JP, Thompson K, Hunt R, Hodgson JG, Cornelissen JHC, et al. (1997) Integrated screening validates primary axes of specialization in plants. Oikos 79: 259–281.

19. Reich PB, Walters MB, Ellsworth DS (1997) From tropics to tundra: Global convergence in plant functioning. P Natl Acad Sci USA 94: 13730–13734.
20. Craine JM, Tilman D, Wedin D, Tjoelker M, et al. (2002) Functional traits, productivity and effects on nitrogen cycling of 33 grassland species. Funct Ecol 16: 563–574.
21. Quested H, Eriksson O, Fortunel C, Garnier E (2007) Plant traits related to whole-community litter quality and decomposition following land use change. Funct Ecol 21: 1016–1026.
22. McIntyre S (2008) The role of plant leaf attributes in linking land use to ecosystem function in temperate grassy vegetation. Agr Ecosyst Environ 128: 251–258.
23. Helsen K, Hermy M, Honnay O (2012) Trait but not species convergence during plant community assembly in restored semi-natural grasslands. Oikos 121: 2121–2130.
24. de Bello F, Vandewalle M, Raitalu T, Lepš J, Prentice HC, et al. (2013) Evidence for scale- and disturbance-dependent assembly patterns in dry semi-natural grasslands. J Ecol 101 1237–1244.
25. Grime JP (1974) Vegetation classification by reference to strategies. Nature 250: 26–31.
26. Hodgson JG, Wilson PJ, Hunt R, Grime JP, Thompson K. (1999) Allocating C-S-R plant functional types: a soft approach to a hard problem. Oikos 85: 282–294.
27. Ranalli M, Harris A, Lundholm J (2008) Evaluating green roof performance in Halifax, Nova Scotia, Canada. 6th North American Green Roof Conference: Greening Rooftops for Sustainable Communities Conference. Baltimore, MD. The Cardinal Group, Toronto.
28. Oberndorfer EC, Lundholm JT (2009) Species richness, abundance, rarity and environmental gradients in coastal barren vegetation. Biodivers Conserv 18: 1523–1553.
29. Köhler M (2006) Long-term vegetation research on two extensive green roofs in Berlin. Urban Habitats 4: 3–26.
30. Floyd D, Anderson J (1987) A comparison of three methods for estimating plant cover. J Ecol 75: 221–228.
31. Harper J (1977) Population biology of plants. London: Academic Press. 900 p.
32. Gleason HA, Cronquist A (1991) Manual of vascular plants of northeastern United States and adjacent Canada. New York: New York Botanical Garden. 900 p.
33. Roland AE, Zinck M, Owen E (1998) Roland's Flora of Nova Scotia. Halifax, Nova Scotia: Nimbus Publishing & Nova Scotia Museum. 1300 p.
34. Westoby M (1998) A leaf-height-seed (LHS) plant ecology strategy scheme. Plant Soil 199: 213–227.
35. Gross N, Suding KN, Lavorel S (2007) Leaf dry matter content and lateral spread predict response to land use change for six subalpine grassland species. J Veg Sci 18: 289–300.
36. Burnham KP, Anderson DR (2002) Model selection and multimodel inference: a practical information-theoretic approach. New York: Springer. 488 p.
37. Burnham KP, Anderson DR, Huyvaert KP (2011) AIC model selection and multimodel inference in behavioral ecology: some background, observations, and comparisons. Behav Ecol Sociobiol 65: 23–35.
38. R Core Team (2013) R: A Language and Environment for Statistical Computing. Vienna, Austria http://www.R-project.org/.
39. Butler C, Orians CM (2011) Sedum cools soil and can improve neighboring plant performance during water deficit on a green roof. Ecol Eng 37: 1796–1803.
40. Reich PB (2014) The world-wide 'fast–slow'plant economics spectrum: a traits manifesto. J Ecol, 102: 275–301.
41. Köhler M, Poll PH (2010) Long-term performance of selected old Berlin greenroofs in comparison to younger extensive greenroofs in Berlin. Ecol Eng 36: 722–729.

Community Interactions Modify the Effects of Pharmaceutical Exposure: A Microcosm Study on Responses to Propranolol in Baltic Sea Coastal Organisms

Hanna Oskarsson[1]*, Ann-Kristin Eriksson Wiklund[2], Gunnar Thorsén[3], Gabriela Danielsson[4], Linda Kumblad[1]

1 Department of Ecology, Environment and Plant Sciences, Stockholm University, Stockholm, Sweden, **2** Department of Applied Environmental Science, Stockholm University, Stockholm, Sweden, **3** Department of Analytical Chemistry, Stockholm University, Stockholm, Sweden, **4** Department of Biochemistry and Biophysics, Stockholm University, Stockholm, Sweden

Abstract

This study investigated the uptake and effects of a common human pharmaceutical, propranolol, on the structure and function of a coastal Baltic Sea model community consisting of macroalga (*Ceramium tenuicorne*), mussels (*Mytilus edulis trossulus*), amphipods (*Gammarus* spp.), water and sediment. The most sensitive species, the mussel, was affected to the same extent as in previous single species studies, while the effects on the amphipod and alga were smaller or even positive compared to experiments performed in less complex test systems. The observed cascade of beneficial effects was a result of inter-specific species interactions that buffered for more severe effects. The poor condition of the mussel led to a feeding shift from alga to mussel by the amphipods. The better food quality, due to the dietary shift, counteracted the effects of the exposure. Less amphipod grazing, together with increased levels of nutrients in the water was favourable for the alga, despite the negative effects of propranolol. This microcosm study showed effects on organisms on different organizational levels as well as interactions among the different components resulting in indirect exposure effects of both functional and structural nature. The combination of both direct and indirect effects would not have been detected using simpler single- or even two-species study designs. The observed structural changes would in the natural environment have a long-term influence on ecosystem function, especially in a low-biodiversity ecosystem like the Baltic Sea.

Editor: Simon Thrush, University of Auckland, New Zealand

Funding: The field sampling was financed by an Askö grant from Stockholm University Baltic Sea Center (http://www.su.se/ostersjocentrum/english/) to H. Oskarsson. Employees of the funder at the Askö Laboratory assisted with the collection of organisms. The funders had no role in the study design, data collection and analyses, decision to publish, or preparation of the manuscript.

Competing Interests: The authors have declared that no competing interests exist.

* E-mail: hanna.oskarsson@su.se

Introduction

Our knowledge on the potential effects of pharmaceuticals on non-target organisms has increased over the last years, yet there is a lack of studies on the effects of pharmaceuticals on ecosystem structure and function, and on ecosystem processes [1]. In complex systems with variations within species as well as inter-specific species interactions, exposure to environmental stressors can result in various indirect and cascading effects, with positive or negative consequences on individual species and communities [2,3]. Differences between species can lead to unexpected effects in response to exposure [4], and compensatory processes in ecological systems can result in both linear and non-linear linkages between effects on individuals and subsequent effects on populations [5]. Ecological buffering capacity and other aspects of ecosystem dynamics and stability can furthermore shield the effects of stressors on an ecosystem [6], resulting in less distinct direct effects than can be detected for species or individuals. This can for instance occur if several species benefit from the same resources, that is, fulfil the same functional roles with the ability to replace one another's function in the system, or as a result of positive indirect effects within the system [6]. Single species experiments could therefore both over- and underestimate the hazards of exposure to toxic substances, and ecotoxicological measures on population level should therefore better predict the effects on ecosystems than measures on individual level [7]. Theories of ecosystem responses to disturbances, unexpected future ecosystem events and the importance of observing change on larger scales derive from the concept of ecosystem resilience, introduced by Holling [8] and reviewed by for example Rönnbäck et al. [9] for the Baltic Sea context. With this in mind, multi-species exposure studies of function and structure are important complements to single-species studies on mechanistic responses, to better understand the possible effects of pharmaceuticals and other environmental contaminants on organisms and ecosystems.

The Baltic Sea is subjected to considerable natural and anthropogenic stress [10], and is considered to be one of the most polluted seas in the world [11]. There is a strong salinity gradient, thermo- and halo-clines and irregular inflows of oxygen rich water from the Atlantic [12]. This contributes to a low biodiversity in the Baltic Sea ecosystem, where each species is of high importance, as the functional redundancy is low [9,13]. Coastal ecosystems function as buffers for nutrient flows from the terrestrial to the open sea ecosystems. In the Baltic Sea, the blue mussels, *Mytilus edulis trossulus*, together with perennial algae, have essential stabilising functions in this process [14,15].

Pharmaceuticals constitute an important group of environmental contaminants due to their biologically active nature, the conservation of drug targets in non-target organisms and their continuous release into the environment via waste water treatment plants. Pharmaceuticals are often relatively water soluble, designed to be effectively absorbed, to induce biological effects [4] and to be biologically persistent [16]. Consequently, this diverse group of substances can pose risks of both bioaccumulation and effects in non-target organisms [17]. Propranolol is a common β-blocker primarily used for treating of hypertension [4]. Propranolol is often detected in aquatic environments, and in the approximate range of 0.01–0.1 μg l^{-1} in the Baltic Sea catchment area [18–21]. Being a non-specific β-blocker, propranolol binds to both β1- and β2-receptors in humans [22], which in turn inhibit the action of catecholamines [23]. β-receptors have been shown to be present in tissues of fish [24], and possibly also in bivalves [25–27] as well as in a range of other organisms [24]. It is hence possible that propranolol exert a similar mode of action as in humans, also in non-target organisms. Previous studies on the effects of propranolol on aquatic organisms have shown effects within a large range of exposure concentrations, dependent on both exposure durations, test species and endpoints. Tests with for example algae, crustaceans and fish have revealed EC50's ranging from 0.5–24300 μg l^{-1} [28–30], and NOEC's for small crustaceans from 1–125 μg l^{-1} [24]. Previous single species experiments have shown effects of propranolol also in aquatic organisms from coastal habitats in the Baltic Sea, in concentrations ranging from 10–10000 μg l^{-1} [31–33].

To study the direct, indirect and possible buffering effects in a community exposed to pharmaceuticals, we performed a laboratory multiple-species microcosm study. We studied the effects of the common human pharmaceutical propranolol on a model community including organisms with different feeding modes and from different biological organisational levels. The setup allowed studies of the propranolol distribution within the community, and served as a comparison to our previous experiments, which showed effects of propranolol, diclofenac and ibuprofen in less complex experimental systems with one and two species [31–33]. By mimicking a coastal Baltic Sea community, we hypothesised that we would find smaller negative effects of exposure than in less complex systems. Based on the detected interactions and indirect effects within the model community, the study provides novel insights into how pharmaceuticals can affect food web structure and ecosystem functions in the field.

Material and Methods

2.1 Overview of the Experimental Set up

In this study, model communities (containing macroalga, omnivorous amphipods, filter feeding mussels, water and sediment) were created in aquaria connected to a flow through system with fresh seawater. Three treatments (control and exposure to two propranolol concentrations) were studied in five replicates (n = 5) during six weeks in a climate chamber laboratory. At the start, during and after the exposure experiment, measurements were conducted to identify potential effects of the propranolol exposure.

2.2 Study Organisms

The model communities studied in this experiment were formed to resemble a habitat of a coastal ecosystem, with natural abundances of the respective organisms at the sampling site [34–36]. Each community was represented by: i) blue mussels (*Mytilus edulis trossulus*, 31 individuals of even size, 2.3 cm±se 0.006, cleaned from epibionts), ii) a red filamentous macroalgae (*Ceramium tenuicorne*, initial mean weight 8.3 g ww±se 0.01), iii) small amphipods (*Gammarus* spp., 30 evenly sized individuals, appr. 1–1.5 cm), as well as iv) sediment (in petri-dishes Ø 10*1.5 cm) and v) water.

Sampling of organisms and sediment was performed by diving, by hand and by a sediment sledge (1–3 m depth), the 20th–21st of June 2011 in the vicinity of Stockholm University's Baltic Sea Center, Askö Laboratory (58°49′3″ N, 17°37′4″ E) in the Baltic Sea, Sweden. The salinity and water temperature at sampling was 6.5 psu and 13.4°C, respectively. The sediment was sieved through a 3 mm net to remove larger organisms and larger inorganic material.

No permits were required for the described study, which complied with all relevant regulations. No protected species were sampled during the collection of organisms and sediment.

2.3 Experimental Conditions

The sampled organisms and sediment were distributed among 15 glass aquaria (LWH 24*19*18 cm, volume 8 l) and acclimated for seven days in a climate chamber at Stockholm University, with a water temperature of 12.5±0.5°C and the light regime 16:8 h (light:dark, 11.11±se 0.72 UM (μmol s^{-1} m^{-1})) before exposure. The same conditions persisted throughout the experimental period. Fresh seawater (salinity 6.52±se 0.0075 psu) was continuously added into each separate aquaria via a flow through system, releasing the water through PVC tubings (inner Ø 8 mm) at approximately 5 cm from the bottom. The daily incoming water corresponded to the aquaria volume, so that the total water volume of each aquarium was exchanged daily. Surplus water was continuously discharged through a hole in the aquaria glass wall at the water surface. The hole was covered with a fine net to avoid escape by amphipods or losses of alga.

Salinity and temperature of the water were measured regularly. A daily addition of 10 ml solution of the microalgae *Isochrysis galbana* (Reed Mariculture) to each aquarium ensured sufficient food supply for the mussels. The microalgae concentration in the aquaria after addition was 7.8×10^{7} cells l^{-1}, or 2–3 mg dw l^{-1}, according to relationships from Lora-Vilchis et al. [37] and Zhue and Lee [38] and measured dry weight of the microalgae. The administered amount (in cells l^{-1}, cells l^{-1} $mussel^{-1}$ or mg dw l^{-1}) was similar to several previous studies [31,39–41].

2.4 Exposure

The model communities were exposed to 0 (control), 100 (P100) and 1000 (P1000) μg l^{-1} of propranolol (nominal concentrations), n = 5. These concentrations have been previously used in single and two-species exposure studies with propranolol [32,33,42]. The propranolol, purchased from Sigma Aldrich, was dissolved in dilute phosphoric acid and the pH was adjusted to 7.1 by the addition of a buffer (bisodium carbonate). This buffer was also added to the control treatment to control for solvent effects, although previous studies showed no difference between solvent control and seawater controls [32]. Propranolol and buffer were continuously added to the aquaria by siphons of PTFE Teflon tubing (length: 30 cm, inner Ø 0.5 mm) connected to fused silica capillaries (length: 15 cm, inner Ø 0.25 mm). The siphons led propranolol or buffer from glass reservoirs placed on shelves above the aquaria, into the water of each experimental aquaria. The aquaria were distributed randomly in the climate chamber to minimize possible effects from placement in the room.

2.5 Measurements of the Model Communities and its Separate Components

2.5.1 Nutrients in water. Water samples from all aquaria were collected at the end of the exposure period and kept cool (4°C) and dark until analysis of ammonium (NH_4^+), nitrite+nitrate ($NO_2^-+NO_3^-$), total amount of nitrogen (Tot-N) and phosphate (PO_4^{3-}). The nutrients were determined by Segmented Flow Analysis (SFA), slightly modified ALPKEM O. I. Analytical Flow Solution IV Methods (#3199527, NO_2+NO_3 – N; #319526, NH_4 – N and #319528, PO_4 – P).

2.5.2 Propranolol in water. Quantification of propranolol in water samples from two selected aquaria of each concentration (0, 100, 1000 μg l^{-1}) was performed twice during the exposure period (day 2 and day 15), and from all aquaria at the end of the exposure, to assess that the concentrations in the aquaria remained constant. Water samples were collected in glass vials and kept cool and dark until analysed according to the method described by Eriksson Wiklund et al. [32].

2.5.3 Propranolol in biota and sediment. Quantification of propranolol in mussels, amphipods and sediment samples from selected aquaria was performed at the end of the exposure period. Analyses were made for 1–3 mussels from three different aquaria per exposed treatment (100 μg l^{-1} and 1000 μg l^{-1}) and for one control aquaria (0 μg l^{-1}); for 1–2 amphipods per treatment (analysed together) and for one sub-sample of sediment from 2–3 aquaria per treatment. The mean±se for mussels in P1000 was based only on data from two of the analysed aquaria. The third replicate (one individual showed a considerably higher propranolol concentration than the other analysed individuals from P1000. It was identified as an outlier (Dixon test, $p = 0.02$) and excluded from the overall mean. Biota and sediment samples were dried (45°C 3 days) and kept cool and dark until analysed. Macroalgae samples could not be analysed due to laboratory circumstances.

The propranolol content in blue mussels was analysed according to the earlier developed methods [30,31] with some minor changes. The instrument used was a Shimadzu gradient liquid chromatography system (LC-10Avp) coupled to a Thermo Scientific ion trap mass spectrometer (LCQ-deca classic).

The following sample preparation steps were used for the mussel tissue and sediment: rewetting and homogenisation of the tissue, addition of surrogate standard (d_7-propranolol, CDN isotopes, Pointe Claire, CAN, product no: D-2386), extraction for more than 3 hours using formic acid (2%) in methanol with 30 minutes ultrasonication, centrifugation to remove solids, buffer exchange to ammonium -acetate buffer (0.1%) containing 20% acetonitrile (ACN), liquid-liquid extraction with n-heptane to remove fat components, solid phase extraction of basic compounds using Oasis MCX columns (Waters, Millford, MA, USA), evaporation of eluate and reconstitution in ammonium-acetate buffer (0.1%) containing 20% acetonitrile. The mussels were analysed individually and approximately 1 g (dry weight) of each sediment sample was weighed on an analytical scale.

Amphipod samples were frozen in liquid nitrogen and ground to a fine powder. Surrogate standard was added after transfer of the powder to Eppendorf tubes for extraction. The extraction was performed as for the mussel samples. Centrifugation was used to remove particulate material and 30 kD molecular weight cut-off filters to remove proteins (Ultracel YM-30, Millipore, Bedford, MA, USA). The solvent was evaporated and the residue dissolved in ammonium-acetate buffer (0.1%) containing 20% acetonitrile.

The injection volumes used for the LC-MS analysis depended on the expected levels in the sample, 1 μl for high concentration samples and 5 μl for low concentration samples. Two fragments were monitored from the ions of the non-labeled and deuterium labeled propranolol in order to assess that the measured signals originated from the desired compound: the fragments at $m/z = 157$ and 183 were monitored for the parent ion 260 and the fragments at $m/z = 163$ and 189 for the parent ion 167. All chemicals used were of analytical grade and all water used was double deionised using a Milli-Q water purification apparatus (Millipore, Bedford, MA, USA), having a resistance greater than 18 MΩ.

Bioconcentration factors (BCFs) for the biota and concentration factors (CFs) for the sediment samples were determined as the ratio between detected amounts of propranolol in biota or sediment, and detected concentrations of propranolol in the water of the respective aquaria. The relative distribution of propranolol in the microcosms was estimated from the measured concentrations in the respective microcosm components (assuming a similar concentration in the alga and the amphipod) and the initial (bio)mass of each component.

2.5.4 Community Gross Production (GP) to respiration (R) Ratio - GP:R. The gross production to respiration ratio (GP:R) of the model communities was measured three times; after the acclimatisation period but before the start of the exposure, as well as after 4.5 weeks of exposure and at the end (6 weeks of exposure). Oxygen measurements were conducted with a micro-sensor connected to a pA-meter (Unisense). After measuring the initial oxygen concentration (t_0), the water surfaces of the aquaria were covered with plastic film to limit oxygen diffusion over the water surface, and the systems were subsequently left in light conditions for a known period of time before the oxygen concentration was measured again (t_1), to attain the net primary production (NP). The water surfaces were subsequently re-covered with plastic film and the systems were left in dark conditions, before the oxygen concentration was measured a third time (t_2), to attain the respiration (R) of the systems. Potential changes in oxygen level not due to organism production and/or respiration were controlled for using blanks containing only seawater. The gross production (GP) was estimated as the sum of NP and R. All production and respiration measurements were related to the water volume of the aquaria.

2.5.5 Measurements of macroalgae. Gross production to respiration ratio (GP:R) of the macroalgae was measured twice, after 5 weeks of exposure and at the end. The measurements were performed in the same manner as for the model communities. During the measurements the macroalgae were removed from the aquaria and kept in 1 l plastic jars with seawater and the respective concentration of propranolol or buffer. After the measurements after 5 weeks of exposure, the algae were replaced in their respective aquaria again. The measurements were normalised to water volume of the aquaria and alga biomass (dw). Dry weight of the macroalgae after 5 weeks was attained from wet weights, normalised by the relation between dry weight and wet weight at the end of the experiment.

2.5.6 Measurements of mussels. The mussel respiration rate was measured only at the end of the exposure period to avoid extra stress from detachment of byssus threads from their substrate. The measurement was performed in the same way as the respiration measurements described for the GP:R measurements for the whole system. During the measurements all mussels were removed from the aquaria and kept in 1 l plastic jars with seawater and the respective propranolol concentration or buffer. Directly after the respiration measurement the mussel feeding rate was also measured. The feeding rate measurements were performed in the same jars, and started with the addition of microalgae solution (1 ml) to each jar, resulting in a concentration of 3.9×10^7 cells l^{-1} and 0.95 mg dw l^{-1}. Initial water samples (50 ml, t_0) were taken with a 60 ml syringe, and transferred into 50 ml Falcon tubes. The water was gently stirred while the water samples were taken to obtain an as homogenous sub-sample as possible. The mussels were then left to filter feed for 21 ± 1 min, before the water was sub-sampled again (t_1), in the same way. The water samples were stored cold and dark until analysed for number of cells 3–5 μm in a particle counter (Beckman Coulter Z1 DT). The respiration and feeding rate measurements were normalised to water volume and biomass (dw).

2.5.7 Measurements of amphipods. The amphipod respiration was also measured at the end of the exposure period and was estimated from the measurement of two individuals from each aquarium. The amphipods were transferred to two 25 ml plastic jars with seawater and the respective concentration of propranolol or buffer. The measurements were conducted in the same way as for the respiration measurements described for the GP:R measurements, and were normalised to water volume and biomass (dw).

2.5.8 Measurements of sediment. Measurements of gross production, respiration, and GP:R, of the sediment were performed at the end of the experiment, in the same way as for the model communities. During the measurements the sediment samples (in their respective Petri dishes) were removed from the aquaria and kept in 0.5 l plastic jars with seawater and the respective concentration of propranolol or buffer. The measurements were normalised to water volume and sediment mass (dw). Dry weight of the sediment was attained from initial wet weights, normalised by the relation between dry weight and wet weight of a subsample at the end of the experiment.

2.5.9 Mortality. The communities were controlled daily to identify possible dead mussels and/or amphipods. Dead organisms were removed and stored in $-20°C$ for later measurements of length and weight.

2.5.10 Dry weight, ash free dry weight and carbon content. The dry weight and ash free dry weight of mussels, amphipods, macroalgae and sediment was attained after the completed exposure. Mussel soft tissue was removed from the shell. All samples were placed in pre-ashed (450°C, 5 h) aluminium cups, and then left to dry (45°C, 3 days) before weighing. Thereafter the samples were ashed (500°C, 2 h) in the same aluminium cups in a muffle furnace before weighed again. The carbon content (%) was determined as the difference in weight between dry samples and ashed samples, divided by dry weight.

2.5.11 Visual observations. Observations of possible changes among the communities were made throughout the exposure, and in detail at the end. Colour, structure and abundance of the macroalga, as well as bacteria/epiphytes on the aquaria walls were recorded.

2.6 Statistical Analyses

To determine differences in measured variables between the three treatments, continuous data was analysed in the statistical software package PRIMER 6 version 6.1.13, with the PERMANOVA+ add-on, version 1.0.3 [43,44], specified for Euclidean distances and 9999 permutations. Prior to PERMANOVA analyses, homoscedasticity was tested by PERMDISP, in the PERMANOVA platform, and transformed values were used to reach homogenous variances if these were insufficient. Differences between treatments were determined by pair-wise PERMANOVA tests with Monte Carlo sampling as the number of unique permutations were low. Categorical data was analysed by generalized linear models (glm), specified for Poisson distribution, and significant effects were analysed with subsequent Tukey HSD post hoc tests, in R version 3.0.1. Correlation analyses were performed using Pearson correlation with logged data to reach homogenous variances (based on Levenes test for homoscedasticity) and pairwise removal of missing data, in SPSS version 20. All data is reported as mean$\pm$se if not otherwise stated.

Results

3.1 Effects on the Model Communities

3.1.1 Gross Production to respiration Ratio (GP:R) in the model communities. The ecological function in each model community, assessed as community gross production to respiration ratio (GP:R), was similar in all treatments before the start of the exposure (PERMANOVA: $F_{2,12} = 0.089$, $p = 0.91$), but different at the end of the exposure (PERMANOVA: $F_{2,12} = 4.7$, $p = 0.024$, Table 1). After the exposure, GP:R was lower in communities exposed to 1000 μg l^{-1} (P1000) compared to the control (67% lower than control, pair-wise PERMANOVA: $p = 0.047$) and compared to communities exposed to 100 μg l^{-1} (P100), although not significantly (64% lower than P100, pair-wise PERMANOVA: $p = 0.075$). At the measurement after 4.5 weeks, GP:R of the community was considerably affected, displaying negative values, but due to large variations, no statistical difference could be determined (PERMANOVA: $F_{2,12} = 0.718$, $p = 0.51$, Table 1).

3.1.2 Nutrient levels in the model communities. No differences in levels of nitrite+nitrate (NO_2^-+NO_3^-, PERMANOVA: $F_{2,12} = 0.68$, $p = 0.52$) were found between the treatments. Although the main test indicated differences in ammonium (NH_4^+) levels among the treatments (PERMANOVA: $F_{2,12} = 4.3$, $p = 0.034$) only close to significantly higher levels pf ammonium were found in P1000, compared to the control (pair-wise PERMANOVA: $p = 0.054$) and to P100 (pair-wise PERMANOVA: $p = 0.061$, Table 1). A similar pattern was found for the total amount of nitrogen (Tot-N, PERMANOVA: $F_{2,12} = 13.8$, $p = 0.003$), where the concentration in P1000 was higher than in both the control (69% higher, pair-wise PERMANOVA: $p = 0.0026$) and P100 (56% higher, pair-wise PERMANOVA: $p = 0.0067$, Table 1).

The concentration of nitrite + nitrate was lower in all treatments compared to seawater blanks (PERMANOVA: $F_{3,14} = 8.2$, $p < 0.003$; pair-wise PERMANOVA: $p < 0.05$ for all treatments) while the levels of ammonium (PERMANOVA: $F_{3,14} = 17$, $p < 0.001$; pair-wise PERMANOVA:, $p < 0.01$ for all treatments) and total nitrogen (PERMANOVA: $F_{3,14} = 19$, $p < 0.001$; pair-wise PERMANOVA: $p < 0.01$ for all treatments) were higher in all treatments compared to seawater blanks.

The concentrations of phosphate (PO_4^{3-}) were several orders of magnitude higher in all experimental aquaria compared to seawater blanks, due to the continuous addition of the phosphoric acid buffer that propranolol was dissolved in (also added to the controls). No difference in phosphate was found between the treatments (PERMANOVA: $F_{2,12} = 1.54$, $p = 0.25$, Table 1).

3.2 Effects on the Separate Community Components

3.2.1 Responses of the macroalga: Gross Production (GP), respiration (R) and GP:R-ratio. The macroalga GP:R-ratio of P1000 was the lowest of the three treatments, at both

Table 1. Results from nutrient analyses and physiological responses of the different components of the model communities.

Component	Variable (time)	Unit	Control	P100	P1000
Community	GP:R (start)	mg l^{-1} h^{-1} g dw^{-1}/mg l^{-1} h^{-1} g dw^{-1}	1.57±0.045	1.53±0.071	1.52±0.119
	GP:R (4.5w)	mg l^{-1} h^{-1}g dw^{-1}/mg l^{-1} h^{-1} gdw^{-1}	0.661±0.169	0.701±0.359	−0.398±0.780
	GP:R (6w)	mg l^{-1} h^{-1} g dw^{-1}/mg l^{-1} h^{-1} g dw^{-1}	1.42±0.062	1.31±0.115	0.469±0.394 *
	NO_2^- +NO_3^- (6w)	µg l^{-1}	17.4±4.14	21.18±4.48	15.09±2.01
	NH_4^+ (6w)	µg l^{-1}	140±15.8	141±25.1	317±75.7
	Tot-N (6w)	µg l^{-1}	452±20	488±34	763±69 **
	PO_4^{3-} (6w)	µg l^{-1}	15048±1051	13132±643	14615±668
Macroalga	GP:R (5w)	mg l^{-1} h^{-1}g dw^{-1}/mg l^{-1} h^{-1} g dw^{-1}	1.44±0.217	1.43±0.335	0.93±0.463
	GP:R (6w)	mg l^{-1} h^{-1}g dw^{-1}/mg l^{-1} h^{-1} g dw^{-1}	1.52±0.271	2.12±.724	1.19±.463
	Weight loss (6w)	g ww	3.82±0.74	3.99±0.69	2.30±0.30
	Carbon content (6w)	g C g dw^{-1}	0.76±0.005	0.79±0.009	0.83±0.007***
Mussel	Respiration (6w)	mg l^{-1} h^{-1}g dw^{-1}/mg l^{-1} h^{-1} g dw^{-1}	−0.515±0.022	−0.549±0.037	−0.690±0.069
	Feeding rate (6w)	cells h^{-1} g dw^{-1}	13179±3920	19177±4358	18148±7089
	Weight (6w)	g dw	0.04±0.001	0.039±0.001	0.038±0.001
	Carbon content (6w)	g C g dw^{-1}	0.880±0.004	0.876±0.004	0.885±0.005
	Mortality (6w)	amount (%) dead	1.3±0.8	3.2±1.0	54±11***
Amphipod	Respiration (6w)	mg l^{-1} h^{-1}g dw^{-1}/mg l^{-1} h^{-1} g dw^{-1}	−2.68±0.31	−3.72±0.41	−2.51±0.28
	Juveniles (6w)	No.	9±4.9	13.4±6.1	25.2±9.7
	Juveniles/Adult (6w)	No.	1.5±0.9	0.81±0.3	1.90±0.9
	Weight (6w)	g dw	0.0074±0.0006	0.0075±0.0009	0.0085±0.004
	Carbon content (6w)	g C g dw^{-1}	0.680±0.005	0.706±0.05	0.752±0.051
	Mortality (6w)	amount (%) dead	77±6.1	64±12	51±9.6***
Sediment	GP:R (6w)	mg l^{-1} h^{-1}g dw^{-1}/mg l^{-1} h^{-1} g dw^{-1}	1.17±0.15	0.94±0.14	0.95±0.0003
	Carbon content (6w)	g C g dw^{-1}	0.0249±0.0008	0.0254±0.0006	0.0248±0.0006

Measurements made at start (start), after five weeks (4.5w and 5w), and/or after six weeks (6w) exposure to propranolol in 0 (Control); 100 µg l^{-1} (P100) and 1000 µg l^{-1} (P1000), mean ± se. dw = dry weight, ww = wet weight, C = carbon. *Denotes significant differences from the respective control treatments (*$p<0.05$, **$p<0.01$, ***$p<0.001$)."

measurement occasions (Table 1). No statistical differences in macroalga GP:R were however found between the treatments after 5 weeks of exposure (PERMANOVA: $F_{2,12} = 0.72$, $p = 0.51$, Table 1), nor after 6 weeks (PERMANOVA: $F_{2,12} = 0.83$, $p = 0.47$, Table 1), possibly due to the large within mean variation. The lower GP:R in P1000 was initially the result of mainly low GP (5 weeks), and in the end a combination of low GP and high R (6 weeks). After 6 weeks the GP:R displayed a non-linear pattern as the decrease GP:R in P1000 remained (22% lower than the control), while the GP:R in P100 was slightly increased (40% higher than the control), indicating what could be a hormesis effect.

3.2.2 Responses of the macroalga: weight loss and carbon content. The macroalga in all replicates lost weight during the exposure period (Table 1). The weight loss seemed smaller in P1000 compared to both the control and P100, although there was no significant difference between the treatments (PERMANOVA: $F_{2,12} = 2.4$, $p = 0.13$). There were differences in macroalgae carbon content among the treatments (PERMANOVA: $F_{2,7} = 22.4$, $p = 0.0015$), where P1000 had higher carbon content than both the control (pair-wise PERMANOVA: $p = 0.0004$) and P100 (pair-wise PERMANOVA: $p = 0.029$, Table 1). P100 had slightly higher carbon content than the control (pair-wise PERMANOVA: $p = 0.057$).

3.2.3 Responses of the mussel: respiration and feeding rate. The mussels' respiration increased slightly by increasing propranolol exposure, and there were only close to significant differences among the treatments (PERMANOVA: $F_{2,12} = 3.68$, $p = 0.060$, Table 1).

There was also a trend of higher filtration rate of mussels in exposed treatments compared to the control, but there were large variances in filtration rate within all treatments and no significant differences were found (PERMANOVA: $F_{2,11} = 0.43$, $p = 0.65$, Table 1).

3.2.4 Responses of the mussel: weight at end. The average mussel weights of both dead and surviving mussels at the end of the experiment were similar among all treatments (PERMANOVA: $F_{2,12} = 1.7$, $p = 0.23$), and neither were there any differences in mussel carbon content between the treatments (PERMANOVA: $F_{2,12} = 1.24$, $p = 0.31$, Table 1).

3.2.5 Responses of the amphipod: respiration. The amphipods' respiration was different between the treatments after the exposure (PERMANOVA: $F_{2,12} = 3.9$, $p = 0.05$). Amphipods exposed to P100 had 39% larger mean respiration than individuals from P1000 (pair-wise PERMANOVA: $p = 0.041$) and slightly, but not significantly, larger respiration than the control (pair-wise PERMANOVA: $p = 0.073$, Table 1).

3.2.6 Responses of the amphipod: weight at end. There were no statistical differences between the weights of surviving amphipods at the end of the exposure (PERMANOVA: $F_{2,12} = 0.76$, $p = 0.50$), and neither were there any differences in amphipod carbon content between the treatments (PERMANOVA: $F_{2,9} = 0.19$, $p = 0.89$, Table 1).

3.2.7 Responses of the amphipod: reproduction (number of juveniles). At the end of the experiment there were amphipod juveniles in 11 of the 15 experimental aquaria. Despite a large variation in number of juveniles within each treatment, there were also differences between the treatments (glm: $p<0.001$) and there was a pattern of increasing number of juveniles with increasing concentration of propranolol. The number of juveniles in P1000 was significantly higher than in both the control (180% higher than control, Tukey: $p<0.001$, Table 1), and in P100 (49% higher than P100, Tukey: $p<0.001$). The number of juveniles was correlated to the number of surviving adults, with a slightly higher number of juveniles per adult in P1000, however, there was no significant difference between the treatments (PERMANOVA: $F_{2,12} = 0.54$, $p = 0.61$, Table 1).

3.2.8 Responses of the sediment: GP:R and carbon content. No statistical differences could be found in sediment GP:R (PERMANOVA: $F_{2,12} = 0.012$, $p = 0.99$), among the treatments, although the results suggested a lower GP:R for the treatments with propranolol, compared to the control (Table 1). Neither were there any differences in sediment carbon content between the treatments (PERMANOVA: $F_{2,12} = 0.21$, $p = 0.81$, Table 1).

3.2.9 Mortality. The mussel mortality in general was relatively low but with large variations (overall mean 12.5%±9.5%). There were however differences between the treatments (glm: $p<0.001$) and the mortality increased with increasing concentration of propranolol (Table 1, Figure 1A). The mortality of P1000 was significantly higher than both the control (Tukey: $p<0.001$) and P100 (Tukey: $p<0.001$). There were differences also in amphipod mortality between the treatments (glm: $p = 0.024$), although amphipod mortality conversely was generally high (overall mean 68.4%±4%). An unexpected decrease in mortality with increasing concentration of propranolol was detected, where the mortality of P1000 was lower than of the control (Tukey: $p = 0.019$, Table 1, Figure 1B). At the end, a relatively large number of amphipods were missing (in total 27%), which was most probably due to cannibalism, as the amphipods were observed to feed on each other. The missing individuals were assumed to have been predated with an even distribution in time over the course of the experiment, and were included in such a manner in the Kaplan Meier amphipod survival curve (1B). The main mussel mortality of P1000 started after 30 days of exposure and continued to the end of the experiment, while the amphipod mortality was evenly distributed over the exposure period.

3.3 Propranolol in Water, Biota and Sediment

Propranolol was detected in all analysed components of P100 and P1000 at the end of the experiment (Table 2). There was an even concentration of propranolol in exposed aquaria over the length of the experiment and the measured concentrations were close to the nominal concentrations (8.2% deviation in P100 and 5.8% deviation in P1000). The highest concentrations and BCF was found in mussels from P1000, but also mussels from P100 and amphipods from both concentrations contained significant concentrations of propranolol. Most of the added propranolol was present in the water (80% in P100 and 77% in P1000), only 14% (P100) and 19% (P1000) in biota, and 6% and 4% respectively, in the sediment. For the sediment, amphipods and algae, the relative uptake declined with increased exposure, whereas for the mussels it was the other way around and the amount of propranolol was 60% higher in P1000, compared to P100.

3.4 Visual Observations

There were clear differences in colour of the macroalga and of bacteria/epiphytes on the aquaria walls, between the treatments. The macroalga turned from red to different shades of red/brown/green in all but one treatment from P100, evenly distributed among the treatments. The aquaria walls of the control and P100 replicates were covered in brown biofilm at the end of the exposure. There was however almost none of this biofilm in P1000, which instead had a much higher ratio of red and green colonies, presumably different bacterial colonies, on the aquaria walls, compared to the controls. Black colonies were found on the aquaria walls in one of the P1000 replicates.

3.5 Interactions among the Separate Components

Based on the detected effects on the separate components and especially the unexpected decrease in amphipod mortality with increasing exposure, as well as the smaller weight loss and increased carbon content in exposed macroalgae, a conceptual model of relationships among the species and possible indirect effects of the exposure was formed (Figure 2). Correlation analyses confirmed several of the conceived links.

The number of dead mussels correlated with the increased total nitrogen concentration in the water (Pearson: $r = 0.52$, $p = 0.05$) and the increased nitrogen levels further correlated well with increased macroalgae carbon content (Pearson: $r = 0.69$, $p = 0.028$). Lower amphipod mortality correlated well with increased number of amphipod offspring (Pearson: $r = 0.59$, $p = 0.02$). Amphipod mortality was found to be nearly significantly correlated to both macroalgae carbon content (Pearson: $r = 0.56$, $p = 0.091$) and mussel mortality (Pearson: $r = 0.48$, $p = 0.078$).

Discussion

Exposure to propranolol in this study resulted in several typical stress responses in mussels (increased mortality, increased respiration and increased excretion) [45], possible hormesis responses (macroalga GP:R, amphipod respiration, sediment carbon content and community nitrite+nitrate concentrations) [31,46], as well as some unexpected indirect positive effects (decreased amphipod mortality, decreased macroalga weight loss and increased macroalga carbon content). The hypothesis that buffering processes would reduce the effects of propranolol in exposed organisms compared to similar doses in single species experiments could, however, only partly be corroborated. The mussels were affected to a similar extent as in the single species experiments [31], while the additional organisms were less, or even positively, affected by the propranolol exposure: the macroalga with regard to biomass and carbon content; and the amphipods with regard to biomass and mortality. In previous single-species experiments, propranolol both reduced chlorophyll fluorescence and GP:R in a macroalga (*Fucus vesiculosus*) [33], and the activity and physiology of amphipods [32,33]. In this multi-species experiment, the toxic effects of propranolol were thus masked or reduced by ongoing interactions, and inter-specific relations between the different components buffered for effects of propranolol previously detected in less complex experimental systems with similar settings [31–33]. We therefore conclude that this is an example of positive indirect effects within communities that reduce or counter direct negative effects of contaminants and other stressors through inter-specific species interactions [47–49].

When scrutinising the conceptual model in detail (Figure 2), there were several interactions and processes responsible for the observed responses, and sometimes lack of expected negative effects. The starting point was the significant relationship between

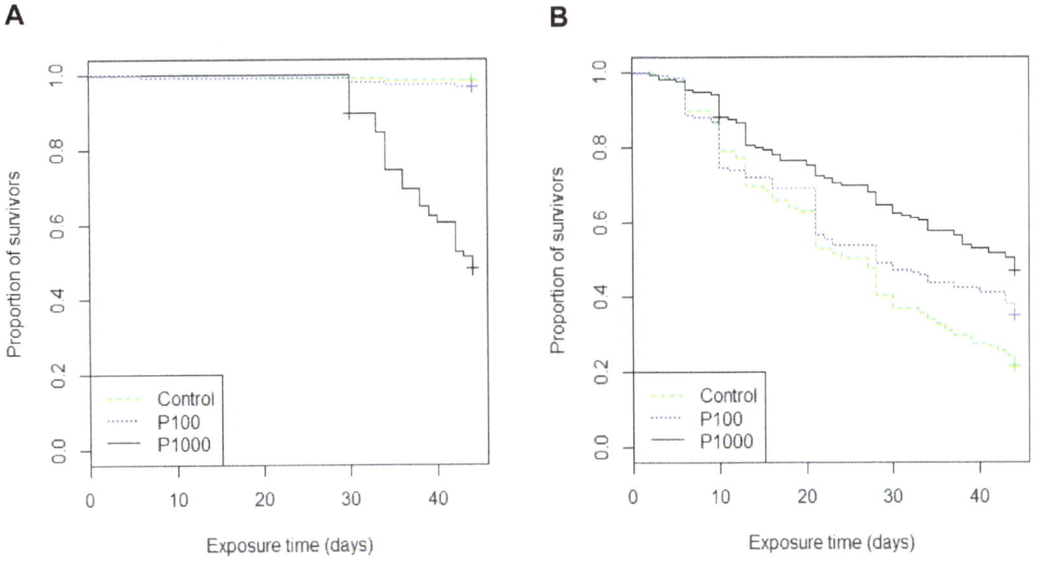

Figure 1. Mussel and amphipod mortality. Kaplan Meier survival curves depicting proportion of surviving individuals: A) mussels (*Mytilus edulis trossulus*), and B) amphipods (*Gammarus* spp.), after 6 weeks of exposure to propranolol in 0 µg l^{-1} (Control); 100 µg l^{-1} (P100) and 1000 µg l^{-1} (P1000).

mussel mortality and pharmaceutical concentration. The poor condition of mussels, regarding the mortality, induced a series of interactions in the microcosms. The mussel degradation increased the nitrogen concentrations in P1000, mainly ammonium, which is a strong indication of degradation of organic matter [50]. The excretion from stressed animals likely made a minor contribution [45]. It was repeatedly observed how amphipods fed on dying mussels, which did not seem to have their normal ability to close their shells in protection. The substantial feeding of amphipods on mussels hence increased the degradation and release of nutrients into the water, at the same time as the amphipods shifted from feeding on the macroalga to the mussel. Less grazing from amphipods, together with increased levels of nutrients in the water, favoured the macroalga, despite the direct negative effects on the alga from the propranolol exposure. Ammonium is furthermore more bioavailable to macroalga than nitrite+nitrate [51], which is also reflected by the good condition of the macroalga in the treatments where the ammonium levels were high, that is, P1000.

The amphipods in all treatments suffered from a generally high mortality, which could have been caused by several reasons, like natural death due to the end of life cycles, which for some of the sampled amphipods probably occurred during the exposure period [52,53], the laboratory conditions, or cannibalism [54]. For the amphipods in the P1000 systems, the dying mussels, however, provided higher availability and better quality of food. This in turn caused higher amphipod weight, lower mortality and slightly increased carbon content and number of offspring, which was corroborated with significant correlations. If considering the observed negative effects of propranolol on macroalga and amphipods in previous studies [32,33], it is reasonable to conclude that the positive effects of propranolol on these organisms were beneficial indirect effects induced by increased mussel mortality. Such interactions make an ecosystem susceptible for this type of unforeseen trophic cascade effect [2,55].

There were no confounding effects due to buffer addition or nutrient availability in the experimental systems. The high

Table 2. Distribution of propranolol in water, biota and sediment.

		Water	Mussels	Amphipods	Sediment	Algae
Control	Concentration	<LOQ	<LOQ	<LOQ	<LOQ	-
P100	Concentration	108±5.8	5.3±0.63	3.2±n.a.	0.61±0.047	-
	BCF or CF		46±6.7	70±n.a.	5.6±0.52	-
	Distribution	80%	11%	0.30%	6%	2%
P1000	Concentration	1058±37	89±11	6.3±n.a.	4.0±0.043	-
	BCF or CF		87±11	45±n.a.	3.9±0.37	-
	Distribution	77%	18%	0.10%	4%	0.50%

Concentration of propranolol in water (µg l^{-1}), biota (µg g ww^{-1}) and sediment (µg g ww^{-1}). Bioconcentration factor (BCF: (mg kg ww^{-1})/(mg l^{-1}))) determined for mussels and amphipods and concentration factor (CF: (mg kg ww^{-1})/(mg l^{-1})) for sediment. Propranolol distribution (%) among the components in the microcosms (assuming a similar concentration in the algae as in the amphipods). Quantifications made after exposure to propranolol in 0 (Control); 100 µg l^{-1} (P100) and 1000 µg l^{-1} (P1000), mean±se. One mussel replicate from P1000 (one individual) showed a considerably higher concentration of propranolol (310 µg g ww^{-1}) and BCF (295) than the other analysed individuals from the same treatment, and was excluded from the overall mean. LOQ = level of quantification, ww = wet weight.

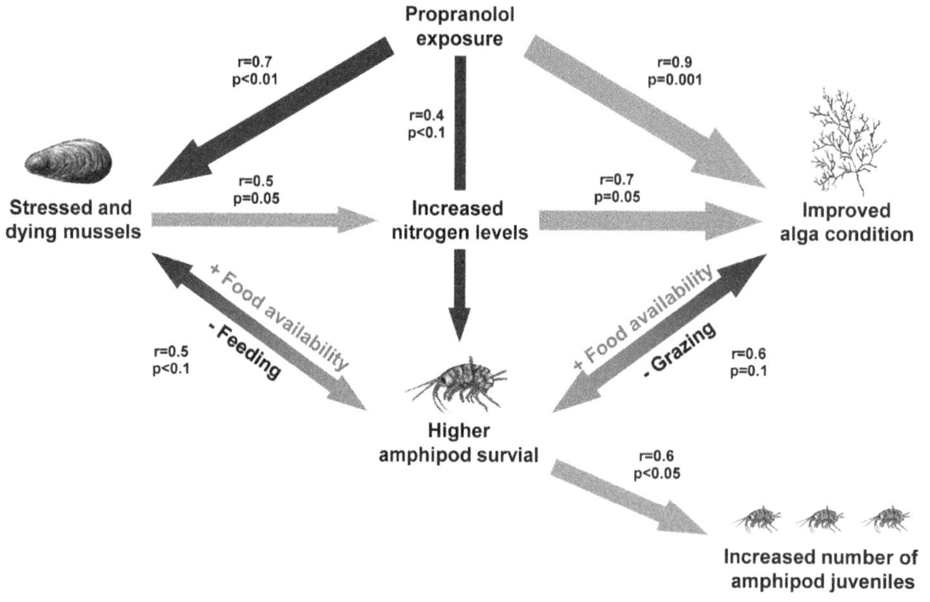

Figure 2. Conceptual model of interactions and indirect effects within the model communities. The interactions are determined by correlation coefficients. The width of the arrow denotes the strength of the correlation. Black arrows illustrate a negative influence on the organism whereas grey arrows illustrate a positive influence.

phosphate concentrations in all aquaria were due to the daily addition of the solvent buffer, which contained phosphoric acid. No aquaria suffered from N-limitation, despite the increased phosphorous levels, as the nitrogen levels were higher in all treatments than in the seawater blanks.

High biodiversity of a community or an ecosystem is considered to provide increased stability, functioning and recovery potential [56], which correlate both to its response to disturbances [1,3] and to its ecological resilience [57]. Communities with low diversity and resilience would hence generally be more sensitive to stressors. Contaminants will thus have a more severe effect in a low-biodiversity ecosystem, like the Baltic Sea, where there is a limited possibility for other organisms to regulate the same ecosystem function, and where the loss or change in function could affect the entire ecosystem [58,59]. The mussel used in this experiment has such a fundamental function in the Baltic Sea, as it is a significant benthic filter feeder. The extensive beds they form exert significant effects on benthic-pelagic coupling and energy flows through their filtration activities [60,61], which influence the structure and abundance of the associated community [62]. The high mussel mortality in this community exposure scenario is likely an effect from their substantial uptake of propranolol. The concentrations in mussel tissues were up to an order of magnitude higher than in the amphipod and sediment, and although they only contributed to 0.3% of the total mass in the experimental system, nearly 20% of the propranolol ended up in the mussels. The uptake also increased with increasing exposure, both in terms of tissue concentration, bioconcentration and relative amount in the system. The high uptake and sensitivity of the mussel is likely due to its filter feeding behaviour. *In situ*, the combination of contaminant exposure and a long life cycle, render a chronic exposure that makes the mussels especially vulnerable.

From a community ecology perspective, contaminant induced mortality has similarities with predation [63,64]. In this case propranolol acted as a predator on the mussels. It has been shown that removal of a competitively dominant species by contaminants induces positive responses in inferior organisms, reviewed by [64].

The blue mussels of the Baltic Sea are important ecosystem engineers and facilitators of high biodiversity by creating both substrate and other livelihood prerequisites for associated species [65,66], and by improving the water quality [61]. In spite of this, their mortality in the microcosm experiment induced positive indirect effects on the alga and amphipod, although such effects are likely smaller in a real Baltic Sea exposure scenario. Previous studies [32,33] have demonstrated negative effects of pharmaceutical exposure also on macroalga. Eutrophication of the Baltic Sea results in increased algae growth, and a subsequent low grazing pressure; factors that this study have shown to counter the toxic effects of propranolol to macroalga, at least temporarily. If the observed positive indirect effects on the amphipod would remain in a real exposure scenario, is more difficult to predict. In a eutrophied environment there will always be high food availability for grazers; one variable that contributed to the positive indirect effects on the amphipods in this microcosm study. However, as amphipods are omnivorous organisms, it might as well be the quality of the food, that is, the higher energy content in the dying mussels compared to the algae, that was the decisive factor providing the prerequisites needed to sustain the stress caused by the propranolol, which was detected in earlier single-species studies [32,33]. Thus, if the coastal area of the Baltic Sea would be exposed to pharmaceuticals to the extent that for example mussels would be severely affected, the exposure would likely not initially exert any negative effects on the amphipods. However, in a longer perspective of the low-biodiversity Baltic Sea, functional changes in the blue mussel population would affect the whole coastal zone. In addition, the detected amounts of propranolol in the sediment imply that also benthic organisms are at risk for exposure if pharmaceuticals reach the recipient in larger quantities. Although the potential effects of pharmaceuticals on sediment organisms are yet unknown, the results of this and previous single species studies [31–33] show that pharmaceutical exposure may have implications on the coastal ecosystem of the Baltic Sea.

Conclusions

The study of effects in microcosms demonstrates a feasible yet relevant way of observing possible environmental effects. This multi-species exposure experiment revealed uptake and effects of the human pharmaceutical propranolol on non-target organisms. The most sensitive species, the mussel, was affected to the same extent as in previous single species studies [31–33], while the effects on amphipods and algae were smaller, or even positive, compared to in less complex test systems. This was likely due to compensatory effects within the model communities, as a result of both negative and positive interactions. Buffering processes resulting in smaller net effects have been observed also in experiments with other variables [47] and contaminant exposure can even induce more indirect than direct effects [63]. The combination of effects of both direct and indirect character would not have been detected if using simpler experimental set-up, with fewer community components. If a more complex experimental set-up is used, buffering effects between organisms having the same or a similar biological role may on the other hand reduce the possibilities of detecting interaction effects. It is, however, the combination of effects with interactions among species that will be found in the environment. This exposure study hence contributes to the understanding of the direct effects of pharmaceuticals in the environment as well as their possible subsequent effects on both community function and structure. The similar response patterns between the treatments indicate that also lower concentrations may affect a community in the field, especially in a more chronic exposure scenario. The structural changes found, in terms of survival and biomass of the mussels, macroalga and amphipods, and especially the increased amphipod propagation, would in the natural environment have a long-term influence on the ecosystem function.

Acknowledgments

We thank Judith Reise, Charlie Persson, Britta Eklund, Lisen Johansson and John Oskarsson for laboratory assistance, Ellen Schagerström, Susann Ericsson and Carl-Magnus Wiltén for assisting with collection of organisms, the chemistry lab at the Department of Ecology, Environment and Plant Sciences at Stockholm University for nutrient analyses, and Nils Kautsky and Clare Bradshaw for comments on the manuscript.

Author Contributions

Conceived and designed the experiments: HO AKEW GT LK. Performed the experiments: HO AKEW GT GD LK. Analyzed the data: HO AKEW LK. Contributed reagents/materials/analysis tools: AKEW GT LK. Wrote the paper: HO. Collection of study organisms: HO AKEW GD. Method development: HO, GT GD. Writing a section of the article (2.4.3): GT. Critically revising article: AKEW GT GD LK.

References

1. Hughes AR (2012) Disturbance and Diversity: An Ecological Chicken and Egg Problem. Nature Education Knowledge 3: 48.
2. Fleeger J, Carman K, Nisbet R (2003) Indirect effects of contaminants in aquatic ecosystems. Science of the Total Environment 317: 207–233.
3. Hooper D, Chapin F III, Ewel J, Hector A, Inchausti P, et al. (2005) Effects of biodiversity on ecosystem functioning: a consensus of current knowledge. Ecological Monographs 75: 3–35.
4. Fent K, Weston A, Caminada D (2006) Ecotoxicology of human pharmaceuticals. Aquatic Toxicology 76: 122–159.
5. Forbes V, Calow P (2012) Promises and problems for the new paradigm for risk assessment and an alternative approach involving predictive systems models. Environmental Toxicology and Chemistry 31: 2663–2671.
6. Sheehan PJ (1984) Effects on community and ecosystem structure and dynamics. In: PJ S, Miller D, GC B, Ph B, editors. Effects of pollutants at the ecosystem level SCOPE 22. New York, USA: John Wiley & Sons Ltd. 51–100.
7. Forbes V, Calow P (1999) Is the per capita rate of increase a good measure of population-level effects in ecotoxicology? Environmental Toxicology and Chemistry 18: 1544–1556.
8. Holling C (1973) Resilience and stability of ecological systems. Annual Review of Ecology and Systematics 4: 1–23.
9. Rönnbäck P, Kautsky N, Pihl L, Troell M, Söderqvist T, et al. (2007) Ecosystem Goods and Services from Swedish Coastal Habitats: Identification, Valuation, and Implications of Ecosystem Shifts. Ambio 36: 534–544.
10. Elmgren R (2001) Understanding human impact on the Baltic ecosystem: Changing views in recent decades. Ambio 30: 222–231.
11. Wulff F, Bonsdorff E, Gren IM, Johansson S, Stigebrandt A (2001) Giving advice on cost effective measures for a cleaner Baltic Sea: A challenge for science. Ambio 30: 254–259.
12. Jansson B-O (1980) Natural Systems of the Baltic Sea. Ambio 9: 128–136.
13. Elmgren R, Hill C (1997) Ecosystem function at low biodiversity - The Baltic example. In: Ormond RFG, Gage J and Angel M, editors. Marine Biodiversity: Patterns and processes, Cambridge University Press. 319–336.
14. Kautsky U (1995) Ecosystem processes in coastal areas of the Baltic Sea. Doctoral thesis, Stockholm University, Stockholm, Sweden.
15. Norling P (2009) Importance of blue mussels for biodiversity and ecosystem functioning in subtidal habitats. Doctoral thesis, Stockholm University, Stockholm, Sweden.
16. Jjemba P (2006) Excretion and ecotoxicity of pharmaceutical and personal care products in the environment. Ecotoxicology and Environmental Safety 63: 113–130.
17. Bengtsson B-E, Gunnarsson B, Wall T, Wennmalm Å, editors (2005) Läkemedel och miljö. Apoteket AB. 148 p.
18. Bendz D, Paxéus NA, Ginn TR, Loge FJ (2005) Occurrence and fate of pharmaceutically active compounds in the environment, a case study: Hoje River in Sweden. Journal of Hazardous Materials 122: 195–204.
19. Wahlberg C, Björlenius B, Paxéus N (2010) Läkemedelsrester i Stockholms vattenmiljö. Förekomst, förebyggande åtgärder och rening av avloppsvatten. Stockholm Vatten AB. 141 p.
20. Andreozzi R, Raffaele M, Nicklas P (2003) Pharmaceuticals in STP effluents and their solar photodegradation in aquatic environment. Chemospere 50: 1319–1330.
21. Falås P, Andersen HR, Ledin A, la Cour Jansen J (2012) Occurence and reduction of pharmaceuticals in the water phase at Swedish wastewater treatment plats. Water Science and Technology 66: 783–791.
22. Alexander BS, Wood MD (1987) Stereoselective Blockade of Central [H-3] 5-Hydroxytryptamine Binding to Multiple Sites (5-Ht1a, 5-Ht1b and 5-Ht1c) by Mianserin and Propranolol. Journal of Pharmacy and Pharmacology 39: 664–666.
23. Escher BI, Bramaz N, Richter M, Lienert J (2006) Comparative ecotoxicological hazard assessment of beta-blockers and their human metabolites using a mode-of-action-based test battery and a QSAR approach. Environmental Science & Technology 40: 7402–7408.
24. Huggett D, Brooks B, Peterson B, Foran C, Schlenk D (2002) Toxicity of select beta adrenergic receptor-blocking pharmaceuticals (B-blockers) on aquatic organisms. Archives of Environmental Contamination and Toxicology 43: 229–235.
25. Adamo SA (2008) Norepinephrine and octopamine: linking stress and immune function across phyla. Invertebrate Survival Journal 5: 12–19.
26. Lacoste A, Malham SK, Cueff A, Poulet SA (2001) Noradrenaline modulates oyster hemocyte phagocytosis via a beta-adrenergic receptor-cAMP signaling pathway. General and Comparative Endocrinology 122: 252–259.
27. Massarsky A, Trudeau VL, Moon TW (2011) beta-Blockers as Endocrine Disruptors: The Potential Effects of Human beta-Blockers on Aquatic Organisms. Journal of Experimental Zoology Part a-Ecological Genetics and Physiology 315A: 251–265.
28. Cleuvers M (2003) Aquatic ecotoxicity of pharmaceuticals including the assessment of combination effects. Toxicology Letters 142: 185–194.
29. Ferrari B, Mons R, Vollat B, Fraysse B, Paxeus N, et al. (2004) Environmental risk assessment of six human pharmaceuticals: Are the current environmental risk assessment procedures sufficient for the protection of the aquatic environment? Environmental Toxicology and Chemistry 23: 1344–1354.
30. Stanley JK, Ramirez AJ, Mottaleb M, Chambliss CK, Brooks BW (2006) Enantiospecific toxicity of the beta-blocker propranolol to Daphnia magna and Pimephales promelas. Environmental Toxicology and Chemistry 25: 1780–1786.
31. Ericson H, Thorsén G, Kumblad L (2010) Physiological effects of diclofenac, ibuprofen and propranolol on Baltic Sea blue mussels. Aquatic Toxicology 99: 223–231.
32. Eriksson Wiklund A-K, Oskarsson H, Thorsén G, Kumblad L (2011) Behavioural and physiological responses to pharmaceutical exposure in macroalgae and grazers from a Baltic Sea littoral community. Aquatic Biology 14: 29–39.
33. Oskarsson H, Eriksson Wiklund A-K, Lindh K, Kumblad L (2012) Effect studies of human pharmaceuticals on *Fucus vesiculosus* and *Gammarus* spp. Marine Environmental Research 74: 1–8.

34. Ankar S, Elmgren R (1978) The benthic macro- and meiofauna of the Askö-Landsort area (northern Baltic proper). A stratified random sampling survey. Contributions from the Askö Laboratory, Stockholm University, Sweden 11: 115 p.
35. Kautsky H (1989) Quantitative distribution of plant and animal communities of the phytobenthic zone in the Baltic Sea. Contributions from the Askö Laboratory, Stockholm University, Sweden 35: 80 p.
36. Jansson B-O, Wulff F (1977) Ecosystem analysis of a shallow sound in the northern Baltic – a joint study by the Askö group. Contributions from the Askö Laboratory, Stockholm University, Sweden 18: 160 p.
37. Lora-Vilchis MC, Cordero-Esquivel B, Voltolina D (2004) Growth of *Artemia franciscana* fed *Isochrysis* sp. and *Chaetoceros muelleri* during its early life stages. Aquaculture Research 35: 1086–1091.
38. Zhu CJ, Lee YK (1997) Determination of biomass dry weight of marine microalgae. Journal of Applied Phycology 9: 189–194.
39. Clausen I, Riisgard HU (1996) Growth, filtration and respiration in the mussel *Mytilus edulis*: No evidence for physiological regulation of the filter-pump to nutritional needs. Marine Ecology Progress Series 141: 37–45.
40. Maire O, Amouroux J-M, Duchêne J-C, Grémare A (2007) Relationship between filtration activity and food availability in the Mediterranean mussel *Mytilus galloprovincialis*. Marine Biology 152: 1293–1307.
41. Prevodnik A, Gardeström J, Lilja K, Elfwing T, McDonagh B, et al. (2007) Oxidative stress in response to xenobiotics in the blue mussel *Mytilus edulis* L.: Evidence for variation along a natural salinity gradient of the Baltic Sea. Aquatic Toxicology 82: 63–71.
42. Ericson H, Thorsén G, Kumblad L (2010) Physiological effects of diclofenac, ibuprofen and propranolol on Baltic Sea blue mussels. Aquatic Toxicology 99: 223–231.
43. Anderson MJ (2001) A new method for non-parametric multivariate analysis of variance. Austral Ecology 26: 32–46.
44. McArdle BH, Anderson MJ (2001) Fitting multivariate models to community data: A comment on distance-based redundancy analysis. Ecology 82: 290–297.
45. Widdows J (1985) Physiological measurements. In: Bayne B, Brown D, Burns K, Dixon D, Ivanovici A et al., editors. The effects of stress and pollution and marine animals. New York: Praeger. 3–45.
46. Calabrese EJ (2008) Hormesis: Why it is important to toxicology and toxicologists. Environmental Toxicology and Chemistry 27: 1451–1474.
47. Alsterberg C, Eklöf JS, Gamfeldt L, Havenhand JN, Sundbäck K (2013) Consumers mediate the effects of experimental ocean acidification and warming on primary producers. PNAS 110: 8603–8608.
48. Shure DJ (1971) Insecticide effects on early succession in an old ecosystem. Ecology 52: 271–279.
49. Relyea R, Schoeppner N, Hoverman J (2005) Pesticides and amphibians: The importance of community context. Ecological Applications 15: 1125–1134.
50. Smith RL, Smith TM (2003) Elements of Ecology. San Francisco: Benjamin Cummnings.
51. Ruist E (2008) Fosfor- och kvävefraktioner i miljöövervakningen – En studie av bohuslänska vattendrag. Länsstyrelsen Västra Götaland län.
52. Kolding S (1981) Habitat Selection and Life-Cycle Characteristics of 5 Species of the Amphipod Genus Gammarus in the Baltic. Oikos 37: 173–178.
53. Kolding S, Fenchel TM (1981) Patterns of Reproduction in Different Populations of 5 Species of the Amphipod Genus Gammarus. Oikos 37: 167–172.
54. MacNeil C, Dick JTA, Elwood RW (1999) The dynamics of predation on Gammarus spp. (Crustacea : Amphipoda). Biological Reviews 74: 375–395.
55. Relyea R, Diecks N (2008) An unforseen chain of events: lethal effects of pesticides on frogs at sublethal concentrations. Ecological Applications 18: 1728–1742.
56. Worm B, Barbier E, Beaumont N, Duffy J, Folke C, et al. (2006) Impacts of biodiversity loss on ocean ecosystem services. Science 314: 787–790.
57. Gunderson L (2000) Ecological resilience – in theory and application. Annual Review of Ecology and Systematics 31: 425–439.
58. Janson B-O (1980) Natural systems of the Baltic Sea. Ambio 9: 128–136.
59. Bonsdorff E, Pearson T (1999) Variation in the sublittoral macrozoobenthos of the Baltic Sea along environmental gradients: A functional-group approach. Australian Journal of Ecology 24: 312–326.
60. Kautsky N, Evans S (1987) Role of Biodeposition by *Mytilus edulis* in the Circulation of Matter and Nutrients in a Baltic Coastal Ecosystem. Marine Ecology Progress Series 38: 201–212.
61. Newell RIE (2004) Ecosystem influences of natural and cultivated populations of suspension-feeding bivalve molluscs: A review. Journal of Shellfish Research 23: 51–61.
62. Norling P, Kautsky N (2007) Structural and functional effects of *Mytilus edulis* on diversity of associated species and ecosystem functioning. Marine Ecology Progress Series 351: 163–175.
63. Clements WH, Rohr JR (2009) Community Responses to Contaminants: Using Basic Ecological Principles to Predict Ecotoxicological Effects. Environmental Toxicology and Chemistry 28: 1789–1800.
64. Rohr JR, Kerby JL, Sih A (2006) Community ecology as a framework for predicting contaminant effects. Trends in Ecology & Evolution 21: 606–613.
65. Norling P, Kautsky N (2008) Patches of the mussel *Mytilus* sp are islands of high biodiversity in subtidal sediment habitats in the Baltic Sea. Aquatic Biology 4: 75–87.
66. Koivisto M, Westerbom M, Riihimaki A (2011) Succession-driven facilitation of macrofaunal communities in sublittoral blue mussel habitats. Marine Biology 158: 945–954.

8

Relationships between Meiofaunal Biodiversity and Prokaryotic Heterotrophic Production in Different Tropical Habitats and Oceanic Regions

Antonio Pusceddu[1]*, Cristina Gambi[1], Cinzia Corinaldesi[1], Mariaspina Scopa[1,2], Roberto Danovaro[1,3]

1 Dipartimento di Scienze della Vita e dell'Ambiente, Università Politecnica delle Marche, Ancona, Italy, **2** Istituto Zooprofilattico Sperimentale dell'Abruzzo e del Molise 'G. Caporale', Experimental Regional Centre for Fishery and Acquacolture, Termoli, Italy, **3** Stazione Zoologica Anton Dohrn, Napoli, Italy

Abstract

Tropical marine ecosystems are among the most diverse of the world oceans, so that assessing the linkages between biodiversity and ecosystem functions (BEF) is a crucial step to predict consequences of biodiversity loss. Most BEF studies in marine ecosystems have been carried out on macrobenthic diversity, whereas the influence of the meiofauna on ecosystem functioning has received much less attention. We compared meiofaunal and nematode biodiversity and prokaryotic heterotrophic production across seagrass, mangrove and reef sediments in the Caribbean, Celebes and Red Seas. For all variables we report the presence of differences among habitats within the same region, and among regions within the same habitat. In all regions, the richness of meiofaunal taxa in reef and seagrass sediments is higher than in mangrove sediments. The sediments of the Celebes Sea show the highest meiofaunal biodiversity. The composition of meiofaunal assemblages varies significantly among habitats in the same region. The nematode beta diversity among habitats within the same region is higher than the beta diversity among regions. Although one site per habitat was considered in each region, these results suggest that the composition of meiofaunal assemblages varies primarily among biogeographic regions, whereas the composition of nematode assemblages varies more considerably among habitats. Meiofauna and nematode biodiversity and prokaryotic heterotrophic production, even after the removal of covariate effects linked with longitude and the quantity and nutritional quality of organic matter, are positively and linearly linked both across regions and within each habitat type. Our results confirm that meiofauna and nematode biodiversity may influence benthic prokaryotic activity, which, in turn, implies that diversity loss could have negative impacts on ecosystem functioning in these systems.

Editor: Maura (Gee) Geraldine Chapman, University of Sydney, Australia

Funding: This study was carried out in the framework of the projects Ricerca di Ateneo 2004–2011 financed by the Polytechnic University of Marche. This study is a contribution to the FIRB Project EXPLODIVE financially supported by MIUR Italy. The funders had no role in study design, data collection and analysis, decision to publish, or preparation of the manuscript.

Competing Interests: The authors have declared that no competing interests exist.

* E-mail: a.pusceddu@univpm.it

Introduction

Marine coastal ecosystems provide crucially important goods and services to the human beings [1,2]. Marine coastal ecosystems are characterized by the presence of a complex and heterogeneous mosaic of habitats, such as sandy beaches, rocky shores, mangrove forests, seagrass meadows, coral reefs and transitional zones linking terrestrial and coastal ecosystems. Nevertheless, because of their proximity with the planet region most dense in human population, the coastal oceans are exposed to multiple anthropogenic stressors, including, among the others, aquaculture, dredging, mining, pollution, species invasion, over-harvesting and destructive fishing practices, large-scale oil and gas operations, watershed and offshore renewable energy development, coastal engineering habitat having strong and direct effects on the marine biodiversity [3,4].

The annual loss rates of five of the most important biogenic marine habitats (seagrass beds, salt marshes, coral and oyster reefs and mangrove forests) range between 1 and 9% [5]. Based on historical evidence, the total global loss of these habitats ranges from ca. 19% for coral reefs (in between 2004–2008) [6], to 29% (since 1879) for seagrass [7], to 35% (since 80ies) for mangrove forests [8] to >85% (in the last 20–130 years) for oyster reefs [9]. Many of these habitats, which are also hot spots of biodiversity, belong to tropical regions.

Ecological theory predicts that biodiversity can control ecosystems' functioning, although outputs of correlative and manipulative investigations have provided at times equivocal or contrasting results [10]. The relationships between biodiversity and functioning of marine ecosystems are most often positive [11], so that any biodiversity loss could result in a decrease of the ecosystem functioning [12] and, consequently, in a lower provision of goods and services to the humans [2,13]. A biodiversity loss can thus potentially impair the ecosystems' capacity to sustain humanity [14]. This is of particular concern in tropical and subtropical environments which host an important fraction of the biodiversity of the coastal oceans and are among the world regions that will experience the earliest emergence of historically unprecedented climates and changes in biodiversity [15].

Either correlative or manipulative approaches for assessing the shape and strength of the relationships between biodiversity and ecosystem functioning in marine ecosystems have been mostly

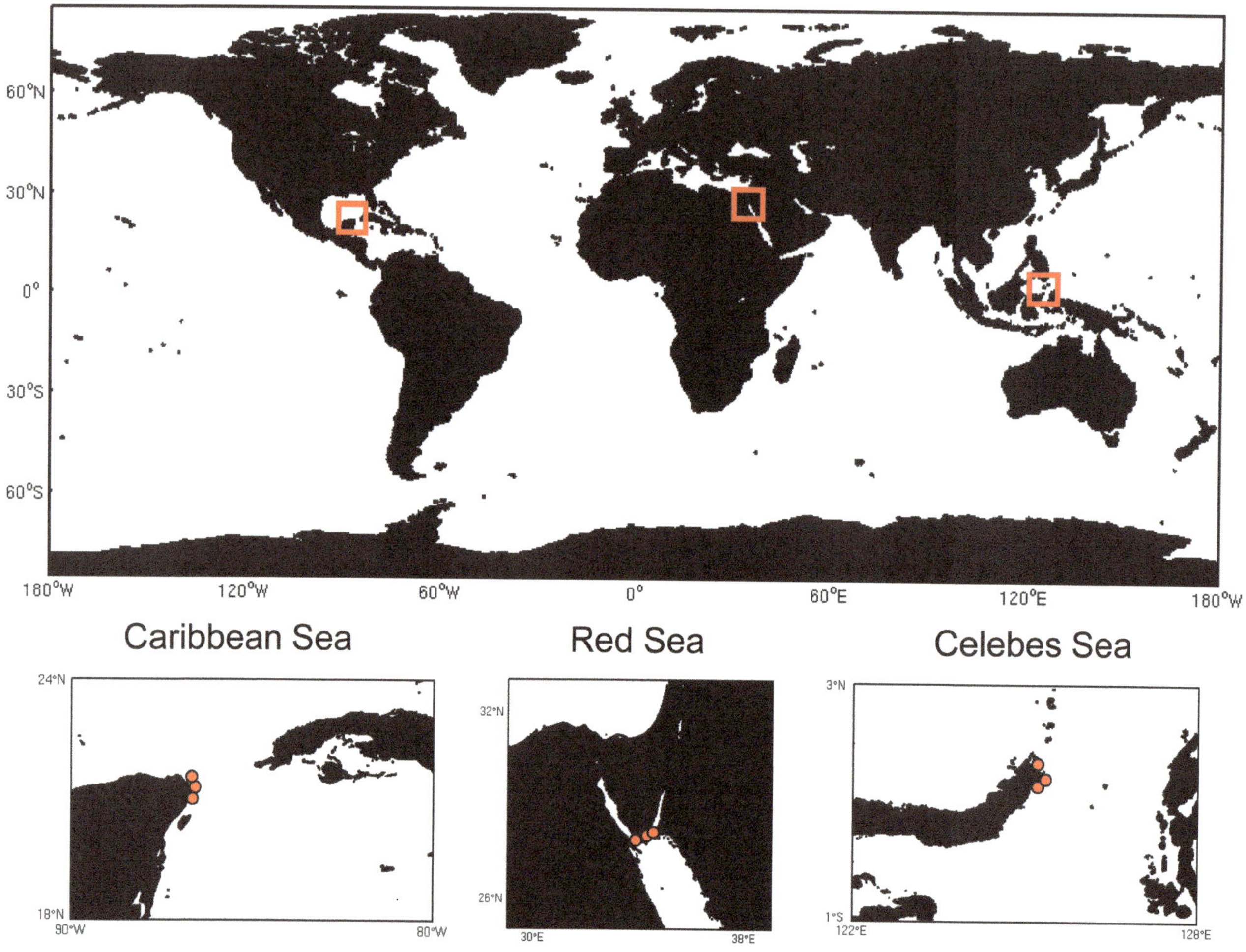

Figure 1. Sampling areas and location of the investigated regions with details of the positioning of each habitat type.

carried out using the diversity of macro-fauna or macro-algae as the independent variable [16]. More recently, meiofauna, the most abundant benthic group of metazoans in marine ecosystems [17], have been utilized for investigating, though only with a correlative approach, the linkages between biodiversity and ecosystem functioning in deep-sea sediments [18,19]. Among metazoan meiofauna, nematodes respond rapidly to many different sources of natural and anthropogenic disturbance affecting sedimentary environments; thus, also due to their high abundance, species and functional (trophic) diversity, nematodes are an ideal tool to investigate the relationships between biodiversity and ecosystem functioning [20]. Meiofauna play also a key ecological role in linking detrital (and prokaryotic) resources with higher trophic levels: in fact most of the meiofaunal taxa eat microalgae, prokaryotes and detritus and, at the same time, it is known that meiofauna are a food source for macrofauna and fishes [21–25]. Meiofauna and nematodes, based on laboratory and *in situ* experiments, are in fact able to influence microbial activities and to graze their production [17,26]. Meiofauna and nematodes are also very sensitive to the broad variations in natural environmental conditions (tidal influence, river inputs and local rainfall, food availability, sediment chemistry, bottom current regimes, habitat heterogeneity, among the others) [17,27] that characterize marine sediments across all spatial scales and water depths [28–31].

The functioning of marine sedimentary ecosystems relies on the rates of organic matter cycling, which is also related to the production of heterotrophic prokaryotes: in turn, these are related to the food quantity and food availability [32,33]. This holds true, in particular for highly detrital ecosystems like seagrass beds [34], mangroves [35] and coral reefs [36], where organic detritus, through the so-called microbial loop, is firstly incorporated into prokaryotic biomass, then enters higher trophic levels through bacterivorous, detritivorous and deposit feeders inhabiting the benthos [32]. Thus, it can be hypothesized that, in those ecosystems, the shape and strength of the relationship between biodiversity and ecosystem function could be influenced by the quantity and bioavailability of food resources.

To provide insights about the relationships between meiofaunal biodiversity and the ecosystem functioning in different tropical sedimentary habitats, we analyzed the richness of meiofaunal taxa and of nematode species (biodiversity) along with sedimentary organic matter quantity and nutritional quality, and prokaryotic heterotrophic production (functioning) across three habitats (i.e., seagrass, mangrove and reef sediments) in the Caribbean, Red and Celebes Seas. We also explored how some of the potential drivers of biodiversity considered in our study (longitude, quantity and quality of trophic sources) could influence the biodiversity-ecosystem functioning linkage. By comparing different habitats and regions we posed the following questions: i) how the different target variables vary among habitats and regions?; ii) whether and how changes the relationship between biodiversity and ecosystem functioning across different habitats?.

Materials and Methods

Ethical Statement

No special permits were requested at the time of sampling as in all regions sediment samples were collected out of protected areas and respecting local legislation. No protected species or taxa were sampled.

Study Areas and Sampling Activities

The sampling strategy included a total of nine sampling sites, from three tropical regions: the Caribbean, Red and Celebes Seas (Fig. 1, Table 1). In each region, three habitat types were visited once each: seagrass, mangrove and coral reef sediments.

In the Caribbean Sea, sediment samples were collected in the east coast of the Yucatan peninsula (Mexico) from a seagrass bed (Isla Mujeres), a coral reef (Xelha) and a mangrove forest (dominated by *Rizophora sp.*). In the Red Sea, sediment samples were collected in the Nabq lagoon from a seagrass bed, a mangrove forest dominated by *Avicennia marina* and a coral reef. In the Celebes Sea (Sulawesi Island, Indonesia), sediment samples were collected from a seagrass bed (dominated by *Enhalus acoroides*), a mangrove forest and a coral reef.

Sampling sites (each representing one type of habitat) within a region were 10–35 kilometers distance apart. At all sampling sites, replicate sediments within each habitat have been collected from three 0.5×0.5 m quadrates randomly selected from one 5×5 m area per habitat. The choice of this sampling area was constrained by the limited capability of operational movement in the three regions. Although this could have possibly reduced the ability to identify precisely the patterns of variability at the habitat scale, this design allowed the collection of sediments from the different habitats in each region within a reasonable time interval (hours-days).

Sediment samples were collected by scuba diving using Plexiglas manual corers (internal diameter 4.6 cm). The top first centimeter of three independent corers was collected for the determination of the quantity and biochemical composition of sedimentary organic matter, prokaryotic abundance, biomass and C production.

All samples for organic matter and biomass determinations were stored at −20°C until the analyses in the laboratory, whereas samples for the determination of prokaryotic heterotrophic production were immediately incubated as described below.

Three additional independent corers, collected for the analyses of meiofaunal abundance, biomass and biodiversity, were preserved in buffered formalin (4%), stained with Rose Bengal (0.5 gL^{-1}) and stored at 4°C until analysis.

Sedimentary Organic Matter

The sedimentary contents of phytopigments and biopolymeric C pools reflect the overall trophic conditions of marine coastal sediments, whereas the algal fraction of biopolymeric C pools reflects the food quality of sedimentary detritus [37]. Chloroplastic pigments (chlorophyll-*a* and phaeopigments) were analyzed fluorometrically [38]. Pigments were extracted from the top centimeter of each core with 90% acetone (24 h in the dark at 4°C). After centrifugation (800×g), the supernatant was used to determine the functional chlorophyll-*a* and acidified with 0.1 N HCl to estimate the amount of phaeopigments. Chlorophyll-*a* and phaeopigment concentrations were summed up and reported as total phytopigment concentrations. Total phytopigments, after

Table 1. Biochemical composition of sedimentary organic matter, prokaryotic biomass and C production in the investigated habitats and regions.

Region	Habitat	Latitude	Longitude	Depth	Chlorophyll-*a*	Phaeopigment	Protein	Carbohydrate	Lipid	Biopolymeric Carbon	Prokaryotic biomass	Prokaryotic heterotrophic production
		° N	°E	m	$\mu g\ g^{-1}$	$\mu g\ g^{-1}$	$mg\ g^{-1}$	$mg\ g^{-1}$	$mg\ g^{-1}$	$mgC\ g^{-1}$	$\mu gC\ g^{-1}$	$\mu gC\ g^{-1}\ d^{-1}$
Caribbean	Seagrass	21.6	−87.0	0.50	1.4±0.2	8.7±1.5	1.44±0.37	1.66±0.42	0.54±0.11	1.77±0.42	1.73±0.42	1.73±0.42
	Mangrove	21.3	−86.8	0.50	4.8±0.1	92.6±5.0	10.46±1.42	21.24±3.29	8.36±1.00	19.89±0.89	71.4±11.1	1.78±0.28
	Reef	20.5	−86.8	3.00	2.7±2.8	34.7±5.9	0.83±0.14	2.49±0.22	0.52±0.01	1.79±0.15	0.47±0.04	0.21±0.02
Red	Seagrass	28.2	34.4	0.50	0.2±0.1	16.4±4.7	0.61±0.15	0.53±0.03	0.03±0.00	0.54±0.06	1.04±0.34	1.04±0.34
	Mangrove	28.2	34.4	0.50	0.1±0.0	13.4±1.4	0.28±0.06	0.07±0.04	0.02±0.00	0.18±0.02	1.71±0.24	1.20±0.17
	Reef	28.2	34.4	2.00	0.4±0.1	49.9±8.0	0.67±0.09	0.10±0.04	0.05±0.03	0.41±0.07	0.36±0.08	0.16±0.03
Celebes	Seagrass	1.6	125.4	0.50	2.2±1.0	20.4±8.9	2.72±0.46	7.23±0.61	1.05±0.26	7.17±0.45	4.04±2.04	4.04±2.04
	Mangrove	1.5	125.2	0.50	0.7±0.3	4.0±1.9	1.78±0.21	1.63±0.22	0.63±0.31	2.35±0.19	6.38±0.53	2.23±0.18
	Reef	1.6	125.5	5.00	1.0±0.3	5.8±1.1	0.44±0.12	0.80±0.10	0.20±0.04	0.90±0.08	8.24±0.59	3.59±0.26

Table 2. Results of PERMANOVA and pairwise test for differences in sedimentary organic matter quantity and biochemical composition, prokaryotic biomass and heterotrophic production.

	Source	DF	MS	Pseudo-F	P	Pairwise tests: Region (Habitat)	Pairwise tests: Habitat (Region)
Phytopigments	**Region**	2	3.8	57.14	***	Car [M>R>S]	S [Car = Red = Cel]
	Habitat	2	0.4	6.07	*	Red [R>S = M]	M [Car>Red>Cel]
	Region × Habitat	4	3.1	46.71	***	Cel [S>M = R]	R [Red = Car>Cel]
	Residuals	18	0.1				
Proteins	**Region**	2	1.8	187.69	***	Cel [S>M>R]	S [Cel>Car>Red]
	Habitat	2	1.2	127.98	***	Car [M>S>R]	M [Car>Cel>Red]
	Region × Habitat	4	1.2	127.14	***	Red [R = S>M]	R [Car>Cel]
	Residuals	18	0.0				
Carbohydrates	**Region**	2	5.8	736.42	***	Car [M>R = S]	S [Cel>Car>Red],
	Habitat	2	1.3	163.86	***	Red [S>R = M]	M [Car>Cel>Red]
	Region × Habitat	4	2.4	301.53	***	Cel [S>M>R]	R [Car>Red>Cel]
	Residuals	18	0.0				
Lipids	**Region**	2	2.2	278.69	***	Car [M>S = R]	S [Cel>Car>Red]
	Habitat	2	1.2	145.26	***	Red [R>S>M]	M [Car>Cel>Red]
	Region × Habitat	4	1.2	144.26	***	Cel [S>M = R]	R [Car>Cel>Red]
	Residuals	18	0.0				
Biopolymeric C	**Region**	2	4.6	1076	***	Car [M>R = S]	S [Cel>Car>Red]
	Habitat	2	1.5	346.56	***	Red [R = S>M]	M [Car>Cel>Red]
	Region × Habitat	4	2.1	501.85	***	Cel [S>M>R]	R [Car>Cel>Red]
	Residuals	18	0.0				
Biochemical composition	**Region**	2	20.2	42.83	***	Car [M≠R = S]	S [Cel≠Car≠Red]
	Habitat	2	10.1	21.43	***	Red [R≠[M = S]	M [Cel≠Car≠Red]
	Region × Habitat	4	15.2	32.14	***	Cel [M≠S≠R]	R [Cel≠Red]
	Residuals	18	0.5				
Prokaryotic biomass	**Region**	2	9.2	138.3	***	Car [S = M>R]	S [Cel>Car = Red]
	Habitat	2	11	163.5	***	Red [S = M>R]	M [Car = Cel>Red]
	Region × Habitat	4	5.9	88.7	***	Cel [R>M = S]	R [Cel>Car = Red]
	Residuals	18	0				
Prokaryotic heterotrophic production	**Region**	2	2.0	64.7	***	Car [S = M>R]	S [Cel> Red = Car]
	Habitat	2	0.5	18.7	***	Red [S = M>R]	M [Car = Cel>Red]
	Region × Habitat	4	0.3	10.6	***	Cel [S = R>M]	R [Cel>Car = Red]
	Residuals	18	0.0				

DF = degrees of freedom; MS = mean squares; Pseudo-F = F statistic; P = probability level; *** = P<0.001; * = P<0.05; ns: not significant. Car = Caribbean Sea; Red = Red Sea; Cel = Celebes Sea; M = mangrove; R = reef; S = seagrass.

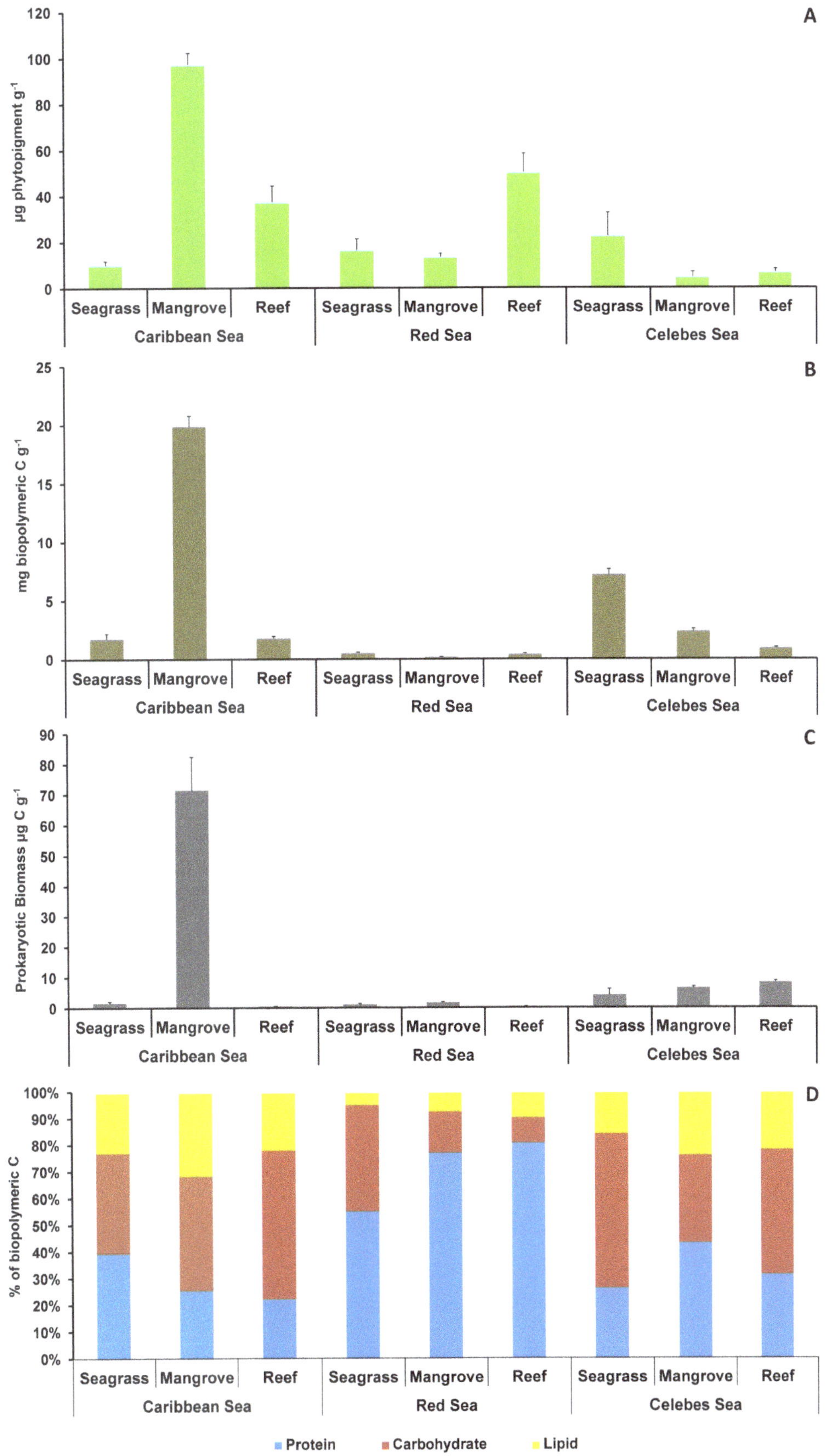

Figure 2. Biochemical composition of sedimentary organic matter. Reported are the concentrations of A) total phytopigments, B) biopolymeric carbon, C) prokaryotic biomass, and D) the biochemical composition (as percentage contribution of the biopolymeric carbon content)

of the sedimentary organic matter in seagrass, mangrove and reef habitats of the Caribbean, Red and Celebes Seas. Reported are mean values ± standard deviation.

transformation into C equivalents using 30 as a conversion factor [39], are thus used as an estimate of the organic material of algal origin including either the living (chlorophyll-*a*) and senescent/detrital (phaeopigment) fractions [40].

Protein, carbohydrate and lipid analyses were carried out on the top 1 cm of three independent corers from each site using photometric protocols [39]. Protein, carbohydrate and lipid sedimentary contents were converted to carbon equivalents by using the following conversion factors: 0.49, 0.40 and 0.75 mg C mg^{-1}, respectively and their sum referred as biopolymeric C (BPC) [37].

Prokaryotic Biomass and Heterotrophic Production

Prokaryotic biomass was determined from the prokaryotic cell biovolume and abundance by epifluorescence microscopy. To estimate the cell biovolume we examined at least 100 cells for sediment sample classifying prokaryotes into three different size classes: small (<0.065 μm^3), medium (0.065–0.320 μm^3) and large (0.320–0.574 μm^3) and the biovolume was converted into carbon content assuming 310 fg C μm^3 as a conversion factor [41]. The total prokaryotic abundance was determined using epifluorescence microscopy [39]. Briefly, sediment subsamples (0.25 g) were added with 1125 µl of 2% formalin and 125 µl of pyrophosphate (final concentration, 5 mM). Then, these were treated with ultrasounds (three times for 1 min; Branson sonifier 2200, 60 W) to increase the extraction efficiency and diluted 250–500 times using autoclaved and 0.2 ml pre-filtered seawater. One ml of the supernatant was stained with Acridine Orange (final concentration, 0.01%), for 5 min in the dark) and filtered under vacuum (< 100 mmHg) using 0.2 µm pore size black Nuclepore Polycarbonate filters. The filters were then washed twice with 3 ml of 0.2 µm pre-filtered and sterilized reagent grade water, mounted on microscope slides, and analyzed under an epifluorescence microscope (Zeiss Axioskop 2; magnification, ×1000). For each slide, at least 10 fields were observed, for a total of at least 400 cells counted per filter. Prokaryotic abundance is expressed as cells g^{-1} dry sediment (after desiccation at 60°C for 24 h).

Benthic prokaryotic heterotrophic production was measured by ^{3}H-leucine incorporation [42]. Sediment sub-samples (200 µl), added to a saturating aqueous solution of ^{3}H-leucine (6-µCi final concentration per sample), were incubated for 1 h in the dark at in situ temperature. After incubation, prokaryotic C incorporation was stopped with 1.7 ml of 80% ethanol before scintillation counting. Sediment blanks were made adding ethanol immediately before ^{3}H-leucine addition. Data were normalized to sediment dry weight after desiccation (60°, 24 h).

Meiofaunal Abundance and Biomass

For metazoan meiofaunal extraction, sediment samples were sieved through a 500 µm and a 30 µm mesh, respectively, to retain the smallest organisms. The fraction remaining on the latter sieve (including organisms with a size of 30–500 µm) was resuspended and centrifuged three times with Ludox HS 40 (diluted with tap water to arrange density to 1.18 g cm^{-3}) for muddy samples, whereas for sandy samples, meiofauna were extracted using decantation (repeated 10 times for 95% of efficiency) [39,43]. After the extraction the sediments have been carefully checked to search for remnant organisms. No organisms were observed in the residual sediments after the treatment with Ludox or decantation. All metazoan animals were counted and classified per taxon under a stereomicroscope using Delfuss cuvettes, after staining with Rose Bengal (0.5 g L^{-1}). For the determination of meiofaunal biomass, we calculated the individual biomass of all animals belonging to different taxa. Nematode biomass was calculated from biovolume (n = 100 per sample) using Andrassy's formula ($V = L \times W^2 \times 0.063 \times 10^{-5}$; body length L in µm and width W in µm). For all other taxa, the biovolume was measured for all of the specimens encountered. Body volume was derived from measurements of body length (L; in mm) and width (W; in mm) using the formula $V = L \times W^2 \times C$; where C is the approximate conversion factor for each meiofaunal taxon [44]. The body volume was multiplied by an average density (1.13 g cm^{-3}) to obtain the biomass (µg DW) assuming that the dry : wet weight ratio was 20–25%, and the C content was considered as 40% of the dry weight [44].

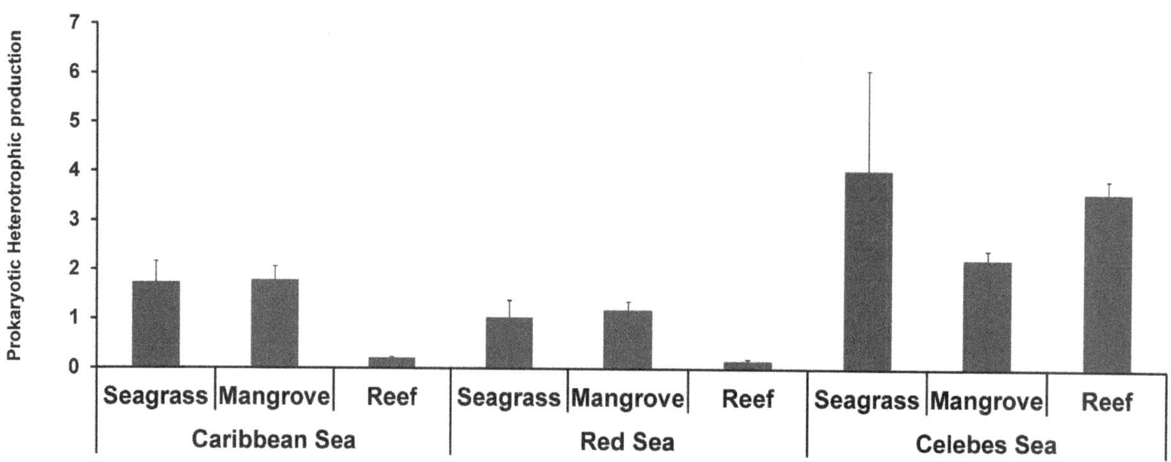

Figure 3. Ecosystem functioning as prokaryotic heterotrophic production (µgC g^{-1} d^{-1}) in seagrass, mangrove and reef habitats of the Caribbean, Red and Celebes Seas.

Table 3. Meiofaunal abundance and richness of taxa, nematode diversity as Species Richness, ES(51), evenness (Pielou's J) and index of trophic diversity in all habitats and regions.

Region	Habitat	Latitude	Longitude	Depth	Meiofaunal abundance	Richness of meiofaunal taxa	Species Richness	Expected species number	Pielou's evenness	Trophic diversity
		° N	°E	m	Ind. 10 cm^{-2}			ES(51)	J	ITD
Caribbean	Seagrass	21.6	−87.0	0.50	1268±731	11.3±4.2	33	14.1±4.3	0.82±0.05	0.49±0.04
	Mangrove	21.3	−86.8	0.50	2474±117	5.0±1.0	13	6.7±0.1	0.48±0.01	0.80±0.02
	Reef	20.5	−86.8	3.00	2871±1307	9.0±1.0	28	10.9±2.0	0.68±0.00	0.38±0.02
Red	Seagrass	28.2	34.4	0.50	1475±166	11.0±1.7	n/a	n/a	n/a	n/a
	Mangrove	28.2	34.4	0.50	343±41	4.0±1.0	n/a	n/a	n/a	n/a
	Reef	28.2	34.4	2.00	488±253	6.0±2.0	n/a	n/a	n/a	n/a
Celebes	Seagrass	1.6	125.4	0.50	1604±267	16.3±2.5	59	24.5±4.0	0.90±0.03	0.44±0.08
	Mangrove	1.5	125.2	0.50	627±192	9.7±1.5	40	16.1±2.9	0.79±0.04	0.43±0.08
	Reef	1.6	125.5	5.00	706±271	17.0±1.0	77	27.5±1.2	0.90±0.05	0.45±0.05

n/a = not analyzed.

Nematode Biodiversity

Nematode species richness and assemblage composition was determined only in the Celebes and Caribbean Seas. For the analysis of nematode diversity, 100 specimens (or all of the retrieved nematodes if <100) were randomly picked up from each of the three independent replicates at each sampling station. Nematodes were mounted on slides (following the formalin-ethanol-glycerol technique to prevent dehydration) [39] and identified to species level. Species identity was not considered in this study but, for the purposes of estimating species richness and other diversity indexes, the different morphotypes belonging to each genus were indicated as sp1, sp2, sp3 and considered as separate species [45]. Nematode species richness was estimated as the total number of species identified in each habitat. Since species richness is strongly influenced by the number of the individuals identified, to standardize the values of nematode diversity, the species-abundance data were converted into rarefaction diversity indices [46,47]. The expected number of species for a theoretical sample of 51 nematode specimens, ES(51), was calculated. Previous studies have shown that this approach enables the provision of robust data on species richness. Although it is far from being perfect [48], the expected species number is the density-independent index most commonly used for the comparison of areas with a non-standardized sample size [31,49]. The Pielou's index (J') was also calculated [50].

We measured also the beta diversity of meiofaunal and nematode assemblages among habitats within each region and among regions within each habitat using the SIMPER analysis based on Bray-Curtis matrixes, and expressed as percentage of dissimilarity [48]. Taxon and species-abundance data were presence/absence transformed prior to the analysis to search for specific assemblages in each habitat and region, irrespectively of the taxon and species relative abundance. Diversity indexes and the dissimilarity estimates were calculated using the routine DIVERSE and SIMPER, respectively included in the PRIMER6+ software [51].

Functional diversity of nematodes was estimated using the Index of Trophic Diversity (ITD) calculated as $ITD = g_1^2+g_2^2+g_3^2 \ldots + g_n^2$, where g is the relative contribution of each trophic group to the total number of individuals and n is the number of trophic groups (with n = 1–4) [52]. Nematodes were divided into four groups as follows: (1A) no buccal cavity or a fine tubular one - selective (bacterial) feeders; (1B) large but unarmed buccal cavity-non-selective deposit feeders; (2A) buccal-cavity with scraping tooth or teeth-epistrate or epigrowth (diatom) feeders; (2B) buccal cavity with large jaws-predators/omnivores [52]. For four trophic guilds, the Index of Trophic Diversity ranges from 0.25 (highest trophic diversity; i.e., the four trophic guilds account for 25% each) to 1.0 (lowest diversity; i.e., one trophic guild accounts for 100% of nematode abundance) [43].

Ecosystem Functioning

Resembling one of the approach used for assessing biodiversity and ecosystem functioning relationships in the deep sea [18], we used prokaryotic heterotrophic production as proxy of the benthic ecosystem functioning and related it to biodiversity, estimated in terms of richness of meiofaunal taxa and nematode expected species number [ES(51)].

Statistical Analyses

To assess separately the differences in the quantity of sedimentary trophic resources, prokaryotic variables (biomass and heterotrophic production), total meiofaunal abundance, biomass, and richness of taxa among habitats within each region,

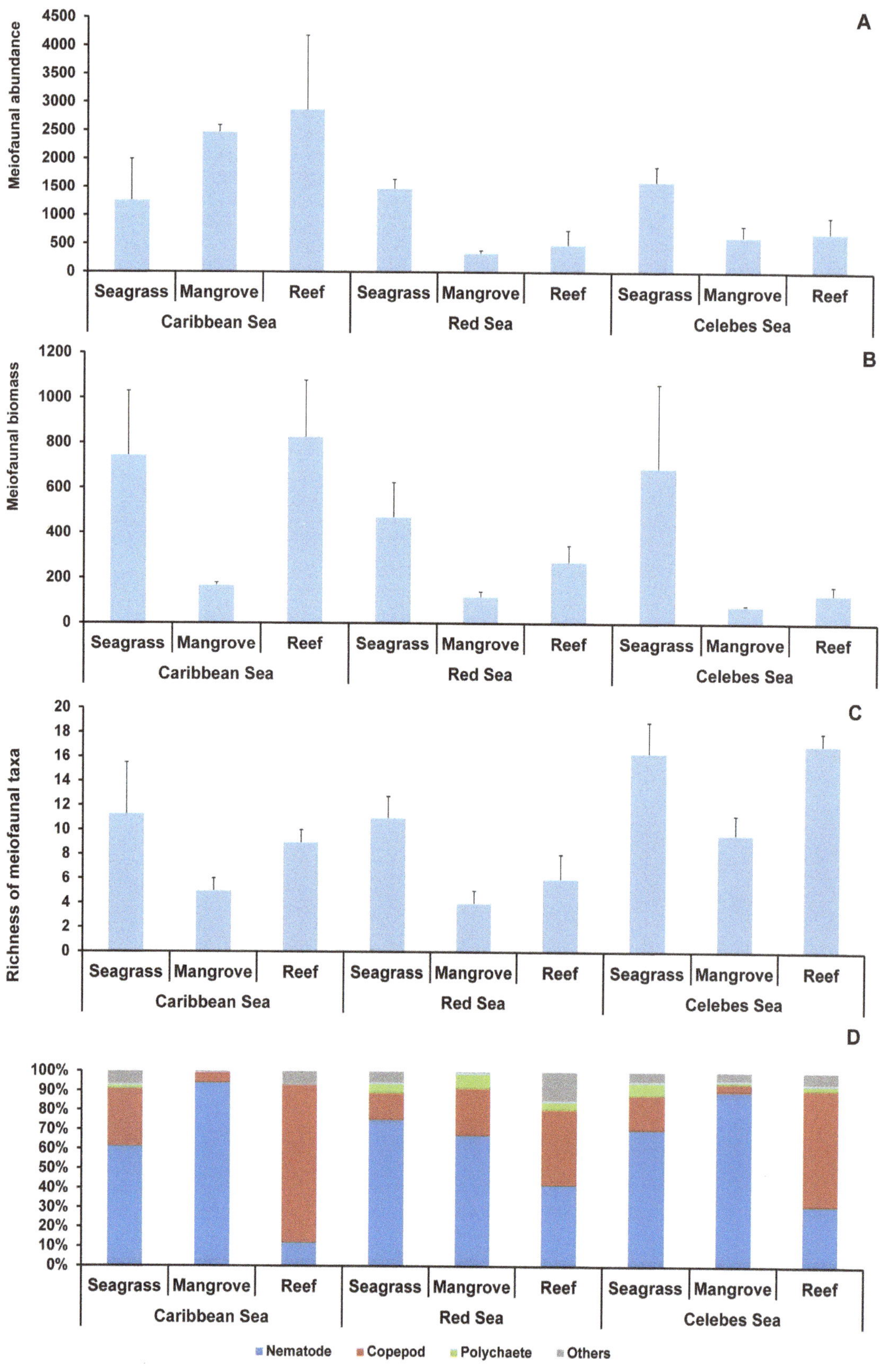

Figure 4. Meiofaunal assemblages. Illustrated are meiofaunal: A) abundance (ind 10 cm^{-2}), B) biomass (µgC 10 cm^{-2}), C) richness of taxa, and D) assemblage composition in seagrass, mangrove and reef habitats of the Caribbean, Red and Celebes Seas. Reported are average values ± standard deviation.

Table 4. Results of PERMANOVA testing for differences in meiofaunal abundance, biomass, richness of taxa and assemblages composition among habitats across regions and among regions across habitats.

	Source	DF	MS	Pseudo-F	P	Pairwise tests	
						Region (Habitat)	**Habitat (Region)**
Meiofaunal abundance	**Habitat**	2	0.74	5.54	**	Car [S=M=R]	S [Car=Red=Cel]
	Region	2	3.31	24.9	***	Red [S>M=R]	M [Car>Cel>Red]
	Habitat × Region	4	1.3	9.8	***	Cel [S>M=R]	R [Car>Cel=Red]
	Residuals	18	0.13				
Meiofaunal Biomass	**Habitat**	2	2.0	6.91	***	Car [R=S>M]	S [Car=Red=Cel]
	Region	2	5.0	17.77	***	Red [S=R>M]	M [Car>Red=Cel]
	Habitat × Region	4	0.6	2	ns	Cel [S>R=M]	R [Car>Red=Cel]
	Residuals	18	0.3				
Richness of meiofaunal taxa	**Habitat**	2	135.81	33.04	***	Car [R=S>M]	S [Cel>Red=Car]
	Region	2	103.70	25.23	***	Red [R=S>M]	M [Cel>Car=Red]
	Habitat × Region	4	7.70	1.87	ns	Cel [R=S>M]	R [Cel>Car=Red]
	Residuals	18	4.11				
Assemblages composition	**Habitat**	2	3430	5.67	***	Car [S=M=R]	na
	Region	2	2759	2.80	ns	Red [S≠M=R]	na
	Habitat × Region	4	985	1.63	ns	Cel [M≠S=R]	na
	Residuals	18	605				

Results of the pairwise tests among levels of the relevant factors are also reported. DF = degrees of freedom; MS = mean squares; Pseudo-F = F statistic; P = probability level; *** = $P<0.001$; ** = $P<0.01$; ns: not significant.
Car = Caribbean Sea; Red = Red Sea; Cel = Celebes Sea; M = mangrove; R = reef; S = seagrass; na = not applicable.

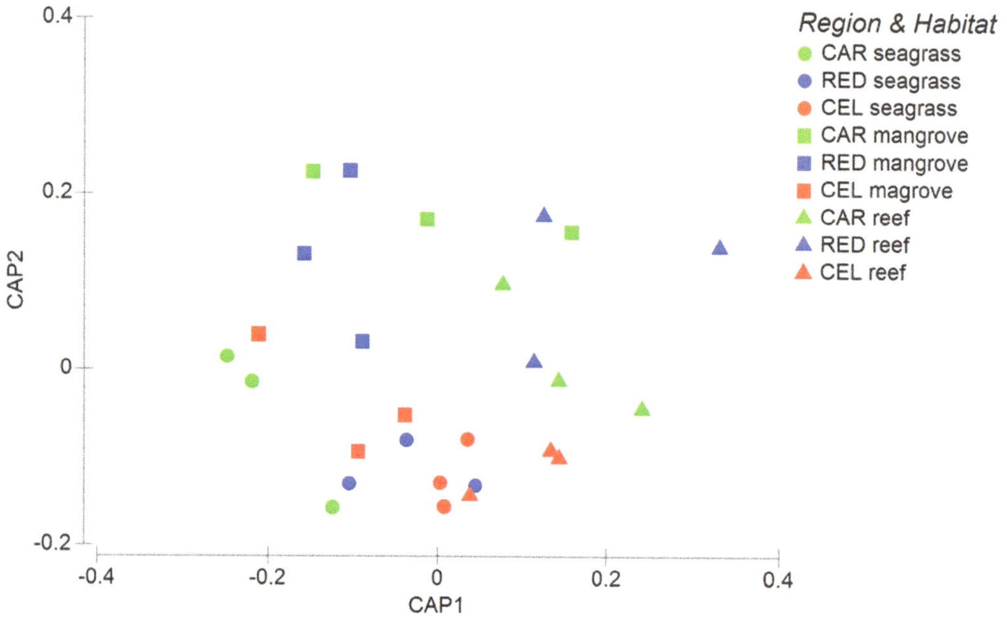

Figure 5. Bi-plots after the canonical analysis of principal coordinates illustrating differences in the composition of total meiofaunal taxa in the sediments of the investigated oceanic regions and habitats. CAR = Caribbean Sea (green); CEL = Celebes Sea (red); RED = Red Sea (blue).

and among regions within each habitat, we used two-way permutational analyses of variance (PERMANOVA) under a reduced model. The analyses were carried out using Habitat (3 fixed levels: mangrove, reef and seagrass; orthogonal to Region) and Region (3 fixed levels: Celebes, Red and Caribbean Seas) as main sources of variance. The same design was used for the analysis of nematode species diversity, but, in this case, the factor region included two levels only (Celebes and Caribbean Seas).

The same design has been used as the basis for a distance-based permutational multivariate analyses of variance (PERMANOVA) [53] to test variations among habitats and regions in the biochemical composition of sedimentary organic matter (based on Euclidean distance matrices), in the meiofaunal and the nematode assemblages (based on Bray Curtis similarity matrices after presence/absence transformations). The analyses were carried out using the permutation of residuals under a reduced model. In both uni- and multivariate tests, significant terms were investigated using a posteriori pair-wise comparisons with the PERMANOVA *t* statistic and 999 permutations.

The variations in meiofaunal communities and nematode assemblages among regions and habitats were illustrated using bi-plots produced after a canonical analysis of principle coordinates (CAP).

Biodiversity vs. ecosystem functioning relationships were assessed using distance-based analyses for a linear model using the routine DISTLM before and after the removal of covariates associated with the geographical location (longitude) and the availability of food resources (here evaluated in terms of biopolymeric C and protein to carbohydrate ratio), as synthetic descriptors of quantity and nutritional quality of food, respectively.

PERMANOVA, CAP and DISTLM tests were carried out using the homonymous routines included in the PRIMER6+ software [51].

Results

Sedimentary Organic Matter

The results of the two-way PERMANOVA carried out including the three habitat types in the Caribbean, Red and Celebes Seas revealed the presence of a significant effect of the interaction Habitat × Region for phytopigment, biopolymeric C and prokaryotic biomass (Table 2).

In the Caribbean Sea, the highest phytopigment and biopolymeric C contents occurred in mangrove sediments; in the Red Sea the highest total phytopigment contents occurred in reef sediments, whereas in the Celebes Sea the highest contents were observed in the seagrass sediments (Fig. 2A–C). Comparing seagrass habitats in different regions, phytopigment and biopolymeric C contents in the Celebes Sea were significantly higher than those in both the Caribbean and the Red Seas, whereas values in the Caribbean mangrove sediments were consistently higher than those in the Celebes and Red Seas.

Prokaryotic biomass in seagrass sediments did not vary significantly across regions (Table 2). In mangrove sediments prokaryotic biomass showed the highest values in the Caribbean Sea and the lowest in the Red Sea, whereas in reef sediments it was significantly higher in the Celebes Sea than in the Caribbean and Red Seas (Table 2).

The pairwise tests revealed also the presence of significant differences in the biochemical composition of sediments among regions (for mangrove and seagrass sediments) (Table 2). The reef sediments of the Celebes and Caribbean Seas exhibited a similar biochemical composition characterized by the dominance of carbohydrates, whereas the protein fraction dominated in the Red Sea reef sediments (Fig. 2D).

Prokaryotic Biomass and Heterotrophic Production

Prokaryotic biomass and heterotrophic production values were characterized by a significant effect of the Habitat × Region interaction (Table 2). In the Caribbean and Red Seas, prokaryotic

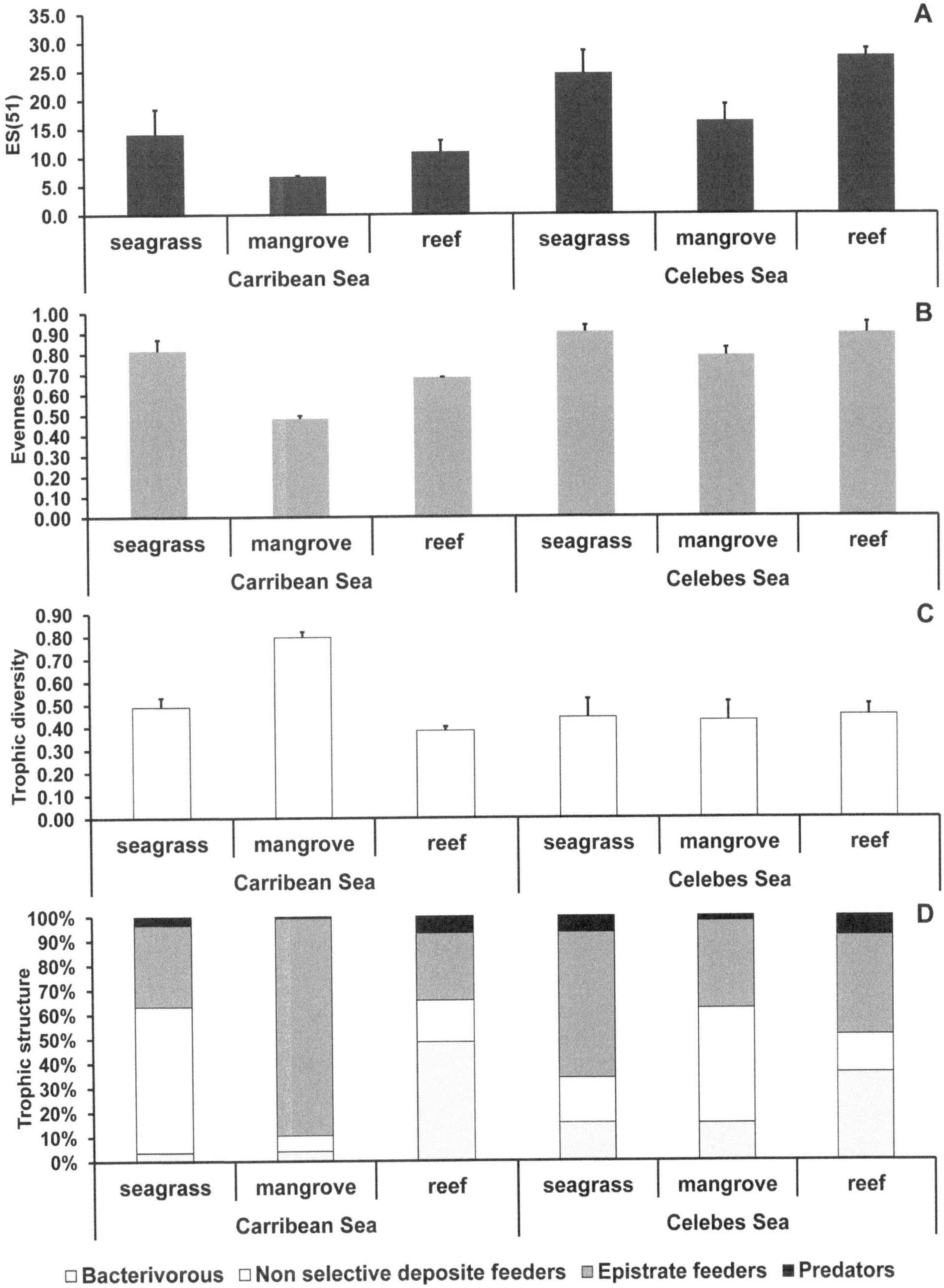

Figure 6. Nematode diversity. Illustrated are A) expected species number [ES(51)], B) evenness (Pielou's J), C) index of trophic diversity (ITD) and D) trophic composition of nematode assemblages in seagrass, mangrove and reef habitats of the Caribbean, Red and Celebes Seas.

Table 5. Results of PERMANOVA testing for differences in nematode biodiversity, as nematode Species Richness, ES(51), Pielou's evenness and index of trophic diversity (ITD), and nematode assemblage composition among habitats across regions and among regions across habitats.

	Source	DF	MS	Pseudo-F	P	Pairwise test	
						Region (Habitat)	Habitat (Region)
Species richness	**Region**	1	2.39	59.13	**	Car [S=R>M]	S [Cel>Car]
	Habitat	2	0.3	7.45	*	Cel [S=R>M]	M [Cel>Car]
	Region × Habitat	2	0.03	0.79	ns		R [Cel>Car]
	Residuals	12	0.04				
ES(51)	**Region**	1	2.44	93.8	***	Car [R=S>M]	S [Cel>Car]
	Habitat	2	0.5	19.36	***	Cel [R=S>M]	M [Cel>Car]
	Region × Habitat	2	0.05	1.79	ns		R [Cel>Car]
	Residuals	12	0.03				
Evenness	**Region**	1	0.06	148.78	***	Car [S>R>M]	S [Car=Cel]
	Habitat	2	0.03	67.18	***	Cel [R=S>M]	M [Cel>Car]
	Region × Habitat	2	0.01	17.29	***		R [Cel>Car]
	Residuals	12	0.00				
ITD	**Region**	1	0.02	16.37	***	Car [M>S>R]	S [Cel=Car]
	Habitat	2	0.02	16.27	***	Cel [M=R=S]	M [Car>Cel]
	Region × Habitat	2	0.03	21.33	***		R [Cel=Car]
	Residuals	12	0.00				
Nematode assemblage	**Region**	1	7074	4.62	**	Car [S≠M≠R]	S [Cel=Car]
	Habitat	2	7267	1.43	ns	Cel [S=M=R]	M [Cel≠Car]
	Region × Habitat	2	5071	3.31	***		R [Cel=Car]
	Residuals	12	1532				

Results of the pairwise tests among levels of the relevant factors are also reported. df = degrees of freedom; MS = mean squares; Pseudo-F = F statistic; *** = $P<0.001$; ** = $P<0.01$; * = $P<0.05$; ns: not significant. Car = Caribbean Sea; Red = Red Sea; Cel = Celebes Sea; M = mangrove; R = reef; S = seagrass.

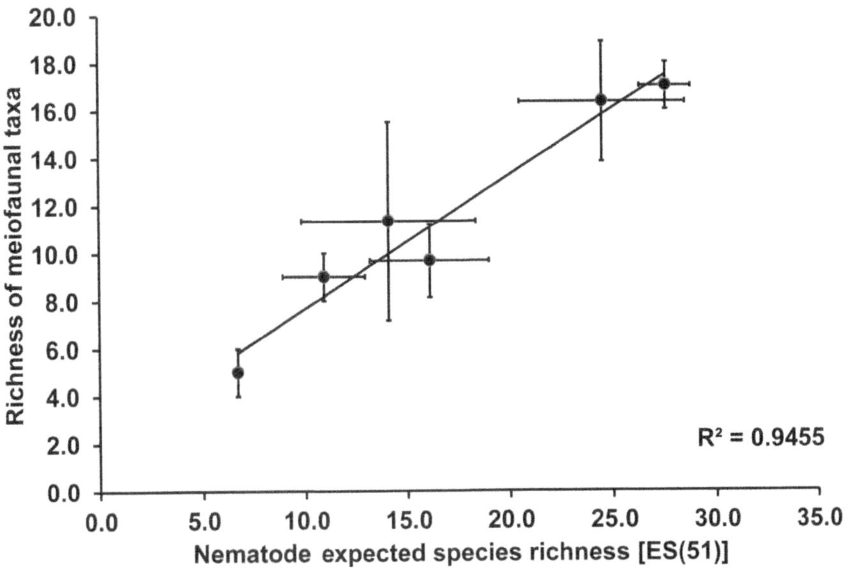

Figure 7. Relationship between richness of meiofaunal taxa and nematode biodiversity as ES(51) in tropical habitats from different oceanic regions. Illustrated are mean values ± standard deviation.

heterotrophic production values in seagrass and mangrove sediments were significantly higher than those in reef sediments, whereas in the Celebes Sea the mangrove sediments were characterized by values significantly lower than those in seagrass and reef sediments (Fig. 3). Overall, prokaryotic heterotrophic production in the sediments of the Celebes Sea were generally and significantly higher than those in the two other regions.

Meiofaunal Abundance, Biomass and Community Composition

Data on meiofaunal abundance, and richness of taxa, nematode diversity as species richness, ES(51), evenness (Pielou's J) and index of trophic diversity in all habitats and regions are reported in Table 3.

The results of the PERMANOVA on meiofaunal abundance revealed a significant effect of the interaction Habitat × Region (Table 4, Fig. 4A). In more detail, the pairwise tests revealed that, in both the Celebes and Red Seas, meiofaunal abundance was significantly higher in seagrass than in mangrove and reef sediments, whereas in the Caribbean Sea values did not vary significantly among habitats. The pairwise tests revealed also that the meiofaunal abundance in both mangrove and reef habitats was significantly higher in the Caribbean than in the Celebes and Red Seas. Meiofaunal abundance in seagrass sediments did not vary significantly across regions.

Meiofaunal biomass varied significantly among habitats (within the same region) and among regions (comparing the same habitat) (Table 4, Fig. 4B). Meiofaunal biomass was generally lower in mangrove sediments (range 79–170 μgC 10 cm^{-2}) than in the other two habitats (range 131–826 μgC 10 cm^{-2}). Significantly higher values of meiofaunal biomass were observed in the Caribbean followed by the Celebes and the Red Seas (Table 4).

The richness of meiofaunal taxa (Fig. 4C) ranged from 4 (in the mangrove sediments of the Red Sea) to 17 (in the reef sediments of the Celebes Sea) and varied significantly among habitats and among regions (Table 4). In more detail, the pairwise tests revealed that in all regions the richness of meiofaunal taxa in seagrass and reef sediments was consistently and significantly higher than in mangrove sediments. The richness of meiofaunal taxa in the Celebes Sea was significantly higher than that in the Caribbean and Red Seas (Table 4, Fig. 4C).

In all habitats and regions, with exception of reef sediments of the Celebes and Caribbean Seas, nematodes were the dominant taxon (52–97%), followed by copepods (3–25%) and polychaetes (0–13%) (Fig. 4D). Only reef sediments of the Caribbean and Celebes Seas were dominated by copepods (78 and 54% of total meiofaunal abundance, respectively). The contribution of all other identified taxa (acarians, amphipods, bivalves, cladocerans, cnidarians, cumaceans, gastrotrichs, gastropods, isopods, kinorinchs, decapods larvae, oligochaetes, ostracods, priapulians, tanaidaceans, tardigrades, termosbanaceans and turbellarians) varied from 0 to 25% of the total meiofaunal abundance (Fig. S1).

The PERMANOVA test showed that the composition of meiofaunal communities varied significantly only among habitats (Table 4). Moreover, the pairwise tests revealed that differences in the composition of meiofaunal communities were not consistent among habitats within each region (Table 4). Consequently, the results of the CAP analyses revealed that the differences among regions in the composition of meiofaunal assemblages were not well defined (Fig. 5).

In the Caribbean Sea, the highest meiofaunal beta diversity occurred between seagrass and mangrove sediments (60%). In the Red Sea, the highest meiofaunal beta diversity occurred between seagrass and reef sediments (44%), whereas in the Celebes Seas the highest beta diversity occurred between mangrove and reef sediments (29%) (Table S1). When contrasting the regions, the highest meiofaunal beta diversity among seagrass sediments is observed between the Red and the Celebes Sea (30%), whereas for mangrove sediments the highest beta diversity is observed between the Caribbean and Celebes Sea (61%). For reef sediments the highest meiofaunal beta diversity is observed between the Red and Celebes Seas (48%).

Nematode Biodiversity

The results of PERMANOVA conducted on nematode species richness and expected species number [ES(51)] revealed significant

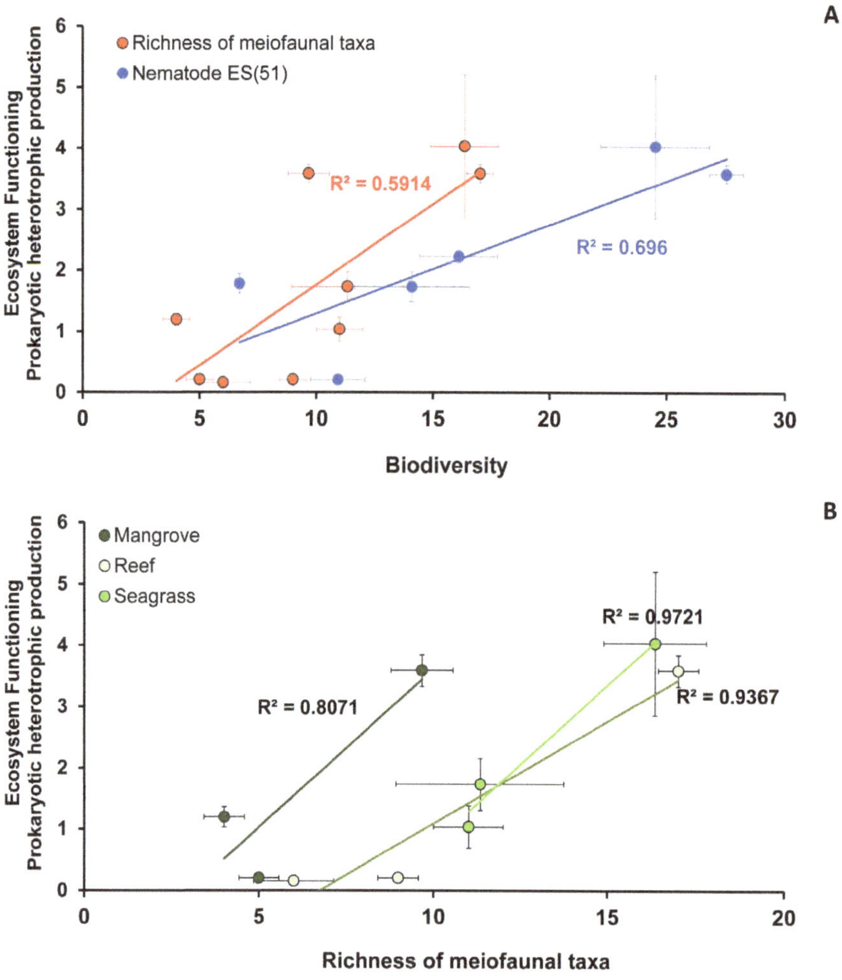

Figure 8. Relationship between biodiversity and ecosystem functioning. Illustrated are the relationships between: A) richness of meiofaunal taxa and prokaryotic heterotrophic production (μgC g^{-1} d^{-1}); nematode diversity as ES(51) and prokaryotic heterotrophic production (μgC g^{-1} d^{-1}); B) richness of meiofaunal taxa and prokaryotic heterotrophic production in different habitat types. Reported are R^2 values. $P<0.01$ for all linear regressions. Error bars indicate standard deviations among replicates (n = 3).

effects of the factors Region and Habitat, but no significant interaction effects (Table 5). Overall, in both the Celebes and Caribbean Seas the mangrove sediments displayed consistently and significantly lower values of nematode species richness, expected species number and evenness values (Table 5; Fig. 6). In all habitats, nematode expected species number and evenness

Table 6. Effects of biodiversity, measured as richness of meiofaunal taxa and nematode expected species number ES(51), on ecosystem functioning, as prokaryotic heterotrophic production before and after the removal of the covariables' effect: longitude, sediment biopolymeric organic C content and the protein to carbohydrate ratio.

Independent variable	R^2	SS	Pseudo-F	P
Richness of Meiofaunal Taxa	0.56	39.25	32.39	***
Covariates	0.63	43.52	12.82	***
Taxa Richness after removal of covariates	0.72	6.53	7.37	*
Nematode ES(51)	0.73	39.97	42.24	***
Covariates	0.77	42.46	15.67	***
Nematode ES(51) after the removal of covariates	0.88	5.78	10.94	***

In the regression analyses, all tests were based on Euclidean distances calculated among observations from untransformed data, using all data from different regions and habitats. The following abbreviations are used: R^2 = regression coefficient, SS = sum of squares, Pseudo-F = statistic F; P = probability level (*** = $P<0.001$; * = $P<0.05$).

were consistently higher in the Celebes than in the Caribbean Sea (with the only exception of evenness in seagrass sediments) (Table 5; Fig. 6A–B). Values of the index of trophic diversity did not vary among habitats in the Celebes Sea, whereas decreased from mangrove to seagrass and reef sediments in the Caribbean Sea. Values of the trophic diversity index varied significantly among regions only in mangrove sediments (Fig. 6C). In seagrass and reef sediments of the Caribbean Sea the selective (bacterial) and non-selective deposit feeders cumulatively represented more than 60% of the total nematode abundance, whereas in mangrove sediments dominated the epistrate feeders (more than 80% of the total abundance, Fig. 6D). In mangrove and reef sediments of the Celebes Sea the selective and non-selective deposit feeders cumulatively represented more than 50% of the total nematode abundance, whereas in seagrass sediments dominated the epistrate feeders (more than 60% of the total abundance). In both the Celebes and Caribbean Seas and in all habitats predator nematodes represented a minor fraction of the total nematode abundance (range 1–7%).

The richness of meiofaunal taxa was significantly and positively correlated with the number of nematode species as ES(51) (Fig. 7).

The results of PERMANOVA showed the presence of a significant effect of the interaction Habitat × Region on the composition of nematode assemblages (Table 5). The pairwise comparisons revealed that the composition of the nematode assemblages varied among regions only in mangrove sediments, whereas differences among habitats were consistently present only in the Caribbean Sea (Table 5). In the Celebes Sea, the nematode beta diversity among habitats was high and ranged from 64% (seagrass vs. reefs) to 79% (reefs vs. mangroves) (Table S1). In the Caribbean Sea the nematode beta diversity among habitats (74–93%) was slightly higher than that in the Celebes Sea (64–71%). Nematode beta diversity among regions was 68%, 77% and 83% in seagrass, mangrove and reef sediments, respectively. Overall the nematode beta diversity among the Caribbean and Celebes was 79%.

Relationship between Biodiversity and Ecosystem Functioning

The richness of meiofaunal taxa and the nematode expected species number [ES(51)] were significantly and positively related with prokaryotic heterotrophic production across all regions and in each habitat (Fig. 8A–B). The linear relationships between biodiversity and ecosystem functioning were not influenced by any of the tested covariates (i.e., longitude, biopolymeric C contents and the values of the protein to carbohydrate ratio) only when all data, irrespectively of region or habitat, were included in the analysis (Table 6).

Discussion

Meiofaunal and Nematode Biodiversity in Tropical Marine Sediments

Meiofaunal abundance and biomass values reported in this study fall within the range of those previously reported for the same regions (Caribbean Sea, Celebes Sea and Red Sea) or from other tropical and subtropical habitats [54–60]. Nonetheless, although our survey is limited to one replicate habitat per region, we report some differences in the structure and biodiversity of meiofaunal assemblages comparing both the same habitat from different regions and different habitats within the same region. For instance, in all investigated regions both seagrass and reef sediments generally host the highest number of meiofaunal taxa. These results suggest that, at both investigated spatial scales (i.e., oceanic region and habitat) the differences in meiofaunal diversity appear more evident than those in terms of abundance or biomass.

Although the use of correlations must be always considered with caution, the presence of a significant positive correlation between the number of higher meiofaunal taxa and the number of the nematode expected species number (Fig. 7) suggests that, at least for the investigated regions and habitats, the analysis of biodiversity based on nematodes could be a good proxy for the analysis of the patterns of the whole meiofaunal biodiversity. Such relationship, however, should be further explored on a much larger set of tropical sedimentary environments.

The analysis of structural nematode biodiversity (either expressed in terms of species richness, expected species number, or evenness) is significantly higher in the Coral triangle (Celebes, Indonesia) than in the Caribbean Sea, whatever the considered habitat is. Moreover, although a proper replication of habitats included in our survey would have shed more light on the actual differences among habitats in each region, we report here that in both the Celebes and Caribbean Seas, seagrass beds and reef sediments apparently host a significantly higher species number than mangrove sediments.

A comparative analysis of nematode diversity among different habitats, including reef sediments [60], temperate seagrass beds [61,62] and mangrove sediments in Australia, Africa, Asia and South America [59,63–65] is difficult as different estimates (indices) of diversity have been used in different studies. Moreover, such a comparison could be biased because of the different geographic locations and sampling efforts [66] as well as by the lack of temporal replication of the surveys. In presence of these potential biases, which altogether could weaken or reinforce the differences among habitats and regions, the results from our study can be nevertheless used at least in comparative terms, as, at all investigated areas, we adopted the same sampling strategy (and sampling effort), sample storage, methodology of extraction and determination/identification procedures.

The comparison of the biodiversity of the different habitats from each region suggests that each single habitat provided an important and significant contribution to the regional (i.e., gamma) nematode diversity only in the Caribbean Sea. In fact, the nematode beta diversity among seagrass, mangrove and reef sediments is >64% in both the Celebes and Caribbean Seas, but the post-hoc tests after the PERMANOVA revealed that the differences in the composition of nematode assemblages were significant only in the Caribbean Sea. This result would suggest that factors regulating nematode gamma diversity in the Caribbean and Celebes Sea are far different each other and leave open this issue to further investigations.

Soft bottoms are typically characterized by environmental variations that operate at the scale of a few centimeters on micro- and meiofaunal assemblages [33]. For instance, a recent study conducted from a single mangrove system from northwestern Brazil reported the existence of significant differences in nematode assemblage structure among micro-habitats [29]. Altogether our results and the evidences available from the literature pinpoint that tropical marine sediments are possibly characterized not only by high levels of nematode biodiversity, but also by high levels of beta diversity at different spatial scales, from the micro-scale to the regional one.

Relationships between Biodiversity and Ecosystem Functioning

Empirical and theoretical studies increasingly argue that biodiversity regulates the ecosystem functions that are responsible for the production of natural goods and services [2,67–70]. Many investigations relating biodiversity and ecosystem functioning have been performed using controlled field experiments that assemble model (and often non-natural) communities to measure the effects of changes in diversity on several ecosystem processes. Meta-analyses have also shown that species diversity generally has a positive effect on ecosystem processes and that this effect is remarkably consistent across trophic levels and different ecosystems [11–12,70].

In this study, the relationships between the richness of meiofaunal taxa and the nematodes species diversity vs. prokaryotic heterotrophic production are positive linear across all investigated regions and habitats (Fig. 8A–B). Our results are slightly different from those recently reported from coral reefs, where using fish as a model for biodiversity, the relationship appears to be positive exponential [71]. Nevertheless, this allows us to hypothesize that positive relationships could be a peculiar characteristic of tropical shallow habitats, but underpin also that the shape of the relationship (linear vs. exponential) could vary when different components of biodiversity (e.g., meiofauna, nematodes and fish) are taken into account. Differences in the shape of the relationships between biodiversity and ecosystem functioning from different marine environments also emerge when comparing habitats from different water depth. For instance, a positive exponential relationship has been observed in the deep Atlantic, Pacific and Antarctic Oceans and in the deep Mediterranean Sea (at depths ranging 200–8200 m) [18], whereas a recent study conducted along the upper slope off New Zealand (at depths ranging 264–1238 m) reported linear negative or null relationships [19]. Moreover, a recent manipulative experiment carried out on natural nematode assemblages response to thermal stress [72] led to hypothesize the same probability of (saturating) rivet-like [73] or idiosyncratic [74] relationships between nematode species richness and ecosystem functioning.

The large variability in the shape of the relationships between marine biodiversity and ecosystem functioning can be due to: i) the relatively limited number of regions and habitats considered so far, ii) the different approaches used (i.e., correlative from the real world and manipulative experiments in the laboratory), and iii) the different environmental characteristics of the investigated regions and habitats (e.g., coastal vs. deep-sea sediments). These discrepancies do not allow us making any robust speculation about the possible mechanisms explaining the different shapes in the relationship between biodiversity and ecosystem functioning observed in different marine ecosystems. Nevertheless, our results corroborate the presence of a pre-eminently positive effect of biodiversity on marine ecosystem functioning, and let us concluding that any loss in marine metazoan biodiversity (whatever the phylum considered) could result in a variably severe impairment of marine ecosystem functions.

Finally, it must be taken into account that the correlative approach used in this and other previous studies leaves open yet the conceptual possibility to interpret the reverse relationship, i.e. addressing whether and how ecosystem functions control biodiversity. This was not among our aims, but we must acknowledge that, in our study, the reverse relationship would remain positive linear and this would confirm that meiofaunal (and nematode) biodiversity are tightly dependent on the functioning of the microbial loop [26].

Acknowledgments

We acknowledge constructive criticism of two reviewers that helped us to significantly improve an early version of our manuscript.

Author Contributions

Conceived and designed the experiments: AP RD. Performed the experiments: AP RD CC. Analyzed the data: CG MS RD. Contributed reagents/materials/analysis tools: AP RD CC. Wrote the paper: AP CG CC MS RD. Species identification: CG MS. Statistical analysis: AP CG. Supported revision and intellectual contribution: AP CG RD.

References

1. Costanza R, d'Arge R, de Groot R, Farber S, Grasso M et al. (1997) The value of the world's ecosystem services and natural capital. Nature 387: 253–260.
2. Worm B, Barbier EB, Beaumont N, Duffy JE, Folke C, et al. (2006) Impacts of biodiversity loss on ocean ecosystem services. Science 314: 787–790.
3. Crain CM, Kroeker K, Halpern BS (2008) Interactive and cumulative effects of multiple human stressors in marine systems. Ecol Lett 11: 1304–1315.
4. Claudet J, Fraschetti S (2010) Human-driven impacts on marine habitats: A regional meta-analysis in the Mediterranean Sea. Biol Conserv 143: 2195–2206.
5. Duarte CM, Dennison WC, Orth RJW, Carruthers TJB (2008) The charisma of coastal ecosystems: addressing the imbalance. Estuar Coast 31: 233–238.
6. Wilkinson CR (2008) Status of the coral reefs of the world: 2008. Global Coral Reef Monitoring Network and Institute of Marine Science, Townsville, 8 pp.
7. Waycott M, Duarte CM, Carruthers TJB, Orth RJ, Dennison WC, et al. (2009) Accelerating loss of seagrasses across the globe threatens coastal ecosystems. Proc Natl Acad Sci USA 106: 12377–12381.
8. Valiela I, Bowen JL, York JK (2001) Mangrove forests: one of the world's most threatened major tropical environments. Bioscience 51: 807–815.
9. Beck MB, Brumbaugh RD, Airoldi L, Carranza A, Coen LD, et al. (2009) Shellfish reefs at risk: a global analysis of problems and solutions. The Nature Conservancy, Arlington, VA, 52 pp.
10. Loreau M (2010) Linking biodiversity and ecosystems: towards a unifying ecological theory. Proc Trans R Soc B 365: 49–60.
11. Cardinale BJ, Srivastava DS, Duffy JE, Wright JP, Downing AL, et al. (2006) Effects of biodiversity on the functioning of trophic groups and ecosystems. Nature 443: 989–992.
12. Loreau M (2008) Biodiversity and Ecosystem Functioning: The Mystery of the Deep Sea. Curr Biol 18: 126–128.
13. Bulling MT, Hicks N, Murray L, Paterson DM, Raffaelli D, et al. (2010) Marine biodiversity-ecosystem functions under uncertain environmental futures. Philos Trans R Soc Lond B Biol Sci. 365: 2107–2116.
14. Cardinale BJ, Duffy JE, Gonzalez A, Hooper DU, Perrings C, et al. (2012) Biodiversity loss and its impact on humanity. Nature 486: 59–67.
15. Solan M, Raffaelli DG, Paterson DM, White PCL, Pierce GJ (2006) Marine biodiversity and ecosystem function: empirical approaches and future research needs. Mar Ecol Prog Ser 311: 175–178.
16. Mora C, Frazier AG, Longman RJ, Dacks RS, Walton MM, et al. (2013) The projected timing of climate departure from recent variability. Nature 502: 183–187.
17. Giere O (2009) Meiobenthology: The Microscopic Motile Fauna of Aquatic Sediments. 2nd Edition, Springer, Berlin.
18. Danovaro R, Gambi C, Dell'Anno A, Corinaldesi C, Fraschetti S, et al. (2008) Exponential decline of deep-sea ecosystem functioning linked to benthic biodiversity loss. Curr Biol 18: 1–8.
19. Leduc D, Rowden AA, Pilditch CA, Maas EW, Probert PK (2013) Is there a link between deep-sea biodiversity and ecosystem function? Mar Ecol 34: 334–344.
20. Danovaro R, Gambi C, Höss S, Mirto S, Traunspurger W et al. (2009) Case studies using nematodes assemblage analysis in aquatic habitats. In: Wilson MJ, Kakouli-Duarte T (Eds) Nematodes as environmental indicators, CABI, Wallingford, UK, 146–171.

21. Watzin MC (1983) The effects of meiofauna on settling macrofauna: meiofauna may structure macrofaunal communities. Oecologia 59: 163–166.
22. Montagna PA (1984) In situ measurement of meiobenthic grazing rates on sediment bacteria and edaphic diatoms. Mar Ecol Progr Ser 18: 119–130.
23. De Morais LT, Bodiou JY (1984) Predation on meiofauna by juvenile fish in a western Mediterranean flatfish nursery ground. Mar Biol 82: 209–215.
24. Heip C, Vincx M, Vranken G (1985) The ecology of marine nematodes. Ocean Mar Biol Ann Rev 23: 399–489.
25. Danovaro R, Fraschetti S, Belgrano A, Curini-Galletti M, Vincx M, et al. (1995) The potential impact of meiofauna on the recruitment of macrobenthos in a subtidal coastal benthic community of the Ligurian Sea: a field result. In: Biology and Ecology of Shallow Coastal Waters, Eleftheriou A, Ansell AD, Smith CJ (eds), Olsen and Olsen, Fredensborg, Denmark, 115–122.
26. van Oevelen D, Middelburg JJ, Soetaert K, Moodley L (2006) The fate of bacterial carbon in an intertidal sediment: Modeling an in situ isotope tracer experiment. Limnol Oceanogr 51(3): 1302–1314.
27. Wilson MJ, Kakouli-Duarte T (2009) Nematodes as environmental indicators. CAB International, UK, 326 pp.
28. Barnes N, Bamber RN, Bennell G, Cornelius N, Glassom D, et al. (2011) Assessment of regional and local biodiversity in tropical and subtropical coastal habitats in the East African Marine Ecoregion. Biodivers Conserv 10: 2075–2109.
29. Pinto TK, Austen MCV, Warwick RM, Somerfield PJ, Esteves AM, et al. (2013) Nematode diversity in different microhabitats in a mangrove region. Mar Ecol 34(3): 257–268.
30. Vanreusel A, Fonseca G, Danovaro R, Da Silva MC, Esteves AM, et al. (2010) The contribution of deep-sea macrohabitat heterogeneity to global nematode diversity Mar Ecol 31(1): 6–20.
31. Gambi C, Pusceddu A, Benedetti-Cecchi L, Danovaro R (2014) Species richness, species turnover and functional diversity in nematodes of the deep Mediterranean Sea: searching for drivers at different spatial scales. Global Ecol Biogeogr 23: 24–39.
32. Pusceddu A, Danovaro R (2009) Exergy, ecosystem functioning and efficiency in a coastal lagoon: the role of auxiliary energy. Estuar Coast Shelf Sci 84: 227–236.
33. Danovaro R, Armeni M, Dell'Anno A, Fabiano M, Manini E, et al. (2001) Small-scale distribution of bacteria, enzymatic activities and organic matter in coastal sediments. Microb Ecol 42(2): 177–185.
34. Danovaro R (1996) Detritus-bacteria-meiofauna interactions in a seagrass bed (*Posidonia oceanica*) of the NW Mediterranean. Mar Biol 127: 1–13.
35. Nagelkerken I, Blaber SJM, Bouillon S, Green P, Haywood M, et al. (2008) The habitat function of mangroves for terrestrial and marine fauna: A review. Aquatic Bot 89: 155–185.
36. Max LM, Hamilton SL, Gaines SD, Warner RR (2013) Benthic processes and overlying fish assemblages drive the composition of benthic detritus on a central Pacific coral reef. Mar Ecol Progr Ser 482: 181–195.
37. Pusceddu A, Dell'Anno A, Fabiano M, Danovaro R (2009) Quantity and bioavailability of sediment organic matter as signatures of benthic trophic status. Mar Ecol Progr Ser 375: 41–52.
38. Lorenzen CJ, Jeffrey SW (1980) Determination of chlorophyll and phaeopigments spectrophotometric equations. Limnol Oceanogr 12: 343–346.
39. Danovaro R (2010) Methods for the study of deep-sea sediments, their functioning and biodiversity. CRC Press, Boca Raton, USA, 428 pp.
40. Stephens MP, Kadko DC, Smith CR, Latasa M (1997) Chlorophyll a and pheopigments as tracers of labile organic carbon at the central equatorial Pacific seafloor. Geochim Cosmochim Acta 61: 4605–4619.
41. Fry JC (1990) Determination of biomass. In: Methods in Aquatic Bacteriology (Austin B, Ed), John Wiley and Sons LTD, 27–72.
42. van Duyl FC, Kop AJ (1994) Bacterial variation in North Sea sediments: clues to seasonal and spatial variations. Mar Biol 120: 323–337.
43. Heip C, Vincx M, Vranken G (1985) The ecology of marine nematodes. Oceanogr Mar Biol Ann Rev 23: 399–489.
44. Feller RJ, Warwick RM (1988) Energetics. In Introduction to the study of meiofauna (Higgins RP, Thiel H Eds) Smithsonian Institution Press, London, 181–196.
45. De Mesel I, Lee HJ, Vanhove S, Vincx M, Vanreusel A (2006) Species diversity and distribution within the deep-sea nematode genus Acantholaimus on the continental shelf and slope in Antarctica. Polar Biol 29: 860–871.
46. Sanders HL (1968) Marine benthic diversity: a comparative study. Am Nat 102: 243–282.
47. Hurlbert SH (1971) The non-concept of species diversity: a critique and alternative parameters. Ecology 52: 577–586.
48. Gray JS (2000) The measurement of marine species diversity, with an application to the benthic fauna of the Norwegian continental shelf. J Exp Mar Biol Ecol 250: 23–49.
49. Soetaert K, Heip C (1990) Sample-size dependence of diversity indices and the determination of sufficient sample size in a high-diversity deep-sea environment. Mar Ecol Progr Ser 59: 305–307.
50. Pielou EC (1975) Ecological diversity, John Wiley, New York, 165 pp.
51. Anderson MJ, Gorley RN, Clarke KR (2008) PERMANOVA+ for PRIMER: Guide to Software and Statistical Methods. PRIMER-E, Plymouth, United Kingdom.
52. Wieser W (1953) Die Beziehung zwischen Mundhöhlengestalt, Ernährungsweise und Vorkommen bei freilebenden marinen Nematoden. Arkiv Zool 2–4: 439–484.
53. Anderson MJ (2001) A new method for non-parametric multivariate analysis of variance. Austral Ecol 26: 32–46.
54. De Troch M, Gurdebeke S, Fiers F, Vincx M (2001) Zonation and structuring factors of meiofauna communities in a tropical seagrass bed (Gazi Bay, Kenya). J Sea Res 45: 45–61.
55. Netto SA, Gallucci F (2003) Meiofauna and macrofauna communities in a mangrove from the Island of Santa Catarina, South Brazil. Hydrobiologia 505: 159–170.
56. Armenteros M, Martín I, Williams JP, Creagh B, González-Sansón G, et al. (2006) Spatial and temporal variations of meiofaunal communities from the western sector of the Gulf of Batabanó, Cuba. I. Mangrove systems. Estuar Coast 29: 124–132.
57. Chinnadurai G, Fernando OJ (2007) Meiofauna of mangroves of the southeast coast of India with special reference to the free-living marine nematode assemblage. Estuar, Coast Shelf Sci 72: 329–336.
58. Xuan QN, Vanreusel A, Vu Thanh N, Smol N (2007) Biodiversity of Meiofauna in the Intertidal Khe Nhan Mudflat, Can Gio Mangrove Forest, Vietnam with Special Emphasis on Free Living Nematodes. Ocean Sci J 42: 135–152.
59. Barnes N, Bamber RN, Bennell G, Cornelius N, Glassom D, et al. (2011) Assessment of regional and local biodiversity in tropical and subtropical coastal habitats in the East African Marine Ecoregion. Biodivers Conserv 10: 2075–2109.
60. Alongi DM (1988) Microbial-meiofaunal interrelationships in some tropical intertidal sediments. J Mar Res 46: 349–365.
61. Danovaro R, Gambi C (2002) Biodiversity and trophic structure of nematode assemblages in seagrass systems: evidence for a coupling with changes in food availability. Mar Biol 141: 667–677.
62. Danovaro R., Gambi C., Mirto S. (2002). Meiofaunal production and energy transfer efficiency in a seagrass *Posidonia oceanica* bed in the Western Mediterranean. Mar Ecol Progr Ser 234: 95–104.
63. Alongi DM (1987) Inter-estuary variation and intertidal zonation of free-living nematode communities in tropical mangrove systems. Mar Ecol Prog Ser 40: 103–114.
64. Ólafsson E (1995) Meiobenthos in mangrove areas in eastern Africa with emphasis on assemblages structure of free-living marine nematodes. Hydrobiologia 312: 47–57.
65. Somerfield PJ, Gee JM, Aryuthaka C (1998) Meiofaunal communities in a Malaysian mangrove forest. J Mar Biol Ass UK 78: 717–732.
66. Benedetti-Cecchi L, Maggi E (2012) The analysis of biodiversity-ecosystem function experiments: portioning richness and density-dependent effects. In (eds: Solan M, Aspden RJ, Paterson DM) Marine biodiversity and ecosystem functioning: frameworks, methodologies and integration, 73–195.
67. Naeem S, Thompson LJ, Lawler SP, Lawton JH, Woodfin RM (1994) Declining biodiversity can alter the performance of ecosystems. Nature 368: 734–736.
68. Tilman D, Reich PB, Knops J, Wedin D, Mielke T, et al. (2001) Diversity and productivity in a long-term grassland experiment. Science 294: 843–845.
69. Heemsbergen DA, Berg MP, Loreau M, van Hal JR, Faber JH, et al. (2004) Biodiversity effects on soil processes explained by interspecific functional dissimilarity. Science 306: 1019–1020.
70. Balvanera P, Pfisterer AB, Buchmann N, He JS, Nakashizuka T, et al. (2006) Quantifying the evidence for biodiversity effects on ecosystem functioning and services. Ecol Lett 9: 1146–1156.
71. Mora C, Aburto-Oropeza O, Bocos AA, Ayotte PM, Banks S, et al. (2011) Global Human Footprint on the Linkage between Biodiversity and Ecosystem Functioning in Reef Fishes. PloS Biology 9(4): e1000606.
72. Gingold R, Moens T, Rocha-Olivares A (2013) Assessing the response of nematode communities to climate change-driven warming: a microcosm experiment. PLoS ONE 8: e66653.
73. Ehrlich PR, Ehrlich A (1981) Extinction: The Causes and Consequences of the Disappearance of Species. Random House, New York. 305 pp.
74. Lawton JH (1994) What do species do in ecosystems? Oikos 71: 367–374.

Global Patterns in Ecological Indicators of Marine Food Webs: A Modelling Approach

Johanna Jacomina Heymans[1]*, Marta Coll[2,3¤], Simone Libralato[4], Lyne Morissette[5], Villy Christensen[6]

1 Scottish Association for Marine Science, Scottish Marine Institute, Oban, Argyll, United Kingdom, 2 Institute of Marine Science (ICM-CSIC), Passeig Marítim de la Barceloneta, Barcelona, Spain, 3 Ecopath International Initiative Research Association, Barcelona, Spain, 4 Department of Oceanography, OGS (Istituto Nazionale di Oceanografia e di Geofisica Sperimentale), Borgo Grotta Gigante, Sgonico, Zgonik (TS), Italy, 5 M – Expertise Marine, Sainte-Luce, Quebec, Canada, 6 Fisheries Centre, University of British Columbia, Vancouver, Canada

Abstract

Background: Ecological attributes estimated from food web models have the potential to be indicators of good environmental status given their capabilities to describe redundancy, food web changes, and sensitivity to fishing. They can be used as a baseline to show how they might be modified in the future with human impacts such as climate change, acidification, eutrophication, or overfishing.

Methodology: In this study ecological network analysis indicators of 105 marine food web models were tested for variation with traits such as ecosystem type, latitude, ocean basin, depth, size, time period, and exploitation state, whilst also considering structural properties of the models such as number of linkages, number of living functional groups or total number of functional groups as covariate factors.

Principal findings: Eight indicators were robust to model construction: relative ascendency; relative overhead; redundancy; total systems throughput (TST); primary production/TST; consumption/TST; export/TST; and total biomass of the community. Large-scale differences were seen in the ecosystems of the Atlantic and Pacific Oceans, with the Western Atlantic being more complex with an increased ability to mitigate impacts, while the Eastern Atlantic showed lower internal complexity. In addition, the Eastern Pacific was less organised than the Eastern Atlantic although both of these systems had increased primary production as eastern boundary current systems. Differences by ecosystem type highlighted coral reefs as having the largest energy flow and total biomass per unit of surface, while lagoons, estuaries, and bays had lower transfer efficiencies and higher recycling. These differences prevailed over time, although some traits changed with fishing intensity. Keystone groups were mainly higher trophic level species with mostly top-down effects, while structural/dominant groups were mainly lower trophic level groups (benthic primary producers such as seagrass and macroalgae, and invertebrates). Keystone groups were prevalent in estuarine or small/shallow systems, and in systems with reduced fishing pressure. Changes to the abundance of key functional groups might have significant implications for the functioning of ecosystems and should be avoided through management.

Conclusion/significance: Our results provide additional understanding of patterns of structural and functional indicators in different ecosystems. Ecosystem traits such as type, size, depth, and location need to be accounted for when setting reference levels as these affect absolute values of ecological indicators. Therefore, establishing absolute reference values for ecosystem indicators may not be suitable to the ecosystem-based, precautionary approach. Reference levels for ecosystem indicators should be developed for individual ecosystems or ecosystems with the same typologies (similar location, ecosystem type, etc.) and not benchmarked against all other ecosystems.

Editor: John F. Valentine, Dauphin Island Sea Lab, United States of America

Funding: JH was funded by Knowseas (EUFP7) and Global Environmental Facility (GEF). MC was funded by the European Commission through Marie Curie CIG grant to BIOWEB and the Spanish Research Program "Ramon y Cajal". SL has been partially funded by PERSEUS (EUFP7) and the Flagship Project RITMARE - The Italian Research for the Sea - coordinated by the Italian National Research Council and funded by the Italian Ministry of Education, University and Research within the National Research Program 2011-2013. The funders had no role in study design, data collection and analysis, decision to publish, or preparation of the manuscript.

Competing Interests: Lyne Morissette employed by M-Expertise Marine. This does not alter the authors' adherence to PLOS ONE policies on sharing data and materials.

* E-mail: sheilaheymans@yahoo.com

¤ Current address: UMR 212 Ecosystèmes Marins Exploités (IRD/IFREMER/UM2), BP 171, Sète, France

Introduction

Natural resource management approaches have been under development for decades, driven by an increasing need to understand the effect of anthropogenic impacts on ecosystems [1,2]. Often, it was assumed that management could be based on population dynamics at the individual species population level [3]. However, it is now clear that there are trade-offs in management

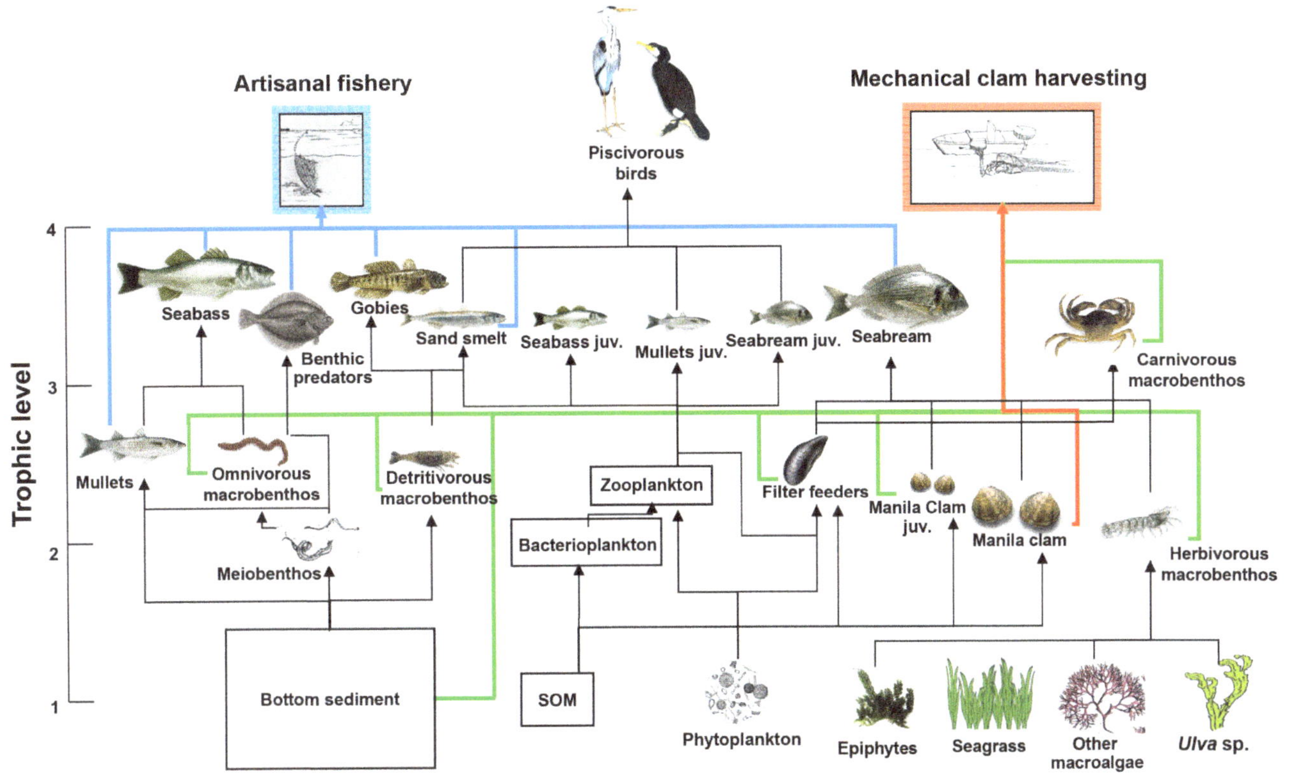

Figure 1. Food web diagram of the Venice lagoon with 27 nodes or funtional groups. Colors of flows depict different fishing target (artisanal fisheries in blue, and clam fishery in red) and non-target species (for clam harvesting, in green). Modified from Pranovi et al. [102].

[4–8] and that community effects therefore must be considered. An example of these trade-offs is food web effects due to competition or predation [2,9,10]. Ecosystems also have emergent properties that call for consideration of ecosystem structure and function in the management of marine resources [11–13].

Although detailed expert knowledge is essential for the management of an ecosystem, a general theory of ecosystem dynamics can help in defining aspects to be considered when conducting ecosystem based management. Such a theory would allow for extrapolation between systems, important given the lack of detailed information available on some systems. Food webs describe the interaction between species at different feeding levels and depict the flow of energy and matter in ecosystems. These predator-prey interactions are considered one of the main regulators of ecosystem dynamics [14,15], and they partially mediate the way ecosystems respond to natural and human perturbations such as fishing, habitat degradation or environmental dynamics. Food web models are simplified representations of natural systems, which help us understand how biodiversity and ecosystems respond to changes. Creating food web models typically calls for quantitative modeling integrating food web dynamics and external factors such as environmental change or fishing.

The study of marine food webs and ecosystems faces difficulties with data collection and quantification of interactions, and the added difficulty of modeling ecosystem processes and dynamics [12]. Therefore, structural and functional properties of marine ecosystems are less known than their terrestrial and freshwater counterparts [12,16]. Moreover, the quantification of many food web properties depends upon the modeling strategy and model structure as they co-vary with model components and links [12,17,18]. However, this lack of knowledge is changing. Ecological modelling applications have increased exponentially and a large body of standardized food web models have been constructed over the last three decades to quantitatively describe marine systems. The quantitative description of food web attributes is essential to advance our understanding of (i) ecosystem structure and functioning; and (ii) how to use ecological indicators to inform policy makers and managers [19–21].

The most widely applied ecosystem modelling approach is Ecopath with Ecosim (EwE, www.ecopath.org) [22–25], developed by Polovina [25] and adapted by Christensen and Pauly [26] and Walters [23,27] into a comprehensive modelling tool [28]. EwE is currently composed of a core mass balance food web module (Ecopath), from which temporal (Ecosim) and spatial (Ecospace) dynamic simulations can be developed [24,29]. EwE models represent complex food webs with non-linear and non-randomly distributed interactions, where each node or functional group of the web may be a species, a group of species, an ontogenetic stage of a species, or a detritus group [24]. Ecopath models are different from cascade models [30] and niche models [31], which are very useful to describe several food web structural properties. Niche and cascade models have been compared to results with EwE [32], but they do not account for the strength of ecological interactions.

The functional groups of EwE models are characterised by specific features such as abundance, biomass, and production, which provide means of measuring biological diversity. The nodes are linked by the strength of trophic (feeding) interactions, while Ecosim can represent non-feeding interactions such as mutualism or parasitism. Therefore, EwE models represent complex ecological networks with a series of properties that characterise food webs and are important for describing ecosystem structure and functioning and for evaluating conservation needs (Fig. 1 shows an illustrated example of a marine food web). Since food web models typically include all trophic levels from the lowest (i.e., primary producers and detritus) to highest (e.g., humans and apex predators), they are able to capture both bottom-up and top-down forcing dynamics.

The structural and functional properties of food webs can be quantified using ecological network analysis [22]. Ecological network analysis stems from the study in 1942 by Lindeman [33]. Trophic interactions and linkages were further conceptualized in food webs of energy transfers by EP Odum [34]. In 1980, Ulanowicz [35] developed indicators of ecosystem development such as ascendency [36].

The currently available EwE food web models provide relevant information to progress the study of a general theory of marine food web traits and dynamics. Taking advantage of the significant number of published marine food web models found in the literature (Table S1), this study investigates whether there are general patterns in ecological traits of marine food webs. To take into account the fact that models vary due to different development strategies, structural properties of the models (such as the number of total, living and non-living, versus only living functional groups, and total number of trophic links, called factors in this study) are used to test the robustness of these patterns by means of covariance analysis.

This paper analyses 105 published EwE models distributed worldwide (Fig. 2) and their emergent ecological network analysis properties to characterise structural and functioning indicators of marine ecosystems. It also includes 26 indicators of ecosystem structure and function as defined in Table 1. The models represent a wide spatial diversity, covering ecosystems from coastal lagoons to the deep sea in all the world's oceans, and large temporal diversity, with ecosystems representing both past and recent timeframes (Fig. 2 & Table S1). To analyse the variation of general patterns in food web indicators, 7 traits were included in the analysis: (i) ecosystem type (coastal, shelf, slope, estuary, bay, lagoon or reef); (ii) latitude; (iii) ocean basin; (iv) depth; (v) size; (vi) period of time represented; and (vii) exploitation rate.

This study is among the first to analyse a large variety of EwE models from different systems in an organised and systematic way. It presents the largest meta-analysis of the structural and functional indicators of marine food webs to date and adds to the general theory of marine food web dynamics and its use for ecosystem conservation and management. It also includes statistical analyses to address co-variance of models and issues of structural uncertainty in these models. The statistical analysis makes this work unique, and the study also includes new and advanced analyses on the key species concept.

Materials and Methods

a) Food web models

A description of the theory and algorithms behind Ecopath with Ecosim (EwE) is given in File S1. The locations of the ecosystems represented by the 105 published EwE models of marine ecosystems from around the world used in this study are shown in Fig. 2. The meta-data (i.e., ecological, network, and synthetic indicators) used to describe the ecosystems are available in Table S1. The models ranged from very simple, with 6 component nodes or functional groups in the Canary Island [Canary Islands 37] to more complex models consisting of 68 groups in the North Sea [North Sea 38], with an average number of 26 groups. In terms of depth, the systems represented ranged from less than a meter deep in Venice Lagoon [Venice Lagoon 39], to the deep sea off the Cape Verde Islands [40], and the West Coast of Scotland [41]. The timeline of the models spanned from 1880 (North Sea) to various models that include recent data from the 2000s. The models represent ecosystems from all continents (except Antarctica), with 17 African systems, 14 Australasian systems, 35 European systems, 31 North American, and 8 South American systems. Mexico (16) has the most models, followed by Italy (8), Canada (8), and the USA (7). The latitude and location of the model areas and references to the papers describing these models are given in Table S1.

Many of the models were previously developed to achieve a first description of the ecosystems using the mass balance approach (42 models), while some had estimates of the ecological network analysis (ENA) parameters (36 models) or were used to compare the Ecopath approach to inverse methods (5 models) (Table S1). In addition, several models had been developed for theoretical dynamic analysis, spatial analysis, or policy analysis (12 models) while 17 models were fitted to time series to hindcast model dynamics, and 6 were used for policy and fisheries impacts. All models used in this analysis were previously peer-reviewed and published in the primary scientific literature (Table S1).

b) Ecological food web- and fishing impact indicators

The food web models representing different marine systems (Table S1) were used to analyse ecosystem structure and function patterns by trait. We calculated several ecological indicators from ecological network analysis (ENA) and performed a meta-analysis across ecosystems applying different statistical multivariate approaches. The indicators were chosen because they had previously been widely applied to highlight ecosystem structure and functioning, were thought to be robust to differences in model construction (such as the number of functional groups) and were independent from currency used for biomass and flows (Table 1) [42,43]. In addition, the robustness of these indicators to model construction was further checked by Factorial Analysis (see section below).

The network analysis indicators are mainly related to the total systems throughput (TST), which is the sum of all flows in the model and considered an overall measure of the "ecological size" of the system [44]:

$$TST = \sum_{i=1,j=1}^{n} Tij \tag{1}$$

where T_{ij} is the flow between any two compartments and includes respiration and export flows.

The more descriptive indicators include the primary production/total systems throughput (PP/TST), flow to detritus/TST (FD/TST), total consumption/TST (Q/TST), total respiration/TST (R/TST), total exports/TST (Ex/TST), total biomass excluding detritus (TBco), and the ratio between primary production and total production (PP/P).

The development capacity (C) of the system is the thermodynamic limit of growth in the system and is calculated as:

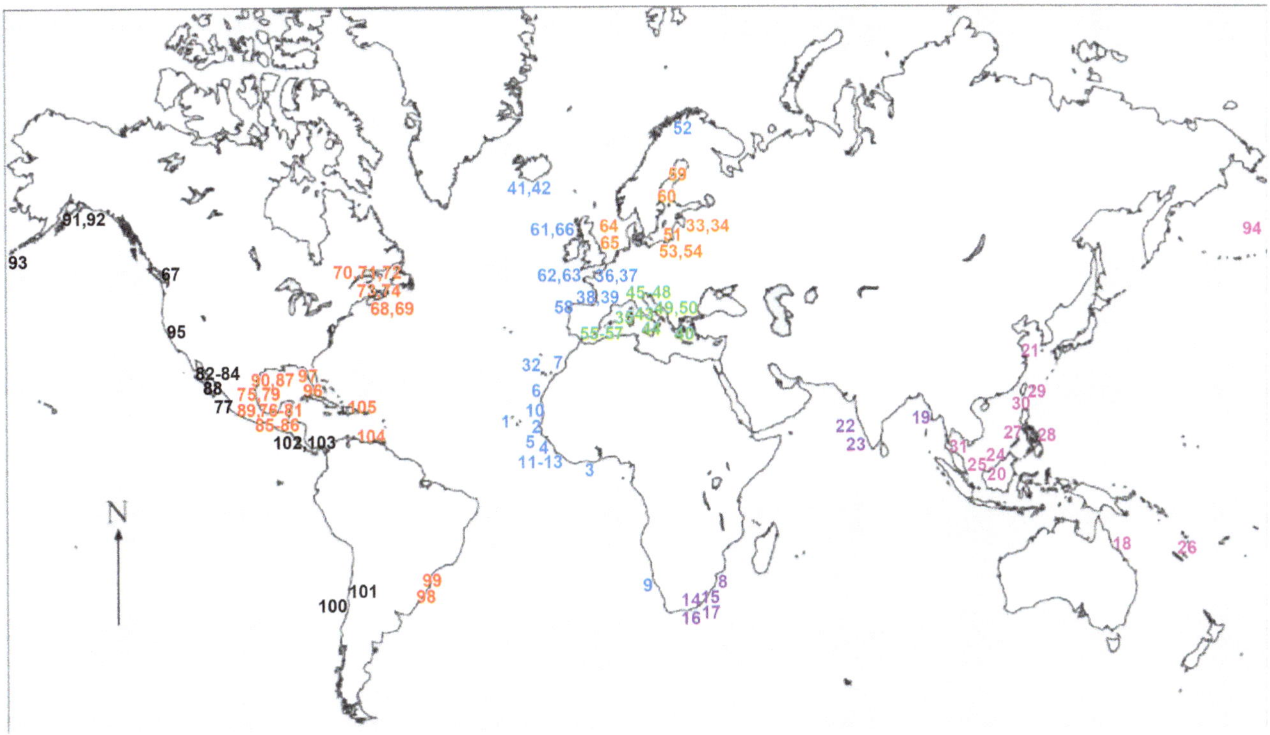

Figure 2. Distribution of food web model used in this study. Models are divided into Eastern Pacific (black), Western Atlantic (red), Eastern Atlantic (blue), North and Baltic Seas (orange), Mediterranean Sea (green), Indian Ocean (purple) and Eastern Pacific (pink). Each model is numbered on the graph according to its number in Table S1, where details and references to each model area are given.

$$C = -TST \cdot \sum\nolimits_{ij} \frac{Tij}{TST} \cdot \log\left(\frac{Tij}{TST}\right) \quad (2)$$

It scales the TST to a measure of the information carried by flows. Capacity is divided between ascendency (A) and the overhead (O). Ascendency is an index of the organisation of a food web [45] and is defined in terms of flow:

$$A = \sum_{ij} (T_{ij}) \cdot \log\left(\frac{Tij \cdot TST}{T_j \cdot T_i}\right) \quad (3)$$

and overhead (O, an indicator of the ecosystem's strength in reserve [45]) is calculated as:

$$O = C - A \quad (4)$$

Overhead and Ascendency are divided into export, dissipation and internal flows [46], and the overhead on internal flows (IFO) has been used as an index of ecosystem redundancy [47]. The redundancy is an indicator of the change in degrees of freedom of the system, or the distribution of energy flow among the pathways in the ecosystem [48], and is calculated as:

$$R = IFO = -\sum_{i=1}^{n} \sum_{j=1}^{n} (Tij) \cdot \log\left(\frac{Tij^2}{\sum_{j=1}^{n} Tij \cdot \sum_{i=1}^{n} Tij}\right) \quad (5)$$

The Finn Cycling Index (FCI) quantifies the amount of recycling relative to TST and is an indication of stress and structural differences [44], and is calculated as:

$$FCI = \frac{TSTc}{TST} \quad (6)$$

where TST_c is the total flow that is recycled.

Other ecological indicators are related to the trophic level (TL) concept, which is the average number of steps for energy to move from primary producers to higher-level consumers and provides an indication of the trophic position of an organism. Thus for a given predator j the TL is calculated as:

$$TL_j = 1 + \sum_i DC_{ij} \cdot TL_i \quad (7)$$

where TL_i is the trophic level of prey i and DC_{ij} is the proportion of prey i in the diet of predator j. The mean trophic level of the community (mTLco) is calculated as the weighted average TL for

Table 1. Ecological and fisheries related indicators used in this comparison.

Acronym	Indicators	Units	Definition	Reference
Ecological indicators				
TST	Total System Throughput	$t \cdot km^{-2} \cdot y^{-1}$	The sum of all the flows through the ecosystem	[45]
PP/TST	Primary production/TST		Primary production over the sum of all the flows through the ecosystem	[28]
FD/TST	Flows to Detritus/TST		Flows to detritus over the sum of all the flows through the ecosystem	[28]
Q/TST	Total consumption/TST		Total consumption over the sum of all the flows through the ecosystem	[28]
R/TST	Total respiration/TST		Total respiration over the sum of all the flows through the ecosystem	[28]
Ex/TST	Total exports/TST		Total exports of the system over the sum of all the flows through the ecosystem	[28]
PP/P	PP/Total Production		Primary production over total production	[28]
MeanPz (MaxPz)	Mean (Max) proportion of total mortality due to predation		The mean (or Maximum) proportion of each group's total mortality that was accounted for by each predator	[103]
meanEE	Mean Ecotrophic Efficiency	%	Ecotrophic efficiency of a group is that proportion of the production that is utilized in the system.	[28]
TBco	Total Biomass (excluding first trophic level)	$t \cdot km^{-2}$	Total biomass of the community excluding detritus and primary producers	[28]
mTLco	Mean Trophic Level of the Community		Weighted average trophic level for functional groups with a TL>2	[28]
TEm	Mean Transfer Efficiency	%	Geometric mean of transfer efficiencies for trophic level II to IV	[33]
A/C	Ascendency/Capacity	%	Relative Ascendency, dimensionless index of ascendency - index of organisation of the food web	[45]
O/C	Overhead/Capacity	%	Relative overhead, dimensionless index of the ecosystem's strength in reserve	[45]
IFO	Internal Flow Overhead or redundancy	%	Indicator of the change in degrees of freedom of the system, or an indicator of the distribution of energy flow pathways in the system	[45,47,48]
FCI	Finn's Cycling Index	%	Quantifies the relative amount of recycling and is an indication of stress and structural differences either among models [44]	[44]
SOI	System Omnivory Index		Variance of trophic levels in the diet	[24,49]
KS	Keystoneness		Index of the ability of a trophic group with low biomass to influence others	[52]
KD	Dominance		Index of both high biomass and high influence	[52]
Fishing indicators				
TC	Total Catches	$t \cdot km^{-2} \cdot y^{-1}$	Total landings and discards exported from the system	[28]
TLc	Mean Trophic Level of the Catch**		Average trophic level of all caught species using weighted by yield	[50]
L_{index}	Loss in production Index		Loss in secondary production due to fishing	[28]
P_{sust}	Probability of being sustainable fished	%	Probability of the system to be sustainably fished adopting [55] ecosystem overfishing definition and criteria	[51,53,54]

*Excluding the cases where the indicator was 0 due to no fishing.

functional groups with TL>2, as:

$$mTLco = \frac{\sum_i TL_i \cdot B_i}{\sum_i B_i} \quad (8)$$

where the weighting factor is the biomass of each functional group (Bi). TL>2 is used to reduce the variability in terms of biomass and production that comes with using low trophic levels.

The system's omnivory index (SOI) is defined on the basis of the omnivory index (OI) of each food web component. It indicates the variance of trophic levels in the diet, and is:

$$OI_i = \sum_{j=1}^{n} (TL_j - (TL_i - 1))^2 \cdot DC_{ij} \quad (9)$$

So from the OI of each functional group, the SOI for the food web is calculated as:

$$SOI = \frac{\sum_{i=1}^{n} [OI_i \cdot \log(Q_i)]}{\sum_{i=1}^{n} \log(Q_i)} \quad (10)$$

where Q_i is the food intake of each consumer [24,49].

The mean transfer efficiency (TEm) for the food web was calculated as the geometric mean of transfer efficiencies for each of the integer trophic levels II to IV. The transfer efficiency of a trophic level is calculated as the sum of the flow transferred from any given level to the next higher level, plus exports (e.g., catches) from the given level relative to the input (or throughput) of the given trophic level.

We also included indicators measuring fishing intensity and impacts in the ecosystems: the mean trophic level of the catch (TLc) [TLc, 50], the primary production required to sustain the catches [PPRc, 28], the loss in production index (L_{index}) [L index, 51] and the probability of an ecosystem being sustainably fished P_{sust} [Psust, 52].

Similar to the mTLco, the mean trophic level of the catches (TLc) is calculated as the weighted average of TL of caught species using catches (Y_i) as the weighting factors:

$$TLc = \frac{\sum_i TL_i \cdot Y_i}{\sum_i Y_i} \quad (11)$$

The proportion of primary production required for the exploited fishery's catch (PPRc) is defined as:

$$PPRc = \sum_{Paths} \left[Y \cdot \prod_{Pred, Prey} \frac{Q_{pred}}{P_{pred}} \cdot DC'_{Pred,prey} \right] \quad (12)$$

where P is production, Q consumption, and DC' is the diet composition for each predator/prey interaction in each path from primary production or detritus through the food web to the catch, with cycles removed from the diet compositions [28]. PPRc can be expressed in percentage terms when it is normalized with the primary production (PPRc/PP = PPRc%).

The Loss in Production Index (L_{index}) and the probability for the ecosystem to be sustainably fished (p_{sust}) are both used to evaluate the ecosystem effects of fishing [51,53,54]. The probability of being sustainably fished was defined by adopting Murawski's [55] ecosystem overfishing definition and criteria, and it includes structural and functional degradation associated with stock collapses and overexploitation of marine resources. L_{index} is defined as:

$$L_{Index} = \frac{PPRc \cdot TEm^{Tlc-1}}{PP \cdot \ln TEm} \quad (13)$$

where the loss in production is estimated on the basis of TLc, TEm and PPRc, that is compared to the primary production at the base of the food web (PP). The probability that such energy loss is sustainable was calculated by comparing L_{index} for a set of overexploited and sustainably exploited ecosystems as reported in Libralato et al. [51]. This allows the definition of a non-linear empirical relationship between the L_{index} and P_{sust} to be used for calculating sustainability of fisheries.

c) Factorial analysis

To analyse marine food webs by the ecological and fishing indicators described above, 7 traits (factors) were chosen: ecosystem type, latitude, ocean basin, depth, size, time period, and exploitation of the ecosystem represented by the model.

Size of ecosystem included six classes (1–10 km^2, 11–100 km^2, 101–1,000 km^2, 1,001–10,000 km^2, 10,001–100,000 km^2, > 100,000 km^2), depth included seven classes (<5 m, <10 m, < 20 m, <50 m, <100 m, <200 m, >200 m), and ecosystem type was divided into six classes (lagoon, estuary, bay, coastal, reef, continental shelf, and upper slope). The classes were developed following from previous analyses [56,57] and were included where there were enough models per class (n>5). The modelled food webs were divided into four latitude classes from the equator to the poles (15°S–15°N and 15–30°, 30–60°, 60–90°, north and south combined). The food webs were also divided from west to east into seven ocean basin classes following a longitudinal gradient (West Atlantic, East Atlantic, North and Baltic Sea, Mediterranean, Indian, West Pacific, East Pacific). The time periods represented by the models were divided into four classes: before 1970, 1970–1980, 1980–1990 and after 1990. With regards to exploitation, the ecosystems were split into three classes: high exploitation, low exploitation (including food webs with recreational fishing only) and no exploitation.

Significant differences between ecosystems were assessed first by comparing all ecological indicators and then comparing individual indicators with each trait and using the non-parametric multivariate permutational analysis of variance (PERMANOVA, in PRIMER with PERMANOVA+ v. 6, PRIMER-E Ltd., Plymouth, UK) on the Euclidean distance matrix. PERMANOVA calculates a pseudo-F statistic that is analogous to the construction of the traditional F-statistic for multifactorial univariate ANOVA models, but uses permutation procedures (here 9999 permutations) to obtain p-values for each term in the ANOVA model [58]. Due to the limited number of observations, and an unbalanced design among traits, we performed a 1-way analysis by trait using first all the indicators together and then each indicator separately. Even if we used indicators less affected by model construction [42,43], the number of functional groups and aggregation used to represent food webs can still influence model outputs and analyses. Therefore we assessed the robustness of indicators to model configuration by including 3 factors as covariates in the PERMANOVA analysis: (i) the number of functional groups or nodes of each food web model; (ii) the number of living groups; and (iii) the total number of trophic links. We used a Type I (or sequential) partition of the sum of squares (SS) since individual SS terms were not independent when including covariates.

When significant, the variation of the different ecological food web indicators and fishing indicators by trait (ecosystem type, latitude, ocean basin, depth, size, time period, and exploitation was plotted using boxplots.

d) Key functional groups

Key functional groups are those with important roles in the food web, and include keystone and structuring functional groups [59]. Keystone groups have relatively low biomass but disproportionately large effects on the food web [52,59], while structuring groups have large effects due to their relatively high biomass [60].

The marine food web models were used to calculate the absolute overall effect of a species on the food web as the sum of all the direct and indirect effects, quantified through the mixed trophic impact analysis (MTI). The MTI analysis quantifies the direct and indirect impacts that each (impacting) group has on any (impacted) group of the food web [61]. The absolute overall effect was then compared with the biomass proportion of each group to identify key species: either keystone (low biomass proportion and high overall effect) or key structuring groups (high biomass proportion and high overall effect [59]). The role of functional groups was assessed through the key role (keystone or structuring functional groups), trophic level (TL) and proportion of top down effects (td; where bottom-up effect is calculated as bu = 100 – td).

On the basis of the work done by Libralato et al. [52], the overall effect of each functional group i on the ecosystem was estimated through:

$$\varepsilon_i = \sqrt{\sum_{j \neq i}^{n} m_{ij}^2} \tag{14}$$

in which the effect of the change in biomass on the group itself (i.e., m_{ii}) is not included. Moreover, accounting for the fraction of positive and negative contributions to the overall effect allows evidence for contribution of bottom-up (positive m_{ij}) and top-down (negative m_{ij}) effects [52]. This was synthesized here by reporting the fraction of top-down contributions (td, as percentage) to the total effect.

The overall effect of each functional group on the ecosystem combined with information on the group's density is useful to identify its role in the ecosystem and to identify key functional groups [59,60]. In particular, groups with a high impact might be identified as keystone or dominant groups if they have a low or high biomass in the ecosystem, respectively [59]. The biomass was found by calculating the contribution of each functional group to the total biomass of the food web, as:

$$p_i = \frac{B_i}{\sum_k B_k} \tag{15}$$

where p_i is the biomass proportion for group i; see Libralato et al. [52].

Using the definition given by Libralato et al. [52], the index of keystoneness was calculated as follows:

$$KS_i = \log[\varepsilon_i \cdot (1 - p_i)] \tag{16}$$

that has the property of attributing high values of keystoneness to functional groups that have both low biomass proportion and high overall effect (as defined above).

The absolute overall effect and the biomass proportion of the functional group were used to identify key functional groups (i.e. species or groups of species with particular roles in the food web), complementing the keystone indicator as proposed by Libralato et al. [52].

The analysis of biomass proportion and overall effect helped to distinguish those key groups that might be difficult to disentangle in terms of keystoneness index only. Therefore, similarly to the keystoneness, an index of dominance of species was calculated from:

$$KD_i = \log[\varepsilon_i \cdot p_i] \tag{17}$$

that assumes high values for functional groups that have both high biomass proportion and high overall trophic effect (as defined above): these groups are considered the dominant functional groups, also known as structural groups. The overall impact 'keystoneness' and 'dominance' proposed here were estimated for each living functional group (thus excluding detritus groups) in the suite of food web models.

For each group, the method provided values for KS and KD: generally groups ranking high in KS were ranking low in KD and *vice versa*. In order to simplify the evaluation, analyses were performed on the most influential functional groups identified as the top 5% ranking groups over all the models. Thus, high ranking keystone functional groups were defined with KS ≥ 0 and dominant species those with KD ≥ -0.7.

Significant differences in the proportion of key functional groups (keystone and structural species) in relation to exploitation level and ecosystem traits (ecosystem type, latitude, ocean basin, depth, size, period represented and exploitation) were evaluated using Correspondence analysis (CA) and performed using Statistica version 6.1 (Statsoft; www.statsoft.com) [62]. CA evaluates deviation from independence between key groups' frequency and traits on the basis of Chi-squared statistic, and decomposes the overall Chi-square in contributions from each combination of trait/key groups, see for example [see for example 63,64]. Therefore, on the basis of the main contributors to overall Chi-square, CA identifies the combination of traits that deviates more from the expected values and the complete independency of factors [complete independency of factors 62].

Results

a) Ecological food web indicators of marine ecosystems

We calculated several ecological indicators from flows and biomasses of food webs (Table 1).

There were significant differences in the food web properties of marine ecosystems by ecosystem type, ocean basin, depth, size, and whether the ecosystem was fished or not (Table 2, values in bold). These results were robust considering differences in model construction since they were corrected for covariance, or the way food webs were described. In contrast, ecological indicators were maintained in ecosystems over time (with no significant difference in food web properties by year, suggesting that the main past and present configurations of marine food webs prevail) and latitudinal gradient (Table 2).

When examining specific ecological indicators by ecosystem type (Table 3), there were significant differences in Total System Throughput (TST, the measure of total trophic flows within an ecosystem) and Total Biomass of the community (TBco), with higher values in reefs, lagoons and shelves (Fig. 3a & 3b). The mean Ecotrophic Efficiency (meanEE, the proportion of overall production used within the system) increased from estuaries to shelves and slopes (Fig. 3c). These results were significant after accounting for covariance.

Some ecological indicators also differed significantly with depth (Table 3), with the Flow to Detritus (FD/TST, the non-living particulate organic matter that returns to the trophic flow as a

Table 2. Significance between ecosystems indicators when grouping data by ecosystem traits.

	TRAITS						
Indicators	**Ecosystem type**	**Latitude**	**Ocean**	**Depth**	**Size**	**Year**	**Fishing**
Ecological indicators	0.01*	0.16 (Fg, Lg, Tl)	0.00* (Fg, Lg, Tl)	0.02* (Fg, Lg, Tl)	0.00* (Fg, Lg, Tl)	0.60 (Fg, Lg, Tl)	0.02* (Fg, Lg, Tl)
Fishing indicators¥	0.41	0.12	0.17	0.00* (Lg)	0.02* (Lg)	0.33	¥

Analyses were performed with PERMANOVA. *: significant difference of the indicator between levels of the factor. Cells with (Fg, Lg, Tl) indicate that the covariate of number of functional groups (Fg), number of living groups (Lg) and number of trophic links (Tl) had a significant effect on the indicator, while cells with (Lg) indicate significant covariance by number of living groups only. Covariate functional groups (Fg). Covariate number of living groups (Lg). Covariate trophic links (Tl) are reported.
¥We excluded the impact of fishing on fishing indicators because for some models there was no fishing, thus fishing indicators could not be calculated.

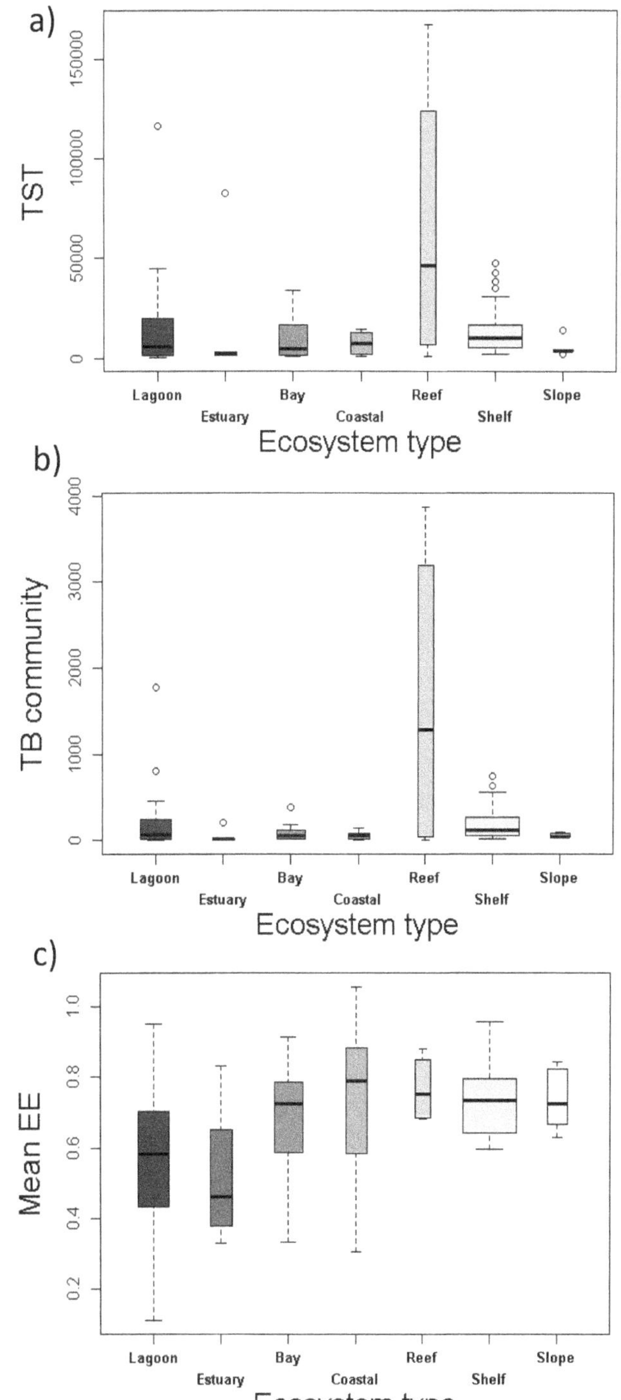

Figure 3. Boxplot of significant differences of food web ecological indicators by ecosystem type. The smallest observation (minimum), lower quartile, median, upper quartile, largest observation (maximum) and outliers are indicated. The boxes are drawn with widths proportional to the square-roots of the number of observations in each class. TST = total systems throughput ($t.km^{-2}.year^{-1}$), TB community = total biomass of the community ($t.km^{-2}$), and mean EE = mean ecotrophic efficiency (proportion).

Table 3. Significance of individual indicator differences between levels of ecosystems traits.

Indicators	TRAITS																				
	Ecosystem type			Latitude			Ocean			Depth			Size			Year			Fishing		
	Fg	Lg	Tl	Fg	Lg	Tl	Fg	Lg	Tl	Fg	Lg	Tl	Fg	Lg	Tl	Fg	Lg	Tl	Fg	Lg	Tl
A/C	0.68	0.68	0.68	0.13	0.13	0.13	0.01*	0.01*	0.01*	0.49	0.49	0.49	0.35	0.35	0.35	0.63	0.63	0.63	0.47	0.47	0.47
O/C	0.60	0.60	0.60	0.11	0.11	0.11	0.00*	0.00*	0.00*	0.53	0.53	0.53	0.22	0.22	0.22	0.75	0.75	0.75	0.57	0.57	0.57
IFO	0.42	0.42	0.42	0.09	0.09	0.09	0.00*	0.00*	0.00*	0.81	0.81	0.81	0.16	0.16	0.16	0.73	0.73	0.73	0.70	0.70	0.70
FCI	0.14	(0.14)	0.14	0.09	0.09	0.09	0.30	(0.30)	0.30	0.67	0.67	0.67	0.01*	(0.01*)	0.01*	1.00	1.00	1.00	0.14	(0.14)	0.14
PP/TST	0.29	0.29	0.29	0.69	0.69	0.69	0.03*	0.03*	0.03*	0.14	0.14	0.14	0.07	0.07	0.07	0.30	0.30	0.30	0.16	0.16	0.16
FD/TST	(0.12)	(0.12)	0.12	(0.73)	(0.73)	0.73	(0.00*)	(0.00*)	0.00*	(0.01*)	(0.01*)	0.01*	(0.90)	(0.90)	0.90	(0.85)	(0.85)	0.85	(0.54)	(0.54)	0.54
SOI	(0.24)	0.24	0.24	(0.43)	0.43	0.43	(0.01*)	0.01*	0.01*	(0.80)	0.80	0.80	(0.07)	0.07	0.07	(0.42)	0.42	0.42	(0.44)	0.44	0.44
TB/TST	(0.52)	0.52	0.52	(0.62)	0.62	0.62	(0.75)	0.75	0.75	(0.39)	0.39	0.39	(0.04*)	0.04*	0.04*	0.50	0.50	0.50	(0.12)	0.12	0.12
TST	0.01*	0.01*	0.01*	0.22	0.22	0.22	0.45	0.45	0.45	0.17	0.17	0.17	0.21	0.21	0.21	0.59	0.59	0.59	0.06	0.06	0.06
Q/TST	0.43	0.43	0.43	0.29	0.29	0.29	0.01*	0.01*	0.01*	0.13	0.13	0.13	0.45	0.45	0.45	0.60	0.60	0.60	0.35	0.35	0.35
R/TST	0.37	0.37	0.37	(0.28)	0.28	0.28	(0.00*)	0.00*	0.00*	(0.01*)	0.01*	0.01*	(0.16)	0.16	0.16	(0.52)	0.52	0.52	(0.48)	0.48	0.48
Ex/TST	0.48	0.48	0.48	0.09	0.09	0.09	0.01*	0.01*	0.01*	0.14	0.14	0.14	0.02*	0.02*	0.02*	0.41	0.41	0.41	0.12	0.12	0.12
PP/P	(0.19)	(0.19)	0.19	0.24	(0.24)	0.24	(0.29)	(0.29)	0.29	0.86	(0.86)	0.86	(0.21)	(0.21)	0.21	0.21	(0.21)	0.21	(0.05*)	(0.05*)	0.05*
Mean Pz	0.88	(0.88)	(0.88)	0.81	(0.81)	(0.81)	0.39	(0.39)	0.39	0.78	(0.78)	(0.78)	0.63	(0.63)	0.63	0.59	(0.59)	(0.59)	0.27	(0.27)	(0.27)
Max Pz	(0.92)	(0.92)	(0.92)	(0.47)	(0.47)	(0.47)	(0.02*)	(0.02*)	0.02*	(0.96)	(0.96)	(0.96)	(0.31)	(0.31)	0.31	(0.77)	(0.77)	(0.77)	(0.17)	(0.17)	(0.17)
Mean EE	(0.01*)	(0.01*)	0.01*	0.04*	(0.04*)	0.04*	0.23	(0.23)	0.23	0.00*	(0.00*)	0.00*	0.00*	(0.00*)	0.00*	0.71	(0.71)	0.71	0.01*	(0.01*)	0.01*
Tbco	0.00*	0.00*	0.00*	0.15	0.15	0.15	0.44	0.44	0.44	0.08	0.08	0.08	0.02*	0.02*	0.02*	0.67	0.67	0.67	0.01*	0.01*	0.01*
mTLco	(0.68)	(0.68)	0.68	(0.72)	(0.72)	0.72	(0.05*)	(0.05*)	0.05*	(0.06)	(0.06)	0.06	(0.07)	(0.07)	0.07	(0.02*)	(0.02*)	0.02*	(0.50)	(0.50)	0.50
TEm	(0.08)	0.08	0.08	(0.30)	0.30	0.30	(0.03*)	0.03*	0.03*	(0.01*)	(0.01*)	0.01*	(0.17)	0.17	0.17	(0.38)	0.38	0.38	(0.91)	0.91	0.91
TLc[¥]	0.07	(0.07)	0.07	0.02*	(0.02*)	0.02*	0.01*	(0.01*)	0.01*	0.00*	(0.00*)	0.00*	0.06	(0.06)	0.06	0.77	(0.77)	0.77	¥	(¥)	¥
TC[¥]	0.22	0.22	0.22	0.61	0.61	0.61	0.28	0.28	0.28	0.00*	0.00*	0.00*	0.02*	0.02*	0.02*	0.92	0.92	0.92	¥	¥	¥
PPR%[¥]	0.97	0.97	0.97	0.09	0.09	0.09	0.35	0.35	0.35	0.11	0.11	0.11	0.10	0.10	0.10	0.40	0.40	0.40	¥	¥	¥
L_{index}[¥]	0.60	0.60	0.60	0.31	0.31	0.31	0.65	0.65	0.65	(0.03*)	0.03*	0.03*	0.03*	0.03*	0.03*	0.48	0.48	0.48	¥	¥	¥
P_{sust}[¥]	0.83	0.83	0.83	0.49	0.49	0.49	0.45	0.45	0.45	0.07	0.07	0.07	0.45	0.45	0.45	0.02*	0.02*	0.02*	¥	¥	¥

Analysis was perform with PERMANOVA. Vaues with a * indicate a significant difference of the indicator between levels of the factor. Values in () indicate that the covariate had a signficiant effect on the indicator. Covariate functional groups (Fg), Covariate number of living groups (Lg), Covariate trophic links (Tl).

[¥]We excluded the impact of fishing on fishing indicators because for some models there was no fishing, thus fishing indicators could not be calculated in those cases, which means we were not comparing like with like.

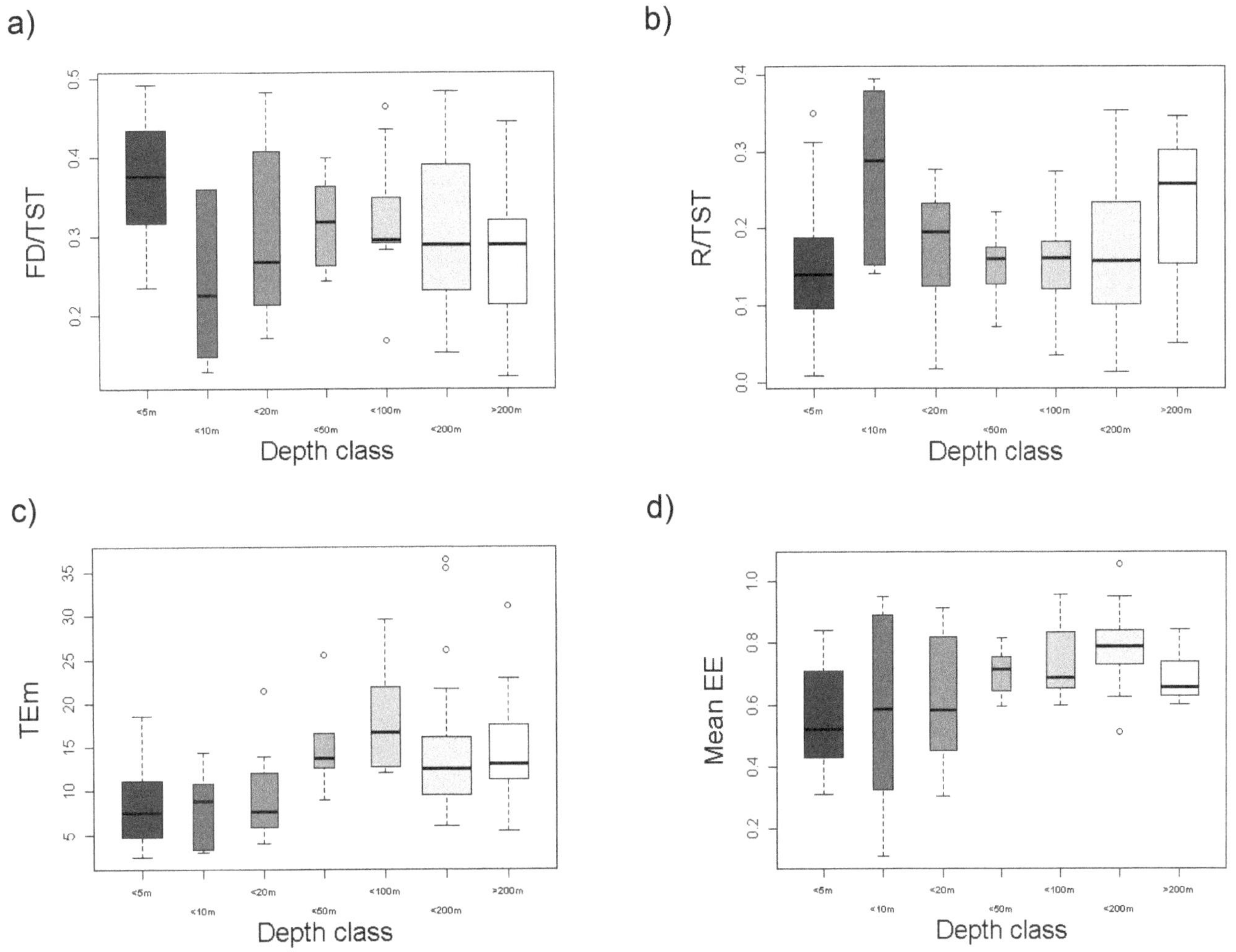

Figure 4. Boxplot of significant differences of food web ecological indicators by depth class. The smallest observation (sample minimum), lower quartile, median, upper quartile, largest observation (sample maximum) and outliers are indicated. The boxes are drawn with widths proportional to the square-roots of the number of observations in each class. FD/TST = flow to detritus/total systems throughput (proportion), R/TST = respiration/total systems throughput (proportion), TEm = mean transfer efficiency (%) and mean EE = mean ecotrophic efficiency (proportion).

ratio of the total flow), total Respiration (per unit of trophic flow, R/TST), mean Ecotrophic Efficiency (meanEE) and mean Transfer Efficiency (TEm, or the conversion of production from lower to higher trophic levels) all showing significant differences (Fig. 4). The flow to detritus (FD/TST, Fig. 4a) was higher in shallower systems, with the trend being reversed in respiration (R/TST, Fig. 4b) and mean transfer efficiency (Tem, Fig. 4c). The trend in mean Ecotrophic Efficiency (meanEE, Fig. 4d) also showed correlation with ecosystem type, where shallower systems (such as lagoons, estuaries and bays) had lower mean Ecotrophic Efficiencies than deeper systems (such as shelves and slopes).

Similarly, ecological indicators were also significantly different by ecosystem size (Table 3). Again, the mean Ecotrophic Efficiency was lower in smaller systems (meanEE, Fig. 5a), and increased with size, while the export from the system (export of matter per unit of flow, Ex/TST) was also lowest in small systems and increased in larger systems (Fig. 5b). This was converse to the trend in the Finn Cycling index (FCI, an index of the relative amount of recycling in the ecosystem), which was highest in smaller systems (Fig. 5c) and declined as size increased.

The ecosystem trait that showed significant differences for most ecological indicators was ocean basin (Ocean, Table 3). Respiration (R/TST, Fig. 6a), export of matter per unit of flow (Ex/TST, Fig. 6b), mean transfer efficiency (TEm, Fig. 6c), flow to detritus (FD/TST, Fig. 6d), the mean trophic level of the community (mTLco, the average trophic level for functional groups with a TL>2 that represents a mean trophic position of organisms in the community, Fig. 6e), total consumption (per unit of flow, Q/TST, Fig. 6f), system omnivory index (SOI, the variance of trophic levels in the diet, Fig. 6g), redundancy or internal flow overhead (IFO, the distribution of energy flow pathways in the system, Fig. 6h), relative ascendency (A/C, an index of organisation of the food web, C being the Development Capacity of the system, Fig. 6i) and relative overhead (O/C, the index of the ecosystem's strength in reserve, Fig. 6j) all showed significant changes with ocean basin. A general decreasing trend between the West and East Atlantic (and the reverse in the West and East Pacific) was observed in respiration (R/TST), consumption (Q/TST) and the redundancy (IFO) in these systems, while an increasing trend was shown in the flow to detritus (FD/TST) and the relative ascendency (A/C). It was also clear that some differences between Ocean basins were reproduced in ecosystem indicators such as the mean transfer efficiency (TEm), mean trophic level of the community (mTLco), and the system omnivory index (SOI): the Pacific Ocean had

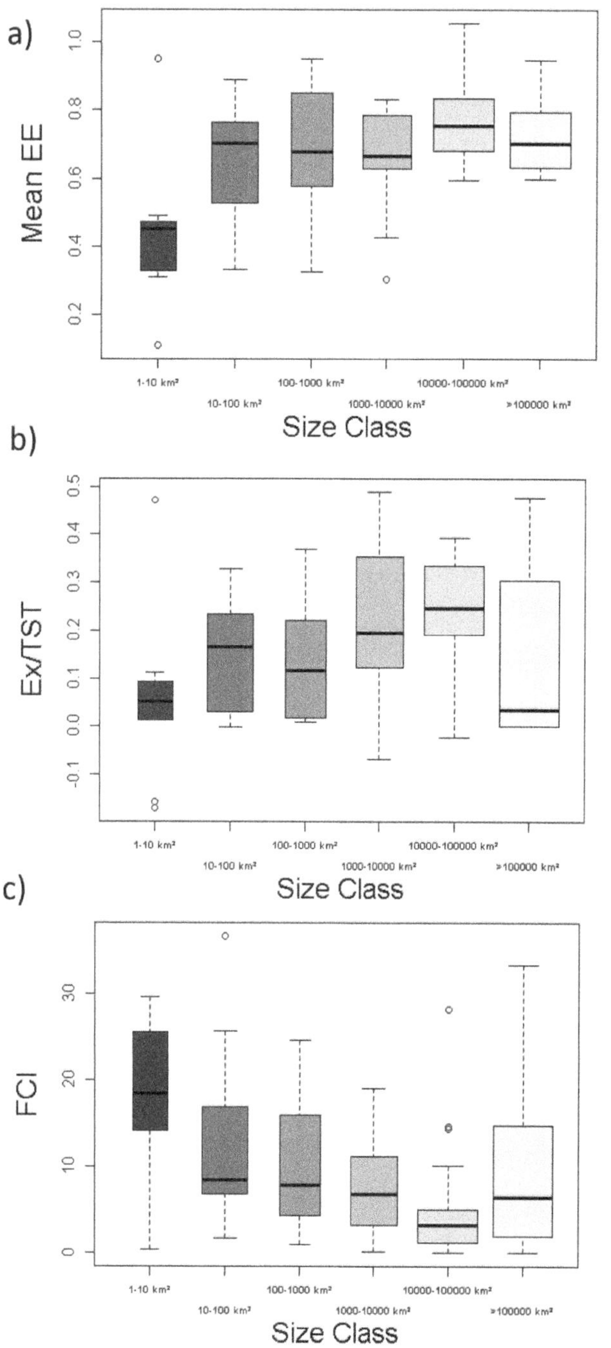

Figure 5. Boxplot of significant differences of food web ecological by size class. The smallest observation (sample minimum), lower quartile, median, upper quartile, largest observation (sample maximum) and outliers are indicated. The boxes are drawn with widths proportional to the square-roots of the number of observations in each class. Mean EE = mean ecotrophic efficiency (proportion), Ex/TST = export/total systems throughput (proportion) and FCI = Finn cycling index.

higher values than the Atlantic and Indian Oceans, and the Indian Ocean generally had the lowest value for all those indicators.

Overall, the traits that showed significant differences between model indicators were ecosystem type, latitude, ocean basin, size, and year. Those indicators that were least prone to co-vary with factors (i.e. functional groups, number of living groups, and number of trophic links) were total export (Ex/TST) and total biomass of the community (TBco, showing no differences in cell colour between covariance in Table 3). Both were only significantly different in two traits each (Ex/TST = ocean basin and size, TBco = ecosystem type and size). Of those ecological indicators analysed, eight were robust to covariates, namely relative ascendency (A/C), relative overhead (O/C), redundancy (IFO), primary production (PP/TST), total systems throughput (TST), consumption (Q/TST), export (Ex/TST), and total biomass of the community (TBco, as highlighted in Table 3). These indicators may be of special interest for future food web studies if correction for covariance is not possible.

b) Key functional groups of food webs

Keystone and key dominant/structuring group indicators have been estimated for all the 2,635 functional groups in the 105 models, and properties of keystone and key dominant/structuring groups were defined on the basis of the top 5% of the ranking groups. The top ranking keystone functional groups had an average TL of 3.28±0.97, an average biomass proportion of 0.022±0.044, and a clear prevalence of top-down effects with an average td effect of 67%±28%. Conversely, top ranking key structuring functional groups had an average TL of 1.53±0.6, an average biomass proportion of 0.43±0.17 and an average td effect of 33%±34% (Fig. 7a & 7b). The last result highlighted a prevalence of bottom-up effects in key structuring groups. However, the large range might be due to predatory groups having high td (i.e. benthic invertebrates), and therefore assumptions about bottom-up effects as a characteristic feature should be made with caution.

The groups' ranking in terms of overall effect from each of the 105 food webs showed that several groups were identified to be keystones (i.e. with low biomass proportion; Fig. 7b, black symbols) and a large proportion of these groups had high trophic levels. Smaller organisms were prevalently key structuring groups (i.e. with high biomass proportion and high impact; grey symbols). The groups that ranked first in terms of overall effect generally had high trophic levels and many of them were larger organisms such as sharks and rays (ten models, n = 10), top predatory fishes (n = 6), marine mammals (n = 7) and seabirds (n = 1). Producers, and especially benthic primary producers such as macroalgae and pleustophytes, were key structuring species (n = 10). Our results highlight groups whose changes in biomass have the largest effect on the food web, and the main distinctive factor between structuring and keystone species is their biomass proportion, although, this is not surprising as biomass proportion is explicit in the definition of key species.

Correspondence analysis indicated that the proportion of keystone and structural functional groups showed significant variation with size, depth, and type of ecosystem at $p<0.001$, as well as significant variability with ocean basin at $p<0.001$ (Fig. 8 and Table 4). In particular, a significantly higher proportion (with the highest Chi-square values by trait) of keystone groups was found in smaller (size <10 km^2, Fig. 8a) and shallower (depth $\leq$ 5 m, Fig. 8b) ecosystems, in estuaries (Fig. 8c) and in the Indian Ocean models (Fig. 8d). This last result is due to the fact that the Indian Ocean models mostly represent bays and estuaries (Fig. 8 & Table 4). The proportion of structural groups never showed high Chi-squared contributions, thus indicating that dominant species did not distribute differently among traits and have a similar ecological role.

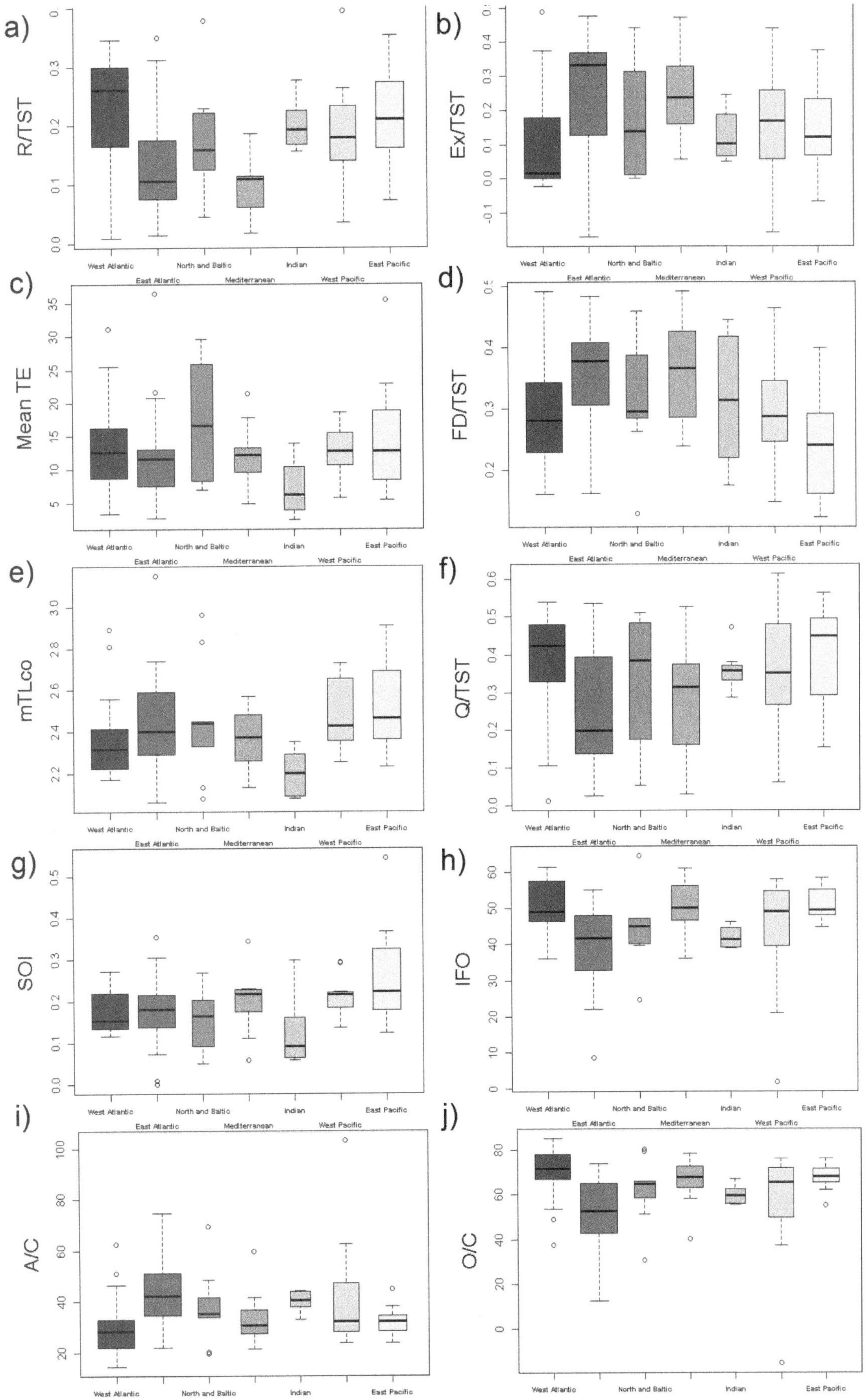
a)
R/TST
b)
Ex/TST
c)
Mean TE
d)
FD/TST
e)
mTLco
f)
Q/TST
g)
SOI
h)
IFO
i)
A/C
j)
O/C
West Atlantic
East Atlantic
North and Baltic
Mediterranean
Indian
West Pacific
East Pacific

Figure 6. Boxplot of significant differences of food web ecological by ocean basin. The smallest observation (sample minimum), lower quartile, median, upper quartile, largest observation (sample maximum) and outliers are indicated. The boxes are drawn with widths proportional to the square-roots of the number of observations in each class. R/TST = respiration/total systems throughput, Ex/TST = export/total systems throughput (proportion), mean TE = mean transfer efficiency, FD/TST = flow to detritus/total systems throughput (proportion), MTLco = mean trophic level of the community, Q/TST = consumption/total systems throughput (proportion), SOI = systems omnivory index, IFO = Internal flow overhead (%), A/C = relative ascendency (%), O/C = relative overhead (%).

c) The impacts of fishing on marine food web indicators

Fishing is at present the main human factor that impacts marine food webs [2,65,66]. Therefore, this study tested if ecological food web indicators also varied among different exploitation levels in marine ecosystems. Our results showed that primary production (PP/P, the unit of primary production over total ecosystem production), mean Ecotrophic Efficiency (MeanEE) and total community biomass (TBco) showed significant differences between exploited and non-exploited food webs (Table 3; Fig. 9a, 9b & 9c).

Moreover, marine food webs also showed differences in the levels of fishing between systems of different depth and size (Table 2), implying different levels of exploitation strategies and impact from fishing with regard to these ecosystem features. Shallower and smaller ecosystems are closer to the coast, thus in general more exposed to fishing (in some cases for centuries), whereas deep and large ecosystems are less accessible, constituting refuges from fishing. The results show, for instance, the mean trophic level of the catch (TLc, which quantifies the mean trophic position of exploited organisms) increased with depth (Fig. 9d). The trophic level of the catch also increased with latitude (TLc, Fig. 9e) and was lowest in the Indian Ocean and highest in the West Atlantic (Fig. 9f), although the TLc was also high in the North and Baltic Seas. Fishing intensity changed with time and we observed a decrease of the probability of being sustainably fished (P_{sust}, Fig. 9i) over time.

Top ranking keystone and structuring species (95th percentile in the KS and KD distribution) were tested for variance according to fishing intensity by analyzing the proportion of these key groups in food webs representing fished, non-fished, or slightly fished (i.e. marine protected areas that included only recreational or artisanal fishing) ecosystems. Of the 2,366 functional groups from the fished ecosystem models, 104 (4.4%) were top ranking keystone groups [52] and 78 (3.3%) were top ranking structuring functional groups. Similar proportions were identified for the non-fished ecosystems: among the 127 functional groups, 6 (4.7%) and 5 (3.9%) groups were identified as top ranking keystone and structural functional groups, respectively. Interestingly, in the 142 functional groups of the lightly exploited food webs, 14 (9.9%) were keystones and only 1 (0.7%) was a structuring functional group. Correspondence analysis between key species and fishing trait was significant at $p = 0.02$ (Chi-squared $= 11.6$; $df = 4$), with a major contribution from the increase of groups with high keystoneness in lightly exploited food webs (Fig. 8e & Table 4).

Discussion

Our approach emphasizes the power of using a large database of ecosystem models to quantify food web indicators at the global level. The ability to generalize from different ecosystems constructed using a common EwE approach is one advantage of having comparable models, based on the same framework.

This study provides a comprehensive analysis for how ecosystems function to show how these indicators might be modified in future with additional human impacts such as climate change, eutrophication, acidification, etc. The impact of model structure and its link with model construction are taken into account by using statistical analyses of co-variance, placing this study among the first to analyse a large variety of EwE models from different systems in an organized, systematic and statistical way.

a) Caveats

This analysis is predicated on a modelling technique that has, like other modelling techniques, its own drawbacks, such as the lack of uncertainty testing in model inputs [67], the difficulty of including non-trophic interactions, and the fact that it does not handle migratory species particularly well [68]. The quality of input data also affects the predictive quality of the outputs. In addition, the suite of models included here range from very early models (created in the early 1990s) with fewer trophic groups, to newer and/or larger models, i.e. models that are more well defined, and therefore have more trophic groups (this is taken into account in our covariance analysis).

At the same time, even models with the same number of trophic groups might have very different topologies; one model with 20 functional groups might have all the herbivorous species combined in one functional group and the top predators defined by species, while another with 20 groups might have one top predator functional group including very different species, and the lower trophic levels defined in more detail. Although this is to some degree encapsulated in the Finn Cycling Index and the Internal Flow Overhead (and this study tested for co-variance in number of trophic links) this still needs to be tested further. The lack of consistency in model construction can be fully addressed only in studies where all models are developed in a comparable and standardized way, as was done in a few other instances [47,69]. However a large scale study such as ours would be very difficult to achieve using those methods.

In addition, only 8 of the models used here did not include fishing. EwE models and -up effects such as eutrophication and environmental changes are not always represented. These drivers might have been included in the models that were fitted to time series (see Table S1 for details), but was not the main drivers for most of the models used. Thus, top-down and bottom-up drivers are not equally represented in our study, resulting in underrepresentation of the impact of bottom-up drivers on the ecosystem. Specifically, very few models had the microbial loop defined, and the impact of detritus and the detritivorous food chain is underrepresented. Future work should include a study of models where the microbial loop is defined to elucidate the importance of this underrepresented part of the food web.

Furthermore, not all indicators are equally robust. Fulton et al. [70] found that some throughput-based indicators are very useful if one has good diet data and knowledge of the ecosystem structure, and similarly trophic level-based indicators are only effective if one has good diet data. They also suggest that the network indicators such as relative ascendency are dependent on the data and model formulation as well as on reliable knowledge of ecosystem structure, concluding that these indicators would be most useful in well-studied systems. Although this large meta-analysis allowed us to disentangle the effects of each trait and covariates, more data will result in better models, and well-validated models will give more reliable indicators. Our study

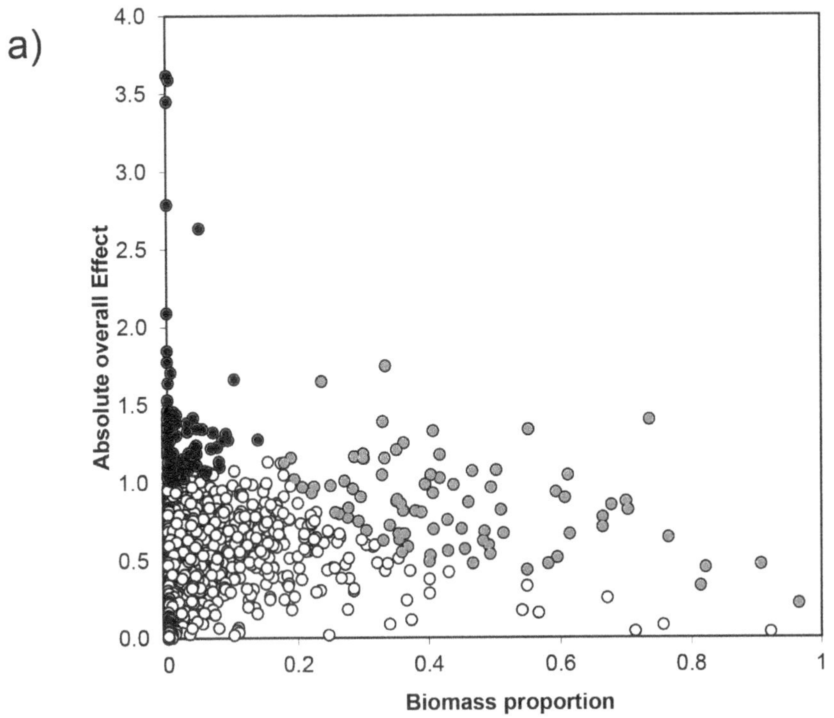

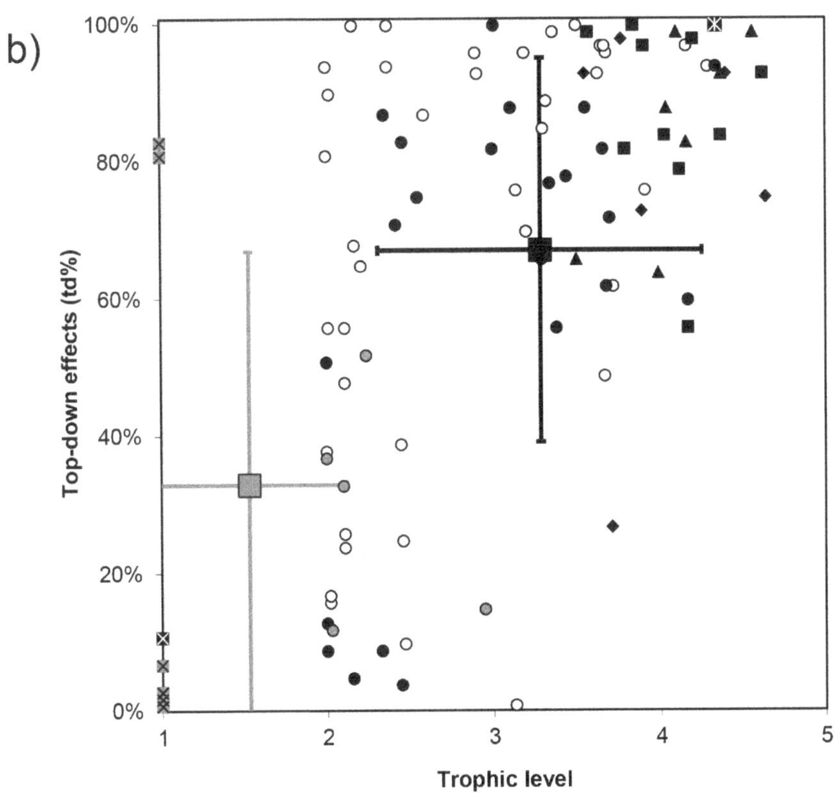

Figure 7. Key ecological roles of functional groups of marine food web models. a) Keystone (KS ≥0; black circles) and dominant groups (KD ≥0.7; grey circles), respectively, in terms of absolute overall effect (ε) for each food web. Open dots represent non key functional groups. b) The 105 trophic groups franking first in terms of absolute overall effect within each food web model (Fig. 2, Table S1). The figure shows the trophic level (TL) vs. the fraction of top-down effect (td%). Groups identified as keystones are represented in black symbols and dominant groups are reported in grey symbols, respectively, whereas open circles represent non key functional groups. Groups are highlighted for both keystones and dominant: birds (star within square), marine mammals (triangles), sharks and rays (squares), top-predators (romboid), primary producers (crossed squares), other groups (circles). Large squares with error bars identify mean+/−SD for all keystones and dominants identified in the 105 models.

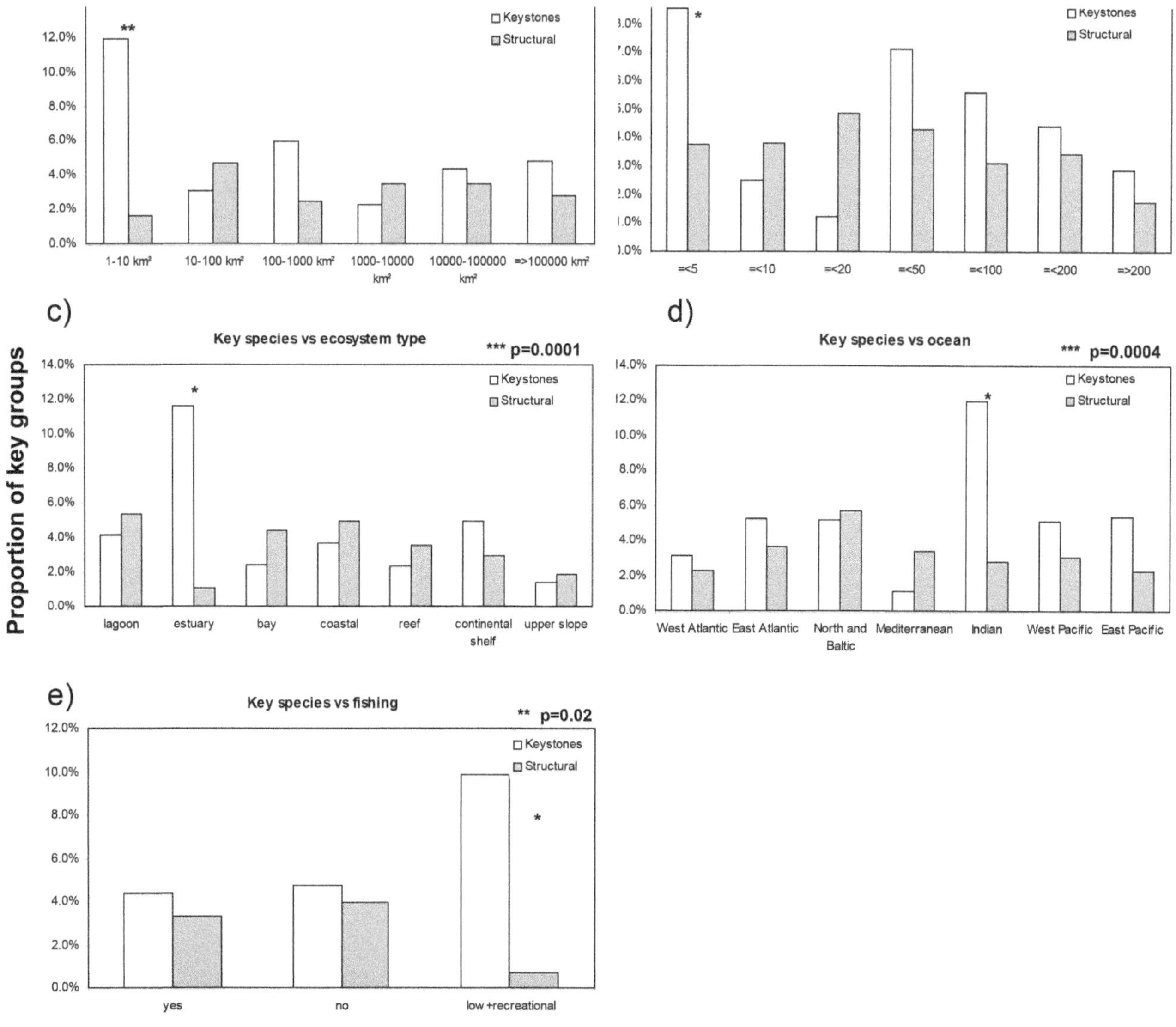

Figure 8. Proportion of key functional groups by food web ecological traits and by exploitation. Graphs report proportion of Keystone (KS ≥0) and dominant groups (KD ≥0.7) by a) ecosystem size, b) depth, c) ecosystem type, d) ocean basin and e) fishing category. Only traits showing significantly different patterns (on the basis of total Chi-squared) are reported (***, ** respectively $p<0.01$, $p<0.05$). Main contribution to Chi-squared are highlighted by asterisk (see also Table 4).

concurs with that of Fulton et al. [70], by showing that not all indicators can be used without taking account of the number of functional groups, the number of living groups or the number of trophic links (Table 3).

Lassen et al. [71] showed that some indicators obtained from EwE models (biomass of important species, food web productivity, etc.) are useful as indicators of good environmental status for the EU Marine Strategy Framework Directive (MSFD) [72]. In addition to the indicators described by Lassen et al. [71], other ecological indicators used here to describe marine food webs have the potential to be indicators of good environmental status as requested by international law and directives such as the MSFD [72], given their capabilities to describe food web changes and sensitivity to fishing. However, our results highlight that some ecosystem indicators vary by ecosystem trait (depth, type, size, etc.), thus implying that the trait effect needs to be accounted for when setting reference levels and thresholds for conservation and management. If food web indicators vary with ecosystem traits, the process of defining and quantifying reference levels and thresholds [73–75] will have to take this into account. Otherwise, management advice could be based on the wrong indicator or on a wrong reference level, or the indicator might be insensitive to the adopted management policies. Our results, therefore, represent a starting point for disentangling the variability of indicators due to ecosystem traits caused by other stressors of interest to managers. For instance, knowing that an indicator is intrinsically lower or higher in certain ecosystem types might help to better understand the locally provided estimates and adapt reference levels and thresholds.

b) Structure and functioning of marine food webs

Our detailed results showed some interesting general trends. For example, coral reefs seem to have the largest energy flow and the largest total biomass per unit of surface area, and as the included reef systems covered large areas [76–79], their mean Ecotrophic

Table 4. Chi-squared obtained from the Correspondence Analysis between traits (fishing, size, ecosystem type, latitude, longitude, depth and year) and key functional groups.

Traits	Classes	Key functional group			total Chi-square	p-value	
		Keystones	Structuring	Other			
fishing (df = 4)	no	0.000	0.278	0.010			
	low/recreational	**{8.013}**	2.645	0.116			
	yes	0.484	0.076	0.011	**11.634**	**0.0203**	**
size (df = 10)	1–10 km²	**{20.556}**	1.298	0.664			
	10–100 km²	1.386	2.003	0.000			
	100–1000 km²	0.654	0.263	0.008			
	1000–10000 km²	5.163	0.185	0.189			
	10000–100000 km²	0.186	0.132	0.001			
	= >100000 km²	0.014	0.430	0.009	**33.142**	**0.0003**	***
ecosystem type (df = 12)	lagoon	0.163	4.073	0.078			
	estuary	**{19.073}**	2.589	0.478			
	bay	2.273	1.046	0.023			
	coastal	0.364	1.690	0.011			
	reef	1.000	0.048	0.035			
	continental shelf	0.203	0.549	0.001			
	upper slope	4.964	1.062	0.480	**40.201**	**0.0001**	***
latitude (df = 6)	= <15	0.105	2.491	0.047			
	15–30	2.191	1.221	0.289			
	30–60	0.903	2.399	0.249			
	= >60	0.100	0.299	0.001	10.294	0.113	
Ocean basin (df = 12)	West Atlantic	2.800	0.148	0.201			
	East Atlantic	0.452	0.054	0.038			
	North and Baltic	0.093	0.000	0.005			
	Mediterranean	9.619	0.088	0.418			
	Indian	**{19.528}**	0.042	0.924			
	West Pacific	0.125	0.098	0.019			
	East Pacific	0.373	0.078	0.008	**35.109**	**0.0004**	***
depth (df = 12)	= <5	**{12.451}**	0.552	0.872			
	= <10	0.794	0.119	0.019			
	= <20	4.280	1.599	0.055			
	= <50	1.767	0.620	0.198			
	= <100	0.554	0.000	0.029			
	= <200	0.152	0.027	0.003			
	= >200	4.516	3.735	0.698	**33.038**	**0.0010**	***

Table 4. Cont.

Traits	Classes	Key functional group			total Chi-square	p-value
		Keystones	Structuring	Other		
year (df = 6)	<1970	0.016	1.287	0.032		
	1970–1980	0.152	1.947	0.119		
	1980–1990	3.219	0.466	0.078		
	>1990	2.775	0.020	0.123	10.235	0.1151

Between parenthesis the degrees of freedom of Covariance analysis (df). Significant total Chi-squared are bold (*** and ** indicate p<0.01 and p<0.05 respectively). Main contributors to Chi-squared indicated by ß.

Efficiencies were also high. However, the smaller, shallower systems such as lagoons, estuaries, and bays had lower mean Ecotrophic Efficiencies. These smaller systems usually have shorter residence times, with a larger proportion of species that either migrate in and out of the system, and so might not be utilized in the system [80], thus increasing the unexplained mortality in that system, and therefore reducing the mean Ecotrophic Efficiency. The shallower systems also had more flow to detritus, which together with the fact that these systems often included more benthic interactions usually at low efficiency [80], explained the low mean transfer efficiency found for the shallower areas [81].

Systems with high transfer efficiencies often have fewer pathways between trophic levels, while systems such as lagoons, estuaries and bays often have more species at the lower trophic levels – detritivores, suspension feeders, etc., therefore reducing the mean transfer efficiency [82–85]. This was also seen in the higher Finn cycling indices in these systems where more energy was recycled [86], while the total export from these systems was significantly reduced. Systems with higher Finn cycling indices often have the ability to recover from perturbations quicker [86], and could therefore be more stable. Conversely, upwelling or pelagic ecosystems, which tend to be deeper, would have higher transfer efficiencies, more export and less recycling [86] and they are often characterised by large fluctuations [83,84,87], although this is not specified in our study and should be examined in future.

Interesting results emerged from the comparisons between ocean basins: there were differences between the eastern and western parts of the Atlantic and Pacific Oceans, which were most noticeable in the Atlantic. From the Western to Eastern Atlantic, we found increased flows to detritus, which was reflected by an increase in ascendency and mean trophic level of the community, while the inverse trend was found in respiration, consumption, redundancy, and overhead. These changes were not necessarily due to differences in ecosystem type, depth, or size categories, as these were all similar for the Eastern and Western Atlantic, nor due to differences in the numbers of groups (average number of groups in the Western Atlantic = 25, Eastern Atlantic = 29). The difference in overhead and redundancy indicates that the Western Atlantic systems seem to have more "strength in reserve" and that the energy in these systems has more pathways to travel from primary producers to top predators than in the systems of the Eastern Atlantic. The differences in the Atlantic could also be due to differences in the biological carbon pumps in these two systems. Helmke et al. [88] showed that there were higher nutrient inputs in the Western Atlantic and more pulsed production events, which accounted for more carbon being produced and fluxed in the west. This higher production and flux explain the higher consumption, respiration, and overhead in that system compared to the East Atlantic.

The difference between the east and west was reversed to some extent in the Pacific, with a higher flow to detritus and lower respiration and consumption in the west than the east. The Pacific models also have similar numbers of compartments (26 for both Eastern and Western Pacific on average), but the Western Pacific models were mostly large shallow systems including reefs and bays, while in the east, models included more deep continental shelves but also estuaries, bays and lagoons. Thus, these differences in the types of systems, depth, and size (which were more similar in the Atlantic) confounded the interpretation as depth, size, and ecosystem type have an impact on ecosystem indicators. Thus, even though the primary production is higher in all four eastern boundary current systems (California and Humboldt in the Pacific and Canary and Benguela in the Atlantic) [89], the increase it created in the ascendency, flow to detritus and export in the

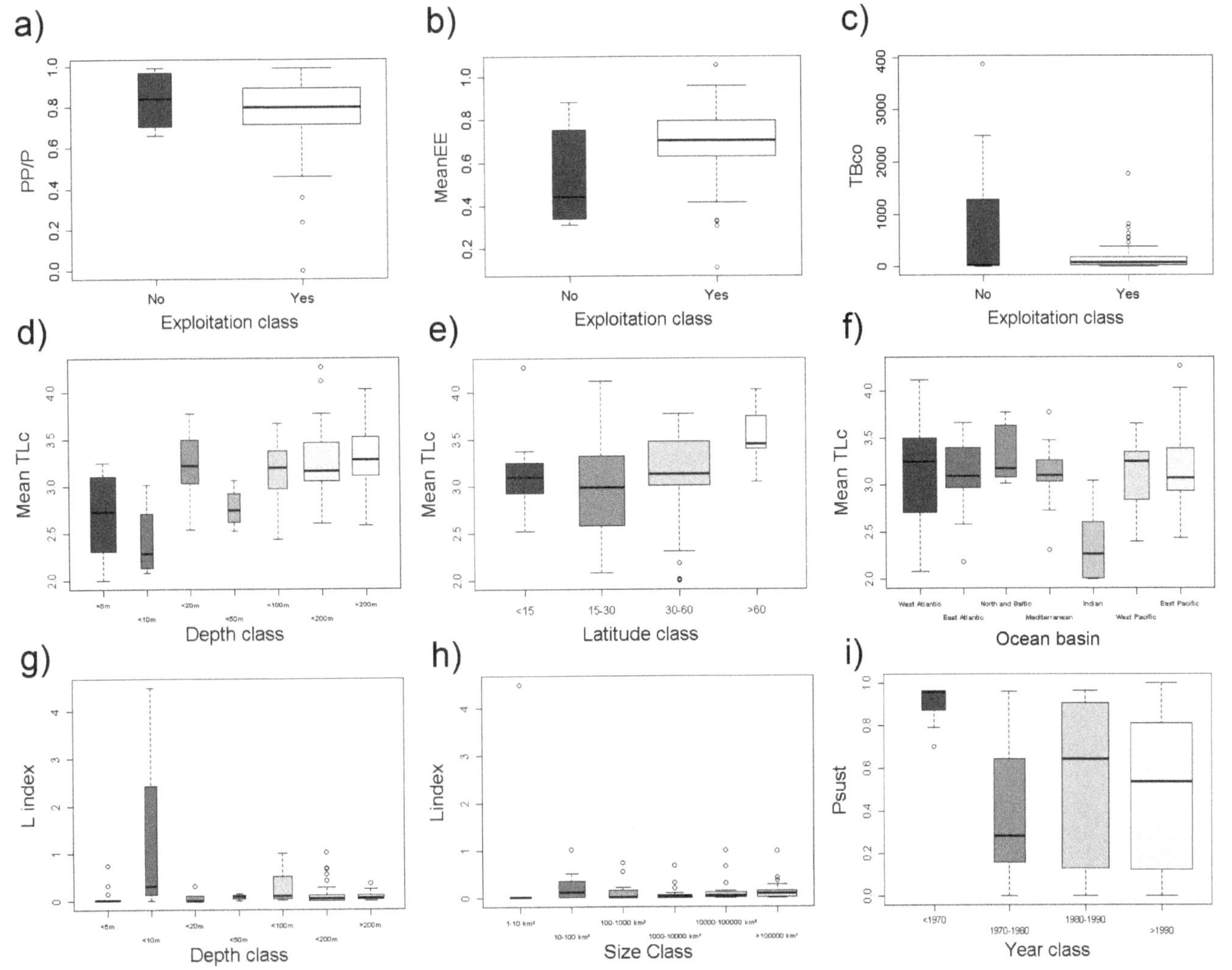

Figure 9. Boxplot of significant differences of food web ecological traits by exploitation and of fishing indicators. The smallest observation (sample minimum), lower quartile, median, upper quartile, largest observation (sample maximum) and outliers are indicated. The boxes are drawn with widths proportional to the square-roots of the number of observations in each class. PP/P = primary production/total production (proportion), meanEE = mean ecotrophic efficiency (proportion), TBco = total biomass of the community (t.km^{-2}), Mean TLc = mean trophic level of the catch, L_{index} = loss in production index, and P_{sust} = probability of being sustainably fished.

Atlantic was not seen in the Pacific. This confirms the results of Carr and Kearns [89], who found that the increase in primary production did not increase the biomass sustained by the available nutrients as much in the Pacific as it did in the Atlantic. They found that the higher iron content, increased physical retention, and differences in plankton community structure accounted for the higher sustained biomass in the Atlantic eastern boundary current (our Western Atlantic), indicated by a higher redundancy (or internal flow overhead, IOF, Fig. 6h).

The mean trophic level of the community and omnivory showed a general increase from the Atlantic westwards to the Pacific, with only the Indian Ocean having much lower values for these indicators. The Indian Ocean models were mainly bays and estuaries, thus mostly shallow and small, with the lowest mean transfer efficiency and high ascendency. These ecosystems in the Indian Ocean are thus rather inefficient in transferring energy up the food chain, with very low omnivory but high organisation of the food webs. In contrast, the models of the North and Baltic Seas have the highest transfer efficiencies, but also have high ascendency, and are thus most efficient at transferring energy up the trophic chain while also being a well-organised system. The high transfer efficiency in the North and Baltic Seas is probably due to the lower species diversity in the Baltic Sea model areas [90], which has translated into high transfer efficiency between fewer species.

Most ecological food web indicators did not show significant differences over time, thus can be considered invariant properties of ecosystems over the broad time ranges and global scales used in our study. Exceptions were the mean trophic level of the consumer community and the probability of the system to be sustainably fished, which both decreased with time. The probability of being sustainably fished was defined by adopting Murawski's [55] ecosystem overfishing definition and criteria, which includes structural and functional degradation associated with stock collapses and overexploitation of marine resources, whereas in sustainably fished ecosystems the main structure and function are preserved. Thus, from our analyses the most relevant changes on these systems through time were due to fishing, which gives such a strong, global signal as to be detectable using these broad time ranges and global scales, supporting claims that fishing is currently the main impacting human factor on marine food webs [2,65,66,91].

This shows that while ecosystems have stable intrinsic properties, human impacts could still be important. However, we did not test in detail for bottom-up processes such as eutrophication and other environmental drivers; therefore these cannot be excluded as important anthropogenic impacts on marine ecosystems. In addition, our indicator of time was relatively crude, by grouping the models in three periods – based on the year(s) that the models represented. To fully take account of changes, a dynamic analysis of the indicators over time should be performed for each calibrated model. Such evaluations have been done for some indicators, e.g. by Heymans et al. [47] and Tomczak et al. [92], but could not be implemented here for all models as it needs models validated with time series analyses. Nevertheless, the analysis conducted by broad time periods at the global scale has the advantage that it identifies important trends at a global level, disregarding local peculiarities that can be studied only when calibrated dynamic simulations are available.

Testing for changes in ecological food web indicators between exploited and non-exploited ecosystems indicates that fishing also induced a decrease in the consumer biomass and a higher use of ecosystem production (increase of mean Ecotrophic Efficiency). The reduction in total biomass of the community confirms work by Worm et al. [93] who found an 11% decline in 144 stock assessment biomass time series since 1977, specifically in pelagic and demersal species in the North Atlantic. In addition, fishing uses the surplus production to some extent, and therefore will increase the explained mortality of the ecosystems. The lower primary production to total production ratio in fished ecosystems could result from fisheries targeting ecosystems with higher secondary production.

Among fishing indicators, we found that some ecosystem traits (latitude, ocean basins, depth) influence trophic level of the catch, while both depth and size affect the loss in production index, thus suggesting the need to account for these confounding traits when evaluating such fishing indicators and using them as ecosystem indicators. Nevertheless, the analysis also showed that, with the exception of trophic level of the catch, fishing indicators were overall less impacted by model covariates (i.e. factors that describe the food webs) than the ecological indicators. These indicators may be of special interest for future food web studies if correction for covariance is not possible. In general, both ecosystem dynamics (such as predator-prey interactions) and external factors (such as fishing) will have an impact on ecosystems and the strength of these impacts depends on a variety of factors [47,94–97]. However, other variations such as economic drivers would also have to be taken into consideration in future studies.

The search for key functional groups in the 105 food webs showed that keystone groups had higher trophic level and mainly affected food webs as predators (top-down), whereas structural functional groups were benthic primary producers, which affected food webs mainly, but not exclusively, through bottom-up effects. Given their high overall impact, modification of the biomass of these key groups through anthropogenic-induced changes may produce important changes in food webs, possibly impairing ecosystem structure and functioning. Reducing top-down impacts exerted through predation by removing or depleting keystone groups (for instance) can cause ecological effects such as the increase of their prey [98], lower predatory mortality for individuals affected by disease or deficiencies, and can result in the decreased transfer efficiency of the ecosystem. On the contrary, modification of structuring groups that exert large effects on food webs through a prevalence of bottom-up effects implies potentially large impacts on the higher trophic levels. Given the prey-predator basis of models analysed, it is very likely that non-predatory roles exerted by structuring species (such as protection, habitat building, interference with physical variables) are underestimated in this analysis and thus the ecological role of these species may be even larger.

Notably, while structuring species appear to be evenly distributed according to ecosystem traits, keystone groups were especially prevalent in estuarine systems and systems smaller than 10 km^2 and less than 5 m in depth. As the Indian Ocean modelled areas were mainly shallow, small estuaries, a significant proportion of keystone groups are identified. However, these results may also be due to the Indian Ocean being historically less impacted by fishing [53] or the fact that mainly bays and estuarine models of the Indian Ocean were available.

These results suggest that coastal and shallow areas with high physical/chemical variability such as estuaries are likely to host a relatively higher proportion of keystone functional groups. Such a result might be important, if further confirmed, for supporting the protection and conservation of these ecosystems, as these groups are often directly implicated with key marine ecosystem services including biodiversity and marine resources [99].

Overall, the disproportionate impacts of keystones and structuring groups [59] implies disproportionate effects if their biomass is modified, thus recommending particular caution when contemplating human impacts. Results of key groups' analysis, therefore, further encourage the protection of estuarine environments, already on the priority list for protection under the Ramsar Convention [100] for the goods and services they provide, for their high ecological value and of importance when valuing ecosystem services [101].

We found a significantly higher proportion of keystone functional groups in lightly exploited ecosystems in comparison to more exploited areas, despite a consistently stable proportion of structuring species. The lack of significant changes in the proportion of keystones in no-fishing models might be due to the low number of unexploited webs available for the analysis. Nevertheless, the significant result for the lightly fished ecosystems may indicate that fishing negatively affects keystones and/or that the keystone role is more prominent and distributed among functional groups in protected environments. The larger abundance of keystone functional groups in lightly exploited ecosystems confirmed previous results for the Mediterranean Sea [57] and highlights a possible effect of fishing in levelling out the species effects. We could not distinguish if the lower proportion of keystones in heavily exploited ecosystems was due to removal of these groups or if the keystone role was hampered by fishing, but we conclude that keystoneness is more clearly pronounced when fisheries exploitation is low. This insightful result merits further study to verify its generality. Overall the keystone groups appear to be more sensitive than structural species to ecosystem properties and exploitation. While their sensitivity might be due to their lower abundance, the large effects they produce in ecosystems through food web interactions make them optimal groups for signalling ecosystem disturbances and impairment, thus being good candidates for ecological indicators.

Conclusions

Our results provide additional knowledge on how marine ecosystems structure and function, and the fact that different patterns occur in different ecosystems pose additional scientific questions and management challenges. For example, significant changes of food web indicators from marine ecosystems highlight the need to set well defined reference levels and thresholds when managing marine resources. It is not possible to set one reference level for all systems regardless of size, depth, or type of ecosystems. Nor it is useful to set reference levels for similar systems in different ocean basins, even if these systems seem to be similar in physical characteristics, i.e. being eastern boundary current systems. Since different baseline references exist and marine ecosystems seem to have intrinsic differences due to ecosystem dynamics, establishing absolute reference values for ecosystem indicators as a whole seems not to be a suitable solution to advance the ecosystem-based and precautionary approach. Reference levels for ecosystem indicators should be developed for individual ecosystems or ecosystems with the same typologies (similar location, ecosystem type, etc.) and not benchmarked against all other ecosystems.

Acknowledgments

We hereby wish to thank Marieke Steuben for proof reading the manuscript. We also thank all the members of the Ecopath Research and Development Consortium and the authors of all the models used in this analysis for their significant contribution to the body of Ecopath models available for this analysis (see Table S1).

Author Contributions

Conceived and designed the experiments: JH MC SL. Performed the experiments: JH MC SL. Analyzed the data: JH MC SL. Contributed reagents/materials/analysis tools: JH MC SL. Wrote the paper: JH MC SL LM VC. Contributed data from models: JH MC SL LM VC.

References

1. McCann K (2007) Protecting biostructure. Nature 446: 29.
2. Jackson JBC, Kirby MX, Berger WH, Bjorndal KA, Botsford LW, et al. (2001) Historical Overfishing and the Recent Collapse of Coastal Ecosystems. Science 293: 629–638.
3. Hilborn R, Walters CJ (1992) Quantitative Fisheries Stock Assessment: Choice, Dynamics and Uncertainty. Boston: Kluwer Academic Publishers. 570 p.
4. Christensen V, Guénette S, Heymans JJ, Walters C, Watson R, et al. (2003) Hundred-year decline of North Atlantic predatory fishes. Fish and Fisheries 4: 1–24.
5. Pikitch EK, Santora C, Babcock EA, Bakun A, Bonfil R, et al. (2004) Ecosystem-Based Fishery Management. Science 305: 346–347.
6. Christensen V, Walters CJ (2004) Trade-offs in Ecosystem-scale Optimization of Fisheries Management Policies. Bulletin of Marine Science 74: 549–562.
7. Heymans JJ, Sumaila UR, Christensen V (2009) Policy options for the northern Benguela ecosystem using a multispecies, multifleet ecosystem model. Progress in Oceanography 83: 417–425.
8. Link J (2010) Ecosystem based fisheries management: Confronting Tradeofs. New York: Cambridge University Press.
9. Christensen V, Walters CJ (2005) Using ecosystem modeling for fisheries management: Where are we? ICES CM 2005: 1–17.
10. Heithaus MR, Frid A, Wirsing AJ, Worm B (2008) Predicting ecological consequences of marine top predator declines. Trends in Ecology and Evolution 23: 202–210.
11. Ulanowicz RE, Wulff F (1991) Comparing Ecosystem Structures: The Chesapeake Bay and the Baltic Sea. In: Cole J, Lovett G, Findlay S, editors. Comparative Analyses of Ecosystems: Patterns, Mechanisms, and Theories. 1 ed. New York: Springer-Verlag. pp. 140–166.
12. Dunne JA (2006) The network structure of foodwebs. In: Pascual M, Dunne JA, editors. Ecological networks: linking structure to dynamics in food webs. Oxford: Oxford University Press. pp. 27–86.
13. Perry RI, Cury P, Brander K, Jennings S, Möllmann C, et al. (2010) Sensitivity of marine systems to climate and fishing: Concepts, issues and management responses. Contributions from Advances in Marine Ecosystem Modelling Research II 23-26 June 2008, Plymouth, UK 79: 427–435.
14. Walters CJ, Martell SJD (2004) Fisheries ecology and management. Princeton: Princeton University Press. 399 p.
15. Allesina S, Pascual M (2008) Network structure, predator–prey modules, and stability in large food webs. Theoretical Ecology 1: 55–64.
16. Belgrano A, Scharler UM, Dunne J, Ulanowicz RE (2005) Aquatic food webs: an ecosystem approach. Oxford: Oxford University Press.
17. Pinnegar JK, Blanchard JL, Mackinson S, Scott RD, Duplisea DE (2005) Aggregation and removal of weak-links in food-web models: system stability and recovery from disturbance. Ecological Modelling 184: 229–248.
18. De Boer RJ (2012) Which of Our Modeling Predictions Are Robust? PLoS Comput Biol 8: e1002593.
19. Shin YJ, Bundy A, Shannon LJ, Blanchard JL, Chuenpagdee R, et al. (2012) Global in scope and regionally rich: an IndiSeas workshop helps shape the future of marine ecosystem indicators. Reviews in Fish Biology and Fisheries DOI 10.1007/s11160-012-9252-z: 1–11.
20. Cury PM, Shannon LJ, Roux J-P, Daskalov GM, Jarre A, et al. (2005) Trophodynamic indicators for an ecosystem approach to fisheries. ICES Journal of Marine Science 62: 430–442.
21. Farnsworth KD, Lyashevska O, Fung T (2012) Functional complexity: The source of value in biodiversity. Ecological Complexity 11: 46–52.
22. Christensen V, Pauly D (1993) Trophic models of aquatic ecosystems. ICLARM Conference Proceedings 26. Manila, Philippines: International Center for Living Resources Management. 1–390 p.
23. Pauly D, Christensen V, Walters C (2000) Ecopath, Ecosim, and Ecospace as tools for evaluating ecosystem impacts of fisheries. ICES Journal of Marine Science 57: 697–706.
24. Christensen V, Walters CJ (2004) Ecopath with Ecosim: methods, capabilities and limitations. Ecological Modelling 172: 109–139.
25. Polovina JJ (1984) Model of a coral reef ecosystem I. The ECOPATH model and its application to French Frigate Shoals. Coral Reefs 3: 1–11.
26. Christensen V, Pauly D (1992) ECOPATH II - a software for balancing steady-state ecosystem models and calculating network characteristics. Ecological Modelling 61: 169–185.
27. Walters C, Pauly D, Christensen V, Kitchell JF (2000) Representing Density Dependent Consequences of Life History Strategies in Aquatic Ecosystems: EcoSim II. Ecosystems 3: 70–83.
28. Christensen V, Walters C, Pauly D (2005) Ecopath with Ecosim: A User's guide. Vancouver, BC: Fisheries Centre, University of British Columbia. 154pp. p.
29. Walters C, Christensen V, Pauly D (1997) Structuring dynamic models of exploited ecosystems from trophic mass-balance assessments. Reviews in Fish Biology and Fisheries 7: 139–172.
30. Cohen JE, Luczak T, Newman CM, Zhou Z-M (1990) Stochastic structure and nonlinear dynamics of food webs: qualitative stability in a Lotka-Volterra Cascade Model. Proceedings of the Royal Society B: Biological Sciences 240: 607–627.
31. Williams RJ, Martinez ND (2000) Simple rules yield complex food webs. Nature 404: 180–183.
32. Coll M, Lotze HK, Romanuk TN (2008) Structural degradation in Mediterranean Sea food webs: Testing ecological hypotheses using stochastic and mass-balance modelling. Ecosystems 11: 939–960.
33. Lindeman RL (1942) The trophic-dynamic aspect of ecology. Ecology 23: 399–418.
34. Odum EP (1969) The strategy of ecosystem development. Science 164: 262–270.
35. Ulanowicz RE (1980) An hypothesis on the development of natural communities. Journal of theoretical Biology 85: 223–245.
36. Ulanowicz RE (1983) Identifying the structure of cycling in ecosystems. Mathematical BioScience 65: 219–237.
37. Moreno T, Castro JJ (1998) Trophic structure of the Maspalomas lagoon (Gran Canaria, Canary Islands), a regenerated ecosystem of brackish water. Boletin do Museu Municipal do Funchal (História Natural) Sup. no.5: 245–261.
38. Mackinson S, Daskalov G (2007) An ecosystem model of the North Sea for use in research supporting the ecosystem approach to fisheries management: description and parameterisation. Lowestoft: CEFAS. Cefas Science Series Technical Report 142 Cefas Science Series Technical Report 142. 200pp. p.
39. Libralato S, Pastres R, Pranovi F, Raicevich S, Granzotto A, et al. (2002) Comparison between the energy flow networks of two habitats in the Venice Lagoon. Marine Ecology 23: 228–236.
40. Stobberup KA, Ramos VDM, Coelho ML (2004) Ecopath model of the Cape Verde coastal ecosystem. In: Palomares ML, Pauly D, editors. West African marine ecosystems: models and fisheries impacts: Fisheries Center Research Reports 12(7). Vancouver, BC: UBC Fisheries Centre. pp. 39–56.
41. Heymans JJ, Howell KL, Ayers M, Burrows MT, Gordon JDM, et al. (2011) Do we have enough information to apply the ecosystem approach to management of deep-sea fisheries? An example from the West of Scotland. ICES Journal of Marine Science 68: 265–280.
42. Pinnegar JK, Polunin NVC (2004) Predicting indirect effects of fishing in Mediterranean rocky littoral communities using a dynamic simulation model. Ecological Modelling 172: 249–267.

43. Dunne JA, Williams RJ, Martinez ND (2002) Food-web structure and network theory: The role of connectance and size. Proceedings of the National Academy of Sciences of the United States of America 99: 12917–12922.
44. Finn JT (1976) Measures of ecosystem structure and function derived from analysis of flows. Journal of theoretical Biology 56: 363–380.
45. Ulanowicz RE (1986) Growth and Development: Ecosystems Phenomenology. Lincoln, NE: toExcel Press. 203 p.
46. Ulanowicz RE (2000) Toward the Measurement of Ecological Integrity. In: Pimentel D, Westra L, Noss RF, editors. Ecological integrity: integrating environment, conservation, and health. Washington DC: Island Press. pp. 99–113.
47. Heymans JJ, Guénette S, Christensen V (2007) Evaluating network analysis indicators of ecosystem status in the Gulf of Alaska. Ecosystems 10: 488–502.
48. Ulanowicz RE (2004) Quantitative methods for ecological network analysis. Computational Biology and Chemistry 28: 321–339.
49. Libralato S (2008) System Omnivory Index. In: Jørgensen SE, Fath BD, editors. Ecological Indicators Oxford: Elsevier. pp. 3472–3477.
50. Pauly D, Christensen V, Dalsgaard J, Froese R, Torres FJ (1998) Fishing down marine food webs. Science 279: 860–863.
51. Libralato S, Coll M, Tudela S, Palomera I, Pranovi F (2008) Novel index for quantification of ecosystem effects of fishing as removal of secondary production. Marine Ecology Progress Series 355: 107–129.
52. Libralato S, Christensen V, Pauly D (2006) A method for identifying keystone species in food web models. Ecological Modelling 195: 153–171.
53. Coll M, Libralato S, Tudela S, Palomera I, Pranovi F (2008) Ecosystem Overfishing in the Ocean. PLoS ONE 3: e3881.
54. Mora C, Myers RA, Coll M, Libralato S, Pitcher TJ, et al. (2009) Management Effectiveness of the World's Marine Fisheries. PLoS Biol 7: e1000131.
55. Murawski SA (2000) Definitions of overfishing from an ecosystem perspective. ICES Journal of Marine Science 57: 649–658.
56. Heymans JJ, Coll M, Libralato S, Christensen V (2012) 9.06 Ecopath theory, modelling and application to coastal ecosystems. In: Wolanski E, McLusky DS, editors. Treatise on Estuarine and Coastal Science: Elsevier. pp. 93–113.
57. Coll M, Libralato S (2012) Contributions of food web modelling to the ecosystem approach to marine resource management in the Mediterranean Sea. Fish and Fisheries 13: 60–88.
58. Anderson MA, Gorley RN, Clarke KR (2008) PERMANOVA+ for PRIMER: Guide to software and statistical methods. Plymouth, UK: PRIMER-E.
59. Power ME, Tilman D, Estes JA, Menge BA, Bond WJ, et al. (1996) Challenges in the Quest for Keystones. BioScience 46: 609–620.
60. Piraino S, Fanelli G, Boero F (2002) Variability of species' roles in marine communities: change of paradigms for conservation priorities. Marine Biology 140: 1067–1074.
61. Ulanowicz RE, Puccia CJ (1990) Mixed trophic impacts in ecosystems. Coenoses 5(1): 7–16.
62. Greenacre MJ (1984) Theory and applications of corespondence analysis. London: Academic Press.
63. Kakarenkov V, Legendre P (2002) Nonlinean redundancy analysis and cononical correspondence analysis based on polynomial regression. Ecology 83: 1146–1161.
64. Fellenberg K, Hauser NC, Brors B, Neutzner A, Hoheisel JD, et al. (2001) Correspondence analysis applied to microarray data. Proceedings of the National Academy of Sciences 98: 10781–10786.
65. Lotze HK, Lenihan HS, Bourque BJ, Bradbury RH, Cooke RG, et al. (2006) Depletion, Degradation, and Recovery Potential of Estuaries and Coastal Seas. Science 312: 1806–1809.
66. Costello MJ, Coll M, Danovaro R, Halpin P, Ojaveer H, et al. (2010) A Census of Marine Biodiversity Knowledge, Resources, and Future Challenges. PLoS ONE 5: e12110.
67. Plagányi ÉE, Butterworth DS (2004) A critical look at the potential of Ecopath with Ecosim to assist in practical fisheries management. African Journal of Marine Science 26: 261–287.
68. Plagányi ÉE (2007) Models for an ecosystem approach to fisheries. FAO Fisheries Technical Paper No. 477. Rome: FAO. 477 477. 108 p.
69. Dame JK, Christian RR (2007) A statistical test of network analysis: Can it detect differences in food web properties? Ecosystems 10: 906–923.
70. Fulton EA, Smith ADM, Punt AE (2005) Which ecological indicators can robustly detect effects of fishing? ICES J Mar Sci 62: 540–551.
71. Lassen H, Pedersen SA, Frost H, Hoff A (2013) Fishery management advice with ecosystem considerations. ICES Journal of Marine Science: Journal du Conseil 70: 471–479.
72. EC (2008) Directive 2008/56/EC, of 17 June 2008 establishing a framework for community action in the field of marine environmental policy (Marine Strategy Framework Directive). 22.p.
73. Shin YJ, Bundy A, Shannon LJ, Simier M, Coll M, et al. (2010) Can simple be useful and reliable? Using ecological indicators to represent and compare the states of marine ecosystems. ICES J Mar Sci 67: 717–731.
74. Samhouri JF, Levin PS, Ainsworth CH (2010) Identifying Thresholds for Ecosystem-Based Management. PLoS ONE 5: e8907.
75. Levin PS, Fogarty MJ, Murawski SA, Fluharty D (2009) Integrated Ecosystem Assessments: Developing the Scientific Basis for Ecosystem-Based Management of the Ocean. PLoS Biol 7: e1000014.
76. Gribble NA (2003) GBR-prawn: modelling ecosystem impacts of changes in fisheries management of the commercial prawn (shrimp) trawl fishery in the far northern Great Barrier Reef. Fisheries Research 65: 493–506.
77. Bozec Y-M, Gascuel D, Kulbicki M (2004) Trophic model of lagoonal communities in a large open atoll (Uvea, Loyalty islands, New Caledonia). Aquatic Living Resources 17: 151–162.
78. Venier JM, Pauly D (1997) Trophic dynamics of a Florida Keys coral reef ecosystem. 8th International Coral Reef Symposium. Panama City: Smithsonian Tropica Research Institute, Balboa, Panama. pp. 915–920.
79. Opitz S (1993) A quantitative model of the trophic interactions in a Caribbean coral reef ecosystem. In: Christensen V, Pauly D, editors. Trophic models of aquatic ecosystems ICLARM Conference Proceedings 26. 1 ed. Manila, Philippines: International Center for Living Aquatic Resources Management. pp. 259–267.
80. Buzzelli C (2012) 9.13 Ecosystem modelling in small subtropical estuaries and embayments. In: Wolanski E, McLusky DS, editors. Treatise on Estuarine and Coastal Science: Elsevier.pp. 331–353.
81. Baird D, Ulanowicz RE (1993) Comparative study on the trophic structure, cycling and ecosystem properties of four tidal estuaries. Marine Ecology Progress Series 99: 221–237.
82. Baird D, McGlade JM, Ulanowicz RE (1991) The comparative ecology of six marine ecosystems. Philosophical Transactions of the Royal Society of London, B 333: 15–29.
83. Jarre-Teichmann A, Christensen V (1998) Comparative modelling of trophic flows in four large upwelling ecosystems: global vs. local effects. In: Cury P, Mendelssohn R, Roy C, Bakun A, Durand MH, et al., editors. Global vs local changes in upwelling ecosystems Proceedings of the first CEOS symposium, 5–9 September 1994 Monterey, CA.
84. Jarre-Teichmann A, Shannon LJ, Moloney CL, Wickens PA (1998) Comparing trophic flows in the southern Benguela to those in other upwelling ecosystems. South African Journal of marine Science 19: 391–414.
85. Heymans JJ, Baird D (2000) A carbon flow model and network analysis of the northern Benguela upwelling system, Namibia. Ecological Modelling 126: 9–32.
86. Vasconcellos M, Mackinson S, Sloman K, Pauly D (1997) The stability of trophic mass-balance models of marine ecosystems: a comparative analysis. Ecological Modelling 100: 125–134.
87. Heymans JJ, Baird D (2000) Network analysis of the northern Benguela ecosystem by means of NETWRK and ECOPATH. Ecological Modelling 131: 97–119.
88. Helmke P, Neuer S, Lomas MW, Conte M, Freudenthal T (2010) Cross-basin differences in particulate organic carbon export and flux attenuation in the subtropical North Atlantic gyre. Deep Sea Research Part I: Oceanographic Research Papers 57: 213–227.
89. Carr M-E, Kearns EJ (2003) Production regimes in four Eastern Boundary Current systems. Deep-Sea Research II 50: 3199–3221.
90. Tomczak MT, Müller-Karulis B, Järv L, Kotta J, Martin G, et al. (2009) Analysis of trophic networks and carbon flows in south-eastern Baltic coastal ecosystems. Progress in Oceanography 81: 111–131.
91. Lotze HK, Coll M, Dunne JA (2011) Historical Changes in Marine Resources, Food-web Structure and Ecosystem Functioning in the Adriatic Sea, Mediterranean. Ecosystems 14: 198–222.
92. Tomczak MT, Heymans JJ, Yletyinen J, Niiranen S, Otto SA, et al. (2013) Ecological Network Indicators of Ecosystem Status and Change in the Baltic Sea. PLoS ONE 8: e75439.
93. Worm B, Hilborn R, Baum JK, Branch TA, Collie JS, et al. (2009) Rebuilding Global Fisheries. Science 325: 578–585.
94. Shannon LJ, Christensen V, Walters CJ (2004) Modelling stock dynamics in the southern Benguela ecosystem for the period 1978–2002. African Journal of Marine Science 26: 179–196.
95. Heymans JJ, Shannon LJ, Jarre A (2004) Changes in the northern Benguela ecosystem over three decades: 1970s, 1980s and 1990s. Ecological Modelling 172: 175–195.
96. Heymans JJ, Guénette S, Christensen V, Trites A (2005) Changes in the Gulf of Alaska ecosystems due to ocean climate change and fishing. ICES CM 2005: 1–31.
97. Heymans JJ (2004) The effects of internal and external control on the northern Benguela ecosystem. In: Sumaila UR, Steinshamn SI, Skog M, Boyer DC, editors. Ecological, economic and social aspects of Namibian fisheries. Netherlands: Eburon. pp. 29–52.
98. Paine RT (1969) A note on trophic complexity and community stability. American Naturalist 103: 91–93.
99. Butchart SHM, Walpole M, Collen B, van Strien A, Scharlemann JPW, et al. (2010) Global Biodiversity: Indicators of recent declines. Science Express 29 April 2010: 1–9.
100. UN (1987) Convention on Wetlands of International Importance especially as Waterfowl Habitat. Ramsar (Iran), 2 February 1971. UN Treaty Series No. 14583. As amended by the Paris Protocol, 3 December 1982, and Regina Amendments, 28 May 1987.
101. Costanza R, d'Arge R, de Groot R, Farber S, Grasso M, et al. (1997) The value of the world's ecosystem services and natural capital. Nature 387: 253–260.
102. Pranovi F, Libralato S, Raicevich S, Granzotto A, Pastres R, et al. (2003) Mechanical clam dredging in Venice lagoon: ecosystem effects evaluated with a trophic mass-balance model. Marine Biology 143: 393–403.
103. Essington TE (2007) Evaluating the sensitivity of a trophic mass-balance model (Ecopath) to imprecise data inputs. Canadian Journal of Fisheries and Aquatic Science 64: 628–637.

10

Effects of Climatic Factors and Ecosystem Responses on the Inter-Annual Variability of Evapotranspiration in a Coniferous Plantation in Subtropical China

Mingjie Xu[1,2], Xuefa Wen[1], Huimin Wang[1]*, Wenjiang Zhang[3], Xiaoqin Dai[1], Jie Song[4], Yidong Wang[5], Xiaoli Fu[1], Yunfen Liu[1], Xiaomin Sun[1], Guirui Yu[1]

1 Qianyanzhou Ecological Station, Key Laboratory of Ecosystem Network Observation and Modeling, Institute of Geographic Sciences and Natural Resources Research, Chinese Academy of Sciences, Beijing, China, **2** University of Chinese Academy of Sciences, Beijing, China, **3** State Key Laboratory of Hydraulics and Mountain River Engineering, Sichuan University, Chengdu, China, **4** Department of Geography, Northern Illinois University, DeKalb, Illinois, United States of America, **5** Tianjin Key Laboratory of Water Resources and Environment, Tianjin Normal University, Tianjin, China

Abstract

Because evapotranspiration (ET) is the second largest component of the water cycle and a critical process in terrestrial ecosystems, understanding the inter-annual variability of ET is important in the context of global climate change. Eight years of continuous eddy covariance measurements (2003–2010) in a subtropical coniferous plantation were used to investigate the impacts of climatic factors and ecosystem responses on the inter-annual variability of ET. The mean and standard deviation of annual ET for 2003–2010 were 786.9 and 103.4 mm (with a coefficient of variation of 13.1%), respectively. The inter-annual variability of ET was largely created in three periods: March, May–June, and October, which are the transition periods between seasons. A set of look-up table approaches were used to separate the sources of inter-annual variability of ET. The annual ETs were calculated by assuming that (a) both the climate and ecosystem responses among years are variable ($V_{cli\text{-}eco}$), (b) the climate is variable but the ecosystem responses are constant (V_{cli}), and (c) the climate is constant but ecosystem responses are variable (V_{eco}). The ETs that were calculated under the above assumptions suggested that the inter-annual variability of ET was dominated by ecosystem responses and that there was a negative interaction between the effects of climate and ecosystem responses. These results suggested that for long-term predictions of water and energy balance in global climate change projections, the ecosystem responses must be taken into account to better constrain the uncertainties associated with estimation.

Editor: Dafeng Hui, Tennessee State University

Funding: This work was funded by the National Key Basic Research Special Foundation Project (2010CB833501), Major Program of National Natural Science Foundation of China (31290221), National Natural Science Foundation of China (31070559, 31210103920) and the National Key Basic Research Program of China (2012CB416903). The funders had no role in study design, data collection and analysis, decision to publish, or preparation of the manuscript.

Competing Interests: The authors have declared that no competing interests exist.

* E-mail: wanghm@igsnrr.ac.cn

Introduction

Evapotranspiration (ET) is the second largest component of the water cycle, consuming a large proportion (60–90%) of annual precipitation [1–3]. ET plays an important role in terms of the energy and nutrient exchange in ecosystems [4–6] and affects many other important ecological processes [7]. Forests cover approximately 30% of the global land surface, but their evapotranspiration accounts for more than 45% of the global ET across terrestrial ecosystems [8]. In the context of climate change, forests are considered to provide important climate forcings and feedbacks [9]. While climate change may adversely affect ecosystem functions, forests could be managed to mitigate climate change [10]. As ET is an integral part of forest biogeophysical processes [9], meteorologists and ecologists require an understanding of how the forest ET responds to environmental drivers so that they can better predict how the biosphere will affect (or will be affected by) climate change [11–16].

The factors that control ET have been well documented for many ecosystems on intra-annual scales [17–20]. Evapotranspiration is controlled by abiotic and biotic factors, many of which are projected to change on multiple spatial and temporal scales [21]. Generally, ET is positively related to net radiation (R_n), air temperature (T_a), and the vapor pressure deficit (VPD) [12,22–24]. However, soil water content (SW) sometimes becomes the main control factor under drought conditions [25]. As for biotic factors, leaf area index (LAI) [26] and additional model parameters that have been used to represent the canopy characteristics have influences on ET. For example, canopy conductance, which can be computed by inverting canopy resistance from the Penman-Monteith equation [27–30], can help estimate ET. These parameters were found to be closely related to ET, especially under drought stress [19,31].

In contrast to intra-annual variability, the inter-annual variability of ET and its driving mechanisms are poorly understood. Although many forest evaporation datasets have been collected worldwide using the eddy covariance technique and compiled over years to decades [32,33], only a small fraction of studies have published their long-term ET records. Furthermore, most of these studies have concentrated on tropical forests [20,34,35], and boreal forests [36,37]. There have also been a few studies reporting the inter-annual variability of ET based on 3–5 year measurements in temperate forests [31,38,39]. Some of these studies have determined that the inter-annual variability of ET was small (coefficient of variation (CV) <10%) in contrast to the considerable year-to-year variation in annual rainfall [35,37,39]. Others who held the opposite views have insisted that the inter-annual variability of ET was significant and could not be neglected [38]. In addition, several models have been developed to study or predict long-term variation in ET [28]. However, these models may be inadequate for evaluating the inter-annual variability of ET because of their dependency on climatic factors [40–42].

The inter-annual variability of ET can be affected by biotic and abiotic factors, similar to those that operate on shorter time scales [21,31]. In addition to the factors mentioned above, the leaf phenology should be given careful consideration. As noted by Xue et al. [38], the reason that Ohta et al. [37] could not fully explain the inter-annual variability of ET in a Siberian larch forest was because they neglected the inter-annual variability in the dates of leaf expansion. Zha et al. [36] determined that the inter-annual variability of ET was controlled by early spring soil temperatures, which are a key factor affecting the growing season length (GSL).

In this study, the sources of the inter-annual variability of ET were separated into two parts: (1) those induced by climatic factors (extreme climatic events included) and (2) those produced by ecosystem responses including the changes in phenology (e.g., GSL), ecosystem structure (e.g., LAI), and physiological changes in plants (e.g., stomatal and canopy conductance) [43,44]. Some studies have proposed that ecosystem responses overwhelm the direct influences of environmental factors on the inter-annual variability of ET [5,21,34], but few researchers have separated the sources of inter-annual variability of ET into 'biological' and 'climatological' components because they are quite difficult to tease apart. However, in recent studies, the inter-annual variations of carbon fluxes were successfully separated into those driven by environmental factors or by biological factors with a novel statistical method [44,46]. These studies suggested that it would be possible to separate the sources of inter-annual variability of ET because water processes are highly correlated with carbon assimilation in an ecosystem.

Southern China is characterized by a humid monsoon climate and has large subtropical coniferous plantations, which account for 41% of the total subtropical forest area [47]. Thus, it is important to conduct research on the carbon and water cycles in this region. Previous studies have indicated that the carbon fluxes are characterized by strong inter-annual variations due to seasonal droughts in the summer or low temperatures in the early spring [43,48]. Considering that water and carbon cycles are directly coupled within the stomata through transpiration, photosynthesis, and respiration [12], evapotranspiration may respond to climatic drivers as do the carbon fluxes. Therefore, the goal of this study is to investigate whether the inter-annual variability of ET in this region is strong and to determine the main source of the inter-annual variability of ET.

Eight years (2003–2010) of continuous eddy-covariance measurements that were recorded at a subtropical coniferous forest site were analyzed. Based on previous research on the carbon cycle [44], a set of look-up tables (LUTs) were employed to separate the sources of the inter-annual variability of ET. The objectives of this study were to (1) characterize the seasonal and inter-annual variation of ET and (2) evaluate the contributions of the two types of driving forces, climatic factors and ecosystem responses, to the inter-annual variability of ET.

Materials and Methods

Site description

This study was conducted in a subtropical coniferous plantation (26°44′29″N, 115°03′29″E, 102 m in elevation above sea level) at Qianyanzhou Ecological Research Station (QYZ), which is owned by the Institute of Geographic Sciences and Natural Resources Research of the Chinese Academy of Sciences. No specific permits were required for the described field studies in the research site. The field studies did not involve endangered or protected species. The site is located in a typical red soil hilly region in south China, with a subtropical monsoon climate. The prevailing wind direction of this climate regime is north-northwest in the winter and south-southeast in the summer. The coniferous trees were planted around 1985. The dominated species are Masson pine (*Pinus massoniana* Lamb.), Slash pine (*Pinus elliottii* Englem.) and Chinese fir (*Cunninghamia lanceolata* Hook.), with a tree density of about 1460 stems ha^{-1} and total biomass of 106 t ha^{-1}. The red soil is weathered from red sand rock, and soil texture is divided into 2.0–0.05 mm (17%), 0.05–0.002 mm (68%) and <0.002 mm (15%). Further details of QYZ site can be referred to Wen et al. [49] and Wang et al. [50].

According to meteorological observations from 1989–2010 at QYZ, the mean annual temperature is 18.0°C and mean annual precipitation is 1509.0 mm, while the growing season (April–October) mean values are 23.9°C and 1050.8 mm, respectively (Fig. 1). Eight years (2003–2010) of flux data were observed with the eddy covariance method in a subtropical forest plantation at QYZ and used for this study. During the eight observation years, the dataset represented much of the typical inter-annual variability at this site, including both typical and extreme climatic events. In 2002–2003, a typical El Niño event occurred, which resulted in an extreme summer drought in 2003 [51], and in 2009–2010, there was a strong La Niña event [52].

Observation and instrumentation

The eddy covariance flux observation system was established in a subtropical coniferous plantation at QYZ site in 2002 (Fig. 2). The above-canopy flux was measured at a height of 39.6 m by instruments loaded on a ventilated tower. The wind velocity was detected by a 3-D sonic anemometer (Model CSAT3, Campbell Scientific Inc., USA), and the variations in CO_2 and water vapor concentration were measured with a LI-7500 open-path CO_2/H_2O analyzer (Licor Inc., USA). All signals were sampled at a frequency of 10 Hz and the CO_2 and H_2O fluxes were calculated and recorded at 30 min intervals by a CR5000 datalogger (Campbell Scientific Inc., USA).

Air temperature and relative humidity sensors (Model HMP45C, Campbell) were mounted in ventilated shield at the heights of 23.6 m and 39.6 m above the ground. Soil temperature and soil water content were measured at the depths of 5 cm with thermocouples (105T and 107-L, Campbell), and TDR probes (Model CS615-L, Campbell), respectively. Radiation measurements were made using a four-component net radiometer (Model CNR-1, Kipp & Zonen, Netherlands), a pyranometer (Model CM11, Kipp & Zonen) and a quantum sensor of photosynthetically active radiation (Model LI190SB, Licor Inc.). Rainfall was monitored with a rain gauge (Model 52203, RM Young Inc., USA). Meteorological variables were sampled at 1 Hz, and 30 min average data were recorded with three CR10X datalog-

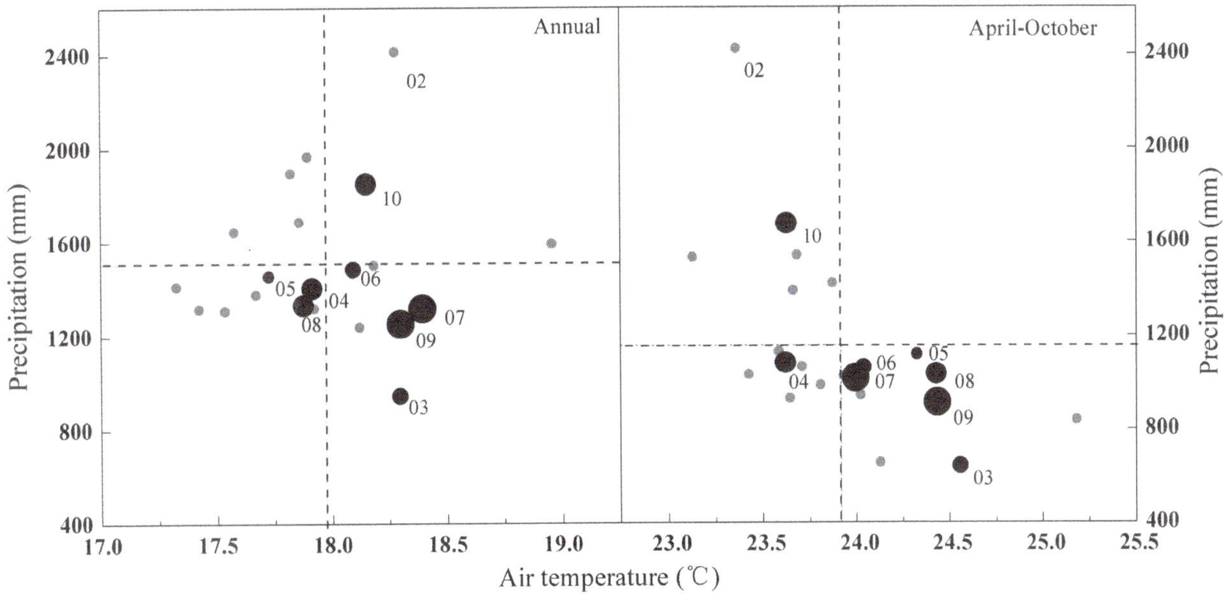

Figure 1. Cumulative precipitation versus average air temperature of the whole year and the growing season (April–October) during the period 1989–2010. For the period of 2003–2010, the size of black dot is approximately proportional to the ET amount. The two-digit numbers (YY) denote the years from 2003 to 2010 (20YY). Dotted lines represent the averages over the 20 year period.

gers, and a CR23X datalogger with a 25-channel solid-state multiplexer (Campbell Scientific Inc., USA).

Flux correction and gap filling

This study adopted the methods of calculating and correcting the carbon dioxide and water vapor fluxes that were presented by Wen et al. [48]. The CO_2 and H_2O fluxes were calculated every 30 minutes from the 10 Hz raw data. Processing of the flux data was performed using routine methods, including three-dimensional rotation [53], a correction for the effects of the fluctuations of air density (the Webb, Pearman and Leuning density correction (WPL correction)) [54], storage calculations and spurious data removal

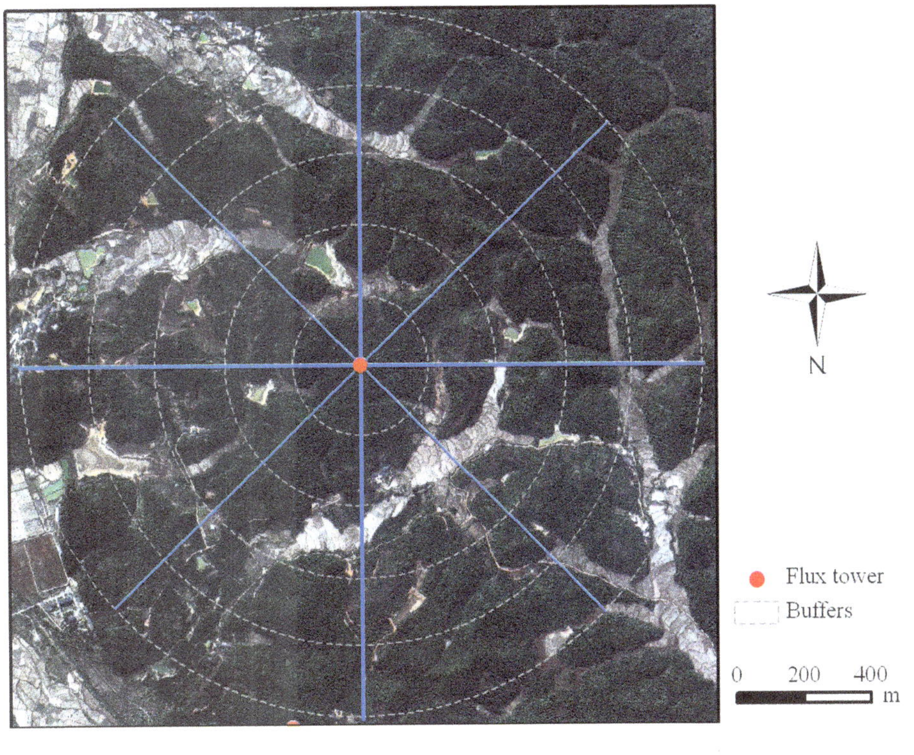

Figure 2. The IKONOS image (Nov. 7, 2003) around the QYZ flux tower (red point). The interval between buffer circles is 200 m [43].

[43,49,55]. Spurious data caused by rainfall, water condensation or system failure were removed from the dataset. To avoid the possible underestimation of the fluxes under stable conditions at night, nighttime data (solar elevation angle<0) were excluded when the value of friction velocity (u_*) was below the relevant thresholds, which were identified with the method described by Reichstein et al. [56]. The threshold values of u_* ranged from 0.16 to 0.22 m s^{-1}, with an average value of 0.20 m s^{-1} for the years from 2003 to 2010. Approximately 23.8% of the half-hour data were excluded for the daytime and 85.4% were excluded for the nighttime. The data gaps were filled using the look-up table method [56]. Any data gaps in meteorological variables were filled using the mean diurnal variation method [57]. Further details of data processing are presented in the previous studies of ChinaFLUX [43,48,55,58]. In this study, the H_2O flux was considered to be ET of the whole ecosystem, as in previous studies [58,59].

Partitioning the sources of inter-annual variation in ET

To separate the effects of climate drivers and ecosystem responses on the inter-annual variability of carbon fluxes, Marcolla et al. [44] established a set of look-up tables (LUTs) according to selected key factors that influence the ecosystem carbon assimilation, such as temperature (T_a) and photosynthetically active radiation (PAR). Using these LUTs, these researchers statistically separated the two sources of inter-annual variability in carbon fluxes. In this study, a similar methodology was applied, but the LUTs for ET were built with the environmental factors R_n and VPD, which have a strong influence on the ET and widely used in combination with Penman-Monteith Equation [12].

To create the LUTs, the half-hour flux values and corresponding meteorological data were integrated to daily scale for each year. Then, the data were classified according to R_n (class width 1 Wm^{-2}) and VPD (class width 0.1 kPa). With the LUTs, we could then use the meteorological data (R_n and VPD) to estimate a corresponding ET value. Specifically, the ETs were calculated in three different ways:

(1) The yearly LUTs were established using the meteorological data (R_n and VPD) and the corresponding ET. Then, the yearly observed meteorological data from the entire time series (DOY (day-of-year) 1–365 (366 for leap year)) was used to reproduce the ET series according to the yearly LUTs. The ET series and the inter-annual variability of ET calculated with this method can be considered to include effects of both climatic factors and the ecosystem responses ($V_{cli\text{-}eco}$).
(2) To isolate the climate factors, an average table that included meteorological data and ET for the entire time series from DOY 1 to 366 (including leap years) was established based on the eight-year data. Then, the LUT_{avg} based on the average table was established. In the LUT_{avg}, the ETs represent the average ecosystem responses for the eight years that were measured. Finally, the LUT_{avg} is looked up using the yearly observed meteorological data from the entire time series to reassign ET series. The inter-annual variability of ET that is calculated through this method can be considered to be the inter-annual variability caused by variable climatic factors (V_{cli}).
(3) To isolate the ecosystem responses, the eight-year averaged daily meteorological data with entire time series (DOY 1–366) can be used to look up the yearly LUTs established in approach (1) to reproduce ET series for each year. The inter-annual variability of ET calculated through this method can be considered to be the inter-annual variability caused by variable ecosystem responses (V_{eco}).

The standard deviations of the yearly values of ET that were obtained through the three approaches ($V_{cli\text{-}eco}$, V_{cli} and V_{eco}) were used as estimates of what portion of the inter-annual variability of ET was generated from which sources. Then the interaction of climate drivers and ecosystem responses was analyzed based on these results.

Correlation between climatic drivers and ecosystem evapotranspiration

To identify the relationship between climatic drivers and ET, we selected five climatic factors, net radiation (R_n), air temperature (T_a), precipitation (P), vapor pressure deficit (VPD) and soil water content (SW), that were important according to the previous studies [31,36,37,60]. We analyzed their correlations with the ecosystem ET on the annual scale, placing emphasis on the time lag effects between the climatic factors and ET.

For this purpose, annual average T_a, SW, VPD and cumulative R_n, precipitation and ET were calculated. Multiple "yearly" statistics (approximately 80 values from the eight years) were obtained using the 12-month intervals and shifting them one month at a time [44,45]. To investigate the lag of the ET's response to the climatic factors, we shifted the climatic series backward one month at a time (up to twelve months) to calculate the correlations between climate drivers and ET. Student's *t*-tests were applied to verify the statistical significance of correlation coefficients.

Results

Seasonal and inter-annual variation of ecosystem evapotranspiration and the corresponding environmental factors

The ET, net radiation (R_n), air temperature (T_a), vapor pressure deficit (VPD), precipitation (P) and soil water contents (SW) are characterized in Figure 3. The ecosystem ET and corresponding R_n, T_a, and VPD values showed a single peak, and reached their maximums in July (Fig. 3). The precipitation also had a strong secondary peak. The precipitation decreased in July as the air temperature reached a maximum. The asynchronous seasonality of temperature and precipitation, an important characteristic of the eastern Asian monsoon climate, often results in summer droughts in southern China. In addition, precipitation in the first half year was usually higher than in the second half of the year. The seasonal pattern in soil water content (SW) (at 5 cm) was not as obvious as the other factors. It remained between 0.12 m^3 m^{-3} and 0.22 m^3 m^{-3} and decreased after July, with values in the second half year generally lower than in the first half year, in accordance with precipitation pattern.

Figure 4 and Table 1 show the environmental condition and ET for each year at QYZ. Over the eight years, the average annual ET was 786.9±103.4 mm (CV = 13.1%) with a substantial variation, from 567.9 mm in 2005 to 876.5 mm in 2009. The years of 2007 and 2009 had the highest ET, whereas 2005 and 2006 had the lowest values (Table 1). Correspondingly, the T_a was higher in 2007 and 2009 in the early spring, and the R_n in 2009 was higher, with positive anomalies for ten months (Fig. 4). In 2004 and 2008, the climatic factors were close to the multi-year averages, except for a lower R_n and T_a in the early spring of 2008. In 2010, although the T_a was lower, the precipitation was abundant. In 2003, the negative anomalies in precipitation were large, and the asynchrony between precipitation and temperature

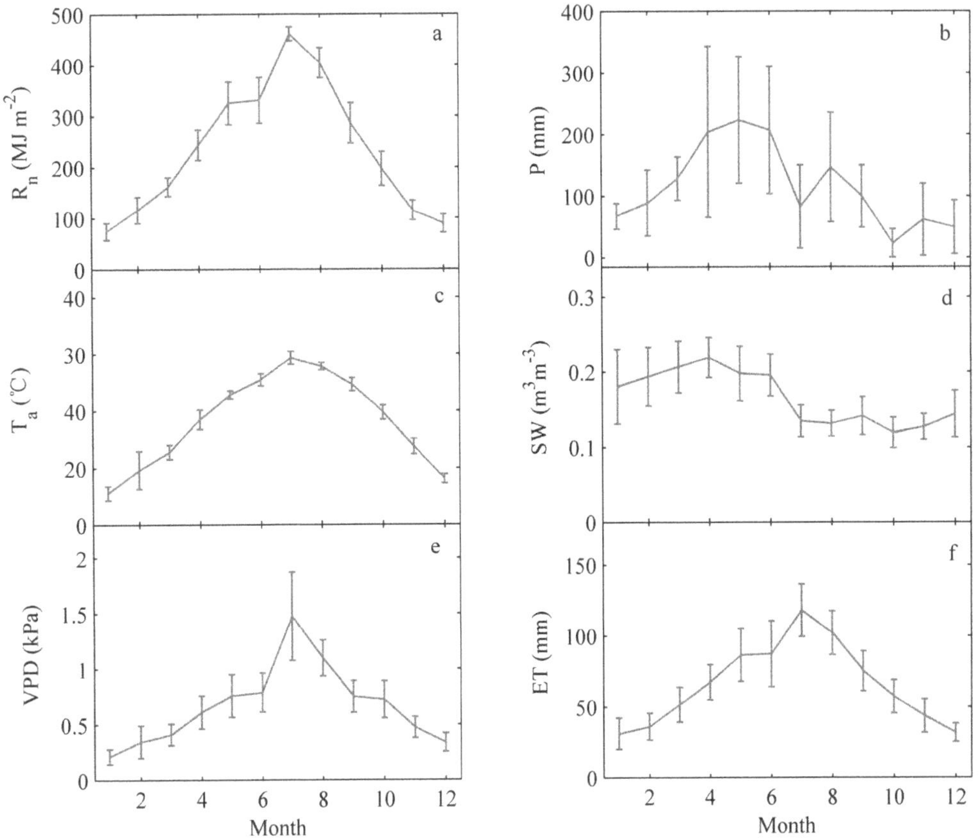

Figure 3. Multi-year (2003–2010) environmental conditions accumulated or averaged monthly: (a) net radiation (R_n); (b) precipitation (P); (c) air temperature (T_a); (d) soil water contents (SW) at 5 cm; (e) vapor pressure deficit (VPD) and (f) ET. The error bars represent the standard deviation.

resulted in a severe summer drought in June and July. In 2005 and 2006, the R_n was low in the first half year, and its annual total was approximately 6.4% lower in both years compared to multi-year average.

The El Niño and La Niña events had significant impacts on the climatic factors. In the El Niño year (2003), the T_a was higher, but the precipitation was lower. In the La Niña year (2010), the precipitation was higher but the T_a was lower (Fig. 4b and c, Table 1). In addition, under the effects of El Niño events, the SW was lower in the second half year of 2003 and the first half of 2004, but the VPD was higher during this period. In contrast, under the influence of La Niña, the pattern in SW and VPD anomalies was opposite to that of the El Niño year (Fig. 4d and e). Correspondingly, in the El Niño year, the anomalies in the ET were negative for ten months, but in the La Niña year, the opposite trend was found, and the anomalies in the ET were positive for eleven months (Fig. 4f).

Partitioning the sources of inter-annual variation of ET

To separate the sources of the inter-annual variability in ET, a statistical method (LUTs) was used to calculate the ET under different conditions. These approaches allowed the climatic drivers and ecosystem responses to change separately or together.

The annual ETs calculated through the three LUT approaches are presented in Figure 5. The average annual ETs were 798.5, 786.9, 754.1 mm for V_{cli}, $V_{cli\text{-}eco}$, V_{eco}, respectively, and the corresponding inter-annual variabilities were 28.13, 103.40 and 93.50 mm (Fig. 6). V_{cli} showed the highest average annual ET and the lowest inter-annual variability. Compared to $V_{cli\text{-}eco}$, which was almost equal to the measured ET, V_{cli} overestimated ET and dampened the inter-annual variability. This result implied that the ecosystem responses may offset climatic effects in the long term. In addition, the inter-annual variability of V_{cli} was only 27.2% of that produced by $V_{cli\text{-}eco}$. At the same time, the correlation coefficient between the ETs of V_{cli} and $V_{cli\text{-}eco}$ was only 0.48 ($p = 0.2316$), which indicated that the ecosystem had a noticeable resistance to direct changes in the external drivers. In contrast, V_{eco} underestimated ET and produced higher inter-annual variability. This indicated that the climatic factors drove the daily function of the ecosystems and maintained the normal ecosystem status, while the ecosystem responses furnished the main source of the inter-annual variability in ET.

Additionally, the inter-annual variability of ET was 28.13 mm as generated by variable climatic factors and was 93.50 mm as generated by ecosystem responses. If there were no interaction, the inter-annual variability of $V_{cli\text{-}eco}$ should be 121.63 mm, but the combined value was 17.6% higher than the real result 103.40 mm (Fig. 6). Therefore, we can infer that the changes in ecosystem responses were counteracting the effects of climatic variations, an inference that agrees with the classical ecological theory.

Discussion

Direct and indirect effects of climatic factors on ET

Previous researchers have confirmed that variations in ET can be attributed to climatic factors and the biotic responses induced

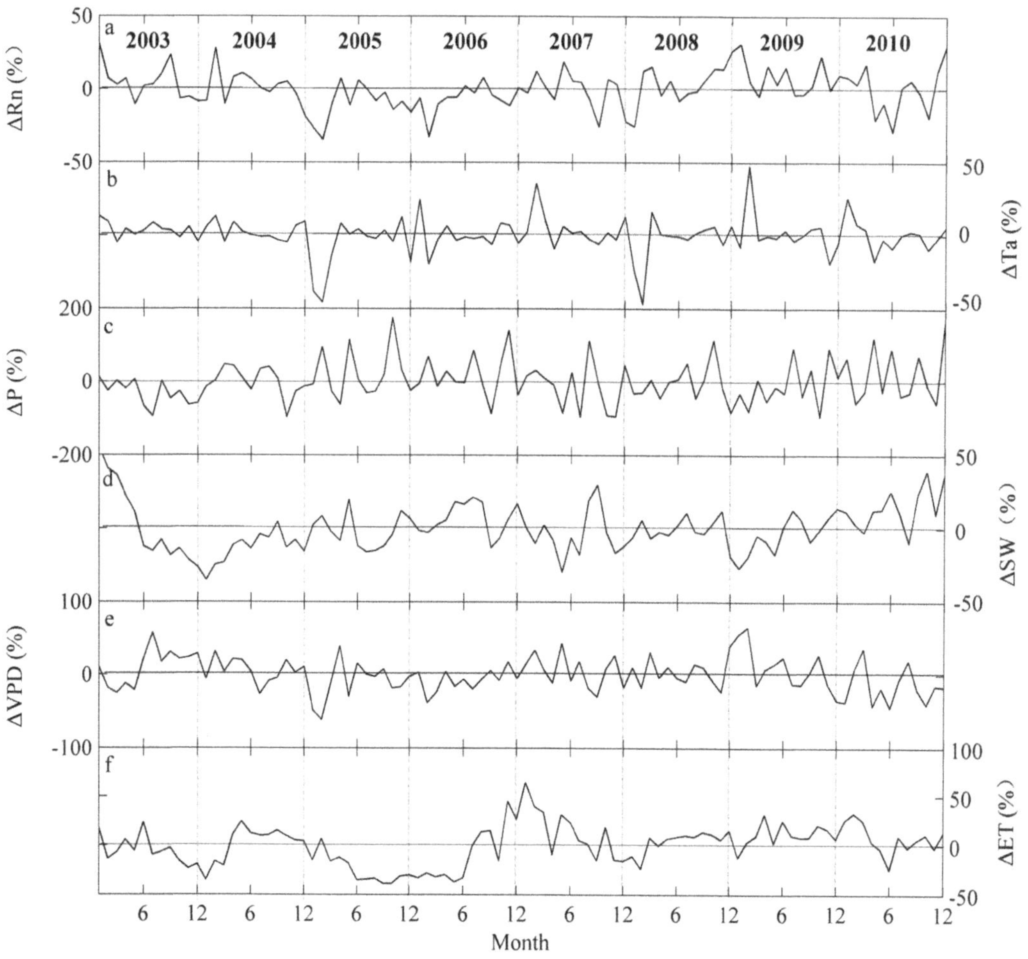

Figure 4. Monthly relative anomalies (%) of environmental factors and ET during 2003–2010: (a) net radiation (R_n); (b) air temperature (T_a); (c) precipitation (P); (d) soil water contents (SW) at 5 cm; (e) vapor pressure deficit (VPD) and (f) ET.

by environmental forcing [12,15,31,61]. Although they are entangled and difficult to separate from each other, there are many studies that refer to the role biotic factors play in driving the variation in ET, especially in the long term [36,62]. In this study, we used a statistical method to separate the effects of climatic drivers and ecosystem responses and quantified their contributions to the inter-annual variability of ET. The results underscore the prominent role that ecosystem responses play in the inter-annual variability of ET (Fig. 5 and Fig. 6). The ecosystem responses include many aspects, such as variation in LAI between years and variability in the photosynthetic ability of plants. These factors are sometimes difficult to measure and quantify over longer time scales. Fortunately, the indirect effects of climatic factors can, to some extent, reflect the ecosystem responses and help us understand the effects of ecosystem responses on the inter-annual variability of ET.

ET is composed of two components: evaporation and transpiration [63,64]. When analyzing how biotic and abiotic factors affect ET, we should consider the effects on each part. Evaporation is mainly a physical process driven by radiation and VPD [65]. Transpiration is a biotic process, which may also be driven by environmental factors but that is mainly controlled by biological variables [66]. In this study, ET was found to be related to R_n, T_a, VPD, SW and precipitation, which is consistent with

Table 1. Annual accumulated ET, R_n, P and annual averaged T_a during 2003–2010.

	2003	2004	2005	2006	2007	2008	2009	2010	mean	CV
R_n (MJ m^{-2})	3036.0	2870.9	2620.8	2621.5	2777.5	2860.8	2939.1	2669.9	2799.6	5.5%
T_a (°C)	18.7	18.3	17.6	18.0	18.5	17.9	18.3	17.8	18.1	2.1%
P (mm)	944.9	1404.5	1455.4	1485.3	1318.7	1332.9	1253.6	1850.2	1380.7	18.4%
ET (mm)	759.3	837.0	567.9	718.5	866.4	832.60	876.6	837.0	786.9	13.1%

The last column presents coefficient of variation (CV).

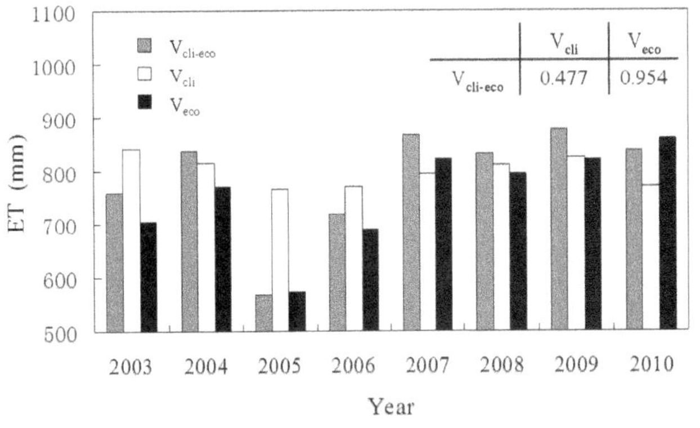

Figure 5. Yearly values of cumulative ET calculated through different approaches: $V_{cli-eco}$ (for variable climate and variable ecosystem responses); V_{cli} (for variable climate); V_{eco} (for variable ecosystem responses). The correlation coefficients of the cumulative values calculated with the three approaches are also reported.

Table 2. Correlation coefficients (R) between ET and some of the climatic drivers (R_n, T_a and VPD) at monthly and annual scale (p represents the probability-value).

Factor	Monthly		Annual	
	R	p	R	p
R_n	0.907	0.000	0.528	0.000
T_a	0.858	0.000	0.261	0.016
VPD	0.791	0.000	0.209	0.055

previous studies [13,67–70]. The lag effects became apparent when the correlograms between annual ET and dominant climatic drivers were drawn. This phenomenon indicated that the ecosystem responses sometimes did not happen simultaneously with the environmental change.

Numerous connections between climatic factors and ET were observed in the short and long term [12]. On a shorter time scale, the direct effects of climatic factors are obvious. The correlation coefficients between climatic factors on monthly time scale were larger than they were on the annual scale (Table 2), which is similar to findings in studies of carbon fluxes [71,72]. In addition, the ET showed a similar seasonal rhythm to the R_n, T_a and VPD (Fig. 3). On the annual scale, the ET is still related to the climatic drivers. The annual ET was sometimes linked to the temperature and precipitation anomalies caused by the extreme climate events such as the El Niño event in 2003, which caused a low ET (27.6 mm lower than the average) (Fig. 4 and Table 1). However, generally the relationships between ET and the climatic factors were not as close as they were on shorter time scales. Moreover, when the statistical methods were employed to further analyze their relationship, important lag effects emerged.

Two mechanisms may explain the lag effects and lower correlation coefficients on longer time scales. First, the rhythm signals of finer time scales are disguised when the analysis is conducted on longer time scales. Second, the ecosystem requires time to respond to climatic factors and its modulating effect is prominent over longer period. This is what we refer to as ecosystem responses. However, the lag effects did not exist for all climatic factors (e.g., R_n and VPD, Fig. 7). The evaporation from the soil and the canopy surface respond quickly to changes of R_n, as demonstrated by Monteith [27]. Meanwhile, R_n and VPD act on transpiration as a pulling force, and their effects were instantaneous [73].

As for T_a, precipitation and SW, there are prominent lag effects as shown in Figure 7. This result agreed with some previous studies [44], and it can be observed as the indirect effect of climatic

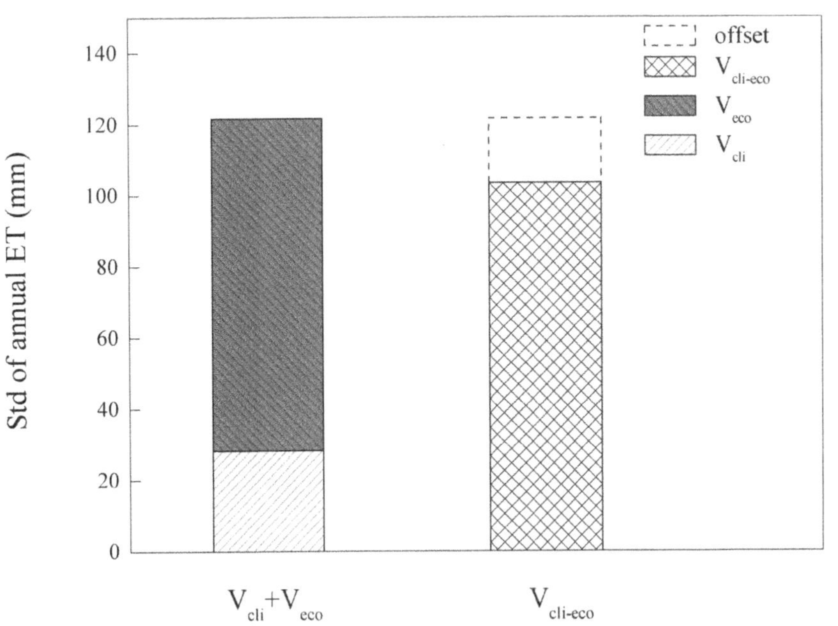

Figure 6. Standard deviation (Std) which represents inter-annual variability of annual cumulative ET calculated through different approaches: $V_{cli-eco}$ (for variable climate and variable ecosystem responses); V_{cli} (for variable climate); V_{eco} (for variable ecosystem responses).

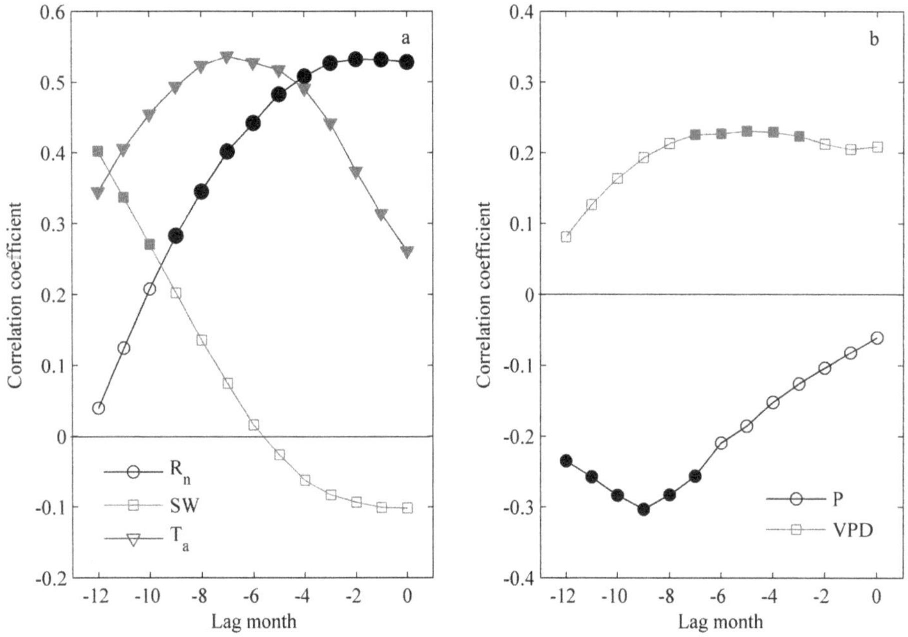

Figure 7. Correlation coefficients between annual ET and climatic drivers (R_n, T_a, P, SW and VPD). Correlograms were calculated by shifting the climatic driver backward one month at a time. The open signs represent that the coefficients are not significant while the closed signs mean significant ($p<0.05$).

factors. The lag may be helpful in understanding the ecosystem responses. Here, we try to interpret the lag effects of T_a, SW and precipitation. The relationship between ET and T_a, showed a parabolic lag effect. The highest correlation coefficient showed up at 8 month lags, which is likely because the growth of forests often requires more time to respond to changes in temperature. This may be partly because of its impact on the growing season length (GSL) [74]. In this study area, the temperatures between April and November were generally above the threshold 5°C. Therefore, the GSL mainly depended on the temperature between December and March (Table 3). Zhang et al. [43] noted that lower temperatures in the early spring greatly reduce the annual net ecosystem exchange (NEE) of carbon by shortening the GSL in this study area. Because transpiration, which is the dominant component of ET, is coupled with photosynthesis, we tested the effect of GSL on ET. The resulting correlation coefficients between ET and GSL ($T_a \geq 5$°C, in days) were as high as 0.56 ($p = 0.1471$, $n = 8$). This may partially explain the lower ET in 2005 and higher ET in 2007, because the GSL was shorter in 2005 and longer in 2007 (Table 3). This result is comparable to the results obtained by Zha et al. [36] which showed that the inter-annual variability of ET was controlled by early spring soil temperatures in three boreal forests and a grassland ecosystem.

The highest correlation coefficients between ET and SW at 5 cm were found at time lags of 10–12 months. This indicated that the soil water conditions in a given year may have significant effects on ET in the following year. This phenomenon may be attributed to the rooting depth and available SW at deeper layers or potentially to plant responses that result in altered water use strategies under the soil water stress [25,75,76]. In our study site, the dominant tree species are Slash pine, Masson pine and Chinese fir. These pioneer species have deep root systems that may be capable of absorbing water from deeper layers, much like the plants in Florida scrub oak and pine flatwoods ecosystems that Bracho et al. [5] described as able to obtain water from deeper sources.

The correlation coefficients between precipitation and ET were negative on the annual scale. When the precipitation was high, forest ET was likely reduced because of the lower solar radiation due to frequent cloud cover. In dry years, as Lundquist and Loheide [22] described, the ecosystem typically has greater potential ET due to higher R_n, higher T_a, higher VPD and longer growing seasons. All of these factors provide strong driving forces for ET. In our study area, the precipitation is abundant on the annual scale. Even in dry years, the trees can access water in deeper layers, which may lead to the relatively high ET. Another possibility as suggested by Pejam et al. [30], is that larger precipitation events would contribute more to the surface run-off than to soil water storage. Because our study site is hilly, with red earth characterized by high clay soils with relatively poor infiltration rates, there may be high runoff when the rain is heavy [47].

Contributions of seasonal variations to the inter-annual variation of ET

During the observed years, the ET of this ecosystem was positive throughout the year (Fig. 8a). Although the ET for all eight years had a similar cumulative trend, large inter-annual variability emerged when comparing the cumulative curves of different years (Fig. 8a). The inter-annual variability of ET in this subtropical coniferous plantation was also relatively large when compared to the other ecosystems (Table 4).

At the beginning of each year, from January to mid-March, the ET increased slowly at less than 2 mm d^{-1}. From mid-March to mid-July, the cumulative ET rose sharply, up to 4 mm d^{-1} in June, and then gradually declined through the end of the year. With respect to the increasing rate of ET, the absolute difference in ET among years was not large in the beginning of the year. Since the growing season, accompanied with the intensive

Table 3. Air temperature (T_a °C) and growing season length (GSL, in days $T_a \geq 5$°C) during 2003–2010.

Factor	Period	2003	2004	2005	2006	2007	2008	2009	2010
	Jan.	6.24	5.82	3.38	6.87	5.62	4.19	5.06	6.94
	Feb.	10.51	10.88	5.21	7.70	13.02	5.10	13.37	10.37
T_a (°C)	Mar.	12.10	12.13	11.00	12.35	14.08	14.79	12.48	13.30
	Apr.-Nov.	23.54	22.80	23.15	22.70	22.53	22.84	22.58	21.74
	Dec.	7.67	8.82	6.56	7.42	9.08	8.58	7.69	8.49
	Jan.	20	19	6	21	19	9	13	21
GSL	Feb.	23	22	11	20	28	14	26	19
(≥5°C)	Mar.	28	31	28	28	28	31	26	27
	Apr- Nov.	244	244	244	244	244	244	239	244
	Dec.	27	22	23	29	30	28	25	25

transpiration, the discrepancies in cumulative curves became larger (Fig. 8a). However, the increases in rates could not fully account for the contributions to the inter-annual variability in the ET. Therefore, to evaluate the contributions of different phenological phases to the variability in annual ET, the difference between the standard deviation of cumulative ET from a given month and its preceding one (DstDev) was calculated to quantify the contribution of each month to the observed inter-annual variability of the ET. Over the whole year, the DstDev had three peaks, in March, May–June and October (Fig. 8b), which coincide with the seasonal transitions.

In subtropical China, March is the beginning of the spring, and new twigs generally sprout as the plant leaves expand. Accompanying this phenological transition, the cumulative ET increased rapidly (Fig. 8a) because of the great increase in transpiration [77]. However, in accordance with variations of early spring temperatures [43], the phenological calendars varied among years. For example, in 2005 the low temperature in spring postponed the twig and leaf emergence and caused phase lags in the enhanced vegetation index (EVI) time series [43]. Therefore, the phenological transition and the advances or lags in growing seasons gave the ET in March a bigger role in the inter-annual variability of ET. Zha et al. [36] found a similar effect in the Siberian forests, years with warmer spring temperatures had a greater overall ET because of earlier leaf emergence. Hollinger et al. [78] also found a similar phenomenon in a carbon flux study in forests. They noted that the climatic conditions in the spring largely influenced the inter-annual variability of NEE. The ecosystem they studied absorbed more atmospheric carbon dioxide in years with warm spring and less carbon dioxide in years with cold spring.

The variations of ET in May and June had the largest contribution to the inter-annual variability of ET in this study.

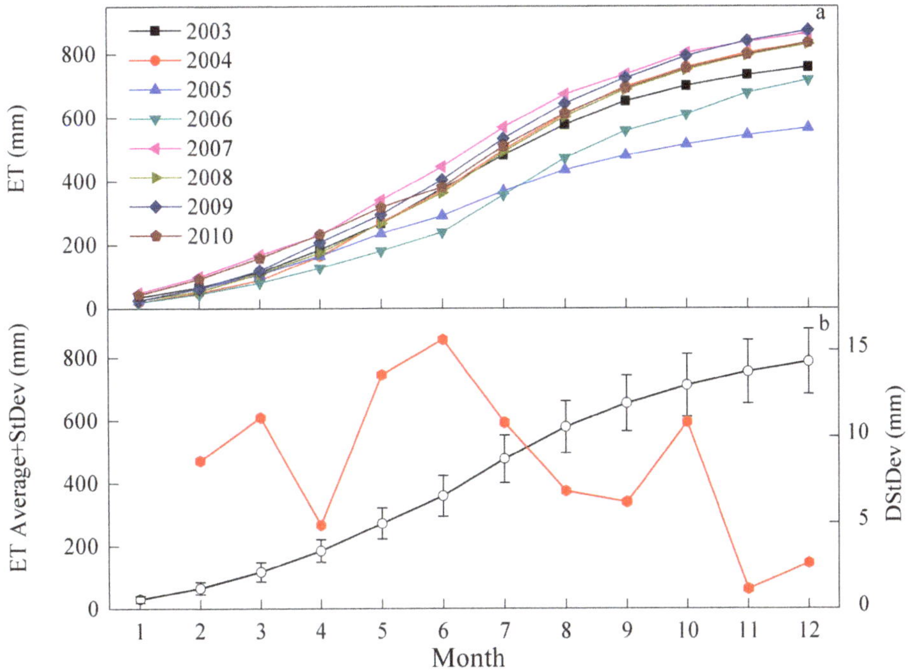

Figure 8. Cumulative ET for each year of 2003–2010 respectively (panel a); cumulative ET averaged over the observation periods and its standard deviation (black line) and monthly variations in standard deviation (red line and the axis on right) (panel b).

Table 4. Coefficient of variation (CV) of ET at flux sites with over 3 years' observations.

Forest type	Location	Mean annual ET (mm)	CV (%)	Observation period	Source
Boreal deciduous coniferous	62°15'18"N, 129°14'29"E	196.1	10	1998–2006	[37]
Boreal broadleaved deciduous	53°37'48"N, 106°11'24"W	405	21	1998–2006	[36]
Boreal coniferous	53°59'24"N, 105°7'12"W	374	9	1998–2006	[36]
	53°55'12"N, 104°41'24"E	300	7	1998–2006	
Temperate broadleaved deciduous	36°58'41"N, 79°05'39"W	633	4.1	2002–2005	[41]
Temperate broadleaved deciduous	35°57'30"N, 84°17'15"W	567.2	13	1995–1997	[31]
Temperate coniferous	34°58'N, 136°00'E	729	4.4	2001–2003	[39]
Subtropical coniferous	26°44'29"N, 115°03'29"E	787	13.1	2003–2010	This study
Tropical rainforest	21°55'39"N, 101°15'55"E	1029	2.8	2003–2006	[20]
	4°20'N, 113°50'E	1323	5.6	2000–2009	

This may have occurred for the following reasons. First, there was substantial variability in the ecosystem structure and function in this period. The ecosystem behaved similarly in this period every year, but the differences were great between years. For example, the LAI or EVI would level off during this period each year, but when compared between years, large differences emerged [43]. Transpiration, which accounts for 70–80% of ET [15,79], became more active in this period [77]. Therefore, under the coactions of these two factors, the ET varied greatly in this period between years. Second, this period marks the transition between wet and dry seasons, which occurs with the seasonal reversal of the monsoon circulation features [80]. The variability in the onset of the monsoon [81,82] led to a high inter-annual variability in the climatic factors. The climatic factors and variable ecosystem responses combined to make this period the most influential on the inter-annual variability of ET.

In October, the DstDev increased abruptly, which might have been caused by the inter-annual variability in senescence time of old needles and the retreat time of the monsoon [81]. After October, the lower Dstdev may have been driven by the lowering activity of the plants. Therefore, the underlying mechanisms that make these seasonal transitions important to the inter-annual variability of ET may be the coactions of climatic factors and ecosystem responses. This is consistent with the view that seasonal transitions have great influence on fluxes and energy exchange and that they play an important role in inter-annual variability [83,84].

The role of ecosystem responses in long term studies

Previous studies have generally focused more on the effects of climatic factors. Although some researchers have considered the biotic factors such as canopy conductance to be as important as climatic factors [31,85,86], the biotic effects on ET have not generally received adequate consideration [12]. In our study, we interpreted all biotic factors as an integrated part of ecosystem responses, and we separated the sources of the inter-annual variability of ET into those that were induced by climate and those produced by the ecosystem responses. The results suggested that ecosystem responses were largely responsible for the inter-annual variability of ET (Fig. 5 and Fig. 6). As discussed above, both of the indirect effects of the climatic factors and the three peaks in variability contributed to the inter-annual variability of ET as shown in Figure 8, underscore the importance of ecosystem responses. By taking ecosystem responses into consideration, more accurate estimations of ET may be made, particularly in years where climatic drivers seem unable to capture the variation in ET (Fig. 5). Our results indicate that the ecosystem ET may respond directly to the climatic forces, but it may also provide some buffering effects because the ecosystems requires time to respond to external stimulations and adjust its functions and structures [87].

Currently, several types of models are used to estimate and predict ET worldwide [12]. However, most of those models tend to underestimate its inter-annual variability [88]. Komatsu et al. [40] suggested that the models can be roughly classified into two groups: simple models that use only climatic factors as input variables [89] and complex models that employ more process-based formulations including the Penman–Monteith equation or the bulk equation [15]. Although the process-based models give limited consideration to biotic factors, they are still limited in their ability to estimate the inter-annual variability of ET. Our results indicate that the ecosystem responses are more important to the inter-annual variability of ET than the environmental drivers, which implies that to more precisely evaluate ecosystem ET, the features of ecosystem responses should be considered alongside the climatic factors, especially for longer time scales.

There is also potential for an effect of CO_2 fertilization on the inter-annual variation of ET under climate change context. It has been reported that the elevated CO_2 concentrations may enhance photosynthesis in forests [90]. They may also affect the transpiration, the dominant component of ET, because photosynthesis and transpiration are coupled in the stomata [12]. In addition, elevated CO_2 concentration may affect the growing season length [91–92] and induce changes in the ecosystem functions [93–94]. These changes in ecosystem responses may in turn increase the inter-annual variability of ET. Recently, the atmospheric CO_2 concentration has been increasing by approximately 2 ppm per year [95], and it rose approximately 14 ppm during our study period (2003–2010) [96]. Compared to CO_2 enrichment experiments and models, this increase is minor, and therefore the effect of CO_2 fertilization on ET might not have been captured by our study. However, in the context of climate change, the rising atmospheric CO_2 concentrations may affect ET in the long term by affecting plants and by magnifying ecosystem responses (e.g., LAI, GSL) [91]. Such scenarios should be discussed in the future.

However, ecosystem responses are complicated and difficult to quantify. To demonstrate the underlying mechanisms, accurate knowledge of how the components of ET (e.g., soil evaporation, canopy evaporation and transpiration) respond to climatic drivers is vital and should be targeted for future investigations of ET. These efforts will require more comprehensive models and additional data. Though some processes respond quickly to variation in climatic factors, other biogeochemical processes may need a longer time to respond to the environmental changes [72]. Thus, long-term physiological and biogeochemical research is needed to elucidate the bioprocesses and the underlying mechanisms.

Conclusions

(1) The ET of a coniferous plantation in subtropical China showed a strong seasonal pattern, consistent with the single peak of climatic factors such as R_n, T_a, VPD,which reached their maximums in July. During the observation period from 2003 to 2010, the mean value of annual ET was 786.9 mm. A relatively large inter-annual variability with a standard deviation of 103.40 mm (CV = 13.1%) was detected.

(2) The main source of the inter-annual variability of ET was ecosystem responses, according to the LUT results. The inter-annual variability of ET caused by climate variability and that caused by changes in ecosystem responses had a negative interaction.

(3) On the annual scale, the main meteorological drivers for ET were R_n and T_a. The R_n affected ET directly without a lag effect. In contrast, the correlation coefficients between ET and T_a, SW and precipitation showed significant lag effects which reflect the ecosystem responses.

(4) The variation during seasonal transitions in March, May–June, and October contributed heavily to the inter-annual variability of ET, which could be attributed to the coactions of climatic factors and ecosystem responses.

Acknowledgments

The authors greatly thank Dr. Xianjin Zhu and Dr. Tao Zhang for their valuable comments and suggestions. We appreciate Dr. M. Luke McCormack for reviewing the English. We are greatly grateful to Dr. Dafeng Hui and the reviewers for their helpful comments that improved this paper.

Author Contributions

Conceived and designed the experiments: MJX XFW HMW. Performed the experiments: MJX HMW WJZ XQD YDW XLF YFL. Analyzed the data: MJX HMW XFW. Contributed reagents/materials/analysis tools: YFL XMS GRY. Wrote the paper: MJX HMW JS.

References

1. Liu SG, Rickerk H, Gholz HL (1998) Simulation of evapotranspiration from Florida pine flatwoods. Ecological Modelling 114: 19–34.
2. Gholz HL, Clark KL (2002) Energy exchange across a chronosequence of slash pine forests in Florida. Agricultural and Forest Meteorology 112: 87–102.
3. Sumner DM, Belaineh G (2005) Evaporation, precipitation, and associated salinity changes at a humid, subtropical estuary. Estuaries 28: 844–855.
4. Powell TL, Starr G, Clark KL, Martin TA, Gholz HL (2005) Ecosystem and understory water and energy exchange for a mature, naturally regenerated pine flatwoods forest in north Florida. Canadian Journal of Forest Research-Revue Canadienne De Recherche Forestiere 35: 1568–1580.
5. Bracho R, Powell TL, Dore S, Li J, Hinkle CR, et al. (2008) Environmental and biological controls on water and energy exchange in Florida scrub oak and pine flatwoods ecosystems. Journal of Geophysical Research 113: G02004.
6. Tian FX, Zhao CY, Feng ZD (2011) Simulating evapotranspiration of Qinghai spruce (Picea crassifolia) forest in the Qilian Mountains, northwestern China. Journal of Arid Environments 75: 648–655.
7. Miao H, Chen S, Chen J, Zhang W, Zhang P, et al. (2009) Cultivation and grazing altered evapotranspiration and dynamics in Inner Mongolia steppes. Agricultural and Forest Meteorology 149: 1810–1819.
8. FAO (2006) Global Forest Resources Assessment 2005: Main Report. Progress Towards Sustainable Forest Management. p. 320. FAO Forestry Paper 147, Rome, Italy.
9. Bonan GB (2008) Forests and climate change: forcings, feedbacks, and the climate benefits of forests. Science 320: 1444–1449.
10. Parry ML, Canziani OF, Palutikof JP, van der Linden PJ, Hanson CE (2007) Climate change 2007: impacts, adaptation and vulnerability: Intergovernmental Panel on Climate Change.
11. Kumagai To, Katul GG, Porporato A, Saitoh TM, Ohashi M, et al. (2004) Carbon and water cycling in a Bornean tropical rainforest under current and future climate scenarios. Advances in Water Resources 27: 1135–1150.
12. Wang K, Dickinson RE (2012) A review of global terrestrial evapotranspiration: Observation, modeling, climatology, and climatic variability. Reviews of Geophysics 50: RG2005.
13. Fisher JB, Whittaker RJ, Malhi Y (2011) ET come home: potential evapotranspiration in geographical ecology. Global Ecology and Biogeography 20: 1–18.
14. Shukla J, Mintz Y (1982) Influence of land-surface evapotranspiration on the earth's climate. Science 215: 1498.
15. Ge Z-M, Zhou X, Kellomäki S, Peltola H, Wang K-Y (2011) Climate, canopy conductance and leaf area development controls on evapotranspiration in a boreal coniferous forest over a 10-year period: A united model assessment. Ecological Modelling 222: 1626–1638.
16. Huntington TG (2006) Evidence for intensification of the global water cycle: Review and synthesis. Journal of Hydrology 319: 83–95.
17. Scott RL, Edwards EA, Shuttleworth WJ, Huxman TE, Watts C, et al. (2004) Interannual and seasonal variation in fluxes of water and carbon dioxide from a riparian woodland ecosystem. Agricultural and Forest Meteorology 122: 65–84.
18. Tognetti R, Giovannelli A, Lavini A, Morelli G, Fragnito F, et al. (2009) Assessing environmental controls over conductances through the soil-plant-atmosphere continuum in an experimental olive tree plantation of southern Italy. Agricultural and Forest Meteorology 149: 1229–1243.
19. Kochendorfer J, Castillo EG, Haas E, Oechel WC, Paw UKT (2011) Net ecosystem exchange, evapotranspiration and canopy conductance in a riparian forest. Agricultural and Forest Meteorology 151: 544–553.
20. Li Z, Zhang Y, Wang S, Yuan G, Yang Y, et al. (2010) Evapotranspiration of a tropical rain forest in Xishuangbanna, southwest China. Hydrological Processes 24: 2405–2416.
21. Stoy PC, Katul GG, Siqueira MBS, Juang J-Y, Novick KA, et al. (2006) Separating the effects of climate and vegetation on evapotranspiration along a successional chronosequence in the southeastern US. Global Change Biology 12: 2115–2135.
22. Lundquist JD, Loheide SP (2011) How evaporative water losses vary between wet and dry water years as a function of elevation in the Sierra Nevada, California, and critical factors for modeling. Water Resources Research 47.
23. Kelliher FM, Hollinger DY, Schulze ED, Vygodskaya NN, Byers JN, et al. (1997) Evaporation from an eastern Siberian larch forest. Agricultural and Forest Meteorology 85: 135–147.
24. Li J, Liu Y, Yang X, Li J (2006) Studies on water-vapor flux characteristic and the relationship with environmental factors over a planted coniferous forest in Qianyanzhou Station. Acta Ecologica Sinica 26: 2449–2456.
25. Alfieri JG, Blanken PD, Yates DN, Steffen K (2007) Variability in the environmental factors driving evapotranspiration from a grazed rangeland during severe drought conditions. Journal of Hydrometeorology 8: 207–220.
26. Maximov T, Ohta T, Dolman AJ (2008) Water and energy exchange in East Siberian forest: A synthesis. Agricultural and Forest Meteorology 148: 2013–2018.
27. Monteith J. Evaporation and environment; 1965. p. 4.
28. Bernier PY, Bartlett P, Black TA, Barr A, Kljun N, et al. (2006) Drought constraints on transpiration and canopy conductance in mature aspen and jack pine stands. Agricultural and Forest Meteorology 140: 64–78.
29. Bucci SJ, Scholz FG, Goldstein G, Hoffmann WA, Meinzer FC, et al. (2008) Controls on stand transpiration and soil water utilization along a tree density gradient in a Neotropical savanna. Agricultural and Forest Meteorology 148: 839–849.
30. Pejam MR, Arain MA, McCaughey JH (2006) Energy and water vapour exchanges over a mixedwood boreal forest in Ontario, Canada. Hydrological Processes 20: 3709–3724.
31. Wilson KB, Baldocchi DD (2000) Seasonal and interannual variability of energy fluxes over a broadleaved temperate deciduous forest in North America. Agricultural and Forest Meteorology 100: 1–18.
32. Baldocchi DD, Hincks BB, Meyers TP (1988) Measuring biosphere-atmosphere exchanges of biologically related gases with micrometeorological methods. Ecology: 1331–1340.

33. Baldocchi D (2008) TURNER REVIEW No. 15. 'Breathing' of the terrestrial biosphere: lessons learned from a global network of carbon dioxide flux measurement systems. Australian Journal of Botany 56: 1.
34. Tanaka N, Kume T, Yoshifuji N, Tanaka K, Takizawa H, et al. (2008) A review of evapotranspiration estimates from tropical forests in Thailand and adjacent regions. Agricultural and Forest Meteorology 148: 807–819.
35. Kume T, Tanaka N, Kuraji K, Komatsu H, Yoshifuji N, et al. (2011) Ten-year evapotranspiration estimates in a Bornean tropical rainforest. Agricultural and Forest Meteorology 151: 1183–1192.
36. Zha TS, Barr AG, van der Kamp G, Black TA, McCaughey JH, et al. (2010) Interannual variation of evapotranspiration from forest and grassland ecosystems in western canada in relation to drought. Agricultural and Forest Meteorology 150: 1476–1484.
37. Ohta T, Maximov TC, Dolman AJ, Nakai T, van der Molen MK, et al. (2008) Interannual variation of water balance and summer evapotranspiration in an eastern Siberian larch forest over a 7-year period (1998–2006). Agricultural and Forest Meteorology 148: 1941–1953.
38. Xue B-L, Komatsu H, Kumagai To, Kotani A, Otsuki K, et al. (2012) Interannual variation of evapotranspiration in an eastern Siberian larch forest. Hydrological Processes 26: 2360–2368.
39. Kosugi Y, Takanashi S, Tanaka H, Ohkubo S, Tani M, et al. (2007) Evapotranspiration over a Japanese cypress forest. I. Eddy covariance fluxes and surface conductance characteristics for 30 years. Journal of Hydrology 337: 269–283.
40. Komatsu H, Cho J, Matsumoto K, Otsuki K (2012) Simple modeling of the global variation in annual forest evapotranspiration. Journal of Hydrology 420–421: 380–390.
41. Oishi AC, Oren R, Novick KA, Palmroth S, Katul GG (2010) Interannual Invariability of Forest Evapotranspiration and Its Consequence to Water Flow Downstream. Ecosystems 13: 421–436.
42. Park H, Yamazaki T, Yamamoto K, Ohta T (2008) Tempo-spatial characteristics of energy budget and evapotranspiration in the eastern Siberia. Agricultural and Forest Meteorology 148: 1990–2005.
43. Zhang W, Wang H, Yang F, Yi Y, Wen X, et al. (2011) Underestimated effects of low temperature during early growing season on carbon sequestration of a subtropical coniferous plantation. Biogeosciences 8: 1667–1678.
44. Marcolla B, Cescatti A, Manca G, Zorer R, Cavagna M, et al. (2011) Climatic controls and ecosystem responses drive the inter-annual variability of the net ecosystem exchange of an alpine meadow. Agricultural and Forest Meteorology 151: 1233.
45. Luyssaert S, Janssens IA, Sulkava M, Papale D, Dolman AJ, et al. (2007) Photosynthesis drives anomalies in net carbon-exchange of pine forests at different latitudes. Global Change Biology 13: 2110–2127.
46. Hui DF, Luo YQ, Katul G (2003) Partitioning interannual variability in net ecosystem exchange between climatic variability and functional change. Tree Physiology 23: 433–442.
47. Wang Y, Li Q, Wang H, Wen X, Yang F, et al. (2011) Precipitation frequency controls interannual variation of soil respiration by affecting soil moisture in a subtropical forest plantation. Canadian Journal of Forest Research 41:1897–1906.
48. Wen X, Wang H, Wang J, Yu G, Sun X (2010) Ecosystem carbon exchanges of a subtropical evergreen coniferous plantation subjected to seasonal drought, 2003–2007. Biogeosciences 7: 357–369.
49. Wen X-F, Yu G-R, Sun X-M, Li Q-K, Liu Y-F, et al. (2006) Soil moisture effect on the temperature dependence of ecosystem respiration in a subtropical Pinus plantation of southeastern China. Agricultural and Forest Meteorology 137: 166–175.
50. Wang Y, Wang Z-L, Wang H, Guo C and Bao W (2012) Rainfall pulse primarily drives litterfall respiration and its contribution to soil respiration in a young exotic pine plantation in subtropical China. Canadian Journal of Forest Research 42: 657–666.
51. McPhaden MJ (2004) Evolution of the 2002/03 El Niño. Bulletin of the American Meteorological Society 85: 677.
52. Kim WM, Yeh SW, Kim JH, Kug JS, Kwon MH (2011) The unique 2009–2010 El Niño event: A fast phase transition of warm pool El Niño to La Niña. Geophysical Research Letters 38: L15809.
53. Zhu Z, Sun X, Zhou Y, Xu J, Yuan G, et al. (2005) Correcting method of eddy covariance fluxes over non-flat surfaces and its application in ChinaFLUX. Science in China Series D: Earth Sciences 48.
54. Webb EK, Pearman GL, Leuning R (1980) Correction of flux measurement for density effects due to heat and water vapor transfer. Quarterly Journal of Royal Meteorological Society 106: 85–100.
55. Zhang LM, Yu GR, Sun XM, Wen XF, Ren CY, et al. (2006) Seasonal variation of carbon exchange of typical forest ecosystems along the eastern forest transect in China. Science in China Series D-Earth Sciences 49: 47–62.
56. Reichstein M, Falge E, Baldocchi D, Papale D, Aubinet M, et al. (2005) On the separation of net ecosystem exchange into assimilation and ecosystem respiration: review and improved algorithm. Global Change Biology 11: 1424–1439.
57. Falge E, Baldocchi D, Olson R, Anthoni P, Aubinet M, et al. (2001) Gap filling strategies for defensible annual sums of net ecosystem exchange. Agricultural and Forest Meteorology 107: 43–69.
58. Yu GR, Song X, Wang QF, Liu YF, Guan DX, et al. (2008) Water-use efficiency of forest ecosystems in eastern China and its relations to climatic variables. New Phytologist 177: 927–937.
59. Wilson KB, Hanson PJ, Mulholland PJ, Baldocchi DD, Wullschleger SD (2001) A comparison of methods for determining forest evapotranspiration and its components: sap-flow, soil water budget, eddy covariance and catchment water balance. Agricultural and Forest Meteorology 106: 153–168.
60. Admiral SW, Lafleur PM, Roulet NT (2006) Controls on latent heat flux and energy partitioning at a peat bog in eastern Canada. Agricultural and Forest Meteorology 140: 308–321.
61. Saugier B, Katerji N (1991) Some Plant Factors Controlling Evapotranspiration. Agricultural and Forest Meteorology 54: 263–277.
62. Yoshida M, Ohta T, Kotani A, Maximov T (2010) Environmental factors controlling forest evapotranspiration and surface conductance on a multi-temporal scale in growing seasons of a Siberian larch forest. Journal of Hydrology 395: 180–189.
63. Iida Si, Tanaka T, Sugita M (2006) Change of evapotranspiration components due to the succession from Japanese red pine to evergreen oak. Journal of Hydrology 326: 166–180.
64. Shuttleworth WJ, Wallace J (1985) Evaporation from sparse crops-an energy combination theory. Quarterly Journal of the Royal Meteorological Society 111: 839–855.
65. Penman HL (1948) Natural evaporation from open water, bare soil and grass. Proceedings of the Royal Society of London Series A Mathematical and Physical Sciences 193: 120–145.
66. Kume T, Komatsu H, Kuraji K, Suzuki M (2008) Less than 20-min time lags between transpiration and stem sap flow in emergent trees in a Bornean tropical rainforest. Agricultural and Forest Meteorology 148: 1181–1189.
67. McCaughey JH, Pejam MR, Arain MA, Cameron DA (2006) Carbon dioxide and energy fluxes from a boreal mixedwood forest ecosystem in Ontario, Canada. Agricultural and Forest Meteorology 140: 79–96.
68. Jassal RS, Black TA, Spittlehouse DL, Brummer C, Nesic Z (2009) Evapotranspiration and water use efficiency in different-aged Pacific Northwest Douglas-fir stands. Agricultural and Forest Meteorology 149: 1168–1178.
69. Si JH, Feng Q, Zhang XY, Chang ZQ, Su YH, et al. (2007) Sap flow of Populus euphratica in a desert riparian forest in an extreme arid region during the growing season. Journal of Integrative Plant Biology 49: 425–436.
70. Chang X, Zhao W, Zhang Z, Su Y (2006) Sap flow and tree conductance of shelter-belt in arid region of China. Agricultural and Forest Meteorology 138: 132–141.
71. Wu J, van der Linden L, Lasslop G, Carvalhais N, Pilegaard K, et al. (2012) Effects of climate variability and functional changes on the interannual variation of the carbon balance in a temperate deciduous forest. Biogeosciences 9: 13–28.
72. Richardson AD, Hollinger DY, Aber JD, Ollinger SV, Braswell BH (2007) Environmental variation is directly responsible for short-but not long-term variation in forest-atmosphere carbon exchange. Global Change Biology 13: 788–803.
73. Butler D (1976) Estimation of the transpiration rate in an apple orchard from net radiation and vapour pressure deficit measurements. Agricultural Meteorology 16: 277–289.
74. Ryu Y, Baldocchi DD, Ma S, Hehn T (2008) Interannual variability of evapotranspiration and energy exchange over an annual grassland in California. Journal of Geophysical Research-Atmospheres 113.
75. Granier A, Reichstein M, Breda N, Janssens IA, Falge E, et al. (2007) Evidence for soil water control on carbon and water dynamics in European forests during the extremely dry year: 2003. Agricultural and Forest Meteorology 143: 123–145.
76. Saleska SR (2003) Carbon in Amazon Forests: Unexpected Seasonal Fluxes and Disturbance-Induced Losses. Science 302: 1554–1557.
77. Tu J, Liu Q, Li H, LIN Y (2008) Dynamics of Sap Flow of Pinus elliottii and Its Relevant Factors in Rapid Growth Season at Qianyanzhou Experimental Station of Jiangxi Province. Scientia Silvae Sinicae 44: 46–51.
78. Hollinger DY, Aber J, Dail B, Davidson EA, Goltz SM, et al. (2004) Spatial and temporal variability in forest-atmosphere CO_2 exchange. Global Change Biology 10: 1689–1706.
79. Granier A, Biron P, Lemoine D (2000) Water balance, transpiration and canopy conductance in two beech stands. Agricultural and Forest Meteorology 100: 291–308.
80. Wang B (2002) Rainy Season of the Asian-Pacific Summer Monsoon*. Journal of Climate 15: 386–398.
81. Wang B, Wu R, Lau K (2001) Interannual Variability of the Asian Summer Monsoon: Contrasts between the Indian and the Western North Pacific-East Asian Monsoons*. Journal of Climate 14: 4073–4090.
82. Ding Y, Chan JCL (2005) The East Asian summer monsoon: an overview. Meteorology and Atmospheric Physics 89: 117–142.
83. Richardson AD, Andy Black T, Ciais P, Delbart N, Friedl MA, et al. (2010) Influence of spring and autumn phenological transitions on forest ecosystem productivity. Philosophical Transactions of the Royal Society B: Biological Sciences 365: 3227–3246.
84. Ham JM, Knapp AK (1998) Fluxes of CO_2, water vapor, and energy from a prairie ecosystem during the seasonal transition from carbon sink to carbon source. Agricultural and Forest Meteorology 89: 1–14.

85. Faver KL, Otoole JC, Krieg DR (1989) Short-Term Estimation of Sorghum Evapotranspiration from Canopy Temperature. Agricultural and Forest Meteorology 48: 175–183.
86. Betts AK, Goulden M, Wofsy S (1999) Controls on evaporation in a boreal spruce forest. Journal of Climate 12: 1601–1618.
87. Odum EP (1969) Strategy of ecosystem development. Science 164: 262–270.
88. Wei H, He H, Liu M, Zhang L, Yu G, et al. (2012) Modeling evapotranspiration and its components in Qianyanzhou plantation based on remote sensing data. Journel of Natural Resources 25: 778–789.
89. Zhang L, Dawes W, Walker G (2001) Response of mean annual evapotranspiration to vegetation changes at catchment scale. Water Resources Research 37: 701–708.
90. Bader MKF, Siegwolf R, Korner C (2010) Sustained enhancement of photosynthesis in mature deciduous forest trees after 8 years of free air CO_2 enrichment. Planta 232: 1115–1125.
91. Piao SL, Friedlingstein P, Ciais P, Zhou LM, Chen AP (2006) Effect of climate and CO_2 changes on the greening of the Northern Hemisphere over the past two decades. Geophysical Research Letters 33.
92. Taylor G, Tallis MJ, Giardina CP, Percy KE, Miglietta F, et al. (2008) Future atmospheric CO2 leads to delayed autumnal senescence. Global Change Biology 14: 264–275.
93. Urban O, Hrstka M, Zitova M, Holisova P, Sprtova M, et al. (2012) Effect of season, needle age and elevated CO_2 concentration on photosynthesis and Rubisco acclimation in Picea abies. Plant Physiology and Biochemistry 58: 135–141.
94. Cernusak LA, Winter K, Dalling JW, Holtum JAM, Jaramillo C, et al. (2013) Tropical forest responses to increasing atmospheric CO_2: current knowledge and opportunities for future research. Functional Plant Biology 40: 531–551.
95. Holtum JAM, Winter K (2010) Elevated CO_2 and forest vegetation: more a water issue than a carbon issue? Functional Plant Biology 37: 694–702.
96. Dai LJ, Cui WH, Jiang YB (2012) Temporal and spatial distribution of tropospheric carbon dioxide from 2003 to 2010 in China. Ecology and Environmental Sciences 21(7): 1266–1270.

Quantifying Process-Based Mitigation Strategies in Historical Context: Separating Multiple Cumulative Effects on River Meander Migration

Alexander K. Fremier[1]*, Evan H. Girvetz[2,3], Steven E. Greco[4], Eric W. Larsen[4]

1 School of the Environment, Washington State University, Pullman, Washington, United States of America, **2** International Center for Tropical Agriculture, Nairobi, Kenya, **3** School of Environmental and Forest Sciences, University of Washington, Seattle, Washington, United States of America, **4** Department of Human Ecology, University of California Davis, Davis, California, United States of America

Abstract

Environmental legislation in the US (i.e. NEPA) requires defining baseline conditions on current rather than historical ecosystem conditions. For ecosystems with long histories of multiple environmental impacts, this baseline method can subsequently lead to a significantly altered environment; this has been termed a 'sliding baseline'. In river systems, cumulative effects caused by flow regulation, channel revetment and riparian vegetation removal significantly impact floodplain ecosystems by altering channel dynamics and precluding subsequent ecosystem processes, such as primary succession. To quantify these impacts on floodplain development processes, we used a model of river channel meander migration to illustrate the degree to which flow regulation and riprap impact migration rates, independently and synergistically, on the Sacramento River in California, USA. From pre-dam conditions, the cumulative effect of flow regulation alone on channel migration is a reduction by 38%, and 42–44% with four proposed water diversion project scenarios. In terms of depositional area, the proposed water project would reduce channel migration 51–71 ha in 130 years without current riprap in place, and 17–25 ha with riprap. Our results illustrate the utility of a modeling approach for quantifying cumulative impacts. Model-based quantification of environmental impacts allow scientists to separate cumulative and synergistic effects to analytically define mitigation measures. Additionally, by selecting an ecosystem process that is affected by multiple impacts, it is possible to consider process-based mitigation scenarios, such as the removal of riprap, to allow meander migration and create new floodplains and allow for riparian vegetation recruitment.

Editor: Maura (Gee) Geraldine Chapman, University of Sydney, Australia

Funding: The authors are grateful for funding for this research that was provided by the California Department of Water Resources (CDWR), Northern District, Red Bluff, California, and the CDWR Planning and Local Assistance Office, Sacramento, California, under Interagency Agreements #4600001950 and #4600000736. The funders had no role in study design, data collection and analysis, decision to publish, or preparation of the manuscript.

Competing Interests: The authors have declared that no competing interests exist.

* Email: alex.fremier@wsu.edu

Introduction

Society's changing perceptions of "impact" to the natural environment influences how natural areas are protected and how project effects are mitigated. In the United States, The National Environmental Policy Act of 1969 (42 U.S.C. §§4321–4370(a) requires evaluation of impact based on 'current condition' [1]; that is, impacts of proposed projects are compared to the current state or baseline of the system and does not include pre-existing alterations. Without properly addressing these cumulative effects over time, environments with a long history of human alteration will incrementally lose natural attributes and move closer to a more completely human-dominated landscape that lacks the structure or function to support natural processes; this phenomenon has been termed a "sliding baseline" [2,3].

Although cumulative effects are being addressed by the Council on Environmental Quality [4,5], a standardized protocol for quantitatively assessing them remains elusive [6,7]. Assessment of cumulative effects therefore requires accurate quantification of past impacts from multiple drivers, anthropogenic or not. Quantifying the legacy of anthropogenic alterations is a significant challenge given natural inter-annual variability and synergisms between multiple forms of impacts [8,9]. Cumulative effects assessment therefore requires a longer term view of environmental impact and typically evaluates changes to ecosystem processes, not just structures that are impacted [10].

In many cases, ecosystem processes (e.g. fire, flooding, climate, vegetation succession, and habitat connectivity) cannot be fully restored because either historical conditions no longer exist (climate regime or flow regulation due to dams) or the political motivation for alternative scenarios is lacking (risk and monetary considerations). Yet, if we are to avoid a sliding baseline, we must consider the history of changes to individual processes and search for efficient ways to mitigate the effect. This is particularly the case in disturbance-dependent systems, such as riverine-riparian areas, where watershed and river channel alterations influence downstream ecosystem processes and functioning [11,12]. For these disturbance-dependent ecological communities, alterations to the processes that create and maintain them can have significant and lasting impact [13], such as wetland loss on water on landscape functions [9] or flow regulation effects on river channel migration and riparian vegetation [14,15].

Quantifying the effects of individual projects on large scale ecosystem processes presents a considerable scientific challenge [6]. Assessment of multiple forms of alterations requires a modeling scientific approach [16]. Where empirical studies are necessary to show impact, retrospective and prospective modeling approaches allows for individual processes and impacts to be separated and quantified. In addition, it allows for development of alternative process-based solutions.

A long-term process-based view is important for understanding riparian vegetation dynamics, restoration and mitigation needs [17]. Natural disturbance processes counteract autogenic forces that drive riparian forest stands into later successional stages through floodplain inundation and by continuously reworking and depositing new lands [18,19]. Therefore, a static view of ecosystem structure and function is insufficient to maintain pre-impact habitat heterogeneity because individual stands are constantly maturing and the recruitment of new seedlings is necessary to take the place of the aging forest [20]. Conservation of riparian vegetation communities requires fostering both flows those that create new land, by channel avulsion or lateral bend migration [21], and secondly, flows that allow riparian recruitment [22,23]. In addition, management of patch dynamics is a critically important concept to create and maintain patches suitable to endangered species on rivers [24].

The overall goal of this research was to use a process-based modeling approach to isolate the specific impacts to lateral river channel migration of flow regulation – past and proposed, and current channel constraints (riprap). Using a mechanism-based numerical model of meander migration allowed us to tease apart the relative effects of riprap and flow regulation separately, as well as the interacting effect of the two together. We discuss our results from a sliding baseline perspective, and identify implications of synergism between impacts, the role of modeling in cumulative impact assessment, and possible process-based mitigation solutions.

Methods

Study area

The Sacramento River flows south through the Central Valley of California, USA (Figure 1a), which has a Mediterranean climate. The main sources of water come from the Sierra Nevada, Coast and Cascade mountain ranges during winter precipitation and spring snow melt events (Nov-April); it is a semi-arid system with a long dry season between May and October. The river channel is partially free to migrate between the towns Red Bluff and Colusa. The study reach was defined from river miles 184–201(river miles or "RM" as location markers defined by the USACE; 30 km in length) (Figure 1b) to investigate the effect of flow changes. This segment was chosen due to the availability of historical river channel records and management interests. Within the study reach, the Sacramento River is primarily a single-thread sinuous channel. Channel cutoff events are infrequent though present with the reach, occurring on average less than once every two years and account for less than 10% of the newly created floodplain [25]. The riverbed material is sand and pebbly gravel with a median grain size that ranges from 20 to 30 mm; the channel banks are composed of sand and gravel with isolated patches of erosion-resistant rock types [26,27,28].

The major man-made structures that have affected the Sacramento River's hydrology in the last 70 years are the Shasta Dam (1943 flow alteration, 1945 completion of construction), located approximately 190 river kilometers upstream from the study site; major flow diversion structures for agriculture (such as the Red Bluff Diversion Dam and the GCID diversion facility); and a number of flood control structures allowing overflow during large flow events into catchment basins and bypass channels. Flow regulation has changed the hydrograph primarily in two ways: 1) winter peak flows were reduced to decrease downstream flooding and increase water storage and 2) summer base flows were increased for dry season agricultural irrigation purposes [29]. Extensive riprap was constructed along the channel margins starting from 1940 up to the present; however, most of the rock revetment in place today was installed between 1970 and 1980 [28]. Approximately one-third of the reach from RM 143.5–194 is constrained by revetment [28]. We selected a model calibration period (1942–1976) and a study reach to avoid calibrating the model in an area with extensive riprap. After dam construction, the valley floor and riparian areas were rapidly cleared and converted to row crops and orchards; and today, 11% of the historic riparian forests remain and only 6% in a non-degraded state [29,30].

Stream hydrology reconstruction

Eight flow datasets were constructed and used in the meander migration model (Table 1). One flow record was created to calibrate the erosion field for the meander model between 1942–1976 at the Hamilton City gauge (Table 1, Figure 1a). Two historical flow records were used to represent the pre-and post-dam flows; these records were built from the USGS Bend Bridge (1892-present) and Hamilton City gauges (1945–1976) (Figure 1a). Simple linear regressions were applied to model historical trends and the relationship between the two gauging stations. Flows at the Hamilton City gauge better represent the actual flows at the study reach, while the Bend Bridge record is the longest record for the Middle Sacramento River. The regression analysis between the Bend Bridge and Hamilton City gauges show a strong correlation at both the daily ($R^2 = 0.90$, $p<0.0001$) and yearly time steps ($R^2 = 0.99$, $p<0.0001$).

The five daily flow management scenarios were simulated using the computer program CALSIM II, as part of the North of Delta Off-stream Storage (NODOS) project, a large planning effort by the California Department of Water Resources (CDWR). CALSIM II is a water resources simulation model designed to evaluate operational alternatives of large, complex river basins. Here, the NODOS project flow scenarios were modeled using quantified knowledge of the entire system. For example, a certain amount of water is necessary year-round both for biological and water quality issues in the upper valley and delta waterways. These scenarios are simulations of the daily river flows that would occur if all current dams and diversions were in operation and are based on the historic record of flows, for the water years 1922–1994.

The NODOS project base hydrograph is a simulation of river flows that would occur under current conditions and is based on flows over the past 70 years. The four scenarios – labeled 4a, 5b, 6, and 7 – are simulations of the flows that would occur with the operation of the NODOS project reservoir and current infrastructure (i.e. management at Shasta Dam and other water diversions and inputs). The four scenarios differ in amount and timing of when water is diverted out of the main channel of the river.

For each hydrology dataset the annual cumulative effective stream power (Ω_{ce}) was calculated as a part of the river meander migration model. This parameter has been shown to be proportional to bank erosion [14,31]. Stream power is used by geomorphologists to quantify the forces that lead to bed and bank movement. Cumulative effective stream power is the summation of stream power over a given threshold value within a given time step (e.g. annually).

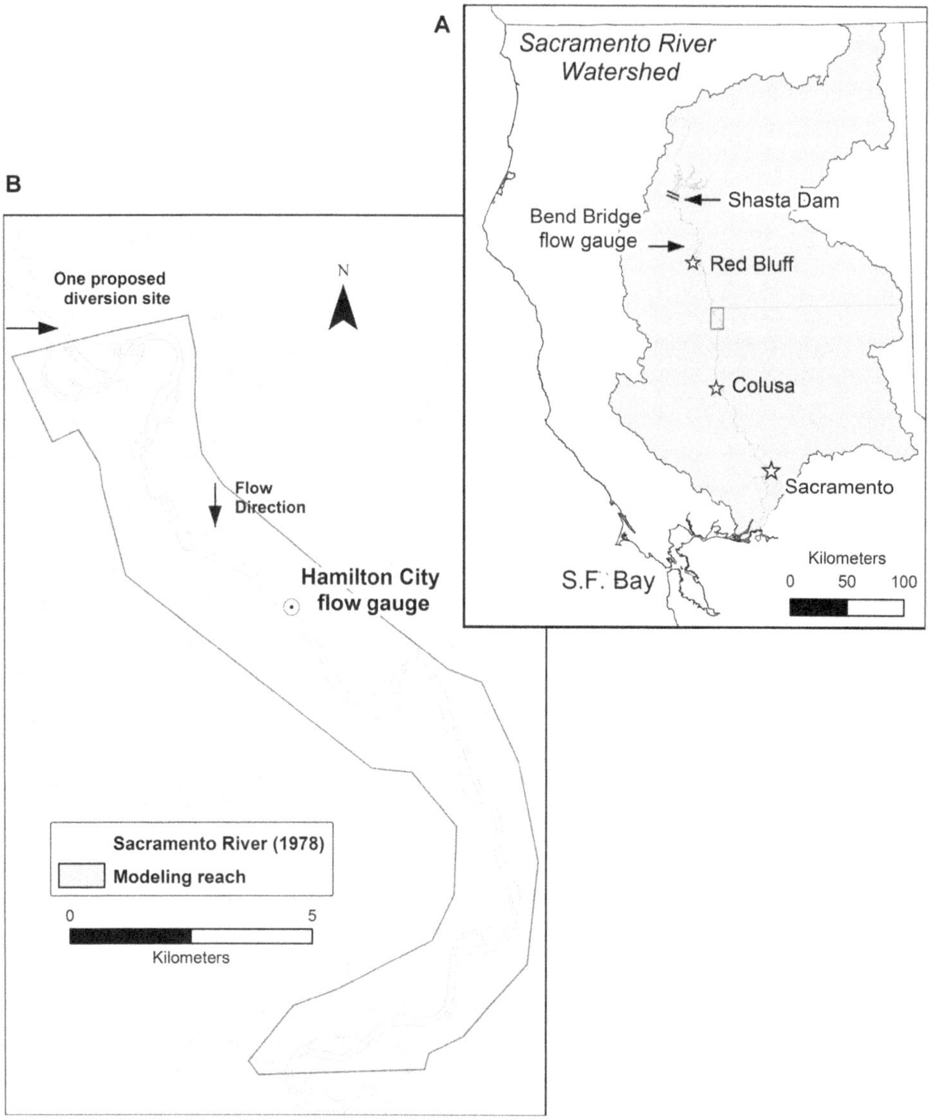

Figure 1. Study site map showing the modeled reach and the historic mapping extent. The meandering reach of the Sacramento River lies between the towns of Red Bluff and Colusa.

$$\Omega = \gamma Q S \tag{1}$$

$$\Omega_{ce} = \gamma S * \Sigma Q_e \tag{2}$$

where, stream power (Ω kg m/s^3) is calculated as the product of discharge (Q m^3/s), slope (S m/m), and the specific weight of water (γ kg/m^{2}s^2). The effective discharges (Q_e) are those flows over a determined threshold. The threshold used in this study was determined using historical data and regression analysis [31]. To analyze the cumulative effects of reduced flows we increased the span of the simulation from 70 to 130 years, the approximate time of the flow record at the Bend Bridge gauge.

In addition, in order to test for long-term changes in climate over the model time frame we compiled historical annual average precipitation data (PRISM) of the Sacramento River from 1895–2002. Long-term climatic influences were not apparent as we observed no significant decrease in precipitation.

River channel meander migration modeling

The river channel meander migration model is a numeric model based on relationships between sediment transport and fluid flow, and has been used to calculate how an alluvial river channel moves over time-scales of years to decades [32,33]. The model assumes that the local bank erosion rate is proportional to a local water velocity factor such that:

$$M = E_o u_b \tag{3}$$

where M is the bank erosion rate (m/s), E_o is a dimensionless bank erodibility coefficient of the order 10^{-8}, and u_b (m/s) is a velocity factor equal to the difference between the velocity near the bank and the reach-average velocity [32]. Higher E_o values result in greater erosion potential. Although the model analytically

Table 1. Description of flow records (daily mean discharge) measured at USGS Bend Bridge Gauge #1137100 and Hamilton City #11388500, and five simulated flows using the CALSIM model.

	Flow Dates	Description
Historic Flows (3)		
Calibration	1942–1976	Observed flows for Hamilton City, 1945–1976. From 1942–1944, flows were modeled based on a regression relationship between Hamilton City and Bend Bridge.
Historical (*pre-dam*)	1892–1942	Predicted flows using the historical record at Bend Bridge using the Hamilton City regression relation for pre-Shasta dam time period.
Historical (*post-dam*)	1943–1992	Predicted flows using the historical record at Bend Bridge using the Hamilton City regression relation for post-Shasta dam time period. These records include the effects of additional infrastructure added over time.
CALSIM II Simulated Flows (5)		
Current (*base*)	1922–1992	Modeled flow record with all current diversion contracts in place. Based on historical inputs at Bend Bridge gauge, but simulates entire record as if the current diversions had been in place for the entire time.
NODOS flow scenarios (4): *4a,5b,6 7*	1922–1992	Modeled flow records with all current diversion contracts and NODOS diversions in place. The four scenarios are based on historical inputs at Bend Bridge gauge and simulate different flow management options in the timing and amount of diversions.

calculates the velocity field in some detail, it represents bank erodibility by an estimated coefficient (E_o) that is calibrated to observed data.

The erodibility potential for the meander migration model was developed in geographic information system (GIS) based on observed geology and vegetation surfaces and known riprap (bank revetment), all of which vary spatially throughout the erosion field. The geology surface dataset was obtained from the California Department of Water Resources [27]. The geology dataset was used to determine areas that were non-erodible due to geologic constraint. The vegetation dataset, used to distinguish between agricultural and riparian land cover, was derived from aerial photography taken in 1997 [34]. Agricultural lands were coded to be twice as erodible as riparian vegetation based on the work of Micheli et al. [35] within the same section of river. These values provided the initial erosion field, which was subsequently adjusted by calibrating the modeled bank erosion to observed data.

To calibrate the erodibility coefficient to observed conditions, simulations require a calibration process. The calibration period for this study used historical channel position data and flow records from 1942 and 1976 (post dam and prior to major channel constraints). Model parameters were adjusted until model erosion rates matched the observed erosion rates between these time periods. The same parameters were then used for each model run of each scenario (Table 2).

The effect of riprap was simulated by modifying the erosion potential raster surface, using a GIS data layer of riprap occurrence from the US Army Corps of Engineers [36]. The spatial error associated with geo-referencing was less than 30 meters; therefore, a 30 meter cell size (0.09 ha) was used as the spatial resolution for the raster surface. The riprap layer was combined with the erosion potential raster surface to simulate areas into which the channel could not erode (ESRI Version 9.1). Migration patterns with channel revetment (i.e. riprap) were modeled separately from the migration with the flow scenarios in order to separate the effect of flow and channel constraints on bank erosion.

To quantify migration patterns, we calculate the 'area reworked' (equivalent to 'depositional area') by lateral channel migration and channel abandonment. We also report migration rates for comparisons with other studies. Migration rates were calculated by dividing the total area reworked by the average stream length of the first and last year and the simulation time (130 years). This represents the average annual lateral movement of the channel per meter of length. In our modeling runs, one cutoff occurred in the first years of each model run. We did not include this cutoff event in our calculations as it was approximately the same for each scenario (Figure 2).

We also present area reworked by individual bend to illustrate the potential number of bends that would be required to mitigate flow and riprap alterations. Using the GIS we separated the modeled migration area by intersecting the first and last centerline and removing polygons created by cutoff events. We show individual bend migration area for each of the NODOS project base scenarios.

Results

Flow regulation and riprap on the Middle Sacramento River have greatly reduced river meander migration potential (Table 3, Figures 2 and 3). Using pre-dam flows, lateral river channel migration reworked 1103 ha of land over the 130 year modeling period without riprap, compared to 832 ha with post-dam flows. This is a 25% loss in migration potential due only to changes in stream power caused by flow regulation to date; this accounting does not include changes in riparian vegetation, riprap or sediment reduction.

Using the CALSIM simulations for current flows and the proposed NODOS project flows, shows a further decrease in area reworked to 687 ha within the same 130 year period. Cumulatively, the current flows have reduced lateral channel migration 38% compared to pre-dam conditions (Table 3, Figure 3). The proposed NODOS project flows would reduce that further to 42–44%. Compared to current conditions, the proposed NODOS project flows would reduce channel migration by 7–10% as compared to the base NODOS project scenario (Table 3, Figure 3).

Synergistic effects of flow regulation and riprap

When riprap is modeled with flow regulation, the NODOS project base flow condition shows a 79% cumulative reduction in lateral river channel migration compared to the pre-dam

Table 2. Meander migration model - geomorphic and hydraulic variables.

Input variable	Description	Value
Discharge (Q_2)	2-year recurrence interval discharge.	2265 m^3/sec
Width (H_2)	Width at Q_2.	235 m
Depth (W_2)	Average depth at Q_2	5.4 m
Slope (S)	Longitudinal water surface slope.	0.00042 m/m
Grain Size (D_{50})	Median bed particle size.	25 mm
Lower threshold	Calculated flow over which significant bank erosion is initiated (Larsen et al. 2006b).	425 m^3/sec
Recurrence Interval	Flow recurrence interval over which cutoff algorithm triggers potential cutoff (Larsen et al. 2006c).	1.5
Sinuosity Threshold	Channel sinuosity threshold used to trigger cutoff simulation. Calculated for each bend between inflection points (Larsen et al. 2006c).	1.9 m/m

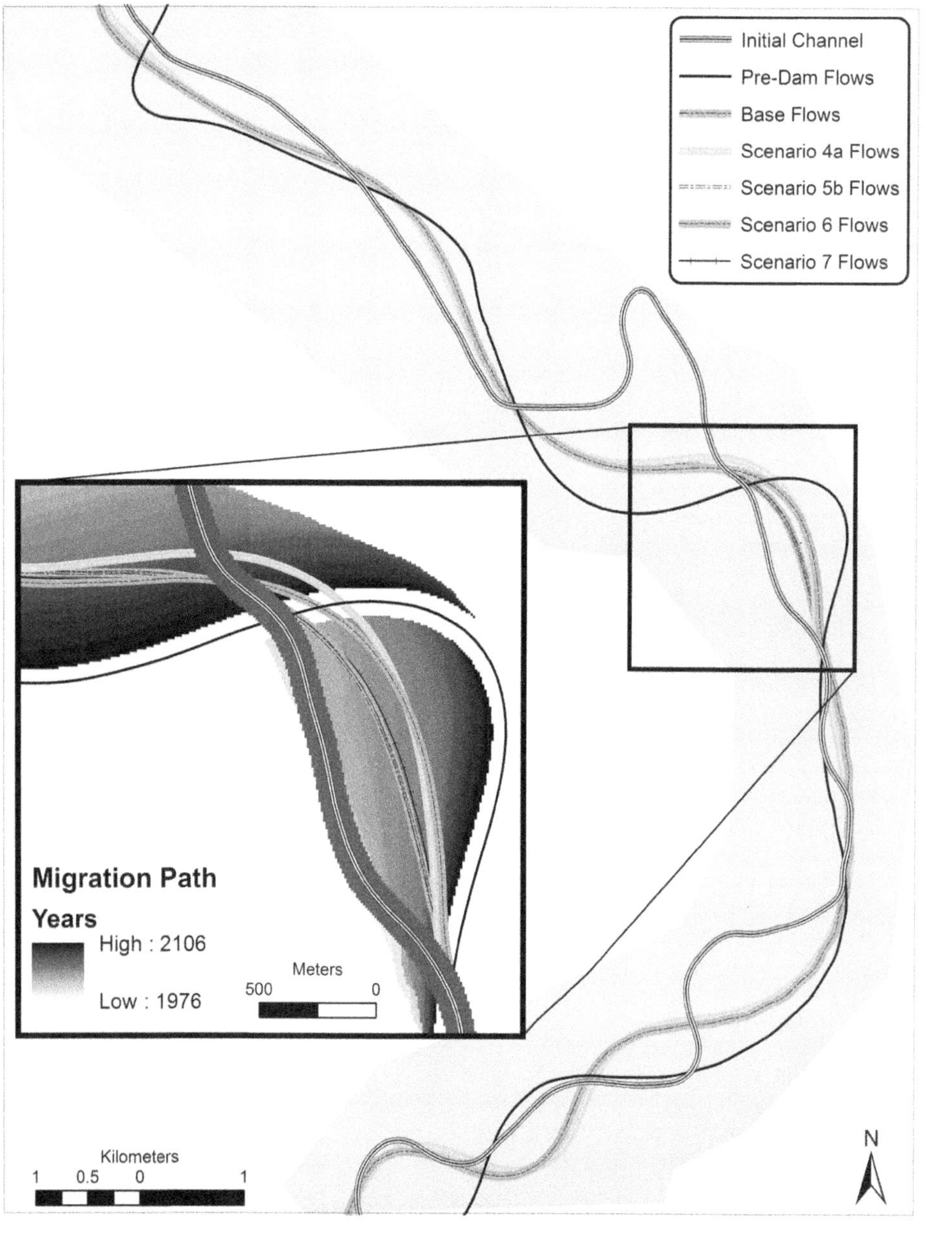

Figure 2. Meander migration model output centerlines for each scenario after 130 year simulation, including the initial channel centerline. The inset map includes a zoomed in migration path of the pre-dam flow channels. Scenarios 5b and 6 are close to overlapping.

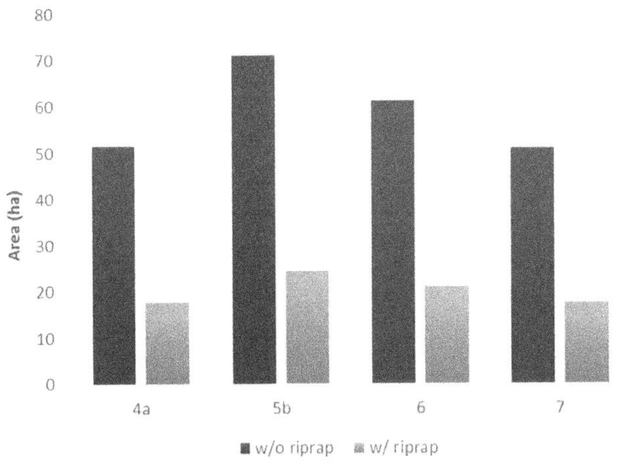

Figure 3. Predicted floodplain reworked by lateral channel migration lost to flow regulation (NODOS scenario flows compared to NODOS base flow) and riprap. These areas represent the area of potential mitigation for the NODOS project. Riprap reduced almost three times (2.89) the area compared to the proposed NODOS flow regulation.

conditions. The percent difference between the NODOS project flow scenario with and without riprap is minimal; however, in absolute terms, the area reworked tells a different story. The NODOS project flow scenarios would reduce the area of floodplain reworked by 51–71 ha without current riprap in place and 17–25 ha with riprap. (Table 3, Figure 3). This difference in modeled impact is almost three times greater (2.9) in absolute area than when channel migration is modeled with riprap in place. Considering mitigation, the effect of riprap, in this case, essentially masks the impacts of flow regulation in absolute area terms. To give perspective to these respective areas, we quantified the area eroded by individual bend. In our 30 km study reach, there were 13 active bends with an average area reworked of 52 ha of land for the NODOS project base flow scenario. This suggests that the removal of riprap and subsequent meander migration from a single bend would approximately mitigate the impact of the proposed NODOS project flow regulation.

Discussion

In this study we applied a modeling tool to compare and help us separate the effects of flow regulation versus channel revetment on lateral river channel migration. If only empirical data were used, the synergistic interacting effect of altered floodplains (e.g. channel constraints and vegetation clearing) and flow regulation taken together would not have been quantifiable, but see Duinker and Greig 2007 for another example [37]. Our modeling procedure allowed us to quantify the impact of channel revetment from flow regulation in an effort to inform management and restoration of an important large-scale ecosystem processes [38]. Without this process-based modeling approach we would have underestimated the mitigation needs by almost a factor of three (see Table 3).

Cumulative effects

Considering cumulative impacts, although the NODOS project would reduce channel migration potential 7–10% from the current base scenario, the changes in cumulative effective stream power would have reduced the migration rate 38% when compared to historical pre-dam conditions. The NODOS water diversion project adds to an overall impact of 40–43% from pre-dam conditions, and 79% including flow alterations and riprap. In this case, the cumulative impacts to channel migration on the Sacramento River are considerable relative to the impact of the proposed flow regulation scenarios (Figure 2 and 3).

Deviations from historical baseline condition, such as the ones described here, are a result of cumulative impacts and are not typically included in the assessment of new projects because of difficulties in defining the appropriate spatial and temporal scale, and identifying the history of direct and/or indirect effects [7].

Table 3. Comparison of area reworked by river channel migration for seven flow scenarios (see Table 1) with and without riprap over a 130-year period.

Modeled scenarios	Area reworked (ha)	% Diff in area from pre dam	Potential Mitigation (ha)
No Riprap			
Pre-dam flow	1103.4		
Post-dam flow	832.4	24.6	
NODOS base flow	687.7	37.7	
flow 4a	636.3	42.3	51.4
flow 5b	616.7	44.1	71.0
flow 6	626.6	43.2	61.1
flow 7	636.8	42.3	50.9
Current Riprap			
NODOS base flow	237.3	78.5	
flow 4a	221.6	80.1	17.7
flow 5b	215.9	80.7	24.5
flow 6	218.3	80.4	21.1
flow 7	223.0	80.1	17.6

The numbers in the Potential Mitigation column are the differences between NODOS base and NODOS scenarios. This can be seen as the total area needed to be mitigated if you assume rip-rap absent (top) or rip-rap present (bottom). The percent difference is the mitigation needs if you assume rip-rap present versus absent (289%). The bottom line is that without rip-rap, almost three times (2.89) increase in the area needed to be mitigated (masked effect).

The presence of multiple interacting human and environmental stressors can act synergistically or additively depending on the environmental setting and character of the alteration [8,12]. Although we might not be able to precisely define the true 'pristine' environment, certainly quantified knowledge of some historical impacts will help in ultimately what we aim to do, which is to assess impacts to the function and structure of the riverine-riparian ecosystem. As illustrated in this study, when the baseline conditions are defined as the system's current conditions and the system has a long history of human alteration, the environment slides farther from historical conditions. Without properly addressing the cumulative effects over time, environments with a long history of human alteration will incrementally lose natural attributes and move closer to a more completely human-dominated landscape that lacks the structure or function to support ecosystem processes.

A process-based mitigation approach

Using a meander migration model we were able to isolate a plausible ecosystem process-based mitigation scenario for the proposed off-stream storage facility. Our results suggest that there are enough geomorphically significant flows to move the channel laterally in the absence of near-channel constraints (riprap), albeit less dynamic and smaller in extent than historical pre-dam conditions [14,31]. One potential management action is targeted removal of channel constraints (riprap) to mitigate the effect of additional water diversions and withdrawals. In our study, the removal of riprap on a single bend would generate enough area to mitigate the impact of the proposed water diversion, assuming lateral channel migration is possible at the site (see Figure 2).

Our goal was to investigate the long-term direction of a single large-scale driver of ecosystem change (i.e. a keystone process), river channel migration. We aimed to isolate the impact of a first-order control, hydrology, on channel migration by holding all else equal. On this river reach, summer base flows (typically less than 425 m^3/s) were found not to cause significant lateral migration through the initiation of bank erosion [15]. Flow scenarios including this lower threshold will show increased impact from flow regulation and diversion because of the redistribution of flows from a few large peak events to many days of flows below the threshold, i.e. dry-season base flows. Without consideration of the lower threshold, the effect of the altered flow magnitude would be missed entirely. Including this geomorphic threshold in the planning process of dam re-operation would be a valuable way to account for the impact of proposed flow management on this vital ecosystem process [39].

Our analysis of system behavior is the beginning of more specific models to quantify the impact of sediment supply loss, riparian vegetation controls on bank erosion rates, adjustments in channel valley slope, and specific riparian vegetation recruitment mechanisms. In selecting the model procedure and parameters, we aimed to minimize model complexity and isolate the impacts of one flow parameter (cumulative effective stream power) on floodplain development. Our study did not model changes in sediment supply due to dam construction which is important to consider as multiple studies have reported reduced lateral migration rates post-impoundment [40]. In addition, to focus the paper, we also excluded a detailed analysis of cutoff dynamics within the reach that might be impacted by flow regulation [16,26]. A more detailed study into flow effects on cutoff dynamics would improve our process-level understanding channel movement in large rivers [41].

Riparian forest extent, pattern and structure have important implications for channel migration rate and bend shape [41,42]. In this study we did not alter the riparian vegetation pattern between scenarios, though we used GIS data to represent the impact of vegetation on bank resistance by increasing bank resistance two-fold between agriculture and riparian vegetation [41]. Previous historical work on lateral channel migration on the Sacramento River suggests that riparian vegetation might have had a large impact on erosion rates post flow regulation [21]. We excluded riparian vegetation in our scenarios, in part, to limit the complexity of the scenarios, but also because it was difficult to envision a management scenario where vegetation would be removed to increase lateral channel migration. Riparian vegetation in the reach is actively being restored to meet multiple conservation objectives [29] due to the legacy of riparian forest loss in this river system.

Summary

Channel migration is an important ecosystem process on low-gradient rivers that structures floodplain environments. We employed a mechanism-based model of river channel migration to quantify the impacts of past, current and proposed flow regulation on lateral river channel migration patterns. We found that channel migration potential would be further reduced with the proposed off-stream storage facility (7–9%), adding to previous flow regulation caused by impoundments upstream (38%). Riprap further reduced channel migration (79%) compared to historical flow patterns. Given competing needs in the state for water, we describe a potential solution to limit conflicts by removing enough riprap to allow lateral migration to occur and offset the impacts of flow regulation. This potential solution focuses on the processes that create and maintain ecosystems and is potentially a quantifiable way to design new approaches to mitigate impacts among multiple stressors in systems with extensive cumulative impacts.

Acknowledgments

We express special thanks to S. A. Cepello and A. Henderson and the staff at Department of Water Resources, Northern District Office; and S. Roberts, J. Wieking, and S. Sou from the Off-stream Storage Investigation and North-of-the-Delta Off-stream Storage Program. The data used in this study were the product of the dedicated efforts of many individuals working in the Landscape Analysis and Systems Research Laboratory in the Department of Human Ecology at the University of California, Davis.

Author Contributions

Conceived and designed the experiments: AKF EHG SEG EWL. Performed the experiments: AKF. Analyzed the data: AKF EHG. Contributed reagents/materials/analysis tools: AKF SEG EWL. Wrote the paper: AKF.

References

1. NEPA (1970) National Environmental Policy Act of 1969.
2. Pauly D (1995) Anecdotes and the shifting baseline syndrome of fisheries. Trends in Ecology & Evolution 10:430.
3. Dayton PK, Tegner MJ, Edwards PB, Risner KL (1998) Sliding baselines, ghosts, and reduced expectations in kelp forest communities. Ecological Applications 8:309–322.
4. CEQ (1987) Council on environmental quality. Regulations for implementing the procedural provisions of the National Environmental Policy Act.
5. CEQ (1997) Considering cumulative effects under the National Environmental Policy Act. Executive Office of the President, Washington, DC.
6. McCold LN, Saulsbury JW (1996) Including past and present impacts in cumulative impact assessments. Environmental Management, 20(5) 767–776.

7. MacDonald LH (2000) Evaluating and managing cumulative effects: Process and constraints. Environmental Management 26:299–315.
8. Gosselink JG, Shaffer GP, Lee LC, Burdick DM, Childers DL, et al. (1990) Landscape conservation in a forested wetland watershed - can we manage cumulative impacts. BioScience 40:588–600.
9. Hunsaker CT (1993) Ecosystem assessment methods for cumulative effects at the regional scale. Pages 480–493 in S. G. Hildebrand and J. B. Cannon, editors. Environmental analysis: The NEPA experience. Lewis Publishers, Boca Raton, FL.
10. Bedford BL, Preston EM (1988) Developing the scientific basis for assessing cumulative effects of wetland loss and degradation on landscape functions: status, perspectives, and prospects. Environmental Management 12:751–771.
11. Hughes FMR, Adams WM, Muller E, Nilsson C, Richards KS, et al. (2001). The importance of different scale processes for the restoration of floodplain woodlands. Regulated Rivers-Research & Management, 17(4–5) 325–345.
12. Michalková M, Piégay H, Kondolf GM, Greco SE (2011) Lateral erosion of the Sacramento River, California (1942–1999), and responses of channel and floodplain lake to human influences. Earth Surface Processes and Landforms 36:257–272.
13. Pickett STA, Rogers KH (1997) Patch dynamics: the transformation of landscape structure and function. In Wildlife and landscape ecology: effects of pattern and scale, Bissonette JA (ed). Springer-Verlag, New York: 101–127.
14. Larsen EW, Fremier AK, Girvetz EH (2006) Modeling the effects of variable annual flow on river channel meander migration patterns, Sacramento River, California, USA. Journal of the American Water Resources Association 42:1063–1075.
15. Larsen EW, Fremier AK, Greco SE (2006). Cumulative effective stream power and bank erosion on the Sacramento River, California, USA. Journal of the American Water Resources Association 42:1077–1097.
16. Larsen EW, Girvetz EH, Fremier AK (2007) Landscape level planning in alluvial riparian floodplain ecosystems: Using geomorphic modeling to avoid conflicts between human infrastructure and habitat conservation. Landscape and Urban Planning 79:338–346.
17. Nilsson C, Svedmark M (2002) Basic principles and ecological consequences of changing water regimes: Riparian plant communities. Environmental Management, 30(4) 468–480.
18. Baker WL, Walford GM (1995) Multiple stable states and models of riparian vegetation succession on the Animas River, Colorado. Annals of the Association of American Geographers 85:320–338.
19. Nilsson C, Berggren K (2000) Alterations of riparian ecosystems caused by river regulation. BioScience 50(9) 783–792.
20. Shin N, Nakamura F (2005) Effects of fluvial geomorphology on riparian tree species in Rekifune River, northern Japan. Plant Ecology 178 (1) 15–28.
21. Greco SE, Fremier AK, Plant RE, Larsen EW (2007) A tool for tracking floodplain age land surface patterns on a large meandering river with applications for ecological planning and restoration design. Landscape and Urban Planning 81:354–373.
22. Mahoney JM, Rood SB (1998) Streamflow requirements for cottonwood seedling recruitment - an integrative model. Wetlands 18(4) 634–645.
23. Stella JC, Hayden M, Battles JJ, Piégay H, Dufour S, et al. (2011) The role of abandoned channels as refugia for sustaining pioneer riparian forest ecosystems. Ecosystems, 14(5) 776–790.
24. Greco SE (2013) Patch change and the shifting mosaic of an endangered bird's habitat on a large meandering river. River Research and Applications 29(6): 707–717.
25. Micheli ER, Larsen EW (2011) River channel cutoff dynamics, Sacramento River, California, USA. River Research and Applications, 27(3) 328–344.
26. Brice J (1977) Lateral migration of the Middle Sacramento River, California. USDI Geological Survey Water Res. Investigations 77–43.
27. CDWR (1995) Memorandum Report: Sacramento River meander belt future erosion investigation. CDWR 155. The Resources Agency, California Department of Water Resources, Sacramento, CA.
28. CDWR (2005) Memorandum report: Bank survey of reach 3 Chico to Colusa, Sacramento River, California. - California Department of Water Resources, Northern District, CA.
29. CALFED (2000) Final programmatic environmental impact statement/environmental impact report. CALFED Bay-Delta Program, Sacramento, CA.
30. Katibah EF (1984) A brief history of riparian forests in the Central Valley of California. In California Riparian Ecosystems: Ecology, Conservation and Productive Management, Warner RE, Hendrix KM (eds). University of California Press, Berkeley: 23–29.
31. Larsen EW, Girvetz EH, Fremier AK (2006) Assessing the effects of alternative setback channel constraint scenarios employing a river meander migration model. Environmental Management 37:880–897.
32. Johanesson H, Parker G (1989) Linear theory of river meanders.in S. Ikeda and G. Parker, editors. River Meandering. American Geophysical Union, Washington, D.C.
33. Larsen EW, Greco SE (2002). Modeling channel management impacts on river migration: a case study of Woodson Bridge State Recreation Area, Sacramento River, California, USA. Environmental Management, 30(2) 209–224.
34. Greco DE, Plant RE (2003) Temporal mapping of riparian landscape change on the Sacramento River, miles 196–218, California, USA. Landscape Research 28:405–426.
35. Micheli ER, Kirchner JW, Larsen EW (2004) Quantifying the effect of riparian forest versus agricultural vegetation on river meander migration rates, Central Sacramento River, California, USA. River Research and Applications 20:537–548.
36. USACE (2002) Sacramento and San Joaquin River Basins Comprehensive Study. US Army Corps of Engineers, Sacramento, CA.
37. Duinker PN, Greig LA (2007) Scenario analysis in environmental impact assessment: Improving explorations of the future. Environmental Impact Assessment Review. 27: 206–219.
38. Florsheim JL, Mount JF, Chin A (2008) Bank erosion as a desirable attribute of rivers. BioScience, 58(6) 519–529.
39. Church M (1995) Geomorphic response to river flow regulation: case studies and time-scales. Regulated Rivers: Research & Management 11:3–22.
40. Shields FD, Simon A, Steffen LJ (2000) Reservoir effects on downstream river channel migration. Environmental Conservation 27:54–66.
41. Constantine JA, Dunne T (2008) Meander cutoff and the controls on the production of oxbow lakes. Geology 36(1):23.
42. Perucca E, Camporeale C, Ridolfi L (2007) Significance of the riparian vegetation dynamics on meandering river morphodynamics. Water Resources Research 43:W03430.

12

Variability of Polychaete Secondary Production in Intertidal Creek Networks along a Stream-Order Gradient

Tianjiang Chu, Qiang Sheng, Sikai Wang, Jihua Wu*

Ministry of Education Key Laboratory for Biodiversity Science and Ecological Engineering, and Institute of Biodiversity Science, Fudan University, Shanghai, China

Abstract

Dendritic tidal creek networks are important habitats for sustaining biodiversity and ecosystem functioning in salt marsh wetlands. To evaluate the importance of creek heterogeneity in supporting benthic secondary production, we assess the spatial distribution and secondary production of a representative polychaete species (*Dentinephtys glabra*) in creek networks along a stream-order gradient in a Yangtze River estuarine marsh. Density, biomass, and secondary production of polychaetes were found to be highest in intermediate order creeks. In high order (3rd and 4th) creeks, the density and biomass of *D. glabra* were higher in creek edge sites than in creek bottom sites, whereas the reverse was true for low order (1st and 2nd) creeks. Secondary production was highest in 2nd order creeks (559.7 mg AFDM m^{-2} $year^{-1}$) and was ca. 2 folds higher than in 1st and 4th order creeks. Top fitting AIC models indicated that the secondary production of *D. glabra* was mainly associated with geomorphological characters including cross-sectional area and bank slope. This suggests that hydrodynamic forces are essential factors influencing secondary production of macrobenthos in salt marshes. This study emphasizes the importance of microhabitat variability when evaluating secondary production and ecosystem functions.

Editor: Jeffrey Buckel, North Carolina State University, United States of America

Funding: This study was financially supported by National Science and Technology Ministry (2013CB430404), and Science and Technology Department of Shanghai (12231204700). The funders had no role in study design, data collection and analysis, decision to publish, or preparation of the manuscript.

Competing Interests: The authors have declared that no competing interests exist.

* E-mail: jihuawu@fudan.edu.cn

Introduction

The primary production in salt marshes can exceed 3900 g C $m^{-2}yr^{-1}$, making them among the most productive ecosystems in the world [1]. This high level of primary production forms the energy base for abundant animal consumers, including benthic macrofauna, nekton, and shorebirds, and thereby sustains a high level of secondary production [2,3]. In the salt marshes of the northern Gulf of Mexico, secondary production exceeds other regions of the United States, which is exemplified by large fishery catches of penaeid shrimps (*Farfantepenaeus aztecus, F. duorarum* and *Litopenaeus setiferus* of 66% U.S.) and blue crabs (*Callinectes sapidus* of 25% U.S.) [4]. Similarly, researchers estimated that the secondary production of crabs (*Chiromantes dehaani, Helice tientsinensis*) in a Yangtze River estuarine salt marsh was 49.4 g AFDM $m^{-2}year^{-1}$, which is higher than that reported from other types of coastal wetlands at a similar latitude [5].

Secondary production is the formation of heterotrophic biomass through time [6]. It represents the cumulative responses of a population to abiotic and biotic stresses, and is an essential variable when quantifying food webs and ecosystem functioning such as material cycling and energy flow [7]. The habitat heterogeneity has been noted when estimating secondary productions. The secondary production of macrobenthos was highest in sandy organic sediment and lowest in muddy sediment in a southern New England salt marsh [8]. Secondary production of macrobenthos differed with vegetation types in the Mobile-Tensaw Delta [9]. A better understanding of the spatial differences in production, especially at local scales, would be helpful to assess the spatial variability of ecosystem functioning.

Intertidal creeks are an important sub-habitat in salt marsh ecosystems and are used by shorebirds and benthic, nektonic, and planktonic animals [10]. These habitats play a key role in promoting material cycle and energy flow between the salt marsh and offshore ecosystem [11]. According to the stream order classification, the smallest headwater stream is defined as a 1st order creek, and creek order increases thereafter when two creeks of the same order connect [12]. Geomorphological characters (e.g. length, width, and depth) and physicochemical properties (e.g. velocity, water temperature and substrate) are usually different between creeks of different orders [13,14,15]. Thus, creek order could be an important factor for evaluating the effects of local habitat heterogeneity on secondary production.

Polychaetes are the dominant macrobenthic component in many tidal creek ecosystems [16]. Their density can exceed 14,500 ind m^{-2} and serve as an important food source for higher trophic level consumers [17,18]. They are also the prominent ecosystem engineers in salt marsh ecosystems. They can affect the organic matter content in sediment via bioturbation [19], accelerate the erosion of intertidal creeks [20], and stimulate the expansion of some plant species [21]. An understanding of the spatial variability in secondary production of this functionally important organism is essential for determining the processes that affect benthic ecosystems across different habitats. As to polychaete, the highest secondary production of *Caulleriella caputesocis* was found in the sediment containing 60 to 100% silt [22]. In the Río de la Plata estuary, secondary production of *Laeonereis culveri* was higher in creeks than in intertidal flat [17]. There are various microhabitats within tidal creek networks, but little is known about the spatial variability in polychaete production.

Dentinephtys glabra (Hartman), which belongs to the family Nephtyidae, is the most numerically dominant macrofaunal species in the creek systems of the Yangtze River estuarine marshes [23,24]. *D. glabra* is a typical infaunal species which has low mobility and is considered an ideal and representative species to study the effects of sediment environmental variability [25]. This species is also a key link between primary producers and higher trophic levels such as nektonic fish and shorebirds, and thus plays an important role in the food web of the marsh ecosystem. To unravel the importance of creek heterogeneity on macrobenthos, we selected this representative polychaete species to evaluate the effects of creek order on benthic population and secondary production. We aim to answer the question how the distribution and secondary production of *D. glabra* differs among creeks of different orders.

Material and Methods

Ethics statement

This study was conducted according to the regulations of local and central government. Sample collections necessary to scientific research were authorized by the Specifications for oceanographic survey - Part 6: Marine biological survey (GB/T 1263.6 - 2007).

Study site

The study site was located in Shanghai Jiuduansha Wetland National Nature Reserve (31°03′–31°17′N, 121°46′–122°15′E) and covered a total area of 420.2 km^2 (Figure 1). Jiuduansha, with a history of about only 60 years, is a young intertidal island in the Yangtze River estuary. It was formed mainly due to the connection of flood tidal currents and ebb tidal currents. Since its formation, it has experienced considerable enlargement, heightening and shape modification because of the dynamic tidal fluxes in the estuary. The salt marsh is dominated by the native plants (*Phragmites australis* and *Scirpus mariqueter*) and an exotic plant (*Spartina alterniflora*). The mean annual temperature, precipitation, and salinity is 15.7°C, 1145 mm, and 11.7, respectively. The intertidal zones are flooded by irregular semi-diurnal tide with the tidal range from 2.45 m to 4.96 m (mean tidal range 2.75) [26].

Sampling design

Monthly samples were collected from a four order creek system during the neap tides between May 2008 to April 2009. We sampled three replicate creeks per stream order (Figure 1). In each creek, we sampled at two locations, the deepest point of the creek (bottom), and an edge position approximately 1 m below the high tide level (edge). All the samples were taken in the intertidal zone and not within the vegetation. At each location, we collected five cores (176.6 cm^2 in area and 20 cm depth for each core) at intervals of 2–3 m. Sediments from these five cores were mixed to form a composite sample for polychaete collection. Thus, we collected a total of 288 composite samples during the study. The sediment samples were washed through a 0.5 mm mesh sieve. *D. glabra* were sorted and fixed in 10% formalin. The top 5 mm of sediment within a 4×8 cm area was collected at each location and preserved at −20°C for chlorophyll *a* (chl *a*) concentration analysis [27]. Additional sediment samples down to a depth of 10 cm (19.6 cm^2) were taken for analysis of water content and grain size. Sediment physical parameters (temperature, pH, and conductivity) were directly measured during each month, except May, June, and July, 2008. A global positioning system (GPS) was used to ensure we sampled the same location each month.

Laboratory procedures

Sediment was dried to constant weight at 80°C to measure water content. Chl *a* content in the sediment was determined using spectrophotometry to represent benthic algae biomass [28]. Algae are an important food source for many primary benthic consumers. Although *D. glabra* do not ingest algae directly, there are indirect relationships between algal biomass and the carnivorous consumer species. Sediment samples for grain size analysis collected from the same location in June, September, and December 2008 and March 2009 were mixed and measured using a LS-POP (VI) laser grain sizer. Geomorphological characters included cross-sectional area, bank slope, bank full elevation, width and depth with the detail measuring methods followed our previous study in [29].

We used this width at setiger 8 (Wd8S) as an indicator of the total length of the polychaetes [30]. The Wd8S of each *D. glabra* individual was measured using a photo imaging system (NIS-Elements D 2.20) to the nearest 0.01 mm. Complete specimens of *D. glabra* were dried at 60°C until constant weight to estimate dry mass (DM). The ash free dry weight (AFDM) was obtained by incinerating at 450°C in a muffle furnace for 8 h.

Data analysis

We used the relationship between AFDM and Wd8S ($AFDM = 0.8082\ Wd8S^{1.1226}$, $N = 32$, $R^2 = 0.9792$) to estimate biomass and secondary production. To estimate secondary production, species were grouped into several size classes at 0.4 mm intervals based on Wd8S.

Secondary production was calculated using the size frequency method [31,32], expressed as:

$$P = [i\sum_{j=1}^{i}(\bar{n}_j - \bar{n}_{j+1}) \times (W_j W_{j+1})^{1/2}] \times 365/CPI$$

where P is the secondary production (g AFDM m^{-2}year^{-1}), i is the number of size classes, $\bar{n}_j$ is the mean density in size class j (ind m^{-2}), W_j is the mean individual weight in size class j (g AFDM), and CPI is the longevity time from hatching to death of the largest size class.

The seasonal VBGF (von Bertalanffy growth function) was used to estimate the longevity of *D. glabra* [33]:

$$L_t = L_\infty\{1 - e^{[-k(t-t_0) + \left(KC/2\pi\pi\right)\sin 2\pi(t - wp + 0.5) - \left(KC/2\pi\pi\right)\sin 2\pi(t_0 - wp + 0.5)]}\}$$

where L_t is the length (Wd8S) at time t (mm), L_∞ is the theoretical maximum length (Wd8S), K is the curvature parameter, C is a constant representing the amplitude of seasonal growth oscillation, t_0 is the age at zero length (Wd8S), wp is the period when growth is slowest. The value for t_0 was estimated on the basis of the smallest width found [30]. Life span was calculated based on the age at which 90% of L_∞ was achieved [17].

Because creeks of the same order have similar width [29], the creek area of each order was calculated based on creek width and its total length. We used ArcGIS 9.3 to measure the length of each creek on Jiuduansha Island based on satellite image data of 2 m resolution.

One-way ANOVA was used to evaluate the effect of creek order on environmental parameters. Two-way or one-way ANOVA was used to evaluate the effects of creek order and sampling location on the density and biomass of *D. glabra*. The effect of creek order on secondary production of *D. glabra* was also evaluated by one-way ANOVA. Post hoc multiple comparisons were performed

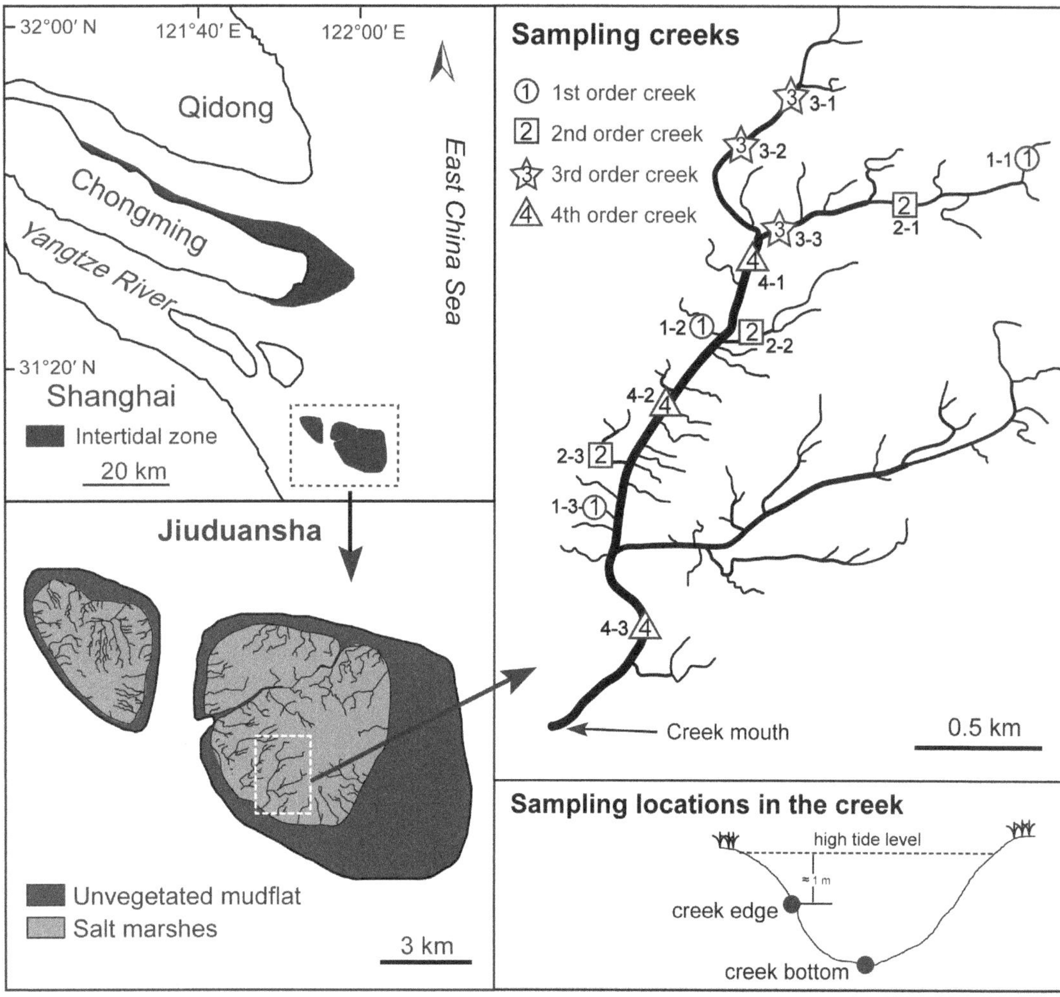

Figure 1. Creeks sampled on the Jiuduansha Island. Samples were collected at two locations (bottom and edge) in each creek.

using the least square differences (LSD) test. Differences were regarded as significant at $P<0.05$. To meet the assumptions of ANOVA, numeric data were log (x+1) transformed before statistical analyses when necessary.

To examine the relationship between secondary production of *D. glabra* and environmental factors, we performed model selection in regression analyses using Akaike's information criterion (AIC) corrected for small sample size (AIC_C) [34]. Spearman correlation were used to exclude the auto-correlated factors with the absolute coefficients larger than 0.7 [35]. Water content, Chl *a* concentration, conductivity, temperature, pH, clay-silt content, bank full elevation, cross-sectional area and bank slope were considered in models. Creek order was a categorical factor. Numeric data were log transformed before analyses. We calculated the Akaike weight (w_i) and R^2 for each model. Akaike weight (w_i) represents the probability that the model *i* is the best model among the given set of alternatives [34]. The proportion of variance explained by the model (R^2) was calculated for each model [36].

Results

Environmental factors by creeks and orders are showed in Table 1. In creeks across 1st to 4th orders, the ranges of sediment water content, Chl *a* concentration, pH, conductivity, temperature and clay-silt content were 52.60–78.91 (%), 0.02–30.60 (μg/g), 6.39–8.55, 1.52–7.97 (ms/s), 0.55–30.95 (°C) and 65.79–87.31 (%), respectively (Table 1). The highest water content, temperature, pH and sand content were in the 4th order creek. The highest Chl *a* concentration and clay-silt content were in 2nd and 1st order creek, respectively. The values of bank full depth, width, and cross-sectional area increased with increasing creek order.

The density and biomass of *D. glabra* are shown in Figure 2, 3 and 4. The density and biomass of *D. glabra* was highest in the 3rd order creeks (332.0 ind m^{-2}, 296.6 mg AFDM m^{-2}), followed by the 2nd order creeks (283.9 ind m^{-2}, 238.5 mg AFDM m^{-2}) and 1st order creeks (205.6 ind m^{-2}, 183.3 mg AFDM m^{-2}), and were lowest in the 4th order creeks (157.1 ind m^{-2}, 136.3 mg AFDM m^{-2}) (Figure 2). The 3rd order creek had significantly higher densities ($P=0.0002$) and biomass ($P=0.0013$) of *D. glabra* than other order creeks. This spatial pattern varied with different sampling months. From May to August, the highest density and biomass of *D. glabra* occurred in the 2nd order creeks, whereas between September to April levels were highest in the 3rd order creeks (Figure 3).

For the 1st and 2nd order creeks, the density (1st: $P=0.0222$; 2nd: $P=0.0160$) and biomass (1st: $P=0.0331$; 2nd: $P=0.0118$) of *D. glabra* were significantly higher at the bottom locations than at the edge locations (Figure 4). However, the trend was reversed in

Table 1. Physicochemical parameters and geomorphological characters in creek of each order with one-way ANOVA results showing the effect of creek order.

Environmental factors	Creek order				ANOVA		
	1st	2nd	3rd	4th	*F*	df	*P*
Physicochemical parameters							
Water content (%)	68.55±0.57	68.70±0.72	67.77±0.73	68.79±0.76	0.444	3	0.722
Chl *a* (μg/g)	5.06±0.82^{ab}	5.63±1.08^{a}	3.52±0.51^{b}	0.51±0.20^{c}	9.849	3	<0.001
Conductivity (ms/s)	4.30±0.20	4.29±0.22	4.78±0.20	4.59±0.34	0.927	3	0.431
Temperature (°C)	15.49±1.67	15.19±1.69	15.25±1.67	15.64±1.71	0.016	3	0.997
pH	7.59±0.08	7.56±0.08	7.46±0.10	7.73±0.07	1.841	3	0.144
Clay-silt (%)	76.10±1.35^{a}	75.17±1.18^{a}	75.69±1.08^{a}	70.39±0.29^{b}	6.315	3	<0.001
Sand (%)	23.90±1.35^{a}	24.83±1.18^{a}	24.31±1.08^{a}	29.61±0.29^{b}	6.315	3	<0.001
Geomorphological characters							
Bank full elevation (m)	3.48±0.15	3.66±0. 27	3.51±0.08	3.55±0.13	0.213	3	0.885
Bank full width (m)	12.43±1.39^{a}	16.00±1.03^{a}	17.93±2.65^{a}	29.33±2.57^{b}	12.865	3	0.002
Bank full depth (m)	1.82±0.11^{a}	2.01±0.16^{a}	2.16±0.17^{a}	2.57±0.07^{b}	6.015	3	0.019
Cross-sectional area (m^2)	12.69±1.33^{a}	17.30±1.97^{ab}	22.00±3.44^{b}	46.88±5.60^{c}	19.042	3	0.001
Bank slope	0.29±0.05	0.26±0.03	0.26±0.05	0.21±0.01	0.850	3	0.505

Shown are Fisher's *F*-ratio (*F*) and *P*-values of ANOVA for testing differences of each variable among creeks of different orders. Different superscript letters (a, b and c) represent a significant difference ($P < 0.05$) among creeks of different orders.

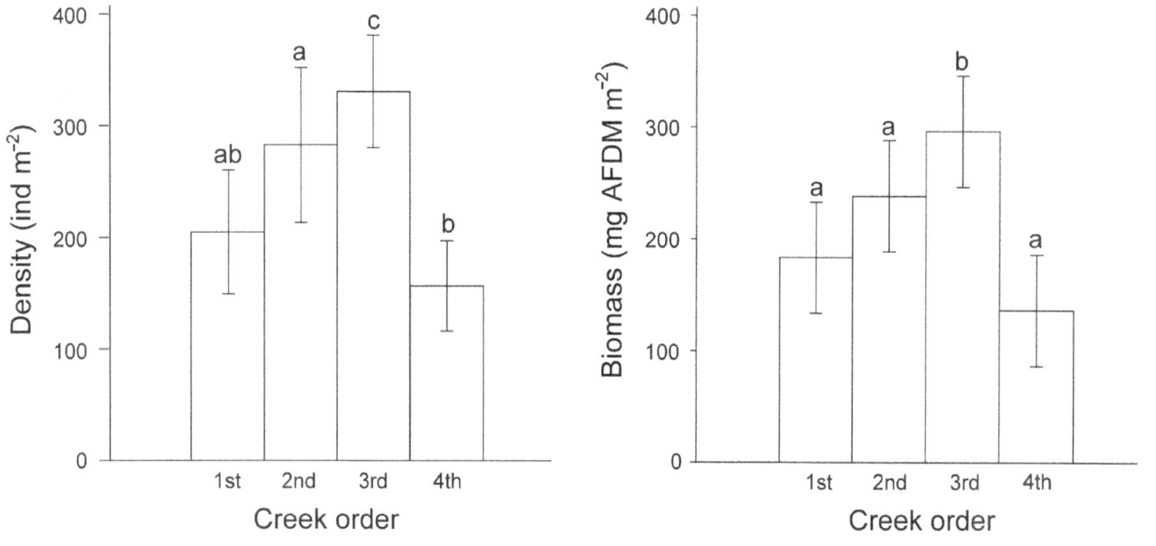

Figure 2. Density and biomass of *Dentinephtys glabra* in creeks of different orders. Error bars represent standard error ($n = 72$). The same letters above the bars denote non-significant differences and different letters represent a significant difference ($P < 0.05$) among creek orders.

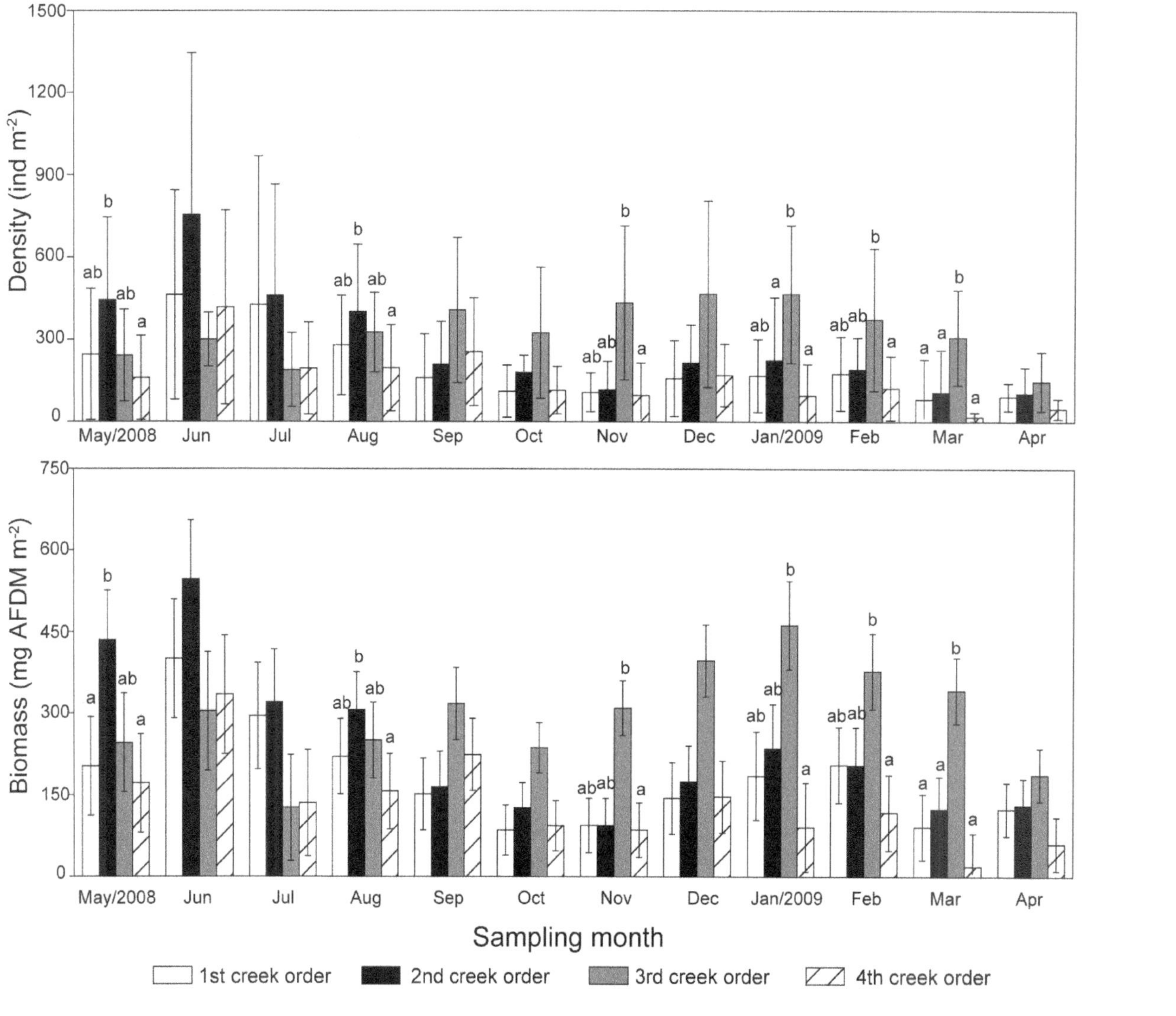

Figure 3. Monthly density and biomass of *Dentinephtys glabra* in creeks of different orders. Error bars represent standard error ($n = 6$). The same letters above the bars denote non-significant differences and different letters represent a significant difference ($P < 0.05$) among creek orders in each month.

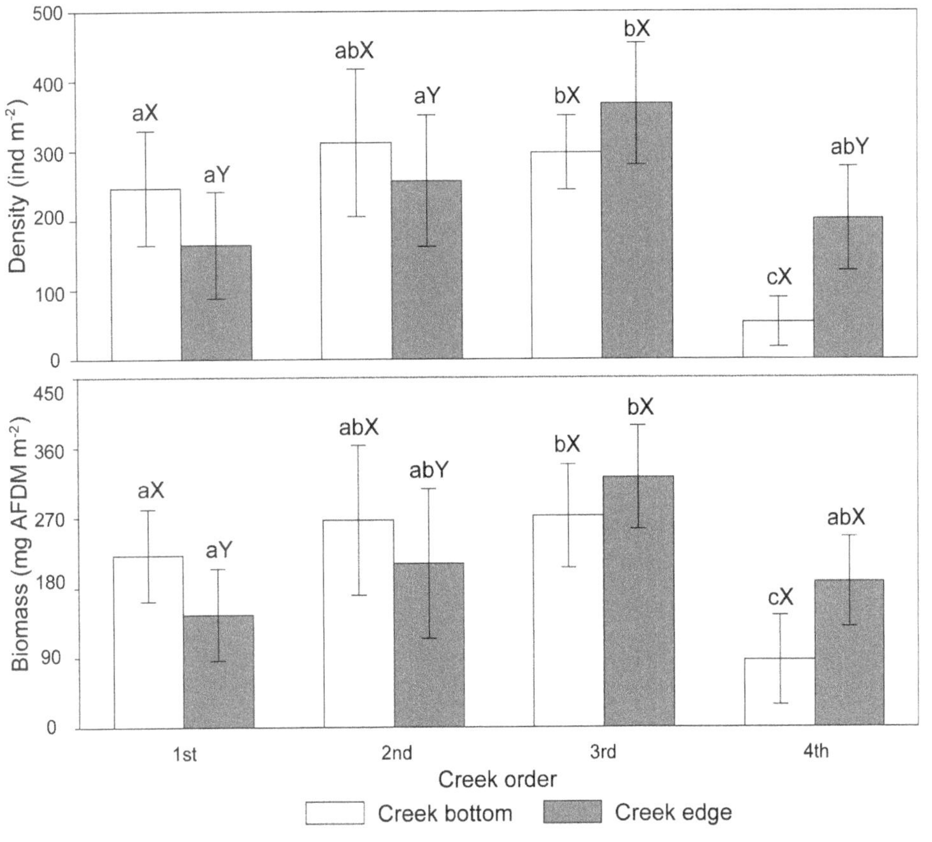

Figure 4. Density, biomass of *Dentinephtys glabra* at bottom and edge sampling locations in creeks of different orders. Error bars represent standard error (n = 36). The same capital letters denote non-significant differences and different capital letters represent a significant difference ($P<0.05$) between bottom and edge locations for creeks of the same order. The same lowercase letters denote non-significant differences and different letters represent a significant difference ($P<0.05$) among creek orders for each sampling location.

the 3rd and 4th order creeks. The density and biomass of *D. glabra* were highest at the edge location of the 3rd order creeks (367.1 ind m^{-2}, 321.3 mg AFDM m^{-2}), and lowest at the bottom location of the 4th order creeks (88.7 ind m^{-2}, 85.9 mg AFDM m^{-2}).

The secondary production of *D. glabra* is shown in Table 2. A total of 6222 individuals were collected during the year-round sampling with the mean Wd8S ranging from 0.84 to 1.59 mm. Using the size frequency method (Table S1–S4), we estimated secondary production to be 286.0, 559.7, 448.2, and 247.4 mg AFDM $m^{-2}year^{-1}$ for 1st to 4th order creeks, with annual P/B ratios of 1.56, 2.39, 1.69, and 1.75, respectively (Table 2). Secondary production differed among creeks of different orders. The secondary production in the 2nd order creeks was 2.26 times greater than in the 4th order creeks.

The secondary production of creek system in Jiuduansha marsh is shown in table 2. The creek system in Jiuduansha marsh was 2952.0×10^3 m^2 in area, with the 1st order creeks accounting for 51.1% of the total area (1507.9×10^3 m^2). The average and total production of *D. glabra* in creeks of the Jiuduansha marsh was estimated to be 387.2 mg AFDM $m^{-2}year^{-1}$ and 1143.8×10^6 mg AFDM $year^{-1}$.

The top five selected AIC models for the relationship between secondary production of *D. glabra* and environmental factors are shown in Table 3. During the analyses, the cross-sectional area was selected as an integrated performance for bank full depth and width. Clay-silt content was selected to represent grain size of the sediment. All five models had the similar w_i value, and included cross-sectional area, bank slope and Chl *a* concentration, suggesting the importance of these variables.

Discussion

Spatial distribution

Our results showed that the highest density and biomass of *D. glabra* occurred in intermediate order creeks. The density of the polychaete *Streblospio benedicti* has been reported to increase with increasing creek order because the creeks of higher order had more constant physical characteristics [37]. Interestingly, the inverse trend of higher polychaete densities in lower order creeks than in higher order creeks has also been documented [11,24,38]. Taken together, these observations suggest that the macrofaunal utilization pattern of creeks along an order gradient differs with species or regions.

Researchers have found that the density of polychaetes was higher in the edge sites than in the bottom sites [23]. Interestingly, the densities and biomass of *D. glabra* were higher in edge locations than in bottom locations in the high order (3rd and 4th order) creeks, and lowest in the edge locations of low order (1st and 2nd order) creeks. Essential abiotic properties including sediment grain size and tidal forces may influence the macrobenthos distributions [39]. For example, the density of polychaete *Capitella* spp. was positively correlated with percent silt [40]. In high order creeks, macrobenthos may be exposed to more intense hydrodynamic

Table 2. Estimation of the total *Dentinephtys glabra* production in Jiuduansha creeks on an order by order basis with one-way ANOVA results showing the effect of creek order.

Creek order	*Dentinephtys glabra*				Total creeks of Jiuduansha			Production of *D. glabra* in total creeks of Jiuduansha (10^6 mg AFDM year^{-1})
	K (year^{-1})	Life span (years)	Production (mg AFDM m^{-2}year^{-1})	P/B	Total length (10^3m)	Average width (m)	Total area (10^3 m^2)	
1st	1.05±0.04	2.39±0.10	286.0±83.6^{ab}	1.56±0.05	121.5	12.4	1506.9	431.0
2nd	1.37±0.37	2.08±0.41	559.7±90.1^{a}	2.39±0.43	54.7	16.0	875.1	489.8
3rd	1.23±0.29	2.22±0.39	448.2±45.1^{ab}	1.69±0.35	22.9	17.9	409.1	183.3
4th	1.07±0.07	2.36±0.12	247.4±76.4^{b}	1.75±0.18	5.5	29.3	160.0	39.6
Total							2952.0	1143.8

Different superscript letters (a and b) represent a significant difference ($P<0.05$) among creeks of different orders.

Table 3. Results of model selection using Akaike's information criterion (AIC) for the relationship between secondary production of *Dentinephtys glabra* and environmental factors.

Model	AICc	△AICc	w_i	R^2	Z values									
					Intercept	Chl *a*	Clay	Cond	Cross	Elevation	pH	Slope	Temp	Water
1	219.49	0	0.07	0.65	0.86	−0.16			−0.47		0.05	−0.24		
2	219.70	0.21	0.06	0.65	0.86	−0.17		0.04	−0.49			−0.24		
3	219.86	0.37	0.06	0.65	0.86	−0.16			−0.46			−0.24	−0.03	
4	220.72	1.22	0.04	0.65	0.85	−0.14	−0.07		−0.49		0.04	−0.26		
5	220.98	1.49	0.03	0.65	0.85	−0.15	−0.07	0.03	−0.50			−0.26		

Only the top five models with Akaike weight (w_i) and R^2 values are shown. Regression coefficients were expressed as Z values (estimate/SE).
Factors are abbreviated as Cond (conductivity), Cross (cross-sectional area), Slope (bank slope), Water (Water content), Chl *a* (Chl *a* concentration), Temp (temperature), Clay (clay-silt content).

forces in the bottom sites than in edge sites, resulting in lower abundance at bottom sites [23].

The pattern of polychaete distribution may also be influenced by other organisms. For example, the presence of burrowing crabs (*Chasmagnathus granulatus*) reduces the abundance of polychaetes [41]. This is because the crabs reduce the burrowing depth of polychaetes, thereby increasing the availability of the polychaetes to shorebirds [41]. In salt marsh creeks of the Yangtze River estuary, the abundance of the crab *Ilyoplax deschampsi* was higher in edge sites than in bottom sites [42]. The effects of *I. deschampsi* on *D. glabra* may be more evident in lower order creeks where *I. deschampsi* is more abundant than in higher order creeks [24]. This may also explain the difference in *D. glabra* abundance between edge and bottom sites between the higher and lower order creeks. Predator is also an important factor to influence the secondary production of polychaete [43]. The secondary production of *D. glabra* was at least two fold higher in intermediate order creeks than in both low and high order creeks. In high order creeks, longer tidal inundation exposed *D. glabra* to carnivorous fishes for longer time than in low order creeks. However, in the low order creeks, benthic polychaete *D. glabra* would be more exposed to shorebirds because of limited tidal inundation [44]. The pattern that higher secondary production of *D. glabra* occurred in intermediate order creeks may reflect the predatory stress to some extent.

Secondary production

Studies evaluating the secondary production of Nephtyidae are very rare, in fact only three species (*Nephtys hombergii*, *N. australiensis* and *N. picta*) have been studied to date. The secondary production of *N. australiensis* was 48.0 mg AFDW $m^{-2}year^{-1}$ [45]. The secondary production of *N. hombergii* ranged from 92.0 [46] to 7335.0 mg AFDW $m^{-2}year^{-1}$ [47], and that of *N. picta* from 95.0 to 628.0 mg AFDW $m^{-2}year^{-1}$ [48]. Compared with these reports, secondary production obtained for *D. glabra* in this study was within the range of 48.0 to 7335.0 mg AFDM $m^{-2}year^{-1}$.

Tidal influences result in wide fluctuations of environmental variables within intertidal creek networks [37]. Turbidity and water temperature decreased, while pH, conductivity, flow, and drainage basin area increased as creek order increased [14]. In our study, Chl *a* concentration and sediment grain size were found to be different among different creek orders. Secondary production is influenced by many physicochemical factors, e.g. temperature, dissolved oxygen, substrate [49]. Secondary production of the polychaete *N. hombergii* differed among areas of different organic matter content [44]. Research found the decrease of secondary production of polychaete *Nereis diversicolor* may be attributed to the increase of salinity and the decrease of organic matter content in the estuary of Oued Souss [50]. In the top fitting AIC models, secondary production of *D. glabra* was mainly associated with geomorphological characters including cross-sectional area and bank slope. This may reflect that the hydrodynamic forces are essential factors for secondary production of macrobenthos in salt marshes. Animals are not evenly distributed across wetland landscapes and the distributions of secondary production change with the high complex landscape topography [51]. Recently, researchers have incorporated the use of GIS analysis to estimate the secondary production for a region. For example, they calculated the area of each sub-habitat (different distances from the marsh edge) using GIS and then calculated the total secondary production of crustaceans for the Galveston Bay salt marsh [52]. In the present study, because of the difference in *D. glabra* production among different order creeks, we think it is necessary to combine area estimation using GIS analysis to estimate total secondary production in the Jiuduansha marsh. Based on this approach, the total production of *D. glabra* in Jiuduansha was 1143.8×10^6 mg AFDM $year^{-1}$. We believe that improved knowledge of the spatial heterogeneity in secondary production combined with landscape ecology is critical to quantify the material cycling and energy flow for the creek networks.

Conclusions

The major findings of this study were: (1) the density and biomass of the polychaete *D. glabra* differed significantly among creeks of different orders, but was highest in intermediate order creeks. The highest density and biomass were found in the 3rd order creeks and the lowest were in the 4th order creeks. (2) In lower order creeks, its density and biomass were higher in the creek bottom than in creek edge sites. However, the reverse was true in higher order creeks. (3) The secondary production of *D. glabra* was much higher in intermediate order creeks than in low and high order creeks. Its secondary production in the 2nd order creeks (559.7 mg AFDM m^{-2} $year^{-1}$) was 2.26 times greater than in 4th order creeks.

Polychaetes are a dominant group and key linkage in the food web of salt marsh ecosystems. The research about polychaete secondary production can be an indicator to evaluate the health of the ecosystems. Our findings indicated the importance of creek order in influencing the abundance and secondary production of macrobenthic animals such as polychaetes. This suggests that future studies of infauna on tidal creek systems should carefully consider the within-system differences. The spatial variability in prey production can, in turn, influence the habitat utilization by predators including nektonic fish and crustaceans, and birds. Thus, the value of different microhabitats should be considered in assessing marsh functions and implementing management and restoration strategies for the purposes of biodiversity conservation.

Supporting Information

Table S1 Annual production of *Dentinephtys glabra* at 1st order creeks estimated by the size-frequency method.

Table S2 Annual production of *Dentinephtys glabra* at 2nd order creeks estimated by the size-frequency method.

Table S3 Annual production of *Dentinephtys glabra* at 3rd order creeks estimated by the size-frequency method.

Table S4 Annual production of *Dentinephtys glabra* at 4th order creeks estimated by the size-frequency method.

Acknowledgments

We thank Mingyao Huang, Li Guo, Bingsong Jin, Haiming Qin, Danqing Huang, Weixing Tang for their help in field samplings.

Author Contributions

Conceived and designed the experiments: JW. Performed the experiments: TC QS SW. Analyzed the data: TC. Wrote the paper: JW TC.

References

1. Mitsch WJ, Gosselink JG (2008) Wetlands, 4th edn. Hoboken: John Wiley & Sons, Inc.
2. Riera P, Montagna PA, Kalke RD, Richard P (2000) Utilization of estuarine organic matter during growth and migration by juvenile brown shrimp *Penaeus aztecus* in a South Texas estuary. Mar Ecol Prog Ser 199: 205–216.
3. Rush SA, Olin JA, Fisk AT, Woodrey MS, Cooper RJ (2010) Trophic relationships of a marsh bird differ between gulf coast estuaries. Estuar Coast 33: 963–970.
4. Zimmerman RJ, Minello TJ, Rozas LP (2000) Saltmarsh linkages to productivity of Penaeid shrimps and blue crabs in the Northern Gulf of Mexico. In: Weinstein MP, Kreeger DA, editors. Concepts and controversies in marsh ecology. Kluwer Academic Press.
5. Chu TJ, Sheng Q, Wang SK, Huang MY, Wu JH (2013) Population dynamics and secondary production of crabs in a Chinese marsh. Crustaceana 86: 278–333.
6. Benke AC, Huryn AD (2007) Secondary production of macroinvertebrates. In: Hauer FR, Lamberti GA, editors. Methods in stream ecology. Academic Press of Elsevier.
7. Benke AC, Huryn AD (2010) Benthic invertebrate production facilitating answers to ecological riddles in freshwater ecosystems. J N Am Benthol Soc 29: 264–285.
8. Sardá R, Foreman K, Valiela I (1995) Macroinfauna of a southern New England salt marsh: seasonal dynamics and production. Mar Biol 121: 431–445.
9. Chaplin GI, Valentine JF (2009) Macroinvertebrate production in the submerged aquatic vegetation of the Mobile-Tensaw Delta: Effects of an exotic species at the base of an estuarine food web. Estuar Coast 32: 319–332.
10. Minello TJ, Able KW, Weinstein MP, Hays CG (2003) Salt marshes as nurseries for nekton: testing hypotheses on density, growth and survival through meta-analysis. Mar Ecol Prog Ser 246: 39–59.
11. Rozas LP, Odum WE (1987) Use of tidal fresh-water marshes by fishes and macrofaunal crustaceans along a marsh stream-order gradient. Estuaries 10: 36–43.
12. Horton RE (1945) Erosional development of streams and their drainage basins, hydrophysical approach to quantitative morphology. Geol Soc Am Bull 56: 275–370.
13. Vlaswinkel BM, Cantelli A (2011) Geometric characteristics and evolution of a tidal channel network in experimental setting. Earth Surf Proc Land 36: 739–752.
14. Harrel RC, Dorris TC (1968) Stream order, morphometry, physico-chemical conditions, and community structure of benthic macroinvertebrates in an intermittent stream system. Am Midl Nat 80: 220–251.
15. Kang SJ, Lin H (2009) General soil-landscape distribution patterns in buffer zones of different order streams. Geoderma 151: 233–240.
16. Lerberg SB, Holland AF, Sanger DE (2000) Responses of tidal creek macrobenthic communities to the effects of watershed development. Estuaries 23: 838–853.
17. Martin JP, Bastida R (2006) Population structure, growth and production of *Laeonereis culveri* (Nereididae: Polychaeta) in tidal flats of Río de la Plata estuary, Argentina. J Mar Biol Assoc UK 86: 235–244.
18. Stehlik LL, Meise CJ (2000) Diet of winter flounder in a New Jersey estuary: Ontogenetic change and spatial variation. Estuaries 23: 381–391.
19. Levin L, Blair N, Demaster D, Plaia G, Fornes W, et al. (1997) Rapid subduction of organic matter by maldanid polychaetes on the North Carolina slope. J Mar Res 55: 595–611.
20. Paramor OAL, Hughes RG (2004) The effects of bioturbation and herbivory by the polychaete *Nereis diversicolor* on loss of salt marsh in south-east England. J Appl Ecol 41: 449–463.
21. Blackburn NJ, Orth RJ (2013) Seed burial in eelgrass *Zostera marina*: the role of infauna. Mar Ecol Prog Ser 474: 135–145.
22. Oyenekan JA (1987) Population dynamics and secondary production in an estuarine population of *Caulleriella caputesocis* (Polychaeta: Cirratulidae). Mar Biol 95: 267–273.
23. Yuan XZ, Lu JJ (2001) Preliminary study on macrobenthic community of the creek in the tidal flat of the Changjiang Estuary. Zool Res 22: 211–215. (in Chinese, with English Abstract)
24. Song CY, Chu TJ, Sheng Q, Wang SK, Wu JH (2011) Macrobenthos communities in salt marshes of the Yangtze River Estuary along a creek-Order gradient. J Fudan Univ 50: 253–259. (in Chinese, with English Abstract)
25. Sun RP, Yang DJ (2004) Fauna Sinica, Polychaeta. Science Press, Beijing.
26. Chen JK (2003) Scientific investigations of the Shanghai Jiuduansha Wetland Nature Reserve. Science Press, Beijing. (in Chinese)
27. Neira C, Levin L, Grosholz ED (2005) Benthic macrofaunal communities of three sites in San Francisco Bay invaded by hybrid *Spartina*, with comparison to uninvaded habitats. Mar Ecol Prog Ser 292: 111–126.
28. Lorenzen CJ (1967) Determination of chlorophyll and pheopigments: spectrophotometric equations. Limnol Oceanogr 12: 343–346.
29. Jin BS, Qin HM, Xu W, Wu JH, Zhong JS, et al. (2010) Nekton use of intertidal creek edges in low salinity salt marshes of the Yangtze River estuary along a stream-order gradient. Estuar, Coast Shelf S 88: 419–428.
30. Otegui MBP, Blankensteyn A, Pagliosa PR (2012) Population structure, growth and production of *Thoracophelia furcifera* (Polychaeta: Opheliidae) on a sandy beach in Southern Brazil. Helgoland Mar Res 66: 479–488.
31. Benke AC (1979) A modification of the Hynes method for estimating secondary production with particular significance for multivoltine population. Limnol Oceanogr 24: 168–171.
32. Hynes HBN, Coleman MJ (1968) A simple method of assessing annual production of stream benthos. Limnol Oceanogr 13: 569–573.
33. Gayanilo FCJ, Pauly D (1997) FAO-ICLARM stock assessment tools. Computerized Information Series: Fisheries 8: 1–262.
34. Burnham KP, Anderson DR (2002) Model selection and multimodel inference: a practical information-theoretic approach. Springer, New York.
35. McGrath MT, DeStefano S, Riggs RA, Irwin LL, Roloff GJ (2003) Spatially explicit influences on northern goshawk nesting habitat in the interior Pacific Northwest. Wildlife Monogr 154: 1–63.
36. Faraway JJ (2006) Extending the linear model with R: generalized linear, mixed effects and nonparametric regression models. CRC Press, Boca Raton.
37. Washburn T, Sanger DM (2013) Microhabitat variability of macrobenthic organisms within tidal creek systems. Hydrobiologia 702: 15–25.
38. Washburn T, Sanger D (2011) Land use effects on macrobenthic communities in southeastern United States tidal creeks. Environ Monit Assess 180: 177–188.
39. Levin LA, Talley TS (2002) Influences of vegetation and abiotic environmental factors on salt marsh invertebrates. In: Weinstein MP, Kreeger DA, editors. Concepts and Controversies in tidal marsh ecology. Kluwer Academic Press.
40. Toomey WA (1997) The effect of soil organic matter quantity on salt marsh functioning and infaunal community structure. Thesis. North Carolina, North Carolina State University.
41. Palomo G, Botto F, Navarro D, Escapa M, Iribarne O (2003) Does the presence of the SW Atlantic burrowing crab *Chasmagnathus granulatus* Dana affect predator–prey interactions between shorebirds and polychaetes? J Exp Mar Biol Ecol 29: 211–228.
42. Wang JQ, Tang L, Zhang XD, Wang CH, Gao Y, et al. (2009) Fine-scale environmental heterogeneities of tidal creeks affect distribution of crab burrows in a Chinese salt marsh. Ecol Eng 35: 1685–1692.
43. Sardá R, Foreman K, Werme CE, Valiela I (1998) Macroinfaunal invertebrate communities of tidal saltmarsh creeks. Estuar, Coast Shelf S 46: 657–669.
44. Jing K, Ma ZJ, Li B, Li J, Chen JK (2007) Foraging strategies involved in habitat use of shorebirds at the intertidal area of Chongming Dongtan, China. Ecol Res 22: 559–570.
45. Robertson AI (1979) The relationship between annual production: biomass ratios and life spans for marine macrobenthos. Oecologia 38: 193–202.
46. Oyenekan JA (1986) Population dynamics and secondary production in an estuarine population of *Nephtys hombergii* (Polychaeta: Nephtyidae). Mar Biol 93: 217–223.
47. Warwick RM, Price R (1975) Macrofauna production in an estuarine mud-flat. J Mar Biol Assoc UK 55: 1–18.
48. Steimle FWJ, Kinner P, Howe S, Leathem W (1990) Polychaete population dynamics and production in the New York Bght associated with variable levels of sediment contamination. Ophelia 31: 105–123.
49. Gong ZJ, Xie P, Yan YJ (2001) Theories and methods of studies on the secondary production of zoobenthos. J Lake Sci 13: 79–88. (in Chinese, with English Abstract)
50. Alla AAA, Gillet P, Deutsch B, Moukrim A, Bergayou H (2006) Response of Nereis diversicolor (Polychaeta, Nereidae) populations to reduced wastewater discharge in the polluted estuary of Oued Souss, Bay of Agadir, Morocco. Estuar, Coast Shelf S 70: 633–642.
51. Kneib RT (2000) Salt marsh ecoscapes and production transfers by estuarine nekton in the southeastern United States. In: Weinstein MP, Kreeger DA, editors. Concepts and controversies in tidal marsh ecology. Kluwer Academic Press.
52. Minello TJ, Matthews GA, Caldwell PA, Rozas LP (2008) Population and production estimates for decapod crustaceans in wetlands of Galveston Bay, Texas. T Am Fish Soc 137: 129–146.

13

Effects of Red-Backed Salamanders on Ecosystem Functions

Daniel J. Hocking*, Kimberly J. Babbitt

Department of Natural Resources and the Environment, University of New Hampshire, Durham, New Hampshire, United States of America

Abstract

Ecosystems provide a vast array of services for human societies, but understanding how various organisms contribute to the functions that maintain these services remains an important ecological challenge. Predators can affect ecosystem functions through a combination of top-down trophic cascades and bottom-up effects on nutrient dynamics. As the most abundant vertebrate predator in many eastern US forests, woodland salamanders (*Plethodon spp.*) likely affect ecosystems functions. We examined the effects of red-backed salamanders (*Plethodon cinereus*) on a variety of forest ecosystem functions using a combined approach of large-scale salamander removals (314-m^2 plots) and small-scale enclosures (2 m^2) where we explicitly manipulated salamander density (0, 0.5, 1, 2, 4 m^{-2}). In these experiments, we measured the rates of litter and wood decomposition, potential nitrogen mineralization and nitrification rates, acorn germination, and foliar insect damage on red oak seedlings. Across both experimental venues, we found no significant effect of red-backed salamanders on any of the ecosystem functions. We also found no effect of salamanders on intraguild predator abundance (carabid beetles, centipedes, spiders). Our study adds to the already conflicting evidence on effects of red-backed salamander and other amphibians on terrestrial ecosystem functions. It appears likely that the impact of terrestrial amphibians on ecosystem functions is context dependent. Future research would benefit from explicitly examining terrestrial amphibian effects on ecosystem functions under a variety of environmental conditions and in different forest types.

Editor: Benedikt R. Schmidt, Universität Zurich, Switzerland

Funding: Partial funding was provided by the New Hampshire Agricultural Experiment Station. This is Scientific Contribution Number 2500. DJH received funding through the University of New Hampshire Graduate School. No additional external funding was received for this study. The funders had no role in study design, data collection and analysis, decision to publish, or preparation of the manuscript.

Competing Interests: The authors have declared that no competing interests exist.

* E-mail: dhocking@unh.edu

Introduction

Ecosystems supply critical services for human societies including food, clean air, and potable water. These services are supported by a variety of ecosystem functions such as primary production, nutrient cycling, and soil formation [1,2]. Despite the importance of these functions in supporting ecosystem services, it remains difficult to predict how organisms contribute to specific ecosystem functions. Predators can affect ecosystem functions indirectly through top-down trophic cascades and directly by altering bottom-up nutrient dynamics (e.g. excretion of soluble nitrogen).

Much of our current understanding of the effects of predators on terrestrial ecosystem function comes from examination of trophic cascades. Predators are able to affect primary producers indirectly through predation on herbivores. In terrestrial ecosystems, carnivores generally reduce herbivore abundance, thereby reducing herbivore damage on plants and increasing plant biomass and reproductive output [3]. However, these patterns are not consistent across all predator species and habitats. No clear pattern has emerged to predict when predators increase or decrease primary production, but it may be related to intraguild predation [4], foraging strategies [5], behaviorally-mediated indirect interactions [6], and the balance of top-down and bottom-up effects [7–9]. Additionally, plants processing anti-herbivore defenses (including ant-tending) and systems with high herbivore diversity attenuate top-down effects on primary producers [3].

Studies in terrestrial systems examining ecosystem functions other than primary production have been even more conflicting. Some predators reduce nutrient availability [5,10], while others increase nutrient availability [5,7,11]. Terrestrial amphibians can increase, decrease, or have no effect on litter decomposition rates depending on the species and habitat [8,12,13]. These idiosyncratic results may arise through differences in predator diversity, functional redundancy of herbivores, indirect effects on behavior or anti-herbivore defenses, or initial productivity and nutrient pools [5,14]. Differences may also arise due to experimental venue with small, controlled experiments often revealing processes not detected in larger field manipulations [15–17]. Variable results may also arise due to the complexity of the detrital food web and the diverse prey consumed by many amphibians. For example, a salamander may feed on predaceous mites that feed on fungivorous collembola, thereby indirectly increasing collembolan abundance, but the salamanders may also prey directly on collembola, directly offsetting the effects of preying on mites (Fig. 1). The balance of these feeding pathways would influence saprotrophic fungi and therefore leaf litter decomposition.

Predators can also affect ecosystem functions by altering nutrient dynamics. Predators ingest energy and nutrients through consumption of prey. Some of these ingested resources are used to produce new tissue and the rest is passed back to the environment through dead tissue, excreted waste, heat, and respiration [9]. The excretion of waste by vertebrate predators often provides nutrients

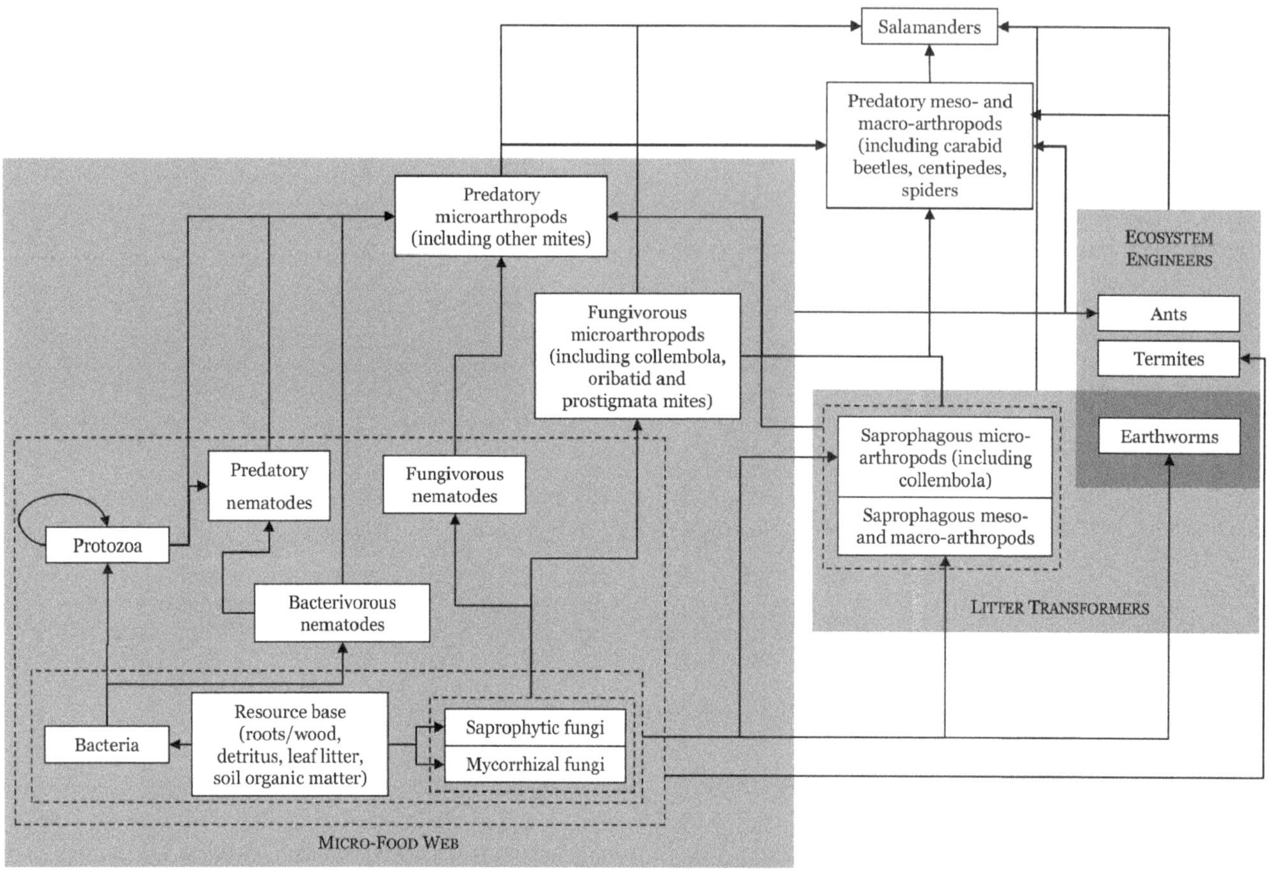

Figure 1. Soil-litter food web including red-backed salamanders as one of the top predators in the system. All organisms in the system contribute nutrients through waste excretion back into the resource base (not shown). The effect of salamanders on ecosystem functions may be a balance of complex trophic cascades through direct predatory effects, indirect behavioral responses of prey to predators, moderated behavior of predators in response to intraguild predators, or changes in nutrient dynamics associated with changes in the food web. Adapted from Coleman and Wall [58] figure 7.2.

in forms readily used by microbes and primary producers and can increase decomposition and plant growth in some systems [7,11] but see [8]. These bottom-up effects on nutrient dynamics may have complementary or opposing effects on particular ecosystem functions when paired with top-down trophic cascades.

To better understand the effects of predators on ecosystem functions, we experimentally manipulated the density of red-backed salamanders (*Plethodon cinereus*), one of the most abundant vertebrate predators in many eastern forests [18]. Lungless salamanders of the family Plethodontidae can occur at densities of 0.2–8.2 m^{-2} [19,20] and can have a biomass twice that of all the passerine birds and equal to all the small mammals in a forest [21]. This tremendous biomass has led many ecologists to speculate on the importance of salamanders in ecosystem functions [22,23]. As vertebrate predators with efficient conversion of food into tissue, woodland salamanders may be important contributors to ecosystem function. Woodland salamanders may influence litter decomposition and potentially net primary production (NPP) [13,23] through their role as abundant predators in the detrital food web [21]. Hairston [23] estimated that, at average densities in the southern Appalachians, 1.165 kcal per m^2 of energy is stored in red-backed salamander biomass, a caloric content greater than all other vertebrate predators combined. He also estimated that salamanders consume more than a complete turnover of the soil invertebrate fauna each year [23]. Additionally, control of herbivorous and leaf-fragmenting invertebrates could alter nitrogen availability and plant growth.

To date, studies on the effect of red-backed salamanders on ecosystem functions have primarily been limited to litter decomposition and remain equivocal [13,24]. We examined the effects of red-backed salamander removal on a variety of ecosystem functions and on a spatial scale an order of magnitude larger than their home range. We conducted this study over 4 years to provide information on the effects of predator removal on the system beyond just the initial perturbation. We also used mesocosm enclosures to more finely manipulate vertebrate predator density in a more controlled experiment. Based on previous research and potential effects on the food web and nutrient dynamics, we expected salamanders to 1) reduce the rate of litter decomposition through predation of litter fragmenting invertebrates [13], 2) increase wood decomposition rates through consumption of fungivorous collembola [18], 3) increase nitrogen immobilization by excretion of waste in soluble forms and increased fungal productivity [7], 4) reduce potential nitrogen mineralization rates due to increased needs of fungi, 5) reduce nitrification rates in response to fungal productivity, 6) increase acorn germination through consumption of herbivorous invertebrates [18], 7) reduce foliar insect damage on red oak seedlings, and 8) reduce the abundance of intraguild predators such as spiders, centipedes, and carabid beetles, primarily through competition [25,26].

Materials and Methods

Ethics Statement

We took measures to reduce handling time and distress of all vertebrate animals in this study. All research was conducted with University of New Hampshire IACUC approval 080301, 091106, and 110403.

Experiment 1: Effects of Salamander Depletion

In April 2008, we established ten 20-m diameter circular plots (314 m^2) in American beech (*Fagus grandifolia*) dominated forest stands at the University of New Hampshire's Kingman Farm property (146 ha.). Plots of this size are approximately 13 times the size of a red-backed salamander home range [27] and similar to the size of depletion plots previously used to study woodland salamander competition [23,28]. We randomly assigned half the plots ($n = 5$) for salamander depletion and the assigned remaining as reference plots ($n = 5$). Plots were all located within 1 km of each other and separated by a minimum of 20 m.

In April - November, we conducted visual encounter surveys (VES) for salamanders on 91 nights from 2008–2011, primarily during or following rain events to remove salamanders from plots and establish experimental treatments. We sampled in all months but more frequently in spring and fall on rainy nights when salamanders were most active. Based on weather conditions, we sampled at infrequent intervals from daily to weeks between surveys. To avoid correlation between plots, we randomly selected a starting plot each night and then proceeded to subsequent plots in the most convenient order, which differed depending on the starting plot. Survey routes within each plot followed concentric rings marked with twine so that an entire plot was covered only once each night. In reference plots, we counted individual red-backed salamanders but did not disturb them. One to four researchers walked the plots at sufficient pace to avoid double counting of wandering salamanders during a survey night (20–30 person-minutes per plot). Survey methods were the same in reference and depletion with the exception that in depletion plots any encountered individual was collected by hand and brought back to the laboratory and euthanized using a 1% solution of MS-222 and preserved in ethanol for potential use in research.

To minimize immigration and edge effects, we delineated a 12 m diameter central "core" of each plot, creating a 4 m wide buffer where we removed salamanders but did not measure ecosystem functions. We flagged the location of each removed salamander and later recorded the distance from the plot center. This provided a way to examine if a greater number of individuals were caught near the plot edge, indicating immigrants from outside the plot were being caught as they entered the plot [29]. We did not conduct intensive mark-recapture estimates on our plots because it would have been impossible to conduct sufficient removals while doing mark-recapture. Additionally, red-backed salamanders are highly fossorial resulting in low detection probabilities, which leads to extremely large uncertainty in population estimates [30–32]. Therefore, to get any reliable population estimates would require the addition of coverboards, fencing, and litter searches [30–32], which would alter ecosystem processes and confound our metrics of interest.

Decomposition. We measured litter decomposition rates using two methods. We used bags (20×20 cm) constructed of fiberglass window screen with 2-mm mesh and filled with 10.2 g of air-dried (to constant mass) beech leaf litter [33] then sewn shut, and we used 1-m^2 leaf litterboxes with larger (1 cm) mesh openings on the top and bottom, and filled with 255 g of air-dried litter. Boxes were surrounded by landscape edging to prevent leaves from blowing out and were staked down to secure the box and leaves. This quantity of leaves used is in the range of annual deciduous leaf fall in the region [33]. We collected freshly fallen leaf litter in late October to early November each year and air dried it in the laboratory for more than a week [33]. Litterbags and boxes were filled and placed in the field at the beginning of each December using a stratified pattern from the same pool of dried leaves to avoid bias when comparing the methods. Three litterboxes and nine litterbags were randomly located within each plot core. We collected one random litterbox and three litterbags from each plot after 6, 12, and 18 months, with the exception of the final year when we only measured decomposition at 6 and 12 months. We then oven-dried the leaves at 60°C to examine the mass lost over the time period. We corrected the initial mass for the difference between air and oven drying [33].

Using both litterbags and litterboxes is important to determine how salamanders affect decomposition of fine litter because the smaller mesh bags exclude many invertebrates. These two methods could help explain previous conflicting results [13,24,35], and test competing hypotheses that salamanders primarily affect decomposition through consumption of large, leaf-fragmenting invertebrates [13] or through consumption of fungivorous collembola [23].

We also examined the effect of red-backed salamanders on woody decomposition. Litter is fragmented by a variety of invertebrates and decomposed by both bacteria and fungi, whereas wood decomposition is almost entirely driven by fungi [33]. Salamanders could affect saprotrophic fungi through predation on collembola, which feed on fungi, allowing us to further test competing hypotheses [13,23]. We used untreated birch dowels (6.35 mm diameter x 30 cm long) to measure woody decomposition. We enclosed each dowel in 2 mm fiberglass mesh sleeves, to enable extraction, and hammered them vertically 20 cm into the ground so that 10 cm of each dowel was above the soil surface [33]. Dowels were placed 10–20 cm from litterbags and were collected at the same time as litterbags and boxes. To determine mass loss over time, we weighed each air-dried dowel and attached a uniquely numbered aluminum tag prior to installation in the field. We oven-dried every tenth dowel to determine a correction for the difference between air and oven drying but did not use these oven-dried dowels in our study as they may have altered hydrophobic properties. Upon collection from the field, we oven-dried the dowels at 60°C and carefully removed any soil and attached fungal hyphae before weighing to determine mass loss.

N-mineralization rate. We used laboratory incubations to measure potential nitrogen mineralization and nitrification rates. In fall 2009, 2010, and 2011 we collected the organic layer from six random locations within each plot. We measured inorganic nitrogen levels from each location immediately and incubated the remaining soil in thin-walled polyethylene bags at a constant temperature (25°C) and humidity (50%) for 28 days [36]. We extracted inorganic nitrogen using 2M KCl, then filtered and froze samples at −20°C until analysis. We measured nitrate and ammonium using an Astoria autoanalyzer [Astoria-Pacific International, Clackamas, OR; 37,38], where ammonium was quantified using the indophenol-blue method [37] and nitrate was quantified using the vanadium (III) reduction color reaction modified for microplate assays [38]. We calculated net nitrogen mineralization and nitrification rates over 28 days from the difference between the initial and incubated samples [36].

Oak germination. We planted red oak acorns in each plot to determine the effect of salamanders on germination rates through the consumption of herbivorous invertebrates or reduced feeding activity of herbivores in response to the presence of a predator [39,40]. We collected freshly fallen acorns in early autumn of 2008–2010 and stored them in moist conditions at 4°C over the winter. In April, we planted 20 acorns per plot under the leaf litter in 2009 and 2010 and 40 acorns per plot in 2011, on the soil surface [41]. We covered the acorns with mesh cages to prevent disturbance from vertebrates. We checked for germination weekly throughout the growing season to record the total number of acorns germinated.

Litter-dwelling macro-invertebrate predators. The densities of litter-dwelling predaceous invertebrates may also affect ecosystem functions compounding or mitigating salamander effects. Therefore, we quantified the abundance of three major macro-invertebrate predator groups: adult carabid beetles, centipedes, and spiders. We collected 0.5 m^2 leaf litter from a random location within the central core of each plot in the spring, summer, and fall of each year. We extracted invertebrates from the litter using large Berlese funnels and enumerated the three predator groups [42].

Statistical analysis. We employed a repeated measures multivariate analysis of variance (rmMANOVA) to test the effect of salamander depletion on ecosystem functions. We used mean decomposition rates for each year, arcsine-transformed proportion of acorn germination, potential nitrogen mineralization rate, and potential nitrification rate as the multivariate response. We used treatment as the primary effect with repeated measures for each of 3 years 2009–2011. Analysis was conducted using the "Manova" function from the car package [43] in R [44].

Experiment 2: Effects of Salamander Density on Ecosystem Functions

Conducting large-scale experiments has the benefit of high realism and a potentially broader scope of inference compared with small-scale experiments, but lacks the precision and control of small-scale experiments. Therefore, we also conducted a smaller scale enclosure experiment to more finely manipulate red-backed salamander density. In May 2010, we constructed 20 mesocosm enclosures (1.41 m × 1.41 m × 1.00 m tall). The mesocosms were enclosed in aluminum (sides) and fiberglass (top and bottom) window screen (2-mm grid) with a secure window screen lid. All enclosures were located within a 40-m radius, in a forest stand dominated by American beech on UNH's Kingman Farm property, within 50 m of two plots from experiment 1. We buried the lower 30 cm of each enclosure belowground. We carefully removed the soil in blocks and replaced it inside the enclosure on top of the mesh screen then added the leaf litter back on top of the soil (following Wyman [13]). Immediately upon addition of the leaf litter, we searched it to ensure no salamanders were added back into the enclosure. During 2010, we allowed soil to settle, fine roots and fungal hyphae to reestablish, and microarthropods and flying insects to recolonize. We left the mesh lids open until April 2011 to allow insect recolonization, while 10 cm window screen baffles overhanging over the top of the enclosure prevented recolonization by salamanders. We added a single hemlock coverboard (1 m x 20 cm x 5 cm) to each enclosure to serve as refuge for salamanders. During 2010 and spring 2011, we conducted nocturnal visual encounter surveys and daytime coverboard searches to remove any salamanders that may have entered during soil or litter replacement.

On 01 May 2011, we collected red-backed salamanders from the forest within 1 km of the enclosures and brought to them the laboratory. All salamanders used in this experiment were adult males or adult, non-gravid females as verified by candling [45]. Within 48 hours of capture, we haphazardly put salamanders in containers one at a time in a stratified pattern until each container had 0, 1, 2, 4, or 8 salamanders. Salamanders were then anesthetized in a 1% solution of MS-222 [46] and given one of eight marks using VIE such that each salamander within an enclosure had a unique mark. Marking was intended to allow for identification of intruders into the enclosures. We then randomly assigned each container to an enclosure, which resulted in salamander density treatments of 0, 0.5, 1, 2, or 4 salamanders per m^2. This range of densities covers the natural variation in red-backed salamander density [18,27] and resulted in four replicates of five treatments.

In enclosures, we measured the same metrics of ecosystem function as described for experiment 1. To accomplish this, we added six litterbags, one litterbox, six birch dowels, and planted 20 red oak acorns in each enclosure. Litterbags and litterboxes were added to the enclosures in December 2010 to coincide with natural beech litterfall. We added the dowels in April 2011. To ensure enclosures were devoid of salamanders at the start of the experiment, we checked the coverboards weekly in April and again when the salamanders were added to the enclosures. We also checked for surface-active salamanders on five rainy nights in April. Any salamanders found were removed and released on the outside of the enclosure. We then added marked salamanders to each enclosure on 03 May 2011. We also checked under the coverboards every 7–10 days during the experiment to check for unmarked individuals. Only one small, juvenile was found unmarked in an enclosure at the beginning of June and was promptly removed. Each week we also recorded the number of newly germinated acorns and marked them with a small zip tie.

In September 2011, we removed all germinated acorn seedlings and measured total leaf area and foliar insect damage using WinFolia (Regent Instruments, v2009a). In addition, we collected soil samples from the organic layer to examine potential nitrogen mineralization rates in October 2011. We used the same techniques to measure ecosystem functions as described in experiment 1 with the exception of litterbag and wooden dowel decomposition. We collected 1 litterbag and 1 dowel from each enclosure monthly beginning in April 2011. From this repeated sampling we were able to calculate the rate of decay using the equation

where M is the mass remaining at time t, M_o is the initial mass, and k is the decay constant.

We also quantified the abundance of macro-invertebrate predators (adult carabid beetles, centipedes, and spiders) at the end of the study. We extracted invertebrates from the litterbox litter using large Berlese funnels and enumerated the three predator groups. Finally, we used coverboard and nighttime visual encounter searches in September – November 2011 to remove salamanders and quantify survival (final density).

We analyzed the effect of salamander density on ecosystem functions using MANOVA. Additionally, we were interested in the potential influence of salamander survival, invertebrate predator abundance, and inorganic nitrogen pools in conjunction with initial salamander density on ecosystem functions. Therefore, we performed Multivariate Analysis of Covariance (MANCOVA) with the addition of salamander final density, amounts of nitrate and ammonium at the start of the incubations, and total abundance of invertebrate predators. For significant multivariate analyses, we used univariate linear regressions to determine the direction and magnitude of effect on each of the ecosystem functions. Analysis was performed in R [44] using Manova in the car package [43].

We analyzed the foliar insect damage separately from the other ecosystem functions because 3 enclosures had zero acorns germinate. Therefore, proportion of foliar insect damage could not be calculated and we did not want to use this reduced sample size for the analysis of all ecosystem functions. We used a linear regression to examine the effect of salamander density on foliar insect damage.

Results

Experiment 1: Effects of Salamander Depletion

We removed red-backed salamanders from all depletion plots on 91 nights from 2008–2011. This resulted in the removal of 3,309 individuals from the five depletion plots (662±32 individuals per plot), an average reduction of 2.1±0.1 salamanders per m^2. This compares with a total of 4,645 salamanders observed on the same occasions on the reference plots (~29% reduction in depletion plots). As we surveyed plots repeatedly, the cumulative number of salamanders observed increased at a greater rate over time in the reference plots compared with the depletion plots (Fig. 1). Although this doesn't elucidate the magnitude of the difference in abundance among treatments, it does suggest that there were fewer salamanders to observe on the depletion plots. The number of salamanders observed per night averaged over each month was consistently greater in the reference plots compared with the depletion plots (Fig. 2).

Despite the 4-m buffer zone around each plot, we were concerned about immigration from surrounding habitat. Using visual implant elastomer, we marked 124 individuals in the 2 meters beyond the edge of the buffer zone (10–12 m from center of the plot) around 2 depletion plots from 09 May –08 July 2009 [47,48]. We subsequently captured 6 of these individuals in the buffer zone and only 1 salamander was found in the plot beyond the buffer zone. This individual was captured 66 cm into the plot past the buffer. Further, the locations of removed salamanders were almost evenly dispersed with respect to distance from the center of each plot. The distribution of salamanders around the plot was fit with a beta distribution as a function of distance from the plot center, accounting for the increasing area with increasing distance from the center (radius^2). The bootstrapped 95% confidence intervals of the beta distribution parameters overlapped or were very near 1, indicating a near uniform distribution of salamanders throughout the plot (Table 1). Therefore, there is no evidence that we caught more salamanders near the outer edge of the plots or that we had significant immigration into the plots.

The means of each ecosystem function and mean density of macroinvertebrate litter predators are summarized in Table 2. Nitrogen mineralization rates were generally low and there was virtually no potential nitrification in fall soil samples from any plot. Red oak acorn germination rates were high in 2009 but very low in 2010 and 2011. Mass loss from litterbags was higher than from litterboxes and was similar to rates of woody mass loss from birch dowels (Table 2). Although dominated by small, juvenile spiderlings, spiders were by far the most abundant of the litter predators followed by centipedes (Table 2).

There was no significant effect of treatment or treatment by year, but there were significant differences among years on ecosystem functions (Table 3). Nitrogen mineralization rates were

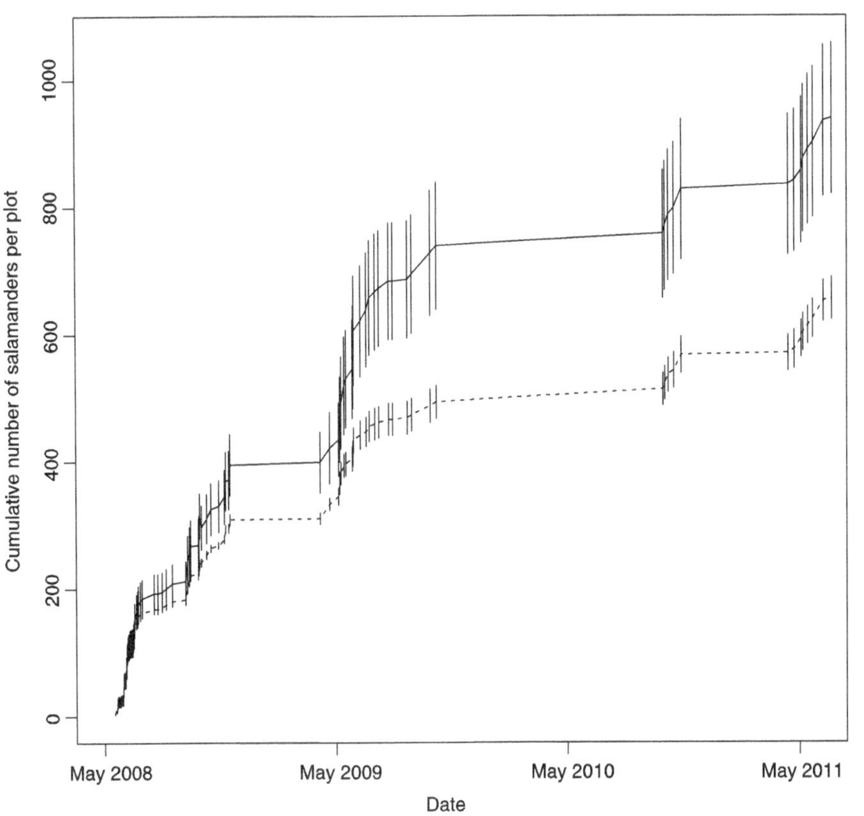

Figure 2. Mean cumulative number of captures per plot (±1 SD) observed in the reference plots (solid line) and removed in the depletion plots (dashed line).

Table 1. Shape parameters defining the estimated beta distributions of salamander removals from the center of the plots.

	Shape 1			Shape 2		
	0.025	**Estimate**	**0.975**	**0.025**	**Estimate**	**0.975**
All Removals	0.98	1.09	1.14	1.02	1.11	1.16
2008	1.02	1.16	1.30	1.06	1.22	1.37
2009	0.95	1.04	1.14	0.97	1.05	1.14
2011	0.98	1.15	1.33	0.99	1.18	1.37

Beta [1,1] indicates the density of salamanders removed is uniform with respect to distance from the plot center. The 95% confidence intervals were estimated from 1000 bootstrap iterations. Salamanders were not removed in 2010.

lower in 2010 and higher in 2011 compared with 2009 (Table 2). There was also a lower rate of decomposition in the litterbags in 2010 than in 2009, while woody decomposition was lowest in 2011 (Table 2). There was also a significant effect of year on invertebrate predator abundance with fewer carabid beetles in 2010. There was no effect of treatment or treatment by year on predator abundances (Table 3).

Table 3. Results of repeated measures MANOVA.

Type II Repeated Measures MANOVA Tests: Pillai test statistic					
Ecosystem Functions					
Factor	**Pillai**	**approx F**	**num df**	**den df**	**P**
Treatment	0.00276	0.022	1	8	0.8854
Year	0.82665	16.691	2	7	0.0022
Treatment*Year	0.17951	0.766	2	7	0.5003
Invertebrate Predators					
Factor	**Pillai**	**approx F**	**num df**	**den df**	**P**
Treatment	0.27791	3.079	1	8	0.1174
Year	0.58595	4.953	2	7	0.0457
Treatment*Year	0.15638	0.649	2	7	0.5515

Treatment tests the difference between red-backed salamander depletion and reference plots. The ecosystem functions nitrogen mineralization, nitrification, acorn germination, and decomposition rates of litterbags, litterboxes, and wood dowels were repeatedly measured in three years 2009–2011. The effect of treatment on the densities of spiders, centipedes, and carabid beetles was also tested using the Pillai test statistic.

Experiment 2: Enclosures

During a total of 19 surveys at the end of the study (01 September –24 October 2011), we captured 40 of the initial 60 animals stocked in the enclosures. Only two enclosures had more individuals than initially stocked. One plot was stocked with zero salamanders but one individual was found. The other was stocked with two but five were found. This anomaly was likely due to a wide, thin, ground-level rip in the windowscreen mesh that occurred during a storm (Hurricane Irene) on 28 August 2011. The rip was missed when checked on 30 August and was not repaired until 03 September. Immigration at this late stage of the study would have been unlikely to influence the ecosystem function metrics. There were likely additional animals that survived in the enclosures but were not captured during these surveys because they remained underground. It is unlikely that many if any salamander escaped except at the end in the one ripped enclosure because they were fully sealed with 1-mm mesh windowscreen on all 6 sides and stapled and glued at all edges except the top, which was secured with 2.5 cm wide strips of industrial strength Velcro.

A total of 177 acorns germinated, a mean (± SE) of 44.25 (±7.2)% germination per enclosure. The mean decomposition rate was 0.524 (±0.035) g g^{-1} $year^{-1}$ in the litterbags and a mean of 39.2% mass was lost from the litterboxes, a mean rate of 0.392

Table 2. Summary of means and standard errors (SE) for ecosystem functions and predator densities for 10 plots over three years across treatments.

	Experiment 1: Plots						**Experiment 2: Enclosures**	
	2009		**2010**		**2011**		**2011**	
Function and Predator Abundance	**Mean**	**SE**	**Mean**	**SE**	**Mean**	**SE**	**Mean**	**SE**
Nitrogen Mineralization Rate	0.268	0.043	−0.245	0.091	0.825	0.109	0.980	0.071
Nitrification Rate	−0.002	0.003	0.008	0.003	−0.001	0.001	−0.020	0.014
Proportion acorns germinated	0.213	0.048	0.085	0.042	0.010	0.006	0.443	0.072
Litterbag Decomposition Rate (g $g^{-1}yr^{-1}$)	0.302	0.011	0.190	0.013	0.263	0.013	0.524	0.035
Litterbox Decomposition (g g^{-1} yr^{-1})	0.125	0.028	0.176	0.038	0.252	0.028	0.392	0.011
Wood Decomposition (g g^{-1} yr^{-1})	0.353	0.038	0.237	0.040	0.144	0.020	0.867	0.091
Proportion Foliar Insect Damage							0.022	0.004
Carabid Beetles (m^{-2})	1.7	0.3	0.3	0.1	1.6	0.5	0.1	0.1
Centipedes (m^{-2})	4.9	0.9	7.8	2.4	8.8	1.9	2.1	0.4
Spiders (m^{-2})	26.9	3.3	52.5	12.3	62.7	11.0	173.5	16.3

Data are from American beech stands in a New Hampshire forest where half the plots had reduced red-backed salamander abundance and from enclosures in similar stands with densities of salamanders from 0–4/m^2.

(0.011) g g^{-1} $year^{-1}$. Mean potential nitrogen mineralization was 0.980 (0.071) μg N g^{-1} dry soil day^{-1} and net nitrification was −0.020 (0.014) μg N g^{-1} dry soil day^{-1}. Of the acorns that germinated, the mean leaf area produced was 690 cm^2 per enclosure (17.8 cm^2 per leaf and 89.7 cm^2 per plant), while the mean proportion of foliar insect damage was 2.2% of the total leaf area produced (Table 2). Of the approximately 175 predaceous invertebrates in the enclosures, 99% were spiders and 93.3% of the spiders were small, mostly hatchlings <3 mm in length (Table 2).

There was no significant effect of salamander density on the ecosystem functions measured (Table 4). We then tested if controlling for the final number of salamanders captured as a covariate in a MANCOVA resulted in a significant effect on ecosystem functions but it did not (Table 4). We performed another MANCOVA to test for the effect when other litter predator density (carabid beetles, centipedes, and spiders) and initial nitrate levels (g g^{-1} dry soil) before incubation were included as covariates. The density of macro-invertebrate predators did not have a significant affect, but initial nitrate levels significantly affected ecosystem functions. However, controlling for these effects did not result in a significant effect of salamander density on ecosystem function (Table 4). Post hoc univariate tests revealed that the significant effect of fall nitrate level on ecosystem functions was driven by its effect on nitrification rates. There was a significant negative effect of initial nitrate levels on potential nitrification rate ($t = -503.15$; $df = 1, 16$; $P<0.0001$). We did not find any significant effect of salamander density, final salamander density, or density of invertebrate predators on the proportion of foliar insect damage on red oak seedlings (Table 4). The estimated magnitude of the effects of each variable can be found in Table 5.

Table 4. The results of three MANOVAs testing the effects of red-backed salamander density, final capture density, soil nitrate levels (g nitrate per g dry soil) on the ecosystem functions: N mineralization rate, Nitrification rate, proportion acorn germination, litterbag decomposition, litterbox decomposition, and woody decomposition.

Type II MANOVA Tests: Pillai test statistic

Ecosystem Functions

Factor	Pillai	approx F	num df	den df	P
Density	0.342	1.127	6	13	1.0000
Density	0.323	0.954	6	12	1.0000
Final Density	0.552	2.465	6	12	0.5191
Density	0.373	1.100	6	11	1.0000
Predator Density	0.660	3.600	6	11	0.1978
Soil Nitrate (g g^{-1})	1.000	31499	6	11	<0.0001

Linear Regression

Foliar Insect Damage

Factor	Estimate	SE	t	P
Intercept	0.0223	0.0052	4.3140	0.0037
Density	−0.0005	0.0023	−0.2060	1.0000
Intercept	0.0174	0.0051	3.3760	0.0271
Density	−0.0049	0.0029	−1.6950	0.6735
Final Density	0.0112	0.0051	2.1950	0.2733
Intercept	0.0331	0.0126	2.6210	0.1272
Density	−0.0021	0.0035	−0.5890	1.0000
Final Density	0.0055	0.0065	0.8580	1.0000
Predator Density	−0.0001	0.0001	−1.3550	1.0000

Additionally, linear regression results testing the effect of salamander and invertebrate predator densities on proportion of foliar insect damage on red oak seedlings (arcsine transformed). P-values are Bonferroni corrected for multiple comparisons.

Discussion

We did not observe any effects of red-backed salamander depletion or density on ecosystem functions. Although researchers have predicted that woodland salamanders are important regulators of ecosystem functions [22,23], we found no evidence that red-backed salamanders affect litter or wood decomposition, nitrogen cycling, acorn germination, herbivory, or the abundance of other litter predators. This is consistent with Homyack et al. [24] who did not find any effect of red-backed salamanders on oak or maple litter decomposition in a Virginia mixed-hardwood forest, but in contrast to Wyman [13] who showed that red-backed salamanders lowered beech-dominated leaf litter decomposition rates by 11–17%. Homyack et al. [49] suggested the conflicting result with Wyman [13] may have been due to differences in litter type. However, we used American beech litter in both of our experiments, similar to Wyman [13], and did not observe an effect of salamanders. Other studies with amphibians have also found conflicting effects on litter decomposition even when nutrient pools were affected [7,8,11,12]. It is likely that the effect of predatory amphibians on terrestrial ecosystem functions is influenced by a variety of biotic and abiotic factors, giving the appearance of idiosyncratic effects when limited to only a few studies.

Some of the variability in results may be due to the complex nature of litter decomposition. Litter decomposition can be influenced by temperature, moisture, microbial community structure, invertebrate community structure, and available nutrient pools; therefore, the effects of amphibians on litter decomposition may be context dependent [32,50,51]. Given this complexity, the effect of amphibians on decomposition is likely influenced by the relative importance of top-down predatory effects on the invertebrate community and the bottom-up effects on available nutrients through ingestion and excretion [14]. For example, the top-down effects of red-backed salamanders on the invertebrate community are known to depend on the community composition and habitat heterogenity [50,51], which suggests that the effects of salamanders on litter decomposition depends on the structural complexity of the environment and the biotic community. Litter decomposition is also likely influenced by the feeding behavior of microarthropods and their behavioral response to predation risk [40,52]. For example, collembola, a primary prey item of red-backed salamanders, can decrease saprotrophic fungal biomass through direct grazing or increase fungal biomass by feeding preferentially on senescent fungal hyphae [53,54]. Given the complex dynamics of forest floor food webs and the variable effect of red-backed salamanders on invertebrates [32,51,55], the effects of salamanders on ecosystem functions should be expected to be highly variable even when top-down effects predominate. The presence of earthworms can further complicate the relationship of salamanders and litter decomposition. Adult nightcrawlers (*Lumbricus terrestris*) burrows provide refuge for red-backed salamanders from extreme temperatures and predators, but

Table 5. The relative effects of salamanders on ecosystem functions estimated from MANOVAs.

	2009		2010		2011		2011: Enclosures	
Metric	Estimate	SE	Estimate	SE	Estimate	SE	Estimate	SE
Nitrogen Mineralization Rate	0.070	0.088	−0.018	0.192	0.178	0.223	0.057	0.050
Nitrification Rate	−0.002	0.006	0.003	0.007	0.001	0.002	0.013	0.010
Proportion acorns germinated	−0.175	0.081	−0.110	0.081	−0.010	0.011	0.110	0.059
Litterbag Decomposition (g $g^{-1}yr^{-1}$)	−0.027	0.021	0.006	0.027	−0.014	0.027	0.013	0.025
Litterbox Decomposition (g g^{-1} yr^{-1})	0.040	0.059	−0.033	0.079	0.052	0.056	−0.005	0.009
Wood Decomposition (g g^{-1} yr^{-1})	−0.067	0.077	0.023	0.085	0.014	0.043	−0.071	0.064
Proportion Foliar Insect Damage							−0.0005	0.002

The estimates for the experimental plots are the effects of salamander depletion relative to the reference plots. The effects in the enclosure experiment represent the change in the ecosystem function with an increase in one salamander per m^2. None of the effects presented are statistically significant (P>0.05).

reduce leaf litter and microarthropod abundance [56]. Juvenile nightcrawlers can be an important prey source for salamanders [56]. In small enclosures, the consumption of juvenile earthworms by adult salamanders reduced the rate of litter decomposition, but the presence of adult earthworms overwhelmed any effect of salamanders by breaking down nearly all leave litter [57]. Our upland, American beech-dominated sites had very few earthworms (*pers. obs.*), so the interactive effects of earthworms and salamanders in more natural settings remains to be tested.

Unlike litter, wood is decomposed almost entirely by saprotrophic fungi. Fungal activity is strongly influenced by temperature, moisture, and available nutrients, especially nitrogen [58]. Therefore, the pathways by which salamanders can influence woody decomposition are slightly more restricted than for litter decomposition. The potential ways salamanders can affect wood decomposition are still numerous and complex as they can derive from top-down effects on the food web or bottom-up effects on nutrient availability. As with litter decomposition, collembola can have variable effects on fungi and salamanders predatory effects on collembola populations may vary [35,53,55]. Therefore, it is difficult to predict the effect of salamanders on woody decomposition.

We did not find an effect of red-backed salamanders on birch dowel decomposition in either our large-scale depletion experiment or in our controlled, density enclosure experiment. There are a number of possible reasons for the lack of an observed effect. First, the top-down and bottom-up effects may roughly balance each other out or create sufficient variability to obscure salamander effects. Second, salamanders prey on wood-chewing invertebrates [18,24], but these taxa may have been restricted from contact with the wood by the mesh sleeves around the dowels. Finally, wood decomposition is strongly influenced by nitrogen availability [58]; therefore, salamanders may have different effects depending on overall pools of available nitrogen. If salamanders make inorganic nitrogen more available for plants and fungi as suggested for abundant, terrestrial frogs [7,11,12], we would expect salamanders to have more influence on wood decomposition in systems with small pools of inorganic nitrogen.

On the contrary, our system had little available inorganic nitrogen and still no observable effect on wood decomposition. Unlike coqui frogs (*Eleutherodactylus coqui* [7]), salamanders may decrease nitrogen availability due to their slow turnover of nutrients, but this would have little effect on woody decomposition in a system already limited in nitrogen. Alternatively, the lack of salamander effect may be due to a limited microbial community structure for mineralizing nitrogen. We found relatively little net potential mineralization and virtually no net nitrification under idealized laboratory conditions. This suggests that nitrogen has already been immobilized by microbes and there is relatively little excess nitrogen available. However, soil N mineralization rates follow a periodic function [59] as does red-backed salamander activity [18]. It is possible that changes in available nitrogen from salamanders in the spring has a greater effect on N cycling and decomposition, since N mineralization remains low in the spring. Variability in salamander density may have obscured this effect in experiment 1 and we did not measure long-term decomposition or springtime decomposition in the better-controlled enclosure experiment.

We expected that salamanders would increase acorn germination through consumption of herbivores that may feed on germinating shoots as they pass through the leaf litter. We observed a trend of higher acorn germination in 2009 in the reference plots compared with the depletion plots. However, there was very little germination in the plots in 2010 or 2011 (Table 2). Low recruitment, likely due to dry, desiccating conditions in the week of germination in those years, may have obscured any effect of salamanders on germination. However, we did have high germination rates in most of the enclosures in 2011 despite the acorns being stratified from the same batch and planted within one week of acorns in reference plots. It is possible that the enclosures kept humidity and moisture levels slightly elevated, at least enough to maintain acorn viability through a mild spring drought. In the enclosures, we did not observe a significant effect of salamander density on germination. Red-backed salamanders might not consistently affect successful red oak germination in American beech-dominated forests depending on other environmental factors. Additionally, salamanders consume large numbers of invasive weevils (*Curculio spp.*) when available [60], which could affect germination in natural systems but was not evaluated in this study because we used only weevil-free acorns. Future studies would benefit from examining the effects of salamanders on recruitment of a variety of plant species in different forest types.

We also did not observe an effect of salamander density on rates of herbivory. However, we only measured foliar insect damage on red oak seedlings. There was very little foliar herbivory in general across density levels. However, lack of foliar damage to red oak seedlings may be limited to beech-dominated stands in southeastern New Hampshire. Homyack et al. [24] found that red-backed salamanders consumed numerous insect larvae, including fungus gnats (Order Diptera, Family Scaiariadae), which feed on plant tissue near the soil surface. Therefore, salamanders may influence herbivory of fine roots and mycorrhizal fungi through predation of fungus gnats and collembola. This could reduce plant growth and survival without generating differences in foliar herbivory. For example, more energy in the forest floor food web may be derived from belowground production, rather than decomposing detritus as previously thought [56]; therefore, salamanders could affect energy and nutrient flow in the system through effects on belowground herbivores and not foliar herbivores. Additionally, the effect would likely depend on the invertebrate community [35,51] and possibly the plant species present.

In addition to ecosystem functions, we did not observe an effect of salamander depletions on the abundances of spiders, centipedes, or carabid beetles. This is in contrast to Hickerson et al. [26] who showed that red-backed salamander depletion resulted increased spider counts and decreased carabid beetle counts. Red-backed salamanders are known to act aggressively toward centipedes, and centipedes are less likely to be found under the same cover objects as salamanders [25] and that reduced centipede abundance can increase salamander counts under cover boards [26]. Additionally, red-backed salamanders are known to prey on spiders and other litter-dwelling invertebrate predators and spiders also prey on amphibians [18]. In the large-scale removal experiment, it is possible that variability in salamander densities obscured any potential effects of red-backed salamanders on these litter-dwelling invertebrate predators; however, in the enclosures the density of these predators did not co-vary with salamander density. There was also no evidence from our enclosures that the abundance of invertebrate predators affected the ecosystem functions. Homyack et al. [24] hypothesized that the lack of effect of red-backed salamanders on litter decomposition was a result of the effect of predatory invertebrates. However, between our two experiments we did not observe an effect of salamanders on predatory invertebrate abundance or an effect of predatory invertebrate abundance on ecosystem functions. This was surprising because spiders are known to reduce herbivore abundance and influence ecosystem functions [3,5]. Spiders tend to reduce plant damage and increase plant biomass and reproduction [3]. However, this depends on the foraging tactics of the spiders. Active-hunting spiders increase primary production and N mineralization rates, whereas sit-and-wait spiders tend to have the opposite effect [5]. It is possible that both the food web structure and types of predatory invertebrate species present dampen ecosystem function effects of salamanders [6,51]. Specifically, salamanders consume a broad range of invertebrates at difference trophic levels in the detrial food web. Thus, abundance of any single prey group may not be significantly reduced and top-down effects related to decomposition could be balanced by multi-trophic level feeding. In old fields, predator functional diversity effects on ecosystem functions can be linearly predicted from the individual effects [61]. Future studies interested in the effects of forest floor predators on ecosystem functions would benefit from explicitly manipulated densities of multiple predatory vertebrates and invertebrates in combination.

The complexity of the forest floor food web and mixture of top-down and bottom-up effects on ecosystem functions makes determining the effects of salamanders difficult, especially since the primary carbon source in forest floor food webs can be detrital or from roots [62]. In addition, there are experimental limitations that create further challenges in elucidating the effects of salamanders. There are always tradeoffs between realism, control, and replication when designing ecological experiments. Effects found in small, highly replicated, well-controlled experiments often do not extrapolate to more complex natural systems. In contrast, large-scale experiments have more realism but can lack the control and replication to detect the effects of specific manipulation.

In this study, we coupled these two approaches to examine the effects of red-backed salamanders on ecosystem functions. Logistics prevented precise determination of salamander abundance on all plots throughout the study. However, the cumulative evidence suggests significant reduction in salamander abundance on the depletion plots. It is possible that large variation among plots could have obscured the effects of depletions and that removals did not surface, subsurface, and age class portions of the population in proportion to their occurrence in the population. This is an inherent challenge with large-scale field manipulations, although we tried to minimize the ratio of between-plot to within-plot variability by using large plots that are variability in microhabitats. We did removal animals ranging in size from recent hatchlings to large adults, including gravid females, over multiple years, making it unlikely that this had a strong influence on the lack of observed effects by salamanders. Despite limitations associated with large-scale field manipulations, the combination of salamander depletions and controlled mesocosm enclosures provide insight into the role of red-backed salamanders in ecosystem functions. We did not find evidence of salamander impacts on decomposition, nitrogen cycling, foliar insect damage, or on predatory invertebrates. The inference from this study is limited to environmental conditions during the study in a beech-dominated forest in New Hampshire, although it may help our understanding of systems where predators influence ecosystem functions. Given complex interactions in soil food webs, the effects of habitat heterogeneity on top-down effects, and the mixture of top-down and bottom-up effects in forest ecosystems, it is likely that predator effects on ecosystem functions are context dependent. Future studies would benefit from additional controlled manipulations of the soil food web and predator densities when examining effects on ecosystem functions. Additionally, the plants and soil properties likely influence salamander effects and explicit study of salamander effects under different soil nutrient conditions and on different plant species would be informative for discerning context-dependent salamander effects. Finally, there may be significant time lags between changes in salamander densities and subsequent changes in ecosystem functions. Therefore, researchers should conduct studies over longer time periods in the future, especially in controlled mesocoms. We did find significant effects of year on ecosystem functions, likely due to natural variation in climatic conditions, and future studies over longer time periods could improve our understanding of this variation and determine if red-backed salamanders influence ecosystem functions under particular conditions.

Acknowledgments

We thank M. Ducey, J. Maerz, J. Aber, A. Kovach, S. Frey, and T. Lee for advice on various aspects of the project. The field and lab work could not have been completed without help from A. Contosta, M. Knorr, J. Vessey, J. Purrenhage, E. Carcagno, S. Coster, J. Walsh, S. Wile, M. Kilpatrick, B. Godboys, J. Bartolotta, J. Bradley, M. Cote, L. Valliere, and E. Willey. The quality of this manuscript was improved by suggestions from L. Hocking, R. Rowe, R. Stephens, K. Jenks, C. Haas, C. Hickerson, and four anonymous reviewers.

Author Contributions

Conceived and designed the experiments: DJH KJB. Performed the experiments: DJH. Analyzed the data: DJH. Contributed reagents/materials/analysis tools: DJH KJB. Wrote the paper: DJH. Provided analytical and editorial advice: KJB.

References

1. Daily GC, Alexander S, Ehrlich PR, Goulder L, Lubchenco J, et al. (1997) Ecosystem services: benefits supplied to human societies by natural ecosystems. Issues in Ecology 2: 1–16.
2. Millenium Ecosystem Assessment (2005) Ecosystems and Human Well-being: Synthesis. Washington, DC: Island Press. 155 p.
3. Schmitz OJ, Hamback PA, Beckerman AP (2000) Trophic cascades in terrestrial systems: a review of the effects of carnivore removals on plants. The American Naturalist 155: 141–153.
4. Mooney KA, Gruner DS, Barber NA, Van Bael SA, Philpott SM, et al. (2010) Interactions among predators and the cascading effects of vertebrate insectivores on arthropod communities and plants. Proceedings of the Academy of Natural Sciences of Philadelphia 107: 7335–7340.
5. Schmitz OJ (2008) Effects of predator hunting mode on grassland ecosystem function. Science 319: 952–954.
6. Schmitz OJ, Beckerman AP, O'Brien KM (1997) Behaviorally mediated trophic cascades: effects of predation risk on food web interactions. Ecology 78: 1388–1399.
7. Beard KH, Vogt KA, Kulmatiski A (2002) Top-down effects of a terrestrial frog on forest nutrient dynamics. Oecologia 133: 583–593.
8. Huang C-Y, Wang C-P, Lucy Hou P-C (2007) Toads (*Bufo bankorensis*) influence litter chemistry but not litter invertebrates and litter decomposition rates in a subtropical forest of Taiwan. Journal of Tropical Ecology 23: 161–168.
9. Burton TM, Likens GE (1975) Energy flow and nutrient cycling in salamander populations in the Hubbard Brook experimental forest, New Hampshire. Ecology 56: 1068–1080.
10. Frank DA (2008) Evidence for top predator control of a grazing ecosystem. Oikos 117: 1718–1724.
11. Sin H, Beard KH, Pitt WC (2008) An invasive frog, *Eleutherodactylus coqui*, increases new leaf production and leaf litter decomposition rates through nutrient cycling in Hawaii. Biological Invasions 10: 335–345.
12. Beard KH, Eschtruth AK, Vogt KA, Vogt DJ, Scatena FN (2003) The effects of the frog *Eleutherodactylus coqui* on invertebrates and ecosystem processes at two scales in the Luquillo Experimental Forest, Puerto Rico. Journal of Tropical Ecology 19: 607–617.
13. Wyman RL (1998) Experimental assessment of salamanders as predators of detrital food webs: effects on invertebrates, decomposition and the carbon cycle. Biodiversity and Conservation 7: 641–650.
14. Milton Y, Kaspari M (2007) Bottom-up and top-down regulation of decomposition in a tropical forest. Oecologia 153: 163–172.
15. Chalcraft DR, Binckley CA, Resetarits WJ (2005) Experimental venue and estimation of interaction strength: Comment. Ecology 86: 1061–1067.
16. Skelly DK (2002) Experimental venue and estimation of interaction strength. Ecology 83: 2097–2101.
17. Morin PJ (1998) Realism, precision, and generality in experimental ecology. In: Resetarits WJ, Bernardo J, editors. Experimental ecology: issues and perspectives. 1998: Oxford University Press. 50–70.
18. Casper GS (2005) *Plethodon cinereus* (Green, 1818) eastern red-backed salamander. In: Lannoo M, editor. Amphibian Declines: The Conservation Status of United States Species. Boston: University of California Press. 796–800.
19. Jung RE, Droege S, Sauer JR, Landy RB (2000) Evaluation of terrestrial and streamside salamander monitoring techniques at Shenandoah National Park. Environmental Monitoring and Assessment 63: 65–79.
20. Bailey LL, Simons TR, Pollock KH (2004) Comparing population size estimators for Plethodontid salamanders. Journal of Herpetology 38: 370–380.
21. Burton TM, Likens GE (1975) Salamander populations and biomass in the Hubbard Brook experimental forest, New Hampshire. Copeia 1975: 541–546.
22. Davic RD, Welsh HH, Jr. (2004) On the ecological roles of salamanders. Annual Review of Ecology, Evolution, and Systematics 35: 405–435.
23. Hairston Sr NG (1987) Community Ecology and Salamander Guilds. New York: Cambridge University Press.
24. Homyack JA, Sucre EB, Haas CA, Fox TR (2010) Does *Plethodon cinereus* affect leaf litter decomposition and invertebrate abundances in mixed oak forest? Journal of Herpetology 44: 447–456.
25. Hickerson C-AM, Anthony CD, Wicknick JA (2004) Behavioral interactions between salamanders and centipedes: competition in divergent taxa. Behavioral Ecology 15: 679–686.
26. Hickerson C-AM, Anthony CD, Walton BM (2012) Interactions among forest-floor guild members in structurally simple microhabitats. American Midland Naturalist 168: 30–42.
27. Petranka JW (1998) Salamanders of the United States and Canada. Washington: Smithsonian Institution Press.
28. Hairston NG (1980) The experimental test of an analysis of field distributions: competition in terrestrial salamanders. Ecology 61: 817–826.
29. Petranka JW, Murray SS (2001) Effectiveness of removal sampling for determining salamander density and biomass: a case study in an Appalachian streamside community. Journal of Herpetology 35: 36–44.
30. Bailey LL, Simons TR, Pollock KH (2004) Comparing population size estimators for Plethodontid salamanders. Journal of Herpetology 38: 370–380.
31. Bailey LL, Simons TR, Pollock KH (2004) Spatial and temporal variation in detection probability of Plethodon salamanders using the robust capture-recapture design. Journal of Wildlife Management 68: 14–24.
32. Bailey LL, Simons TR, Pollock KH (2004) Estimating detection probability parameters for plethodon salamanders using the robust capture-recapture design. Journal of Wildlife Management 68: 1–13.
33. Harmon ME, Nadelhoffer KJ, Blair JM (1999) Measuring decomposition, nutrient turnover, and stores in plant litter. In: Roberston GP, Coleman DC, Bledsoe CS, Sollins P, editors. Standard soil methods for long-term ecological research. New York: Oxford University Press. 202–240.
34. Gosz JR, Likens GE, Borman FH (1972) Nutrient content of litter fall on the hubbard brook experimental forest, New Hampshire. Ecology 53: 769–784.
35. Walton BM, Steckler S (2005) Contrasting effects of salamanders on forest-floor macro- and mesofauna in laboratory microcosms. Pedobiologia 49: 51–60.
36. Roberston GP, Wedin DA, Groffman PM, Blair JM, Holland EA, et al. (1999) Soil carbon and nitrogen availability: nitrogen mineralization, nitrification, and soil respiration potentials. In: Roberston GP, Coleman DC, Bledsoe CS, Sollins P, editors. Standard soil methods for long-term ecological research. New York: Oxford University Press. 258–271.
37. Sims GK, Ellsworth TR, Mulvaney RL (1995) Microscale determination of inorganic nitrogen in water and soil extracts. Communications in Soil Science and Plant Analysis 26: 303–316.
38. Braman RS, Hendrix SA (1989) Nanogram nitrite and nitrate determination in evnironmental and biological-materials by vanadium(III) reduction with chemi-luminescence detection. Analytical Chemistry 61: 2715–2718.
39. Schmitz OJ, Krivan V, Ovadia O (2004) Trophic cascades: the primacy of trait-mediated indirect interactions. Ecology Letters 7: 153–163.
40. Hawlena D, Strickland MS, Bradford MA, Schmitz OJ (2012) Fear of predation slows plant-litter decomposition. Science 336: 1434–1438.
41. Auchmoody LR, Smith HC, Walters RS (1994) Planting northern red oak acrons: is size and planting depth important? In: US Department of Argiculture FS, Northeastern Forest Experiment Station, editor. Radnor, PA. 5.
42. Southwood TRE, Henderson PA (2000) Ecological Methods. Oxford: Blackwell Science Ltd. 575 p.
43. Fox J (2002) An R and S-Plus companion to applied regression. Thousand OaksCalifornia: Sage Publications. 312 p.
44. R Development Core Team (2012) R: A language and environment for statistical computing. 2.14 ed. Vienna, Austria: R Foundation for Statstical Computing.
45. Gillette JR, Peterson MJ (2001) The benefits of transparency: candling as a simple method for determining sex in Red-Backed Salamanders. Herpetological Review 32: 233–235.
46. Peterman WE, Semlitsch RD (2006) Effects of tricaine methanesulfonate (MS-222) concentration on anesthetization and recovery in four plethodontid salamanders. Herpetological Review 37: 303–304.
47. Davis TM, Ovaska K (2001) Individual recognition of amphibians: effects of toe clipping and fluorescent tagging on the salamander *Plethodon vehiculum*. Journal of Herpetology 35: 217–225.
48. Heemeyer JL, Homyack JA, Haas CA (2007) Retention and readability of visible implant elastomer marks in the eastern red-backed salamander (*Plethodon cinereus*). Herpetological Review 38: 425–428.
49. Homyack J, Haas C, Hopkins W (2010) Influence of temperature and body mass on standard metabolic rate of eastern red-backed salamanders (*Plethodon cinereus*). Journal of Thermal Biology 35: 143–146.
50. Walton BM (2005) Salamanders in forest-floor food webs: environmental heterogeneity affects the strength of top-down effects. Pedobiologia 49: 381–393.
51. Walton BM, Tsatiris D, Rivera-Sostre M (2006) Salamanders in forest-floor food webs: invertebrate species composition influences top-down effects. Pedobiologia 50: 313–321.
52. Irmler U (2000) Changes in the fauna and its contribution to mass loss and N release during leaf litter decomposition in two deciduous forests. Pedobiologia 44: 105–118.
53. Hanlon RDG (1981) Influence of grazing by collembola on the activity of senescent fungal colonies grown on media of different nutrient concentration. Oikos 36: 362–367.
54. Lenoir L, Persson T, Bengtsson J, Wallander H, Wirén A (2007) Bottom-up or top-down control in forest soil microcosms? Effects of soil fauna on fungal biomass and C/N mineralisation. Biology and Fertility of Soils 43: 281–294.
55. Rooney TP, Antolik C, Moran MD (2000) The impact of salamander predation on collembola abundance. Proceedings of the Entomological Society of Washington 102: 308–312.
56. Ransom TS (2011) Earthworms, as ecosystem engineers, influence multiple aspects of a salamander's ecology. Oecologia 165: 745–754.

57. Ransom TS (2012) Comparison of direct, indirect, and ecosystem engineering effects of an earthworm on the red-backed salamander. Ecology 93: 2198–2207.
58. Paul EA, editor (2007) Soil Microbiology, Ecology, and Biochemistry. Third ed. Boston: Academic Press. 532 p.
59. Contosta AR, Frey SD, Cooper AB (2011) Seasonal dynamics of soil respiration and N mineralization in chronically warmed and fertilized soils. Ecosphere 2: 1–21.
60. Maerz JC, Karuzas JM, Madison DM, Blossey B (2005) Introduced invertebrates are important prey for a generalist predator. Diversity and Distributions 11: 83–90.
61. Schmitz OJ, Sokol-Hessner L (2002) Linearity in the aggregate effects of multiple predators in a food web. Ecology Letters 5: 168–172.
62. Pollierer MM, Langel R, Korner C, Maraun M, Scheu S (2007) The underestimated importance of belowground carbon input for forest soil animal food webs. Ecology Letters 10: 729–736.

14

Relationships of Biomass with Environmental Factors in the Grassland Area of Hulunbuir, China

Miao Liu[1,2], Guohua Liu[1]*, Li Gong[1,2], Dongbo Wang[4], Jian Sun[3]

1 State Key Laboratory of Urban and Regional Ecology, Research Center for Eco-environmental Sciences, Chinese Academy of Sciences, Beijing, China, **2** University of Chinese Academy of Sciences, Beijing, China, **3** Key Laboratory of Ecosystem Network Observation and Modeling, Institute of Geographic Sciences and Natural Resources Research, Chinese Academy of Sciences, Beijing, China, **4** Monitoring Station for Eco-environment of Hulunbuir, Hulunbuir, China

Abstract

Many studies have focused on the relationship between vegetation biomass and environmental factors in grassland. However, several questions remain to be answered, especially with regards to the spatial pattern of vegetation biomass. Thus, the distributed mechanism will be explored in the present study. Here, plant biomass was measured at 23 sites along a transect survey during the peak growing season in 2006. The data were analyzed with a classification and regression tree (CART) model. The structural equation modeling (SEM) was conducted to explicitly evaluate the both direct and indirect effects of these critical environmental elements on vegetation biomass. The results demonstrated that mean annual temperature (MAT) affected aboveground biomass (AGB) scored at −0.811 ($P<0.05$). The direct effect of MAT on belowground biomass (BGB) was −0.490 ($P<0.05$). The results were determined by SEM. Our results indicate that AGB and BGB in semi-arid ecosystems is strongly affected by precipitation and temperature. Future work shall attempt to take into account the integrated effects of precipitation and temperature. Meanwhile, partitioning the influences of environmental variations and vegetation types are helpful in illuminating the internal mechanism of biomass distribution.

Editor: Keping Ma, Institute of Botany, Chinese Academy of Sciences, China

Funding: This study was supported by the "Strategic Priority Research Program" of Chinese Academy of Sciences (Grant No. XDA05060100), and the National Natural Science Foundation of China (Grant No. 41105117). The funders had no role in study design, data collection and analysis, decision to publish, or preparation of the manuscript.

Competing Interests: The authors have declared that no competing interests exist.

* Email: ghliu@rcees.ac.cn

Introduction

Vegetation biomass, which is usually illustrated by aboveground biomass (AGB), belowground biomass (BGB) and ratio of root to shoot (R/S), is regarded as an important parameter in indicating the carbon cycles of terrestrial ecosystems and in global climate models [1–3]. In particular, the majority of CO_2 in the atmosphere was fixed by grassland ecosystems, which account for 1/4 of the land surface and 1/10 of global carbon storage [4,5]. Changes in environmental conditions will produce rapid and profound influences on vegetation biomass [6,7]. Furthermore, the spatial patterns of biomass shape the dynamics of the grassland's carbon cycles [8]. Therefore, it is necessary to explore the mechanisms of how biomass distributed patterns respond to environmental variation and climate change [9]. Hulunbuir grassland, with one hundred thousand square kilometers of total grassland area, its large tracts of grassland make this area very suitable for research along the different ecological gradients.

The spatial pattern of vegetation biomass is severely shaped by environmental elements. There are many studies examining the relationship between vegetation biomass and environmental conditions. For instance, Peng et al. [10] found that net primary production responds non-linearly to increased vs. decreased rainfall in semi-arid grassland ecosystem, Inner Mongolia. Chang et al. [11] also documented that annual precipitation and air temperature are the key factors affecting the aboveground net primary productivity (ANPP) in temperate grasslands, and hold that ANPP increased with the increasing of precipitation, and declined with the air temperature. Fan et al. [12] reported that the proportion of belowground biomass increased as temperature decreased in Inner Mongolia, and was distributed more deeply in desert grassland owing to the aridity of the grassland types. Meanwhile, changes of AGB and BGB were examined in relation to gradients of temperature and precipitation. Total biomass was negatively correlation with temperature, but positively weak correlation with mean annual precipitation [13]. As a matter of fact, the complex change of biomass was caused by the meteorological factors and soil properties [14]. However, the studies regarding the magnitude of the concrete influence of the main environmental elements on vegetation biomass, especially along the gradients of precipitation and temperature across Hulunbuir grassland, remain few. The principal purposes of the present study were to explore the spatial distribution pattern of AGB, BGB and R/S along the gradients of precipitation and temperature using data investigated from 23 sites across Hulunbuir grassland; examine the relationships of AGB, BGB and R/S with the relevant environmental factors and to screen out the main factors using the classification and regression tree (CART) model; and to identify the concrete effect magnitude of the key elements on the AGB, BGB and R/S using Structural Equation Modeling (SEM). The final objective was to indicate the internal controlling factors of vegetation biomass along the gradients of precipitation and temperature.

Materials and Methods

Ethics Statement

For each site, no specific permits were required for the sample collected and the field studies did not involve endangered or protected species in Hulunbuir, Inner Mongolia.

Study Area

The study area is situated in the western portion of Mt. Daxing'anling, Hulunbuir, Inner Mongolia, China (115°31'–126°04'E, 47°05'–53°20'N). The mean annual precipitation is 339 mm and the mean annual temperature is 2.2°C [15,16]. The topographic features in the area are relatively changeless and the maximal difference in elevation is not more than 50 m. Chernozem and chestnut soil are the main soil types [17]. It is provided with broad precipitation and temperature gradients that are strongly related to the spatial patterns of grassland biomass in the Hulunbuir zone [18].

Data Collection and Sample Analysis

In 2006, 23 sites crossing from west to east and from south to north were surveyed in Hulunbuir grassland during the vegetation growth period (June, July and August). Sampling sites were established along the gradients of precipitation and temperature (Fig. 1). From each site, we harvested AGB and BGB in five plots (1 m×1 m) of similar topography and environmental conditions every 10 m intervals along a transect. AGB was determined by clipping the plants at ground level and oven-drying them at 65°C until they reached a constant weight. BGB was collected from soil depths of 0 cm to 30 cm, where most of the belowground biomass is located [19]. The root samples were obtained from the blocks using 5 cm diameter soil cores, soaked in water to remove the residual soil via a 0.5 mm sieve, and dried at 65°C to a constant weight. Meanwhile, soil samples were collected from five replicate soil profiles to determine soil properties at the 30 cm soil depth. After being air-dried and sieved (using 2 mm mesh), the soil samples were carefully handpicked to extract the surface organic materials and fine roots for soil chemical properties analysis. Each mixed soil sample was divided into two parts. One sub-sample was oven-dried at 105°C to a constant weight to measure gravimetric soil water content (SWC) and soil bulk density (BD). The remaining soil was ground in a ball mill for soil total organic carbon (TOC), total nitrogen (TN), total phosphorous (TP) and available phosphorous (AP) analysis. Soil properties were determined following all standard protocols [20]. Meteorological factors were obtained using spatially interpolated methods from the records of 28 weather stations (Fig. 1). Annual mean temperature (MAT)and annual mean precipitation (MAP) (2003–2006) were regarded as climate factors, and the data of longitude and latitude for each sampling site was determined using the Global Positioning System (GPS).

Data Analysis

The entire document was analyzed according to the followings three steps. First of all, the data regarding MAP, MAT, TOC, TP, AP, TN, BD, SWC, AGB, BGB and R/S were handled through Spearman correlation analysis after normal distribution testing. Secondly, we used classification and regression tree (CART) [21] to screen out the crucial variables influencing vegetation biomass from all samples of environmental factors along gradients of precipitation and temperature according to the data collected across various studied sites. The CART method facilitates determination of the possible interactions and adjustments necessary for making decisions [22], as it can identify the critical variables that significantly influence the response variables. The specific operating process has been described in Sun's report [8] and the R guideline book (R Development Core Team, 2011). Finally, structural equation modeling (SEM) [23] was conducted to explicitly evaluate both the direct and indirect effects of these critical environmental elements on vegetation biomass (AGB, BGB and R/S). SEM has been used in recent studies to exactly assess the causal relationships among multiple interacting variables [24–26]. In this study, statistical analysis and plotting were performed using the R software (version 2.15, R Development Core Team, 2011).

Results

Descriptive Statistic of AGB, BGB and R/S

AGB, BGB and R/S all exhibited large variations along the sampled transect, ranging from 8.37 to 201.96 g m^{-2} for AGB, 105.99–2586.50 g m^{-2} for BGB and 1.01–131.98 for R/S (Fig. 2). The mean values were 62.01 g m^{-2}, 974.21 g m^{-2} and 28.67 for AGB, BGB and R/S, respectively. The standard deviation values of AGB, BGB and R/S were 48.87, 513.72 and 27.46, in sequence.

Correlationship of AGB, BGB and R/S with Environmental Factors

The Spearman correlation analysis method was adopted to analyze the correlation of AGB, BGB and R/S with environmental factors (Table 1). The results demonstrated that AGB was positively related to MAP, TOC and TN across the sampling sites, with correlation coefficients of 0.562, 0.613 and 0.511, respectively. By contrast, AGB was negatively related to MAT, AP and BD, with correlation coefficients of 0.661, 0.579 and 0.650, respectively, while its relation with SWC and TP was non-significant. BGB was positively related to the MAP, TOC and TN at the 0.05 level, with correlation coefficients of 0.406, 0.434 and 0.359, respectively. The relationships of BGB with SWC, TP, AP and BD were non-significant. Meanwhile, R/S was positively related to AP and BD, with correlation coefficients of 0.446 and 0.557, while it had negative correlations to MAP, TOC and TN, with correlation coefficients of −0.328, 0.368 and 0.312 ($P<0.05$). The relationships of R/S with SWC and AP were non-significant.

Identification of Critical Factors by the CART Model

The impacts of all kinds of environmental factors on vegetation biomass (AGB, BGB and R/S) were observed using the CART model. The optional tree was developed. As shown in Fig. 3A, the analysis indicated that MAP, MAT, AP, TN, SWC and BD are most closely associated with large-scale variations in AGB. For the second tree (Fig. 3B), three critical environmental factors containing MAT, AP and TOC were obtained, having a significant influence on BGB. For the third tree (Fig. 3C), R/S was influenced mainly by MAP, TOC and AP.

Structural Equation Modeling to Explain the Effects of Environmental Variables on AGB, BGB and R/S

The critical variables (MAP, MAT, AP, TN, SWC and BD) were inter-correlated, and these apparent relationships combined both direct and indirect correlations (Fig. 3). Thus, we further used SEM to explicitly evaluate the causal relationships among these interacting variables.

From the SEM (Fig. 4), it is evident that, MAT, TN, SWC and BD affected AGB (direct effect), scored at −0.811 ($P<0.05$), −0.608, 0.424 and 0.532, respectively (Table 2, Fig. 4A). The direct

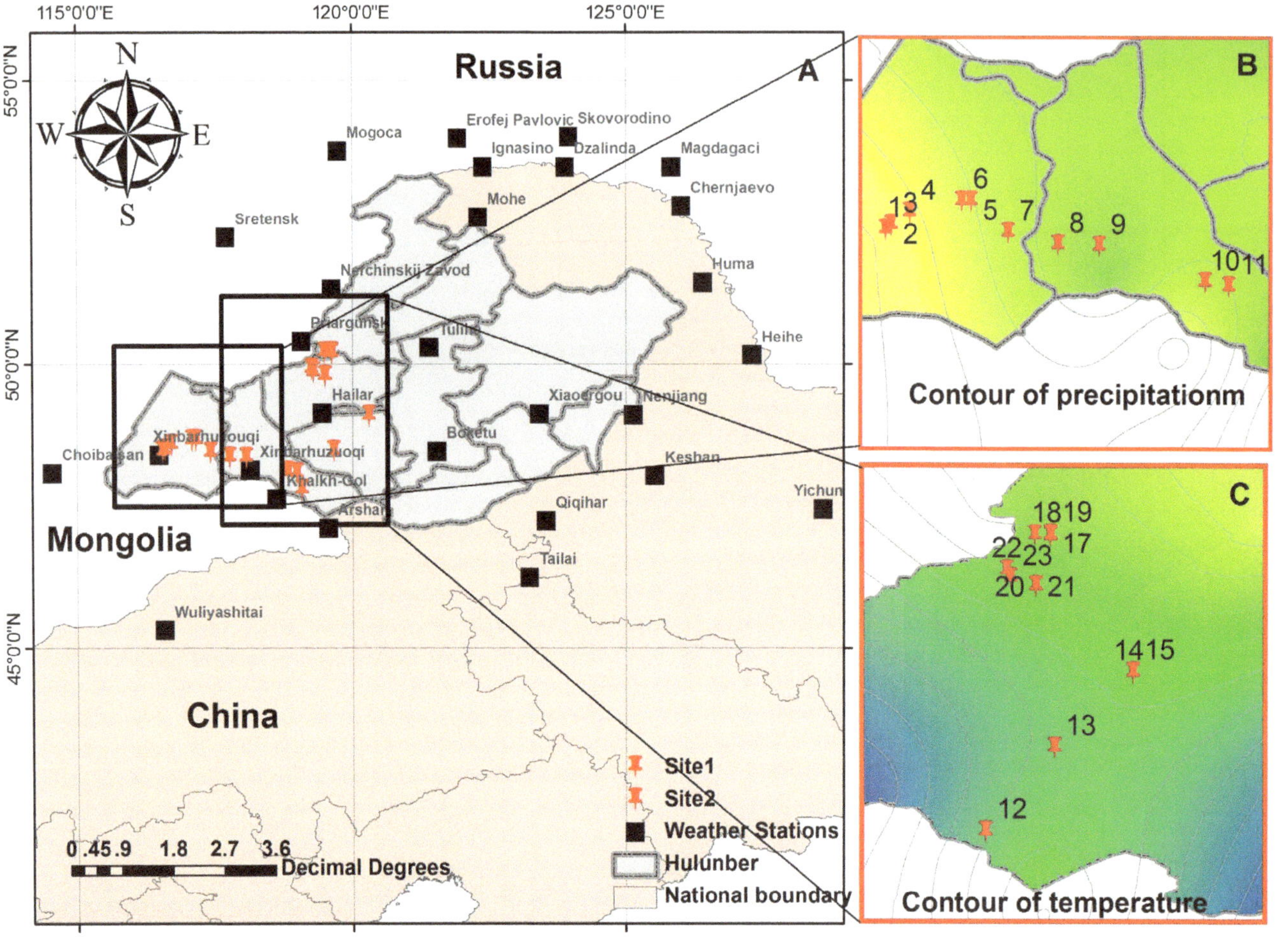

Figure 1. Spatial distribution of the sampling sites across the Hulunbuir grasslands, China. Samples (A) were collected along the gradients of precipitation (B) and temperature (C).

effects of AP and MAT on BGB were 0.280 and −0.490 ($P<0.05$), respectively (Table 2, Fig. 4B). AP had significantly positive direct effects on R/S at 0.408 (Table 2, Fig. 4C). The rank of total effects on AGB, in decreasing order, was: TN, MAT, SWC, BD and MAP. MAT had the strongest negative effects on both AGB and BGB among all predictors (Table 2, Figs. 4A, B), whereas no significant effect of MAP on AGB was found. AGB and BGB responded similarly to MAT. The standardized total effect of MAT on AGB was −0.381, consisting of direct effects (path coefficient = −0.811, $P<0.05$) and indirect effects through soil variables (path coefficient = 0.43). Among soil variables, TN and BD had significant negative effects on AGB, with path coefficients of −0.608 and −0.532 (Table 2, Fig. 4A), whereas SWC had a positive effect on AGB (path coefficient = 0.424). The total effect of AP on R/S was 0.389, consisting of direct effects (path coefficient = 0.408, $P<0.05$) and indirect effects through soil variables (path coefficient = −0.019). However, neither the direct nor indirect effects of MAP on R/S were significant (Table 2, Fig. 4C).

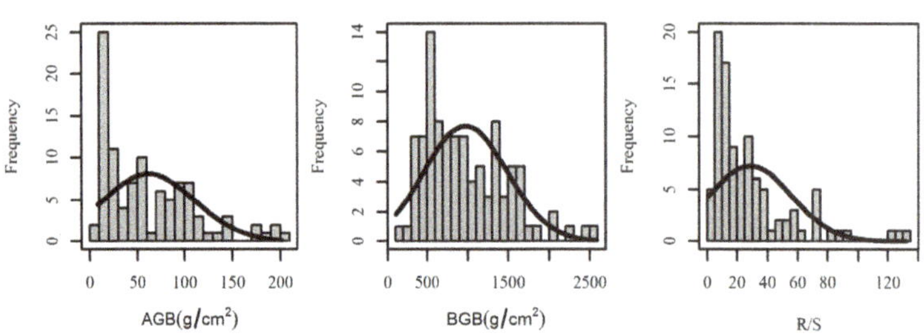

Figure 2. Frequency distribution curves of the AGB, BGB and R/S; the samples were collected across the Hulunbuir grasslands. All designations are the same as those in the footnotes below Table 1.

Table 1. Coefficient of correlation between biome and environmental factors (Spearman).

	MAP	MAT	TOC	TP	AP	TN	BD	SWC	AGB	BGB	R/S
MAP	1										
MAT	−0.923**	1									
TOC	0.610**	−.682**	1								
TP	−0.010	−.053	0.473**	1							
AP	−0.618**	.619**	−0.259**	0.331**	1						
TN	0.537**	−.600**	0.915**	0.553**	−0.185	1					
BD	−0.555**	.638**	−0.763**	−0.417**	0.437**	−0.783**	1				
SWC	−0.034	.147	0.621**	0.396**	0.165	0.620**	−0.312	1			
AGB	0.562**	−.661**	0.613**	−0.076	−0.579**	0.511**	−0.650**	0.167	1		
BGB	0.406**	−.429**	0.434**	0.139	−0.179	0.359**	−0.272	−0.097	0.255*	1	
R/S	−0.328**	.421**	−0.368**	0.138	0.446**	−0.312**	0.557**	−0.139	−0.826**	0.306**	1

Note: **Correlation is significant at the 0.01 level.*Correlation is significant at the 0.05 level (2-tailed). MAP (Mean annual precipitation), MAT (Mean annual temperature), TOC (Total organic carbon), TP (Total phosphorous), AP (Available phosphorous), TN (Total nitrogen), BD (Soil bulk density), SWC (Soil water content), AGB (Aboveground biomass), BGB (Belowground biomass) and R/S (the ratio of root and shoot). The sample size is 100.

Discussion

Characteristics of Vegetation Biomass across all Samples in the Temperate Grassland

The values of vegetation biomass exhibited large variations along the sampled transect. The mean value of AGB in the present study was found to be lower than that in the temperate grasslands of Inner Mongolia (135.3 g m^{-2}) [27]. By contrast, BGB and R/S in the temperate grassland of Hulunbuir was found to be higher than it is in the temperate grasslands of Inner Mongolia (775.2 and 6.3 g m^{-2}). Meanwhile, BGB was far larger than AGB across Hulunbuir grassland, which was approximately 15 times greater than AGB. This result indicated that vegetation biomass in the temperate grassland of Hulunbuir was present to a greater degree below ground than above. Similar results were detailed in Ma's report [27], which described how BGB was approximately 6 times greater than AGB in the temperate grasslands of Inner Mongolia. The average value of R/S in Hulunbuir moderate grassland is far larger than that observed in global temperate grasslands (4.2) [1]. The higher R/S in Hulunbuir grasslands might be associated with the comparatively slow depletion of carbohydrates in roots, resulting from low respiration rates in the extremely cold winters there, and might also be due to slower root turnover in the colder environment relative to the mean level of the global temperate grasslands [28]. In addition, R/S values found in the present study ranging from 1.01 to 131.98 were consistent with results reported by Yang et al. [13], who performed a field investigation from 2001 to 2004, discovering that R/S values have the potential to vary greatly as a result of climate change and anthropogenic activities [29–31]. For example, long term grazing in grasslands may result in the reduction of AGB and ultimately higher R/S.

Effect of critical factor on Spatial Pattern of AGB, BGB and R/S

Moreover, based on the Spearman correlation analysis between biomass and environmental factors, we found that most environmental factors showed a significant correlation at 0.05 levels (Table 1). This may be a result of the degree of multicollinearity with respect to the effect of environmental factors on vegetation biomass, consistent with Sun's report [8].

We used the CART model and SEM [23] as new approaches to conduct variable selection and quantify its effect so as to screen out the critical factors associated with AGB, BGB and R/S and then to identify direct and indirect factors while determining the extent to which these factors may constrain vegetation biomass. To our knowledge, the efficiency of these approaches has not been evaluated empirically in vegetation biomass research simultaneously. Traditionally, stepwise selection and linear regression are used to identify and rank the limiting factors in grassland biomass studies. However, when performing stepwise selection, closely covariate parameters cannot be selected simultaneously in the final model, because the explanatory power does not increase when a closely related variable is included [32]. Field studies examining ecosystem responses to climatic and other environmental changes typically use naturally occurring climatic gradients. Nevertheless, Burke et al. [33] raised the issue that there are inherent problems with utilizing simple statistical relationships of spatial variability as foundations for understanding ecosystem function, because complex covariance along the gradient occurs across large spatial scales, leading to the problem that actual and apparent controlling factors may be confounded. SEM, however, is an appropriate option [34]. The quantitative procedure in the current study demonstrated that the strongest direct factors influencing AGB at the regional scale were MAT, TN, SWC and BD. While the

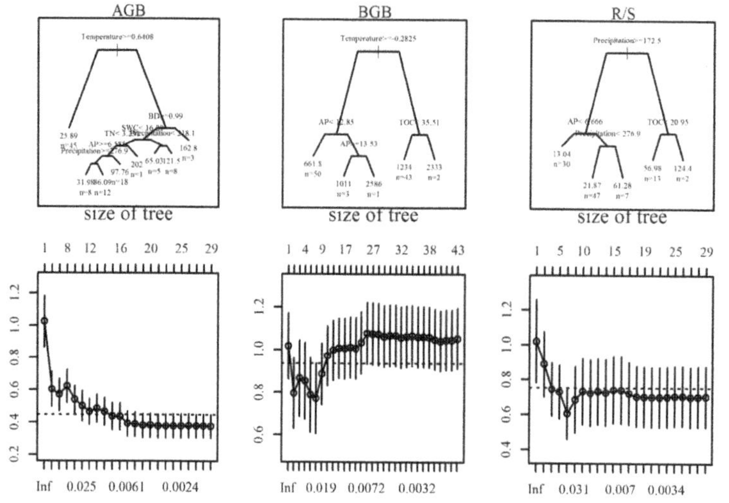

Figure 3. CART analyses of the relationships between biome and environmental factors along the gradients of precipitation and temperature in the Hulunbuir grasslands. The key environmental factors were screened in panels A (AGB), B (BGB) and C (R/S). Branches are labeled with criteria used to segregate data. Values in terminal nodes represent mean vegetation biomass of sites grouped within the cluster. n = number of plots in the category. The below corresponding panels were structural complexity (cp value) of trees. All designations are the same as those in the footnotes below Table 1.

largest total element impacting R/S was AP, AP and MAT both had major direct influences on BGB. This holistic approach is appropriate in across-site comparisons of ecosystem structure and function. Additionally, we discovered that longitude and latitude, which indicated the gradient of precipitation and temperature accordingly, were almost the main determinants on biome with a comparatively larger magnitude of total effect than that of all other relevant environmental factors. Therefore, it is necessary to explore the mechanism of biomass distribution along the gradient of precipitation and temperature.

Precipitation and temperature are considered to be the limiting factors for the growth and distribution of vegetation over the long term [35]. Previous studies precipitation has large impacts on grassland ecosystems biomass in Inner Mongolia [11,27] However, in the present study, AGB and BGB showed a decreasing trend with increasing temperature (Fig. 4A, B), while R/S demonstrated a reverse change (Table 1). The spatial pattern of vegetation biomass in Inner Mongolia in this dry area is associated with temperatures [13]. Under drought conditions high temperatures may further restrain plant photosynthesis. This was shown by Xu and Zhou [36], who reported that high temperatures resulted in the reduction of photosynthetic rate and biomass of *L. chinensis* steppe. Furthermore, higher temperatures will result in increased evaporation, intensifying drought and reducing biomass [11].

In our case, when MAP is retained in the model, MAT will not be selected. Therefore, our results showed temperature to be a significant influence on AGB and BGB, while precipitation was not shown in Fig. 4. As a matter of fact, precipitation also had the strongest positive effect on AGB and BGB (Table 1). This phenomenon indicates that there is an integrated effect between

Table 2. Standardized direct effect, indirect effect and total effect of critical environmental factors on aboveground biomass (AGB), belowground biomass (BGB) and root to shoot ratio (R/S).

Endogenous variable	Exogenous variable	Direct effect	Indirect effect	Total effect
AGB	MAP	−0.047	−0.028	−0.075
	MAT	−0.811**	0.430	−0.381
	AP	0.097	−0.277	−0.180
	TN	−0.608**	0.156	−0.452
	SWC	0.424**	−0.111	0.323
	BD	−0.532**	0.261	−0.271
BGB	MAT	−0.490**	0.056	−0.434
	AP	0.280**	0.01	0.291
	TOC	0.062	0.052−	0.114
R/S	MAP	−0.049	0.183	0.134
	AP	0.408**	−0.019	0.389
	TOC	−0.142	0.055	−0.087

Note: The "**" represented the significant. All designations are the same as those in the footnotes below Table 1.

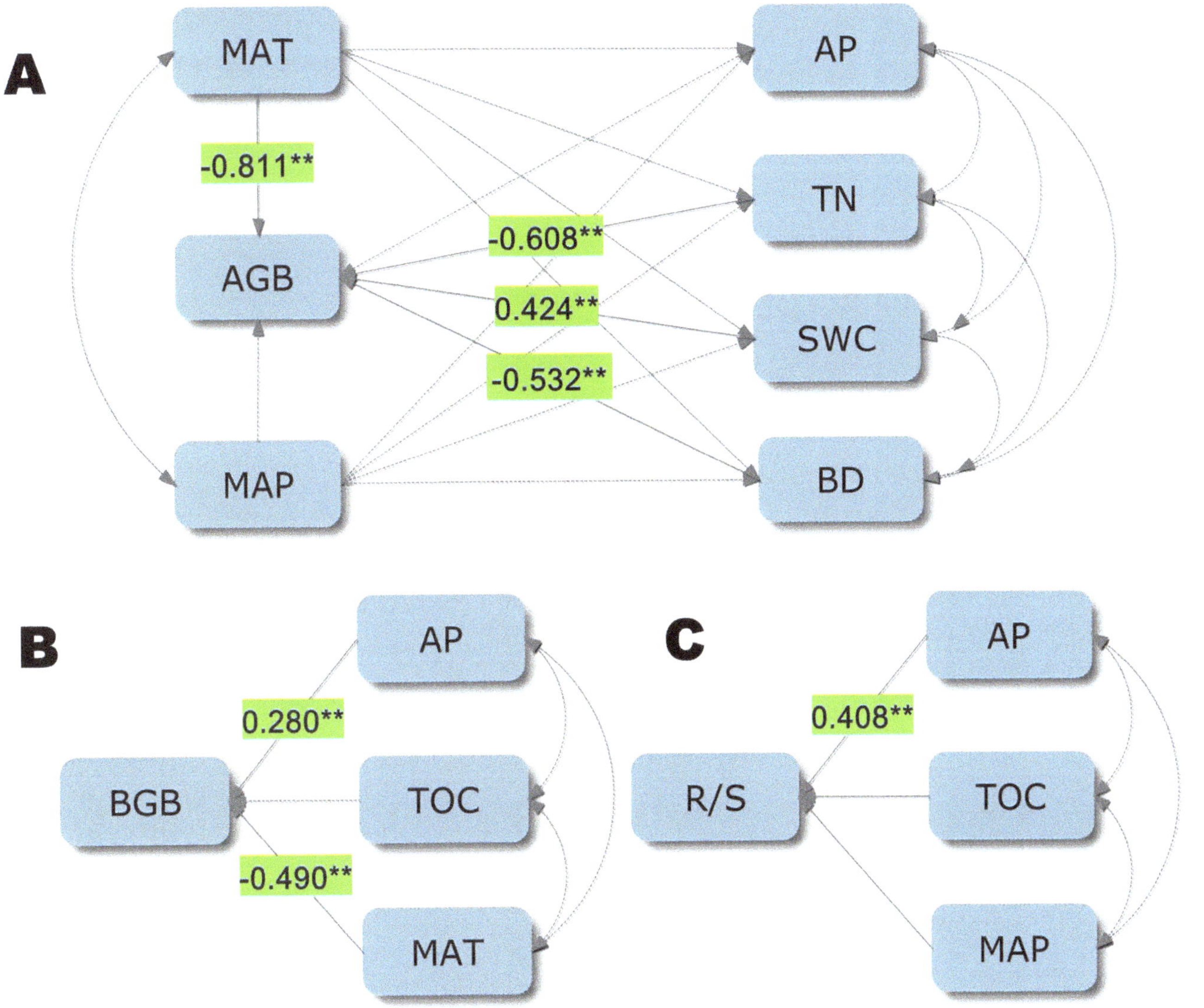

Figure 4. The establishment of a structural equation model for A) aboveground biomass (AGB), B) belowground biomass (BGB) and C) root to shoot ratio (R/S). Each line represents a direct linear causal relationship. The arcs show the correlation between two variables. Values on lines are path coefficients. The asterisks are significant at $P=0.05$ level. The coefficients that are not statistically significant are shown by dashed arrows. All values are standardized. All designations are the same as those in the footnotes below Table 1.

temperature and precipitation on plant growth [37]. Our findings reinforce the idea that precipitation is major limiting factor that control the functions of ecosystems in terrestrial biomes, particularly in arid and semi-arid ecosystems [38–41]. In addition, we found that TOC and TN are also positively related to precipitation [42] and negatively related to temperature (Table 1). Kirschbaum [43] holds that warming will have the effect of reducing soil organic carbon by stimulating decomposition rates more than NPP. On the other hand, some scholars hold the opposite view [44], suggesting that adequate precipitation and temperatures that are not too high can be used to create good soil nutrient conditions for plant growth [33]. Relevant to the above statement, the spatial pattern of vegetation biomass was influenced by soil nutrient level (Fig. 4B, C), and soil nutrient levels were affected by the conditions of water and heat. Soil nutrient levels can be improved by more litter being returned to soil resulting from more plant biomass with the addition of well water and ideal heat conditions.

Soil nitrogen has been found to be a limiting factor for AGB and BGB in most terrestrial ecosystems [37,45]. Nevertheless, others have reported limited or no influence of soil nitrogen on ecosystem production [46,47]. In the present study, we found that there was negative relationship of AGB with total nitrogen content. Thus, we concluded that the different responses of plants to nitrogen content are a result of different nitrogen deficiency levels of the local systems [37]. Soil phosphorous, generally regarded as the most critical factor for plant growth [48], the biomass was related with available phosphorous (Table 1, Figure 4). Meanwhile, the previous studies suggested that soil water content play an important role on biomass allocation [49], and the soil physical structure (e.g soil bulk density) might be more important for determining BGB than other factors [50].

Conclusion

Our results indicate that precipitation and temperature strongly affect aboveground and belowground biomass in semi-arid ecosystems. Future work shall attempt to take into account the integrated effects of precipitation and temperature, which could possibly explain the spatial variance of AGB, BGB and R/S better than MAP or MAT alone. Meanwhile, partitioning the influences of environmental variations and vegetation types is helpful in illuminating the internal mechanism of biomass distribution along the gradient of precipitation and temperature.

Acknowledgments

The authors are grateful to two anonymous reviewers all gave very helpful editorial comments.

Author Contributions

Performed the experiments: ML LG DBW. Analyzed the data: ML. Contributed reagents/materials/analysis tools: GHL ML JS. Wrote the paper: ML GHL.

References

1. Mokany K, Raison RJ, Prokushkin AS (2006) Critical analysis of root: shoot ratios in terrestrial biomes. Global Change Biology 12: 84–96.
2. Titlyanova AA, Romanova IP, Kosykh NP, Mironycheva-Tokareva NP (1999) Pattern and process in above-ground and below-ground components of grassland ecosystems. Journal of Vegetation Science 10: 307–320.
3. Jackson RB, Canadell J, Ehleringer JR, Mooney HA, Sala OE, et al. (1996) A global analysis of root distributions for terrestrial biomes. Oecologia 108: 389–411.
4. Scurlock JMO, Johnson K, Olson RJ (2002) Estimating net primary productivity from grassland biomass dynamics measurements. Global Change Biology 8: 736–753.
5. Hui DF, Jackson RB (2006) Geographical and interannual variability in biomass partitioning in grassland ecosystems: a synthesis of field data. New Phytologist 169: 85–93.
6. Farrar JF (1999) Acquisition, partitioning and loss of carbon. Physiological Plant Ecology: 25–43.
7. Maherali H, DeLucia EH (2001) Influence of climate-driven shifts in biomass allocation on water transport and storage in ponderosa pine. Oecologia 129: 481–491.
8. Sun J, Cheng GW, Li WP (2013) Meta-analysis of relationships between environmental factors and aboveground biomass in the alpine grassland on the Tibetan Plateau. Biogeosciences 10: 1707–1715.
9. Cairns MA, Brown S, Helmer EH, Baumgardner GA (1997) Root biomass allocation in the world's upland forests. Oecologia 111: 1–11.
10. Peng SS, Piao SL, Shen ZH, Ciais P, Sun ZZ, et al. (2013) Precipitation amount, seasonality and frequency regulate carbon cycling of a semi-arid grassland ecosystem in Inner Mongolia, China: A modeling analysis. Agricultural and Forest Meteorology 178: 46–55.
11. Chang J, Wang Z, Li Y, Han G, Wang ZW, et al. (2010) Relationship between aboveground net primary productivity and precipitation and air temperature of three plant communities in Inner Mongolia grassland. Acta Scientiarum Naturalium Universitatis NeiMongol 41: 689–694.
12. Fan JW, Zhong HP, Harris W, Yu GR, Wang SQ, et al. (2008) Carbon storage in the grasslands of China based on field measurements of above- and below-ground biomass. Climatic Change 86: 375–396.
13. Fan JW, Wang K, Harris W, Zhong HP, Hu ZM, et al. (2009) Allocation of vegetation biomass across a climate-related gradient in the grasslands of Inner Mongolia. Journal of Arid Environments 73: 521–528.
14. Sun J, Cheng GW, Li W, Sha YK, Yang YC (2013) On the Variation of NDVI with the Principal Climatic Elements in the Tibetan Plateau. Remote Sensing 5(4): 1894–1911.
15. Liu ZF, Fu BJ, Zheng XX, Liu GH (2010) Plant biomass, soil water content and soil N:P ratio regulating soil microbial functional diversity in a temperate steppe: A regional scale study. Soil Biology & Biochemistry 42: 445–450.
16. Zheng XX, Liu GH, Fu BJ, Jin TT, Liu ZF (2010) Effects of biodiversity and plant community composition on productivity in semiarid grasslands of Hulunbeir, Inner Mongolia, China. Ecological Complexity and Sustainability 1195: E52–E64.
17. Liu ZF, Liu GH, Fu BJ, Zheng XX (2008) Relationship between plant species diversity and soil microbial functional diversity along a longitudinal gradient in temperate grasslands of Hulunbeir, Inner Mongolia, China. Ecological Research 23: 511–518.
18. Zhang ZX, Zhang HY, Zhou DW (2010) Using GIS spatial analysis and logistic regression to predict the probabilities of human-caused grassland fires. Journal of Arid Environments 74: 386–393.
19. Li XJ, Zhang XZ, Wu JS, Shen ZX, Zhang YJ, et al. (2011) Root biomass distribution in alpine ecosystems of the northern Tibetan Plateau. Environmental Earth Sciences 64: 1911–1919.
20. Bao SD (2000) Soil Chemical Analysis of Agriculture. Beijing: Chinese Agriculture Press. (in Chinese).
21. Put R, Perrin C, Questier F, Coomans D, Massart DL, et al. (2003) Classification and regression tree analysis for molecular descriptor selection and retention prediction in chromatographic quantitative structure-retention relationship studies. Journal of Chromatography A 988: 261–276.
22. Toschke AM, Beyerlein A, von Kries R (2005) Children at high risk for overweight: A classification and regression trees analysis approach. Obesity Research 13: 1270–1274.
23. Grace JB, Pugesek BH (1997) A structural equation model of plant species richness and its application to a coastal wetland. American Naturalist 149: 436–460.
24. Grace JB, Kelley JE (2006) A structural equation model analysis of postfire plant diversity in California shrublands. Ecological Applications 16: 503–514.
25. Lamb EG (2008) Direct and indirect control of grassland community structure by litter, resources, and biomass. Ecology 89: 216–225.
26. Shipley B, Lechowicz MJ, Wright I, Reich PB (2006) Fundamental trade-offs generating the worldwide leaf economics spectrum. Ecology 87: 535–541.
27. Ma WH, Yang YH, He JS, Hui Z, Fang JY (2008) Above- and belowground biomass in relation to environmental factors in temperate grasslands, Inner Mongolia. Science in China Series C-Life Sciences 51: 263–270.
28. Gill RA, Jackson RB (2000) Global patterns of root turnover for terrestrial ecosystems. New Phytologist 147: 13–31.
29. Yu HY, Luedeling E, Xu JC (2010) Winter and spring warming result in delayed spring phenology on the Tibetan Plateau. Proceedings of the National Academy of Sciences of the United States of America 107: 22151–22156.
30. Shi FS, Chen H, Wu Y, Wu N (2010) Effects of Livestock Exclusion on Vegetation and Soil Properties under Two Topographic Habitats in an Alpine Meadow on the Eastern Qinghai-Tibetan Plateau. Polish Journal of Ecology 58: 125–133.
31. Li W, Huang HZ, Zhang ZN, Wu GL (2011) Effects of grazing on the soil properties and C and N storage in relation to biomass allocation in an alpine meadow. Journal of Soil Science and Plant Nutrition 11: 27–39.
32. De'ath G, Fabricius KE (2000) Classification and regression trees: A powerful yet simple technique for ecological data analysis. Ecology 81: 3178–3192.
33. Burke IC, Lauenroth WK, Parton WJ (1997) Regional and temporal variation in net primary production and nitrogen mineralization in grasslands. Ecology 78: 1330–1340.
34. Geng Y, Wang YH, Yang K, Wang SP, Zeng H, et al. (2012) Soil Respiration in Tibetan Alpine Grasslands: Belowground Biomass and Soil Moisture, but Not Soil Temperature, Best Explain the Large-Scale Patterns. Plos One 7.
35. Stephenson NL (1990) Climatic Control of Vegetation Distribution - the Role of the Water-Balance. American Naturalist 135: 649–670.
36. Xu ZZ, Zhou GS (2005) Effects of water stress and high nocturnal temperature on photosynthesis and nitrogen level of a perennial grass Leymus chinensis. Plant and Soil 269, 131–139.
37. Kang M, Dai C, Ji W, Jiang Y, Yuan Z, et al. (2013) Biomass and Its Allocation in Relation to Temperature, Precipitation, and Soil Nutrients in Inner Mongolia Grasslands, China. PLoS ONE 8(7): e69561. doi:10.1371/journal.-pone.0069561.
38. Heisler-White JL, Knapp AK, Kelly EF (2008) Increasing precipitation event size increases aboveground net primary productivity in a semi-arid grassland. Oecologia 158: 129–140.
39. Loik ME, Breshears DD, Lauenroth WK, Belnap J (2004) A multi-scale perspective of water pulses in dryland ecosystems: climatology and ecohydrology of the western USA. Oecologia 141: 269–281.
40. Munkhtsetseg E, Kimura R, Wang J, Shinoda M (2007) Pasture yield response to precipitation and high temperature in Mongolia. Journal of Arid Environments 70: 94–110.
41. Epstein HE, Burke IC, Lauenroth WK (2002) Regional patterns of decomposition and primary production rates in the US Great Plains. Ecology 83: 320–327.
42. Trumbore SE (1997) Potential responses of soil organic carbon to global environmental change. Proceedings of the National Academy of Sciences of the United States of America 94: 8284–8291.
43. Kirschbaum MUF (2000) Will changes in soil organic carbon act as a positive or negative feedback on global warming? Biogeochemistry 48: 21–51.
44. Giardina CP, Ryan MG (2000) Evidence that decomposition rates of organic carbon in mineral soil do not vary with temperature. Nature 404: 858–861.
45. LeBauer DS, Treseder KK (2008) Nitrogen limitation of net primary productivity in terrestrial ecosystems is globally distributed. Ecology 89: 371–379.

46. Heggenstaller AH, Moore KJ, Liebman M, Anex RP (2009) Nitrogen Influences Biomass and Nutrient Partitioning by Perennial, Warm-Season Grasses. Agronomy Journal 101: 1363–1371.
47. Wu J, Hong J, Wang X, Sun J, Lu X, et al. (2013) Biomass Partitioning and Its Relationship with the Environmental Factors at the Alpine Steppe in Northern Tibet. PLoS ONE 8(12): e81986. doi:10.1371/journal.pone.0081986.
48. Hejcman M. Szaková J. Schellberg J. Tlustoš P (2010) The Rengen Grassland Experiment: relationship between soil and biomass chemical properties, amount of elements applied, and their uptake. Plant and Soil 333, 163–179.
49. Zhao CY, Feng ZD, Chen GD (2004) Soil water balance simulation of alfalfa (*Medicago sativa L.*) in the semi-arid Chinese Loess Plateau. Agricultural Water Management 69: 101–114.
50. Schenk HJ, Jackson RB (2005) Mapping the global distribution of deep roots in relation to climate and soil characteristics. Geoderma 126: 129–140.

15

Facilitation Drives the Positive Effects of Plant Richness on Trace Metal Removal in a Biodiversity Experiment

Jiang Wang[1], Yuan Ge[2]*, Tong Chen[1], Yi Bai[1], Bao Ying Qian[1], Chong Bang Zhang[1]

1 School of Life Science, Taizhou University, Linhai, China, **2** Bren School of Environmental Science and Management/Earth Research Institute, University of California Santa Barbara, Santa Barbara, California, United States of America

Abstract

Background: Phytoextraction is an environmentally acceptable and inexpensive technique for mine tailing rehabilitation that uses metallophyte plants. These plants reduce the soil trace metal contents to environmentally acceptable levels by accumulating trace metals. Recently, whether more trace metals can be removed by species-rich communities of these plants received great attention, as species richness has been reported having positive effects on ecosystem functions. However, how the species richness affects trace metals removal of plant communities of mine tailing is rarely known.

Methodology/Principal Findings: We examined the effects of species richness on soil trace metal removal in both natural and experimental plant communities. The root lengths and stem heights of each plant species were measured in order to calculate the functional diversity indices. Our results showed that trace metal (Cu, Cd, Pb and Zn) concentrations in mine tailing soil declined as species richness increased in both the natural and experimental plant communities. Species richness, rather than functional diversity, positively affected the mineralomass of the experimental plant communities. The intensity of plant-plant facilitation increased with the species richness of experimental communities. Due to the incremental role of plant-plant facilitation, most of the species had higher biomasses, higher trace metal concentrations in their plant tissues and lower malondialdehyde concentrations in their leaves. Consequently, the positive effects of species richness on mineralomass were mostly attributable to facilitation among plants.

Conclusions/Significance: Our results provide clear evidence that, due to plant-plant facilitation, species richness positively affects the removal of trace metals from mine tailing soil through phytoextraction and provides further information on diversity conservation and environmental remediation in a mine tailing environment.

Editor: Jin Chen, Key Laboratory of Tropical Forest Ecology, Xishuangbanna Tropical Botanical Garden, Chinese Academy of Sciences, China

Funding: This research was financially supported by the National Natural Science Foundation of China (Project 31000256), the Natural Science Foundation of Zhejiang Province, China (Y13C030015) and the Ecology Key Disciplines of Zhejiang Province in Taizhou University (EKD2013-04). The funders had no role in study design, data collection and analysis, decision to publish, or preparation of the manuscript.

Competing Interests: The authors have declared that no competing interests exist.

* E-mail: gyuan@bren.ucsb.edu

Introduction

Research into understanding how species diversity affects ecosystem function has increased over the past two decades [1,2]. Many studies have shown that species richness has positive effects on ecosystem functions and services, particularly in stressful environments [3–9]. Mine tailings are stressful environments characterized by high concentrations of trace metals and low pH, fertility and water retention capacity [10]. Phytoextraction is an environmentally friendly and inexpensive technique for mine tailing rehabilitation that uses metallophyte plants. They accumulate trace metals through their roots, which reduces trace metal content in the soil to environmentally accepted levels [11]. Li et al [12] found that increased algal species richness led to a rise in cadmium removal in an algal microcosm experiment. However, surprisingly few studies have investigated whether species-rich plant communities remove soil trace metals more efficiently than species-poor communities on mine tailing sites.

Previous studies have suggested that the positive effects of species richness on ecosystem functions generally arise through three primary mechanisms: selection effect, niche complementarity and facilitation [13]. Selection effect refers to the increased probability that a randomly assembled species-rich community will contain a species or combinations of species that have the higher than average biomass trait [14]. Niche complementarity and facilitation affect the collective performance of a community through either resource partitioning or positive interactions between species [15]. The two latter mechanisms are called the niche complementarity effect [16]. The selection and complementarity effects are not exclusive and operate simultaneously to bring about changes in ecosystem functions [17]. The principal mechanism behind the positive effects of biodiversity on ecosystem function has been actively debated [18–20].

Severely polluted environments, such as mine tailing sites, put ecosystems under constant stress [21]. The central role of facilitation among higher plants in maintaining ecosystem functions in stressful environments has been widely recognized [22–25]. However, the importance of this facilitative role may depend, to a large degree, on the characteristics of the species being tested, the stress intensity and the ecosystem function measure being considered [23,26–27]. Moreover, recent studies suggest that the role of facilitation may actually decrease in exceptionally severe environments [28–29]. So far, the role played by facilitation in plant communities growing on mine tailings and the underlying mechanism driving it remains unknown.

In a recent study, we found that more diverse natural plant communities led to lower trace metal (Cu, Cd, Pb and Zn) concentrations in mine tailing soil [30]. The results suggested that species-rich plant communities would be more efficient at removing toxic trace metals from mine tailing soils than those with fewer species. Although natural plant communities are more realistic, they hardly allow identifing why the species-rich plant communities having high efficiency as natural plant communities have non-uniform initial conditions, such as different seed pools, soil trace metal concentrations and other soil characteristics. Compared to the natural plant communities, experimental plant communities are more suitable for exploring the underlying mechanism. In this study, we combined both approaches. We aimed to determine: (1) whether plant species richness has a positive effect on the removal of trace metals in mine tailing soil; (2) if a positive effect exists, whether it is mainly attributable to facilitation among plants and (3) what is the mechanism driving the facilitation effects.

Materials and Methods

Ethics Statement

No specific permissions were required for this research. The natural plant community experiment was conducted on Huangyan Pb/Zn mine tailings. The exploitation and processing of minerals from the Huangyan Pb/Zn mine stopped about 20 years ago and now plants spontaneously grow on the mine tailings. Consequently, this site is open and no permissions were required to conduct the experiment. The experimental plant community experiment was located at Taizhou University. Some of the authors of the manuscript are Taizhou University staff and the site used for the study was at the authors' experimental research center. Consequently, no specific permissions were needed to conduct the experimental trial on this site. The plant species used in this study were not endangered or protected species.

Natural Plant Communities

A plant species survey was conducted at the Huangyan Pb/Zn mine tailing site (28°34′23″N, 120°53′44″E) on April 14, 2008 (Table S1 in File S1). Based on the natural species composition data, we designed four species richness levels (1, 2, 4 and 8) for the natural plant communities on the Huangyan Pb/Zn mine tailings. For each species richness level, five plots (1 m×1 m) with different species compositions were randomly selected (Table 1). In each plot, the aboveground and belowground plant material was collected and sorted into species. The plant materials were thoroughly washed with running tap water and rinsed with distilled water to remove any soil particles attached to the plant surfaces. The aboveground and belowground plant materials were separated and oven dried (80°C) to a constant weight and then weighed.

In each plot, five soil samples were randomly collected using a soil core sampler (64 mm in diameter×100 mm in length). Each soil sample was collected at a depth of 0–20 cm, air-dried at room temperature, homogenized and sieved (<2 mm). They were then used to determine the soil trace metal concentrations (Cu, Cd, Pb and Zn) and total N. After digestion with HCl and $HClO_4$ (4:1, v/v) [31], the soil trace metal concentrations were measured using an Inductively Coupled Plasma Optical Emission Spectrometer (ICP/OES, Optima 2100DV, Perkin Elmer, USA). Total N in the filtrates was determined using the Berthelot reaction method [32]. Briefly, 1.0 g of K_2SO_4 catalyst mixture and 5 ml of concentrated H_2SO_4 were added to 0.5 g of air-dried ground tailings in 10 ml digestion tubes. Exactly 20 ml of distilled water was then added to the tubes, which were then filtered. After full digestion, the filtrate and the contents of the tubes were transferred to 50 ml volumetric flasks for the Berthelot reaction.

Table 1. The species composition of each plot at the four different species richness levels for the natural communities growing on the Huangyan Pb/Zn mine tailings.

Species richness			
1	2	4	8
MF	MF+RB	LC+RB+RP+SC	GP+PA+XS+CC+MF+BP+SC+RP
RB	RP+SN	SC+RL+XS+MF	VH+LP+DL+RP+MF+XS+LF+RL
LC	MF+WJ	SC+MF+CP+LC	SC+RL+XS+RP+LF+RC+MF+RB
XS	GP+RB	LC+MF+GP+LF	XS+RP+LJ+AE+LC+MF+RB+CA
DL	SC+GP	MF+RL+GP+XS	SC+LC+WJ+VH+RP+MF+LF+RB

Footnotes: Species abbreviations and their corresponding full names are listed in Table S1.

Experimental Plant Communities

In order to construct the parallel experimental plant communities, eight species (*Bidens pilosa* Linn, *Phytolacca americana* Linn, *Commelina communis* Linn, *Mirabilis jalapa* Linn, *Chenopodium ambrosioides* Linn, *Solanum nigrum* Linn, *Brassica campestris* Linn and *Xanthium sibiricum* Patrin ex Widder) were randomly selected from the species pool of plants found growing on Huangyan Pb/Zn mine tailings (Table S1 in File S1). *P. americana* is a perennial herbaceous plant and the other seven species are annual herbaceous plants. Seeds from the eight species were collected from the plants growing on the Huangyan Pb/Zn mine tailings, then sown in trays between the 5th and 7th April, 2009 and transplanted about two months after germination. Mine tailing soil samples were randomly collected from the Huangyan Pb/Zn mine tailings and then mixed together to make a composite sample (Tailing properties: organic matter, 330±71.2 $mg\cdot kg^{-1}$; total P, 50.3±10.6 $mg\cdot kg^{-1}$; total N, 120.7±41.4 $mg\cdot kg^{-1}$; Cu, 31.29±3.81 $mg\cdot kg^{-1}$, Cd, 8.92±2.32 $mg\cdot kg^{-1}$; Pb, 924.57±61.23 $mg\cdot kg^{-1}$ and Zn, 1312.45±52.44 $mg\cdot kg^{-1}$).

The seedlings from the eight species were transplanted into plastic containers (pots, 80×80×60 cm^3) filled with mine tailing soil between June 2nd and 4th, 2009 and placed outdoors. We constructed 23 different communities and created four species richness levels (Table 2): eight monocultures, five polycultures of two species, five polycultures of four species and five polycultures of eight species, which were replicated five times. Thirty-two plant seedlings were transplanted into each pot. The plant density was similar to that of the natural plant communities growing on Huangyan Pb/Zn mine tailings. The species in the polycultures were planted at equal densities. A random selection of species from the total pool was not used because, with a limited number of communities, equal representation of the species at each species richness level was not guaranteed. Instead, the species compositions were chosen to guarantee that every species would be selected at least once at each of the four species richness levels.

Table 2. The species composition of each pot at the four species richness levels for the experimental plant communities.

Species richness			
1	2	4	8
BP	BP+XS	BP+CC+CA+SN	BP+PA+CC+MJ+CA+SN+BC+XS
PA	PA+MJ	PA+MJ+CA+BC	BP+PA+CC+MJ+CA+SN+BC+XS
CC	CC+CA	XS+PA+SN+MJ	BP+PA+CC+MJ+CA+SN+BC+XS
MJ	BP+PA	PA+BP+CC+BC	BP+PA+CC+MJ+CA+SN+BC+XS
CA	SN+BC	BP+CC+MJ+CA	BP+PA+CC+MJ+CA+SN+BC+XS
SN			
BC			
XS			

Footnotes: Species abbreviations and their corresponding full names are listed in Table S1.

The pots were weeded weekly. In addition to natural rainfall, water was added by spraying the pots during dry periods.

The experimental plant communities were harvested on May 10th, 2010. All the aboveground and belowground biomasses (living plants) were sorted into species, dried and weighed separately. Before biomass harvesting, a water treatment experiment was conducted. The water content of each pot was determined 5 days after it received the same amount of water through artificial spraying (no water was added during the final 5 days before harvesting). Five soil cores, to a depth of 0–20 cm, were randomly collected from each pot and then combined into one sample. Soil water content was calculated by comparing the wet and dry soil masses (mass of wet soil − mass of dry soil)/mass of dry soil). A proportion of each soil sample was air-dried at room temperature, homogenized and sieved (<2 mm) in order to determine the soil trace metal concentrations (Cu, Cd, Pb and Zn) and total N. The soil trace metal concentrations and total N were measured using the same method employed in the natural plant communities experiment.

At each species richness level, five surviving plants were chosen to determine the malondialdehyde (MDA) concentration in the leaves and the trace metal (Cu, Cd, Pb and Zn) concentrations in plant tissues. Generally, one surviving plant of each species was selected in each pot. However, because some species were not present in all the pots, two or three surviving plants were selected from one pot to guarantee five surviving plants per species richness level. Three surviving plants of *X. sibiricum* were selected at species richness levels two and four, as only three plants of *X. sibiricum* survived. *B. campestris* was not involved in the MDA and trace metal analyses because it did not survive in most of the pots. MDA concentration is an indicator of lipid peroxidation, which reflects the damage to plant tissues caused by oxidative stress [33]. The uppermost five leaves of each surviving plant were used to determine the MDA concentration by using the method followed by Heath and Packer [34]. The dried plant tissues were ground and mixed fully in order to determine the trace metal concentrations. After digestion with HNO_3 [31], the trace metal concentrations in the plant tissues were measured using an Inductively Coupled Plasma Optical Emission Spectrometer (ICP/OES, Optima 2100DV, Perkin Elmer, USA). The biomass used to measure the MDA was included in the plant biomass of each pot during data analysis.

Statistical Analysis

The total trace metal content accumulated in the plants (mineralomass) was used to reflect the trace metal removal efficiency by the plants. The mineralomass of a species in one pot was calculated as follows:

$$\text{Mineralomass} = \text{biomass of the species} \times \text{trace metal concentrations in plant tissues}$$

The mineralomass of the plant communities in each pot was the sum of the mineralomasses of all the component species.

Functional diversity indices were represented by the community-weighted mean trait values [35] and the functional diversity Q index [36]. For each species, five plants in monoculture were randomly selected and their root lengths and stem heights were measured with a ruler. The community-weighted mean trait values for each trait were calculated for every pot ($n = 15$) using species trait values and species relative abundances (calculated from the species biomass) following Garnier et al. [35]:

$$CWM = \sum_{i=1}^{n} p_i trait_i$$

where n is the number of species in the pot, p_i is the relative abundance of species i and trait$_i$ is the trait value of species i.

The average values of the root lengths and stem heights for each species were used to calculate the Euclidean distance between species [37]. *B. campestris* disappeared from all the plots so its root lengths and stem heights were not measured. Following the method used by Heemsbergen et al. [38], the trait values of seven species were put into a number of classes (the highest value was defined as the seventh class and the lowest value was defined as the first class). The Euclidean distance was calculated as follows:

$$d_{ij} = \left[\sum_{k=1}^{K} \left(A_{ki} - A_{kj}\right)^2\right]^{1/2}$$

where A_{ki} and A_{kj} are the class number of species i and j for trait k and$K$ is the number of traits measured ($K = 2$).

The functional diversity Q index was calculated as follows:

$$FD_Q = \sum_{i=1}^{S-1} \sum_{j=i+1}^{S} d_{ij} p_i p_j$$

where S is the number of species observed in the plot and p_i and p_j are the relative abundance of species i and j, respectively.

We used the additive partitioning method to quantify the selection and complementarity effects [16]. The complementarity effect and the selection effect were calculated as follows:

$$\Delta Y = N\overline{\Delta RYM} + N\text{cov}(\Delta RY, M)$$

where ΔY is the net diversity effect, $N\overline{\Delta RYM}$ is the complementarity effect and $N\text{cov}(\Delta RY,M)$ is the selection effect. N is the number of species in the polyculture, $\overline{\Delta RY}$is the average change in relative mineralomass for all species in the mixture and $\overline{M}$ is the average mineralomass of the monocultures. The cov($\Delta RY,M$) term is the covariance between the mineralomass of the species grown in monocultures (*M*) and the change in their relative mineralomass when grown in polycultures.

The relative interaction intensity index (RII) was defined as follows [39]:

$$RII = \frac{B_W - B_O}{B_W + B_O}$$

where *Bo* is the mass potentially achieved by the target plant growing in the absence of inter- or intraspecific interactions, i.e. the expected biomass which is the monoculture biomass divided by the species richness of the polycultures. and *Bw* is the mass observed when the target plant was growing with other plants, i.e. the observed biomass of species in the polycultures. This index represents the ratio of the net mass loss/gain due to the interaction relative to the mass affected by only facilitative interaction and competitive interactions. RII has values ranging from −1 to 1, is symmetrical around zero and is negative for competition and positive for facilitation.

The data of soil trace metal concentration in the natural communities were log transformed to meet the assumptions for linear regression. We first used simple linear regression to examine the effect of species richness on soil trace metal concentrations (Cu, Cd, Pb and Zn) in the natural and experimental plant communities, respectively, and to examine the effect of species richness on the RII. Correlation analysis (Spearman coefficient) was used to determine the dependence of the complementarity effect, the selection effect or the MDA concentrations of each species in the experimental plant communities on the actual species richness, and to determine the dependence of biomass on the actual species richness, FD_Q, CWM_{root} or CWM_{stem}. These analyses were performed using the SPSS 11.5 program (SPSS Inc, Chicago, IL, USA) for Windows and significance levels for all analyses were set at $P<0.05$. The multiple stepwise regressions were performed to examine the effects of actual species richness, FD_Q, CWM_{root} and CWM_{stem} on mineralomass in the polycultures. The optimal model was selected with respect to its value of AIC (Akaike Information Criterion). The model with the lowest value of AIC was selected as the optimal model. For the selected optimal model, the normality of residuals (Shapiro-Wilk test), homosedasticity (Breusch-Pagan test) and multicollinearity among predictors (Variance Inflation Factors) were tested to check for multiple regressions preliminary assumptions. The multiple stepwise regressions were performed using R 2.11.1.

Results

In the natural plant communities growing on the Huangyan Pb/Zn mine tailings, soil trace metal (Cu, Cd, Pb and Zn)

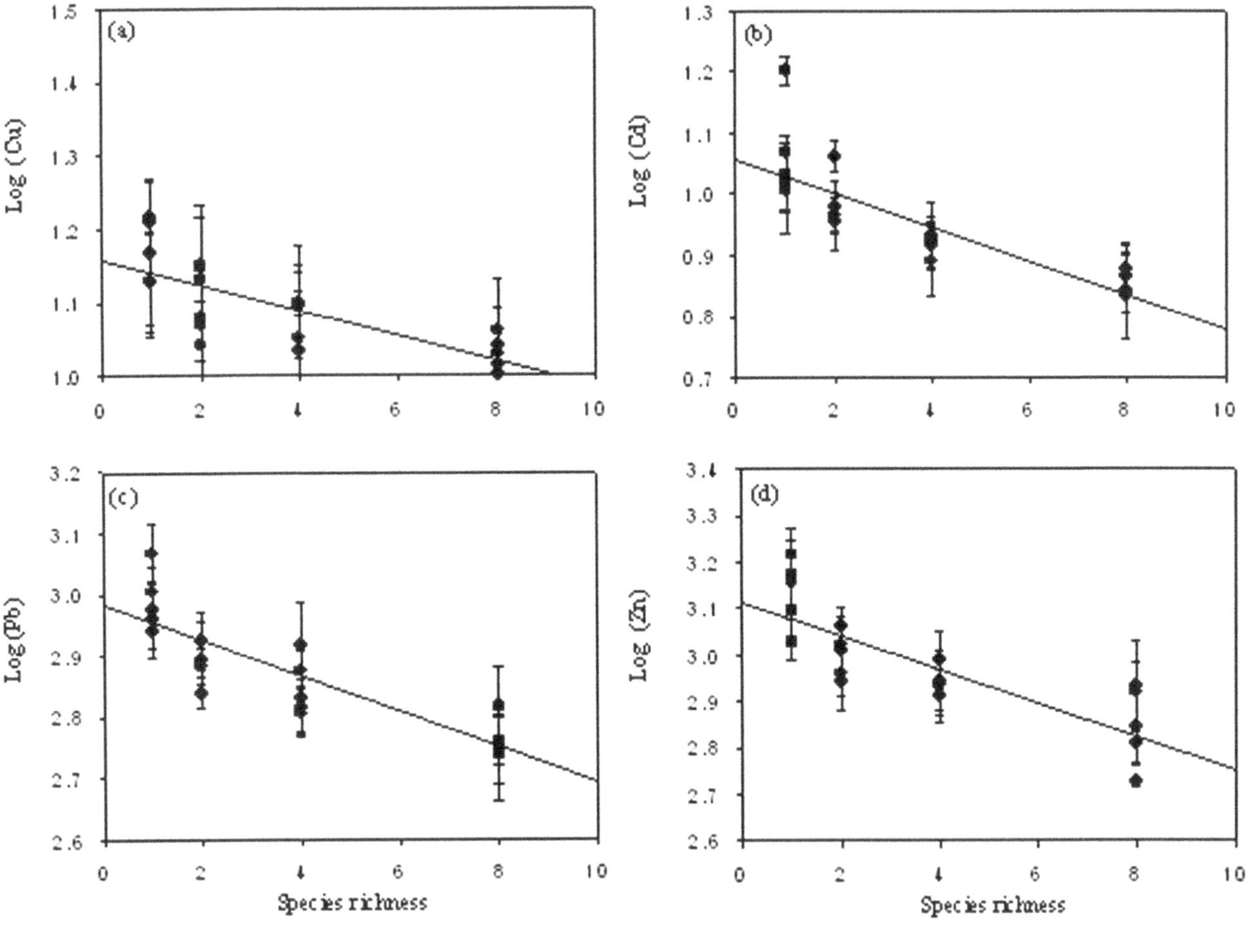

Figure 1. The dependence of soil trace metal concentration on the actual species richness (SR) in the natural plant communities. (a) Fitted line: Log (Cu) = 1.180−0.688 SR, $n = 25$, $P = 0.001$. (b) Fitted line: Log (Cd) = 1.058−0.832 SR, $n = 25$, $P<0.001$. (c) Fitted line: Log (Pb) = 2.981−0.862 SR, $n = 25$, $P<0.001$. (d) Fitted line: Log (Zn) = 3.115−0.815 SR, $n = 25$, $P<0.001$.

concentrations were negatively correlated with species richness (Fig. 1). In the experimental plant communities, even if some plants disappeared in some pots due to the oxidative stress caused by trace metal toxicity, the actual species richness was still significantly and positively correlated with the initial species richness ($r = 0.966$, $n = 23$, $P<0.001$). The trace metal (Cu, Cd, Pb and Zn) concentrations in the soils from the experimental plant communities were negatively correlated with the actual species richness, which was consistent with the results from the natural plant communities (Fig. 2). These results indicated that more diverse plant communities led to lower soil trace metals in the mine tailings.

Actual species richness was always included in the optimal regression model in the multiple stepwise regressions analysis (Table 3), while FD_Q was always excluded. CWM_{root} was included for Cu, Cd and Zn, and CWM_{stem} was included for Cd and Pb. Moreover, actual species richness was positively correlated with the biomass of the polycultures ($r = 0.876$, $n = 15$, $P<0.001$). However, FD_Q ($r = 0.430$, $n = 15$, $P = 0.110$), CWM_{root} ($r = 0.494$, $n = 15$, $P = 0.061$) and CWM_{stem} ($r = 0.485$, $n = 15$, $P = 0.067$) had no significant relationship with biomass. These results suggested that the significant effect of actual species richness on mineralomass was probably closely linked to biomass production.

We also investigated whether the complementarity effect or the selection effect led to the positive effects that an increase in actual species richness had on the mineralomasses of Cu, Cd, Pb and Zn. The selection effect had no significant relationship with actual species richness in the polycultures (Fig. 3). The selection effect values varied, ranging from positive to negative across all the experimental plant communities. In contrast, the complementarity effect significantly rose with the increase in actual species richness in the polycultures, compared to the other cultures. Moreover, the complementarity effect values were positive across all the experimental plant communities. Our results suggested that the positive effects of species richness on trace metal accumulations by plants were mainly attributable to the complementarity effect.

The relative interaction intensity index (RII) was used to reflect interactions among plants. RII was positive in most polycultures (Fig. 4). Moreover, they significantly increased when actual species richness rose. Moreover, with the exception of *B. campestris*, the increase of survival ratio of other seven plant species increased with the actual species richness (Table 4). The results proved that plant-plant facilitation dominated in the polycultures and increased when the actual species richness rose.

The malondialdehyde (MDA) concentration directly reflects the oxidative damage to plant tissues caused by trace metals. In the

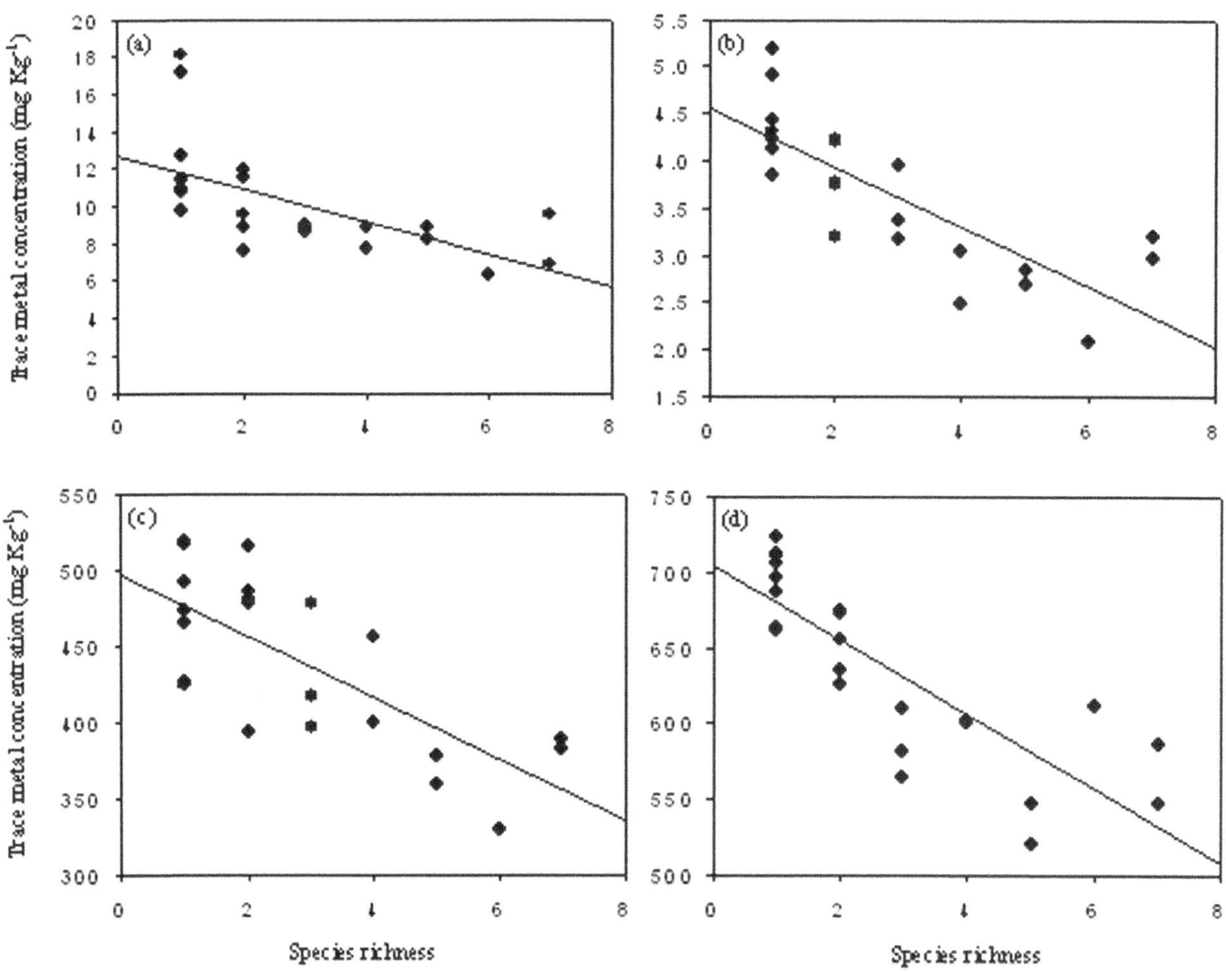

Figure 2. The dependence of soil trace metal concentration (TY) on the actual species richness (SR) in the experimental plant communities. (a) Fitted line: $TY_{Cu} = 12.655 - 0.610$ SR, $n = 23$, $P<0.001$. (b) Fitted line: $TY_{Cd} = 4.547 - 0.796$ SR, $n = 23$, $P<0.001$. (c) Fitted line: $TY_{Pb} = 496.414 - 0.735$ SR, $n = 23$, $P<0.001$. (d) Fitted line $TY_{Zn} = 703.839 - 0.818$ SR, $n = 23$, $P<0.001$.

Table 3. Multiple stepwise regression analyses between mineralomass and explanatory variables from a model that initially contained four variables (actual species richness (SR), FD_Q, CWM_{root} and CWM_{stem} values for Cu, Cd, Pb and Zn).

Mineralomass	Included variables	VIF	AIC	W	BP
$Mineralomass_{Cu}$	SR	1.13	−55.514	0.919(0.188)	3.706(0.157)
	CWM_{root}	1.13			
$Mineralomass_{Cd}$	SR	1.14	−49.798	0.944(0.437)	5.429(0.143)
	CWM_{root}	4.86			
	CWM_{stem}	4.53			
$Mineralomass_{Pb}$	SR	1.13	−52.161	0.912(0.145)	5.902(0.118)
	CWM_{stem}	1.13			
$Mineralomass_{Zn}$	SR	1.13	−52.667	0.903(0.094)	6.032(0.096)
	CWM_{root}	1.13			

Footnotes: the values in the brackets are p-value.

experimental communities, the MDA concentration in the leaves of seven species consistently decreased as actual species richness rose (Table 5). Moreover, the trace metal (Cu, Cd, Pb and Zn) concentrations in the plant tissues consistently increased as actual species richness rose (Figure S1 in File S1). These results suggested that a reduction in oxidative damage to plant tissues probably promoted the accumulation of trace metals.

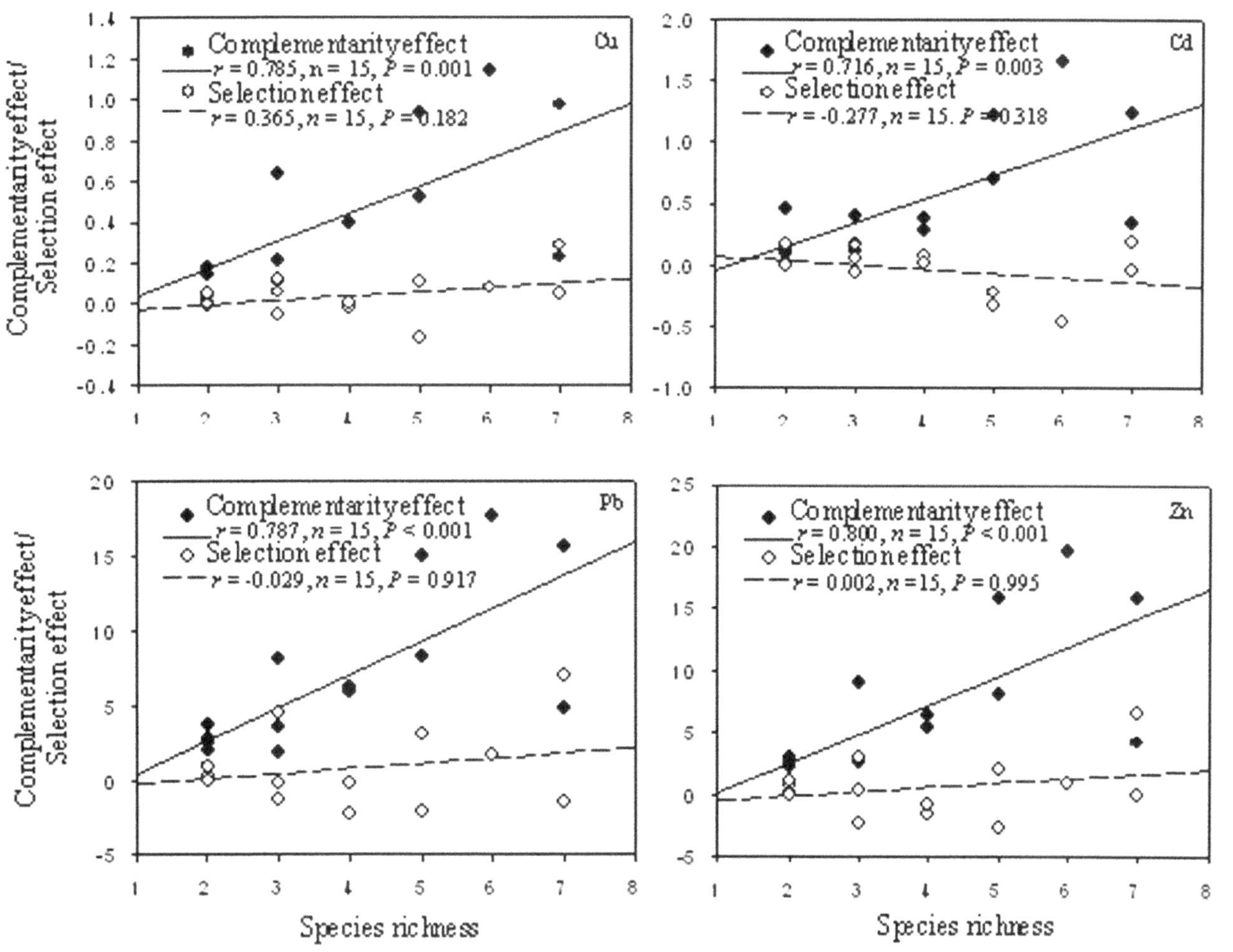

Figure 3. The dependence of the complementarity/selection effect on the actual species richness in the experimental plant communities. a) Cu, b) Cd, c) Pb and d) Zn. The relevant Spearman's correlation coefficient and *P*-value are shown in each panel.

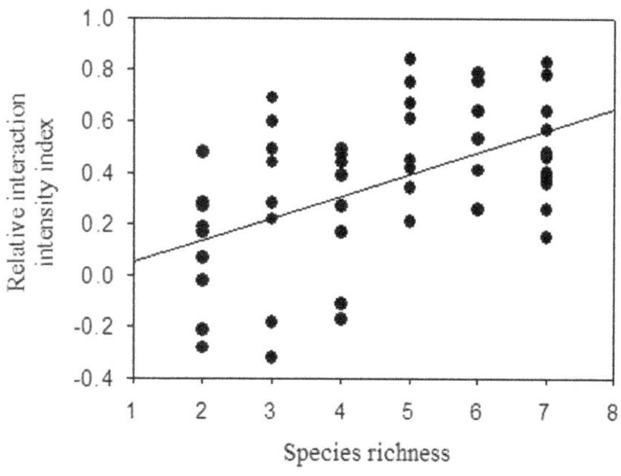

Figure 4. The dependence of relative interaction intensity index (RII) on the actual species richness (SR) in the experimental plant communities. Fitted line: RII = 0.032+0.519 SR, $n = 57$, $P < 0.001$.

Discussion

Our results demonstrated that species-rich plant communities had lower soil trace metal (Cu, Cd, Pb and Zn) concentrations in both the natural and the experimental plant communities than the species-poor communities. More importantly, we found that species richness had significantly positive effects on mineralomass in the experimental plant communities. Consequently, it is almost certain that the positive effects of species richness on mineralomass led to the lower soil trace metal concentrations seen in the more diverse plant communities.

In theory, the observed positive biodiversity effect could be further partitioned into the selection effect and the complementarity effect [16]. Across all the experimental plant communities, the grand mean value of the selection effect was nearly zero and made little contribution to the positive effect. However, the grand mean value of the complementarity effect was significantly positive and positively correlated with actual species richness. Therefore, the positive effect of actual species richness on mineralomass can mainly be attributed to the complementarity effect. Our results suggest that the positive complementarity effect could be attributed to the increase in biomass (see [40] for more details)

Table 4. The survival ratio of each planted species in certain species richness level.

	Planted species richness			
	1	2	4	8
Bidens pilosa Linn	0.531	0.563	0.667	0.800
Phytolacca americana L.	0.156	0.188	0.250	0.500
Commelina communis Linn.	0.438	0.563	0.625	0.900
Mirabilis jalapa L.	0.219	0.438	0.500	0.750
Chenopodium ambrosioides L.	0.344	0.563	0.500	0.700
Solanum nigrum L.	0.156	0.500	0.563	0.650
Brassica campestris L.	0.125	0.125	0.063	0.100
Xanthium sibiricum Patrin ex Widder	0.250	0.188	0.500	0.550

Survival ratio = the survival number of plants in a certain species richness level/ the total planted number in the corresponding species richness level.

Table 5. The correlations between malondialdehyde (MDA) concentration for each species and the actual species richness in the experimental communities.

Species	*r*	*n*	*P*
BP	−0.900	20	$P<0.001$
PA	−0.660	20	$P=0.002$
CC	−0.886	20	$P<0.001$
MJ	−0.829	20	$P<0.001$
CA	−0.744	20	$P<0.001$
SN	−0.935	20	$P<0.001$
XS	−0.892	16	$P<0.001$

Footnotes: Species abbreviations and their corresponding full names are listed in Table S1.

and the increases in trace metal concentrations in plant tissues (Figure S1 in File S1) could be due to the increase in actual species richness.

Distinguishing the effects of niche differentiation and facilitation is very difficult in practice [16]. Niche complementarity or facilitation may have important roles in driving the positive complementarity effect seen in the polycultures. Niche differentiation directly linked to interspecific trait differences that are determined by the functional diversity of communities [41–42]. In this study, the pot experimental design limited environmental fluctuations and spatial heterogeneity, which are important for resource partitioning between species [17]. Consequently, in this study, functional diversity probably made only a limited contribution to the increased mineralomass. However, environmental fluctuations and spatial heterogeneity are ubiquitous in natural conditions [17]. Consequently, we speculate that niche complementarity may provide more contribution to the positive effect of species richness on mineralomass in natural communities of mine tailings. In the experimental communities, most of species in the polycultures had positive RII values; Most of the polycultures had a larger biomass than their most productive component species in the monocultures (see [38] for more details); Most of plant species (with the exception of *B. campestris*) had higher survival ratio in the polycultures. All of the above results proved plant-plant facilitation dominated in the polycultures, as overyielding [43] and the increase of survival ratio [44] are strong indicators of facilitation. Moreover, the yield and survival ratio increased with the actual species richness. Consequently, the positive complementarity effect seen in the polycultures should be ascribed to the facilitation.

The plants used in the experimental communities are all the established species of the Huangyan Pb/Zn mine tailing. Moreover, their seeds were also collected from the Huangyan Pb/Zn mine tailings. Consequently, the plants used should have high tolerance to the stress of mine tailing. Although mine tailings are characterized by high concentrations of trace metals and low pH, fertility and water retention capacities [10,30] and mechanically, physically, chemically and biologically restrict plant growth [10,30], the experimental mine tailing environment may not exert the exceptionally severe stress for these plants. Choler et al. [45] also found, due to lack the experiment at the altitude of >3000 m where should be exceptionally severe for nival communities, the facilitation of the nival communities consistent increased with altitude. Consequently, the role of facilitation decreasing in exceptionally severe environments [28–29] was not seen in this experiment.

In this study, the main driving mechanism behind the facilitative processes could be the amelioration of mine tailing soil by the collective performance of all plant species, as soil total N and soil water increased as species richness rose in the experimental communities (see [40] for more details). Plants can improve the soil physical, chemical and biological properties of mine tailings by preventing erosion, adding nutrients and promoting microbial growth and activity [46]. The species-rich experimental communities showed increased biomass production and surface area coverage (data not shown), which likely decreased light penetration and water evaporation and enhance water retention capacity. The root biomass increased as species richness rose (see [40] for more details), which likely increased soil nitrogen due to root exudates. Additionally, soil microbial biomass and activity increased as species richness rose (data not shown), which likely accelerated the decomposition of plant litter and increase the soil nitrogen levels.

Trace metals stress leads to the production of free radicals in plants, which causes lipid peroxidation and tissue damage [47]. MDA is a cytotoxic product of lipid peroxidation [48]. The ameliorated soil properties probably reduced the trace metal stresses, which led to a decrease in MDA concentrations in the plant tissues. Consequently, as actual species richness increased, the plant stress caused by trace metals declined, which may be the reason why trace metal concentrations in plant tissues increased as actual species richness rose.

Taken together, our results provide clear evidence that increased species richness in plant communities exerts a positive effect on soil trace metal removal through increased facilitation among plants present. In comparison to conventional techniques for mine tailing remediation, phytoextraction is cost-effective, safe for humans and environmentally friendly [10,11]. However, the factors limiting the wider application of this technology involve its relatively low trace metal removal efficiency [49] and the small biomass of most of the hyper-accumulators identified so far [50–52]. Our results indicated that diverse plant communities can promote biomass production and trace metal accumulation by plants. However, the selection of species to be used in the construction of plant communities for phytoremediation should be prudent. In this study, *B. campestris* was sensitive to the mine tailing environment and the facilitative effect was not beneficial. Although similar results were found in our natural and experimental plant communities, future studies should be expanded to include realistic plant communities established on different mine tailings in order to determine whether the implications of our results can be extrapolated to realistic phytoremediation programs for mine tailings.

Supporting Information

File S1 This includes Table S1 and Figure S1. Table S1 The name of species identified in the species survey of Huangyan Pb/Zn mine tailing. Figure S1 The heavy metal concentrations in plant tissues of seven species in pots with different actual sepcies richness. a) Cu, b) Cd, c) Pb and d) Zn. Species abbreviations are seen in Table S1. The lines shown in each panel are the regression lines (All the P values of Cu, Cd, Pb and Zn for seven species are lower than 0.05).

Acknowledgments

We thank Xiu Bai and Xing Luo for help in the field and two anonymous referees for their helpful comments.

Author Contributions

Conceived and designed the experiments: JW, YG. Performed the experiments: TC, YB, BYQ, CBZ. Analyzed the data: JW. Wrote the paper: JW, YG.

References

1. Hooper DU, Adair EC, Cardinale BJ, Byrnes JEK, Hungate BA, et al. (2012) A global synthesis reveals biodiversity loss as a major driver of ecosystem change. Nature 486: 105–108.
2. Midgley GF (2012) Biodiversity and ecosystem function. Science 335: 174–175.
3. Kennedy TA, Naeem S, Howe KM, Knops JMH, Tilman D, et al. (2002) Biodiversity as a barrier to ecological invasion. Nature 417: 636–638.
4. Cardinale BJ, Srivastava DS, Duffy JE, Wright JP, Downing AL, et al. (2006) Effects of biodiversity on the functioning of trophic groups and ecosystems. Nature 443: 989–992.
5. Cadotte MW, Cardinale BJ, Oakley TH (2008) Evolutionary history and the effect of biodiversity on plant productivity. PNAS, 105: 17012–17017.
6. Kaneryd L, Borrvall C, Bergl S, Curtsdotter A, Eklöf A, et al. (2012) Species-rich ecosystems are vulnerable to cascading extinctions in an increasingly variable world. Ecol Evol 2: 858–872.
7. Lei PF, Scherer-Lorenzen M, Bauhus J (2012) The effect of tree species diversity on fine-root production in a young temperate forest. Oecologia 169: 1105–1115.
8. Maestre FT, Quero JL, Gotelli NJ, Escudero A, Ochoa V, et al. (2012) Plant species richness and ecosystem multifunctionality in global drylands. Science 335: 214–217.
9. Thompson PL, Shurin JB (2012) Regional zooplankton biodiversity provides limited buffering of pond ecosystems against climate change. J Anim Ecol 81: 251–259.
10. Meeinkuirt W, Pokethitiyook P, Kruatrachue M, Tanhan P, Chaiyarat R (2012) Phytostabilization of a Pb-contaminated mine tailing by various tree species in pot and field trial experiments. Int J Phytoremediat 14: 925–938.
11. Novo LAB, Covelo EF, González L (2013): Phytoremediation of amended copper mine tailings with Brassica juncea, Int J Min, Reclamat & Environ 27: 215–226.
12. Li SP, Li JT, Kuang JL, Duan HN, Zeng Y, et al. (2012) Effects of species richness on cadmium removal efficiencies of algal microcosms. J Appl Ecol 49: 261–267.
13. Hector A, Bazeley-White E, Loreau M, Otway S, Schmid B (2002) Overyielding in grassland communities: testing the sampling effect hypothesis with replicated biodiversity experiments. Ecol Lett 5: 502–511.
14. Cardinale BJ, Gross K, Fritschie K, Flombaum P, Fox JW, et al. (2013) Biodiversity simultaneously enhances the production and stability of community biomass, but the effects are independent. Ecology. In press.
15. Tilman D, Lehman C (2001) Biodiversity, composition, and ecosystem processes: theory and concepts. In: Kinzig AP, Pacala SW, Tilman D, editors. The functional consequences of biodiversity. Princeton: Princeton University Press. pp. 9–41.
16. Loreau M, Hector A (2001) Partitioning selection and complementarity in biodiversity experiments. Nature 412: 72–76.
17. Špaèková I, Lepš J (2001) Procedure for separating the selection effect from other effects in diversity–productivity relationship. Ecol Lett 4: 585–594.
18. Huston MA, Aarssen LW, Austin MP (2000) No consistent effect of plant diversity on productivity. Science 289: 1255a.
19. Naeem S (2000) Reply to Wardle et al. Bull Ecol Soc Am 81: 241–246.
20. Fargione J, Tilman D, Dybzinski R, Lambers JHR, Clark C, et al. (2007) From selection to complementarity: shifts in the causes of biodiversity–productivity relationships in a long-term biodiversity experiment. Proc R Soc B 274: 871–876.
21. Kondratyev KY, Krapivin VF, Phillips GW (2002) Global Environmental Change: Modeling and Monitoring. Springer, Berlin.
22. Baumeister D, Callaway RM (2006) Facilitative effects of *Pinus flexilis* during succession: a hierarchy of mechanisms benefits other plant species. Ecology 87: 1816–1830.
23. Maestre FT, Bowkerl MA, Escolar C, Puche MD, Soliveres S, et al. (2010) Do biotic interactions modulate ecosystem functioning along stress gradients? Insights from semi-arid plant and biological soil crust communities. Phil Trans R Soc B 365: 2057–2070.
24. Odadi WO, Karachi MK, Abdulrazak SA, Young TP (2011) African wild ungulates compete with or facilitate cattle depending on season. Science 333: 1753–1755.
25. He Q, Bertness MD, Altieri AH (2013) Global shifts towards positive species interactions with increasing environmental stress. Ecol Lett 16: 695–706.
26. Maestre FT, Valladares F, Reynolds JF (2005) Is the change of plant-plant interactions with abiotic stress predictable? A meta-analysis of field results in arid environments. J Ecol 93: 748–757.

27. Kawai T, Tokeshi M. (2007) Testing the facilitation-competition paradigm under the stress-gradient hypothesis: decoupling multiple stress factors. Proc Royal Soc B 274: 2503–2508.
28. Bruno JF, Stachowicz JJ, Bertness MD (2003) Inclusion of facilitation into ecological theory. Trends Ecol Evol 18: 119–125.
29. Michalet R, Brooker RW, Cavieres LA, Kikvidze Z, Lortie CJ, et al. (2006) Do biotic interactions shape both sides of the humped-back model of species richness in plant communities? Ecol Lett 9: 767–773.
30. Wang J, Zhang CB, Ke SS, Qian BY (2011) Different spontaneous plant communities in Sanmen Pb/Zn mine tailing and their effects on mine tailing physico-chemical properties. Environ Earth Sci 62: 779–786.
31. Mcgrath SP, Cunliffe CH (1985) A simplified method for the extraction of the metals Fe, Zn, Cu, Ni, Cd, Pb Cr, Co and Mn from soils and sewage sludges. J Sci food Agr 36: 794–798.
32. Page AL, Miller RH, Keeney DR (1982) Methods of Soil Analysis, Chemical and Microbiological Properties. Wisconsin: Madison.
33. Dhir B, Sharmila P, Saradhi PP (2004) Hydrophytes lack potential to exhibit cadmium stress induced enhancement in lipid peroxidation and accumulation of praline. Aquat Toxicol 66: 141–147.
34. Heath RL, Packer L (1968) Photoperoxidation in isolated chloroplasts 1 kinetics and stoichiometry of fatty acid peroxidation. Arch Biochem Biophys 125: 189–198.
35. Garnier E, Cortez J, Billès G, Navas M, Roumet C, et al. (2004). Plant functional markers capture ecosystem properties during secondary succession. Ecology, 85: 2630–2637.
36. Rao CR (1982) Diversity and dissimilarity coefficients – a unified approach. Theor Popul Biol 21: 24–43.
37. Walker B, Kinzig A, Langridge J (1999) Plant attribute diversity, resilience, and ecosystem function: the nature and significance of dominant and minor species. Ecosystems 2: 95–113.
38. Heemsbergen DA, Berg MP, Loreau M, van Hal JR, Faber JH, et al. (2004) Biodiversity effects on soil processes explained by interspecific functional dissimilarity. Science 306: 1019–1020.
39. Armas C, Ordiales R, Pugnaire FI (2004) Measuring plant interactions: anew comparative index. Ecology 85: 2682–2686.
40. Wang J, Zhang CB, Chen T, Li WH (2013) From selection to complementarity: the shift along the abiotic stress gradient in a controlled biodiversity experiment. Oecologia 171: 227–235.
41. Petchey OL, Gaston KJ (2002) Functional diversity (FD), species richness and community composition. Ecol Lett 5: 402–411.
42. Petchey OL, Gaston KJ (2006) Functional diversity: back to basics and looking forward. Ecol Lett 9: 741–758.
43. Drake JM (2003) Why does grassland productivity increase with species richness? Disentangling species richness and composition with tests for overyielding and superyielding in biodiversity experiments. PNAS 270: 1713–1719.
44. Mulder CPH, Uliassi DD, Doak DF (2001) Physical stress and diversity-productivity relationships: the role of positive interactions. Proc Natl Acad Sci USA 98: 6704–6708.
45. Choler P, Michalet R, Callaway RM (2001) Facilitation and competition on gradients in alpine plant communities. Ecology 82: 3295–3308.
46. Gómez-Sagasti M, Alkorta I, Becerril JM, Epelde L, Anza M, et al. (2012) Microbial Monitoring of the Recovery of Soil Quality During Heavy Metal Phytoremediation. Water Air Soil Pollut (2012) 223:3249–3262.
47. Šuran J, Prišć M, Rašić D, Srebočan E, Crnić A P (2013) Malondialdehyde and heavy metal concentrations in tissues of wild boar (*Sus scrofa* L.) from central Croatia. J Environ Sci Heal Part B 48: 147–152.
48. Choudhary M, Jetley UK, Khan MA, Zutshi S, Fatma T (2007) Effect of heavy metal stress on proline, malondialdehyde, and superoxide dismutase activity in the cyanobacterium Spirulina platensis-S5. Ecotoxicology and Environmental Safety 66: 204–209.
49. Cooke SJ, Suski CD (2008) Ecological restoration and physiology: an overdue integration. BioScience 58: 957–968.
50. Parraga-Aguado I, Gonzalez-Alcaraz MN, Alvarez-Rogel J, Jimenez-Carceles FJ, Conesa HM (2013) The importance of edaphic niches and pioneer plant species succession for the phytomanagement of mine tailings. Environ Pollut 176: 134–143.
51. Zhang SJ, Li TX, Huang HG, Zou TJ, Zhang XZ, et al. (2012) Cd accumulation and phytostabilization potential of dominant plants surrounding mining tailings. Environ Sci Pollut Res 19: 3879–3888.
52. Sheng XF, He LY, Wang QY, Ye HS, Jiang CY (2008) Effects of inoculation of biosurfactant-producing *Bacillus* sp. J119 on plant growth and cadmium uptake in cadmium-amended soil. J Hazard Mater 155: 17–22.

16

Using a Multi-Trait Approach to Manipulate Plant Functional Diversity in a Biodiversity-Ecosystem Function Experiment

Conrad Schittko*, Mahmoud Hawa, Susanne Wurst

Freie Universität Berlin, Functional Biodiversity, Dahlem Center of Plant Sciences, Berlin, Germany

Abstract

A frequent pattern emerging from biodiversity-ecosystem function studies is that functional group richness enhances ecosystem functions such as primary productivity. However, the manipulation of functional group richness goes along with major disadvantages like the transformation of functional trait data into categories or the exclusion of functional differences between organisms in the same group. In a mesocosm study we manipulated plant functional diversity based on the multi-trait Functional Diversity (FD)-approach of Petchey and Gaston by using database data of seven functional traits and information on the origin of the species in terms of being native or exotic. Along a gradient ranging from low to high FD we planted 40 randomly selected eight-species mixtures under controlled conditions. We found a significant positive linear correlation of FD with aboveground productivity and a negative correlation with invasibility of the plant communities. Based on community-weighted mean calculations for each functional trait, we figured out that the traits N-fixation and species origin, i.e. being native or exotic, played the most important role for community productivity. Our results suggest that the identification of the impact of functional trait diversity and the relative contributions of relevant traits is essential for a mechanistic understanding of the role of biodiversity for ecosystem functions such as aboveground biomass production and resistance against invasion.

Editor: Andrew Hector, University of Oxford, United Kingdom

Funding: Funded by Freie Universität Berlin. The funders had no role in study design, data collection and analysis, decision to publish, or preparation of the manuscript.

Competing Interests: The authors have declared that no competing interests exist.

* E-mail: conrad.schittko@fu-berlin.de

Introduction

Biodiversity is declining rapidly [1], which may negatively affect ecosystem functions [2] and economically important ecosystem services at the same time [3]. The role of plant functional diversity on ecosystem properties has been a research focus in recent years [4–6]. Several different experiments have shown that positive diversity effects on plant community biomass production mostly result from niche complementarity in combination with a species-specific sampling effect [7–10]. The number of functional types or groups (e.g. C_4 grasses, C_3 grasses, N-fixing plants, and non-N-fixing plants) represented by the species in a local community is a common measure of functional diversity that is often used to manipulate functional diversity in experimental set-ups [4–6,8]. However, the manipulation of functional group richness in biodiversity-ecosystem function experiments goes along with major disadvantages like the exclusion of functional differences that occur between organisms in the same group [11]. Further, the assumption that ecosystem processes are governed by the abundance and distribution of functional traits in a community [5,12,13] has attracted growing attention on the use of functional trait composition, rather than species or functional group richness, in the explanation of biodiversity-ecosystem functioning relationships.

The effects of functional traits on ecosystem properties are quantified by two conceptually different approaches. On the one hand, community-weighted means of trait values (CWM) are commonly computed as mean trait values weighted by the species relative abundances in a given community [14,15]. This single trait index is interpreted as a trait value expected from a random sample of the community and is theoretically consistent with Grime's [16] mass-ratio hypothesis. On the other hand, a number of new multi-trait approaches such as functional attribute diversity (FAD), functional diversity (FD) or Rao's Q has been introduced which assess functional diversity of a community by quantifying the distances of trait values of species in a multi-dimensional trait space [11,17]. The concept of functional trait diversity is based on the assumption that greater differences between species' trait values represent greater trait complementarity and larger functional diversity. Since the introduction of the continuous multi-trait Functional Diversity (FD) index by Petchey & Gaston [11,18] it has been shown that FD explains variation in ecosystem functions, most notably plant aboveground productivity of grassland species [19,20]. However, multi-trait approaches were solely used for measuring functional diversity either on experimental communities that were compiled by another approach (e.g. species richness or functional group richness) or on natural communities.

So far, no manipulative study has used trait-based functional diversity measures to create different functional diversities during the composition of experimental communities; thus knowledge about their practicability in manipulating functional diversity in experimental set-ups is still missing. Several properties of these measures predict that they will explain greater variance in ecosystem function than either species richness or functional group richness [21]. And, already seven years ago Petchey & Gaston [11] requested experiments that manipulate trait distributions in local assemblages while species richness is held constant, mainly due to their suggested advantage of eliminating the sampling effect.

In the mesocosm study presented here we used the revised version of the FD index by Petchey & Gaston [11] to generate a gradient of functional diversity among plant mixtures while species richness was kept constant. We used a pool of 20 plant species from the urban European flora as our model plant community. Our interest in urban floras is attributed to the fact that cities are remarkably rich in species – usually species richness is greater than in the surrounding areas – and may harbour one of the hot spots of biodiversity [22]. As a novel aspect of this study we treated the attribute of a plant species to be alien or native (hereafter called the 'floristic status' after Klotz et al. [23]) as an additional trait that characterizes a species such as life-history or resource capture traits do. We assume that species with different biogeographical histories differ in the number of associations they form with their biotic environment and that this will not only affect the species' own performance but also community-wide ecosystem properties like traditionally used functional traits do. We are aware that the floristic status does not fulfill the definition of a functional trait in its narrowest sense as being any morphological, physiological or phenological plant feature [24]. However, in case of considering a functional trait as a characteristic that is relevant to ecosystem functioning [5,25,26] or influences the own performance of an organism [12], the addition of the floristic status as additional trait to the classical functional traits appears reasonable.

In this study we explored firstly whether a positive relationship between functional diversity and aboveground productivity is detectable in plant mixtures that were compiled on the basis of the FD index. This new approach may give insight into the effectiveness and appliance of trait-based approaches for manipulative studies. Secondly, we calculated community-weighted means (CWM) of the selected plant functional traits to determine the relative importance of each trait in explaining the relationship between functional trait diversity and aboveground productivity. In a second phase of the experiment we moved the focus to the resistance to invasion, which is an ecosystem function that was linked with species diversity and functional diversity since the early days of invasion ecology [27]. Experimental studies following a functional approach showed that functional group identity rather than functional group richness limits the invasion process of alien plant species in resident communities [28,29], which is a finding that is consistent with the sampling effect hypothesis, but in a functional context. In our study, it was the third goal to elucidate the question whether functional trait diversity affects community resistance to invasion. Therefore we added seeds of the exotic Canada goldenrod *Solidago canadensis* to each plant mixture and monitored seed germination and seedling survival.

Methods

In this study we used a single factor regression design with functional diversity being the single continuous factor that was experimentally manipulated. Functional diversity was calculated for all possible eight-species mixtures from a pool of 20 naturally co-occurring plant species, based on the FD-index introduced by Petchey & Gaston [11,18]. This measure of functional diversity is defined as the total branch length of a functional dendrogram, which is created from data of plant functional traits. From the total number of possible eight-species mixtures (125,970), 40 mixtures were randomly selected and planted out for the experiment. This design was chosen to cover a broad range of functional diversity (from low FD to high FD).

Species pool and functional traits

Plant species selection for the species pool of the experiment was conducted with the purpose of reflecting the composition of a ruderal plant community, which is a typical attribute of temporary abandoned fields and disturbed habitats in urban areas. Specific abiotic conditions and the proximity to human settlements and transport connections support the establishments of non-native plant species in these areas [30]. The species pool contained 20 angiosperm species (13 natives and 7 non-natives), of which 15 were forbs, 3 grasses and 2 tree species (see Table S1). The major part of the plant seeds we used was collected on an urban abandoned site in Berlin (N52°28′30", E13°21′46"). This site was open to the public and it was no permission required to enter that site or to collect plant seeds from there. Regarding to the red list of endangered vascular plant species of Berlin [31] none of our experimental plant species were endangered or protected. A minor part of seeds was provided by Appels Wilde Samen (Darmstadt, Germany) or by the Botanical Garden Berlin-Dahlem. To classify each plant species functionally we used data of seven functional traits (four continuous and three categorical) plus the species' floristic status. We aimed to include traits that are related to plant productivity, resource use, coexistence and competitive strength. The categorical traits were: life form, leaf form and resource use strategy (i.e. root colonization with N-fixing bacteria and/or arbuscular mycorrhizal fungi, AMF). The continuous traits were: specific leaf area (SLA), canopy height, seed mass, leaf dry matter content and floristic status. Trait values for each trait were exclusively obtained from trait-databases (BiolFlor [23] and LEDA [32]) or in some cases of unavailability from approved publications. The trait floristic status defines whether a plant is native or alien to Germany. For the specification of the floristic status we followed the classification from the BiolFlor database [23], which is based on the classification system by Schroeder [33]. This classification acts exclusively on a temporal basis excluding information about the evolutionary history, origin or mode of introduction. Alien species are classified according to their time of human introduction: before the discovery of America (archaeophytes, trait value '2' in our data table) and after the discovery of America in 1492 (neophytes, trait value '1'). Species that colonized the focus area after the end of the last glacial period (about 12,500 years ago) without human assistance are classified as indigenous or native (trait value '3'). Our approach is backed on the aspect that native species have longer evolutionary histories with other local organisms than species that established later in the course of time (see Text S1 for additional information).

Calculation of Functional Diversity (FD)

The calculations of the FD index for each possible eight-species mixture (125,970 in total) were performed by the R-based software FDiversity [34]. After calculating the FD values we ended up with a range of FD between 1.69 (lowest value) and 4.08 (highest value, see Text S2 for additional information on the calculation). From the 125,970 possible mixtures we randomly chose 40 mixtures with the Excel function 'randbetween' (Microsoft Excel 2007).

Experimental set-up and harvest

In November 2011 we started the experimental part of the study: We used 40 plastic pots (7.5 l volume, 21 cm height, 26 cm upper diameter and 19 cm lower diameter) and filled them with 7 kg soil each. The soil was a loamy sand (pH 8.1, C/N ratio of 28:1, determined with an elemental analyzer Euro EA, HEKAtech GmbH, Germany) that was collected from an urban abandoned site in Berlin (N52°28′30″, E13°21′46″) and sieved before the experiment (4 mm mesh size) but remained untreated otherwise. Each of our experimental plant species could be found on that site. 16 equally sized seedlings (two individuals per species) that had been grown on sterilized soil were transplanted carefully into each mesocosm according to the drawn species compositions. Each seedling was assigned by chance to one of 16 possible positions in the mesocosms. The mesocosms were distributed within the greenhouse, watered regularly and their position was randomized twice a week. During the whole experiment the pots were kept in the greenhouse with 16 hours light a day (~180 μE m-2 sec-1 PAR) and 22/18°C day/night temperature. Nine weeks after transplanting the seedlings into the mesocosms, plant shoots were harvested. Shoots were cut 3.5 cm above ground level to enable regrowth. Shoot weight was determined after drying at 50°C for three days.

Invasion by *S. canadensis*

To simulate a biological invasion into an established ruderal plant community after a disturbance by cutting we added 100 achenes (hereafter called seeds) of *Solidago canadensis* L. (Asteraceae) evenly to the soil surface of each mesocosm. The Canada goldenrod *S. canadensis* is an invader of North American origin, was introduced in the 18th century in Europe and began to spread in the 19th century in Central Europe where it may become a highly dominant species on abandoned fields and disturbed habitats in urban areas [30]. *S. canadensis* seeds were collected on the same field described above and each seed was checked with a dissecting microscope for its color and size to prevent that they may differ in the ability to germinate. We executed the seed addition after the cutting to prevent that the seeds and seedlings of *S. canadensis* will get strong shading from the established plants. We checked the mesocosms every second day for numbers of newly emerged seeds and survived seedlings until three months after seed addition. Germination and survival rate were calculated per pot.

Data analysis

Aboveground community biomass at the time of harvest was calculated for each eight-species mixture as the sum of biomass of the 16 individuals per mixture. The influence of FD on aboveground community biomass was tested with a linear regression with FD as the predictor and biomass as response variable. Community-weighted means of trait values (CWM) were calculated for each mixture based on species biomass proportions according to the equation $CWM = \sum_{i=1}^{S} p_i \times t_i$ where S is the number of species in the community, p_i is the species biomass proportion and t_i are the given trait values. To elucidate the relative contribution of CWMs of traits to the amount of explained variation in aboveground biomass production we performed multiple regressions considering statistical models with different combinations of included explanatory variables (traits). We used the Akaike information criterion (AIC) and R^2 values to select for the model with the best fit. Afterwards we performed a standardized principle components analysis (PCA) with the CWMs of the traits from the selected model to determine multiple relationships between them [13]. The influence of FD on *S. canadensis* seedling germination was tested with a regression analysis. The regression function of the curve with the best fit to our data was determined by comparing R^2 and AIC values of different regression functions (linear, exponential, logarithmic, power, polynomial). The percentage of survived seedlings was analyzed with a logistic regression model which is an instance of a generalized linear model. The logistic function avoids predictions below and above 0 and 100% and is widely used to analyze percentage data [35–37].

All models were tested for the underlying assumptions (homoscedasticity, normality). All statistical analysis were done in R (version 3.0.1, [38]).

Results

Statistical analysis of aboveground biomass indicated a positive relationship between FD and biomass production (Fig. 1). Altogether, FD explained 32% of the total variation of biomass among the 40 mesocosms.

The model selection procedure led to a model that included the traits N-fixation, floristic status, AMF and SLA and explained 93% of the variation in aboveground biomass (Table 1). N-fixation, AMF and SLA were positive correlated with aboveground biomass whereas the floristic status showed a negative correlation. In terms of floristic status, high biomass proportions of non-native species in regard to community biomass (hence a low CWM for the trait floristic status) were accompanied with a high community biomass production.

The two leading axes of a standardized principle component analysis (PCA) on the four selected traits N-fixation, floristic status, SLA and AMF explained about 80% of variation (Fig. 2). The first principal component accounted for 50.1% and had a high positive loading for floristic status as well as a high negative loading for N-fixation. Linear regression between total community aboveground biomass and the first PCA component showed a strong significant negative correlation ($R^2 = 0.75$, $P<0.001$) between both variables. The second principal component accounted for 29.9% of variation. SLA had a positive and AMF a negative loading on this axis. Linear regression between total community aboveground biomass and the second PCA component showed no significant correlation ($R^2 = 0.01$, $P>0.05$) between both variables.

Invasion of *S. canadensis*

Seedling germination rate of *S. canadensis* in the 40 mixtures is best explained by a polynomial function (Fig. 3A). Between FD values 1.7 (lowest FD) and 3.5 the number of germinated seeds decreased with increasing FD value. From 3.5 to 4.1 (highest FD) the number of germinated seeds increased again, but less steep and without reaching germination rates from the lowest FD values. The seedling survival rate showed a negative linear relationship with increasing FD value (Fig. 3B).

Discussion

A few approaches and tools have been developed for measuring and assessing community's functional diversity based on functional traits which led to our first research question, whether the proposed positive relationship between functional diversity and aboveground productivity is also prevalent when communities were assembled on the basis of the FD index by Petchey & Gaston [11,18]. A strength of our experimental approach is that species richness is held constant which may lead to the result that a positive functional diversity-productivity relationship would be

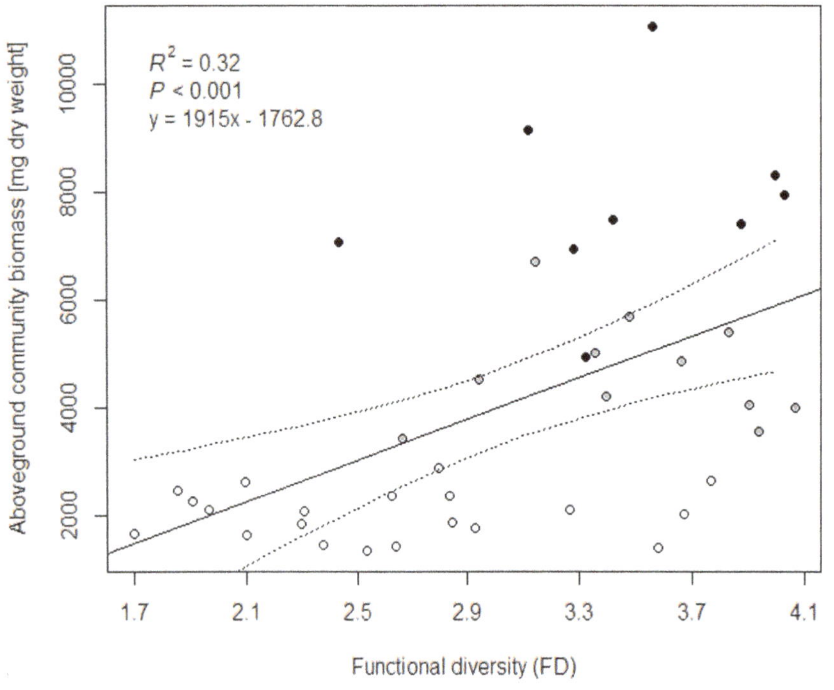

Figure 1. Aboveground community biomass plotted against the FD index of each eight-species mixture. Increasing FD index values represent increasing functional diversity within the mixtures. (intercept = −1762.8, SE = 1577.7, t = −1.1, slope = 1915.0, 95% confidence intervals for slope 889.2–2940.9, SE = 506.8, t = 3.8, n = 40). The regression line is solid with 95% confidence bands (dashed lines). Black points represent mixtures that contained the two most productive species of the experiment (*Melilotus albus* and *Medicago* x *varia*). Grey points represent mixtures that contained either *M. albus* or *M.* x *varia*. Empty points represent mixtures that contained neither of these two species.

explained by species' complementarity only, whereas the sampling effect would have been eliminated, as suggested by Petchey & Gaston [11].

Biomass production is related to functional trait diversity

Our finding that aboveground biomass production is positively related to functional diversity is consistent with the well-known experimental grassland studies like the one carried out at the Cedar creek ecosystem science reserve [4], in California [8], the BIODEPTH sites in Europe [9] or the Jena experiment [10]. By comparing the amount of variation in aboveground biomass production between our small-scale and short-term greenhouse study and these large-scale and long-term field experiments it is notable that functional diversity explained a higher amount of variation in our experiment (32%) than in the other studies (14% at Cedar Creek, 13% at the California experiment, and 15% at the Jena experiment; there is no comparable R^2 value reported from the BIODEPTH experiment), even though Tilman et al. [39] showed that diversity effects strengthen through time. Spaekova & Leps [40] who manipulated plant species richness in a greenhouse pot experiment also found more pronounced effects of diversity in the greenhouse compared to similar field experiments.

However, the biomass production of the 20 plant species was very uneven in our experiment (Fig. 4). Our most productive

Table 1. Summary for the best statistical model based on 4 predictor variables explaining the variation of aboveground community biomass.

Functional trait	*t*-value	*P*-value	Partial R^2
Nitrogen fixation	11.718	<0.001	positive 0.893
Floristic status	2.115	0.0416	negative 0.337
AMF	1.726	0.0933	positive 0.280
SLA	−1.966	0.0573	positive 0.315
Degrees of freedom	35		
Multiple R^2	0.9304		
***P*-value**	<0.001		

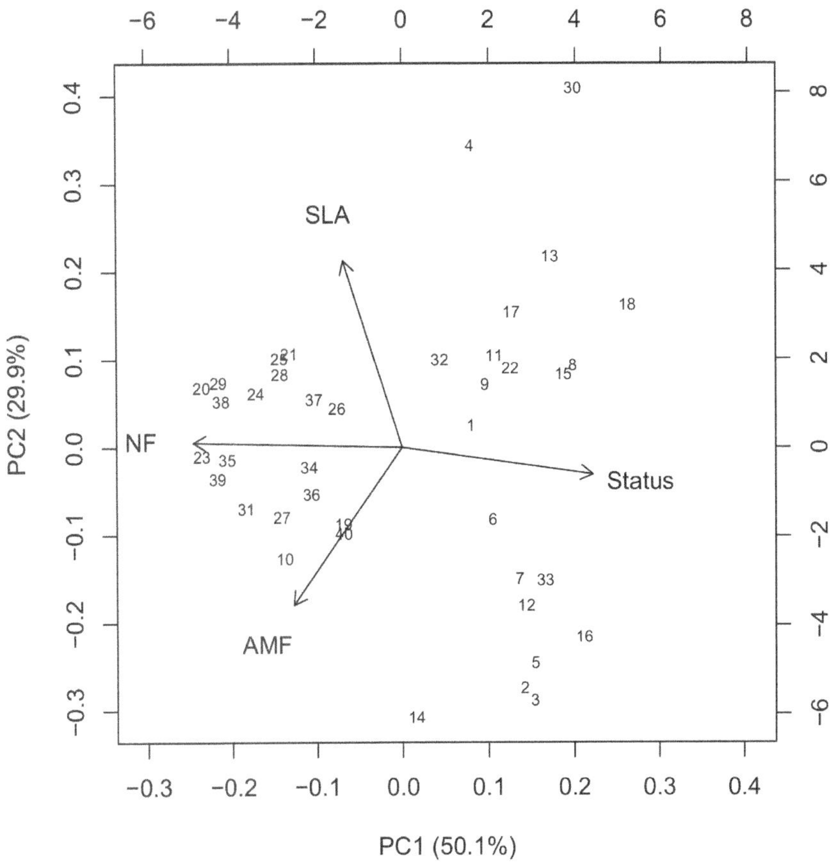

Figure 2. Standardized principal components analysis (PC1 vs. PC2) of the four most important traits. Plotted are the community weighted means (CWM) of the trait values of the functional traits N-fixation (NF), floristic status, specific leaf area (SLA) and arbuscular mycorrhizal fungi (AMF). CWMs of trait values were calculated for each of the 40 mixtures (plotted numbers) based on species biomass proportions.

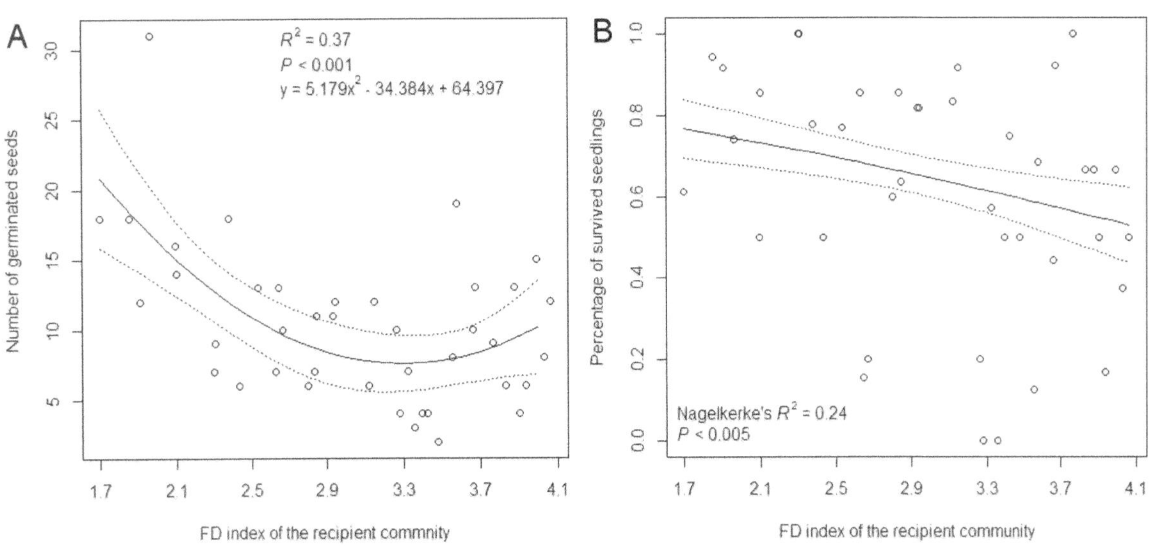

Figure 3. Invasion of *S.canadensis*. (A) Number of germinated seeds of *S. canadensis* from 100 sown seeds in each mesocosm (linear regression, intercept =64.4, SE=14.4, t=4.5, linear coefficient =−34.4, SE=10.1, t=−3.4, quadratic coefficient =5.2, SE=1.7, t=3.1, n=40) and (B) the percentage of survived seedlings plotted against the increasing FD index (logistic regression, intercept =1.9, SE=0.4, z=4.5, slope =−0.5, SE=0.1, z=−3.2, n=40, note that Nagelkerke's R^2 is a pseudo R^2, comparison to standard R^2 from ordinary least square regression is not appropriate). Regression lines are solid with 95% confidence bands (dashed lines).

species (*Melilotus albus*) was on average over 45 times more productive than our least productive species (*Cirsium arvense*). That the two most productive species (*M. albus* and *Medicago* x *varia*) were both legumes is most likely caused by the N-poor soil we used (which had a C/N ratio of 28:1). Although we did not manipulate the composition of the randomly chosen mixtures that were planted out for the experiment, the two most productive species were not distributed evenly over the range of the FD index values. Both species entered the experiment at a FD index value at 2.4 which is the point where the community biomass of mixtures that contained one or both of the two species is markedly increased (Fig. 1, black and grey points). Combining this random compositional effect with our finding that the trait N-fixation played the most important role in explaining aboveground productivity leads to the indication that a sampling effect contributed to the positive relationship between functional diversity and productivity in our experiment. The sampling effect is generally defined as an increased probability of including a species of greatest inherent productivity in a plot that is more diverse. That the sampling effect may not only be important at the species level, but also at the trait level, confirms the suggestion of Lepš [41] that life history strategies of constituent species are more important for ecosystem function than species richness. It is also consistent with a recently conducted trait analysis of the Jena experiment by Roscher et al. [13], where the authors showed that increasing community biomass production was best explained by CWMs and suggested that traits of the dominant species were most important for high productivity. In our case, the differences between high productive and less productive species had such a huge dimension that it becomes obvious that the inherent traits of the dominant species are most influential on community productivity when the traits are separately analyzed via the CWM method. Further, we conducted two separated linear regressions, one that included the mixtures without legumes and a second one with all legume-mixtures (data not shown), and found no relationship between FD and productivity ($P>0.05$) within these subsamples, which is another hint that there may have been a sampling effect present in our experiment. However, having a sampling effect in our set-up does not necessarily mean that there was no complementarity among species present. One common form of complementarity in plant communities (which involves both resource partitioning and facilitation) develops between legumes, which have the ability to fix atmospheric N, and other plants, which have access only to soil nitrogen [42]. The sum of both complementarity and sampling effect is the net biodiversity effect [42], and furthermore it is assumed that both may interact with each other [43]. However, within our set-up, it is not possible to identify their individual contributions to the net biodiversity effect since we had no monocultures in our set-up and cannot apply an additive partitioning method.

Data analysis via CWMs also revealed that the species' floristic status was the second most important trait after N-fixation. In our study the loadings of the traits N-fixation and floristic status are opposite in direction on the first PCA axis which tells that the two CWMs have opposite effects, i.e. N-fixation is positively and the floristic status negatively related to productivity. In the case of floristic status it is unlikely that this effect is driven exclusively by a few high productive species, since the three most productive species of our experiment represent all three trait characteristics (native, archaeophyte and neophyte). Moreover, this result is consistent with the finding of a meta-analysis by Vilà et al. [44] that investigated the ecological impacts of exotic plant species and found that exotics have a general negative impact on native species' productivity, abundance and diversity, but they increased total community production by 56.8%.

Our results also showed that not only weighted means of N-fixing plants and exotic species, but also CWMs of a larger set of traits (including the traits SLA and AMF), explained variation in community biomass. But given the facts that *P*-values for both traits are not significant and the second PCA component (on which both traits have high loadings) is not significantly correlated with community biomass indicates that their influence in explaining variation in community biomass is rather negligible [45]. Concerning SLA, this result is consistent with findings from the Jena experiment [13] where the CWM of SLA also did not contribute to the explanation of aboveground productivity.

Invasion of *S. canadensis*

In terms of seedling survival rate of the invading species *S. canadensis* our results are consistent with previous invasion experiments at small scales, which have shown that increased functional group diversity decreased invasion [46]. The invasion resistance effect has been attributed to the ability of diverse communities to occupy more space, generate more biomass, and use more resources which are than not available to the invading species [39,47]. But also the trait characteristics of resident species seem to play an important role in the establishment of invading species, i.e. negative and positive effects on the invaders have been associated with specific functional traits of single species within the resident community [5]. For example, strong negative effects on invaders are reported from C_4 grasses [48], whereas legumes are generally known to have positive effects on invaders [49]. Especially the latter finding is inconsistent with the results from our experiment. We found in the first phase of our study that the dominance of the two productive legumes was highest at intermediate levels of FD, which indicates, that their strong effect on overall community productivity has been responsible for the minimum of the *S. canadensis* seed germination rate at an intermediate FD level. The beneficial role mediated by N-supply, which legumes may have for the invaders in long-term studies and under natural conditions, seems to be not relevant in our short-term study. Maybe this negative short term effect of legumes on invaders also explains why our focal plant community has a low proportion of exotic plant species and a high proportion of legumes at a very early successional stage (i.e. in the first year after abandonment) in the field (Schittko, personal observation). It is also known that alfalfa (*Medicago sativa*, which is one of the parent species of *M.* x *varia*) produces root exudates with strong allelopathic agents that may reduce seed germination of other species by 50% [50]. Allelopathic effects and the strong dominance of legumes at intermediate FD levels may have led to the minimum of the *S. canadensis* seed germination. That the germination rate increased again slightly at high FD levels may be explained by the fact that legume dominance and thereby community biomass was decreased at these levels.

Conclusions

The approach of using a multi-trait measure for the manipulation of functional diversity can be the superior method because it may explain greater variance in ecosystem functioning than species richness or functional group richness. Our results emphasize the need to incorporate different aspects of functional composition (functional traits, functional diversity) in studies of biodiversity-ecosystem functioning relation-

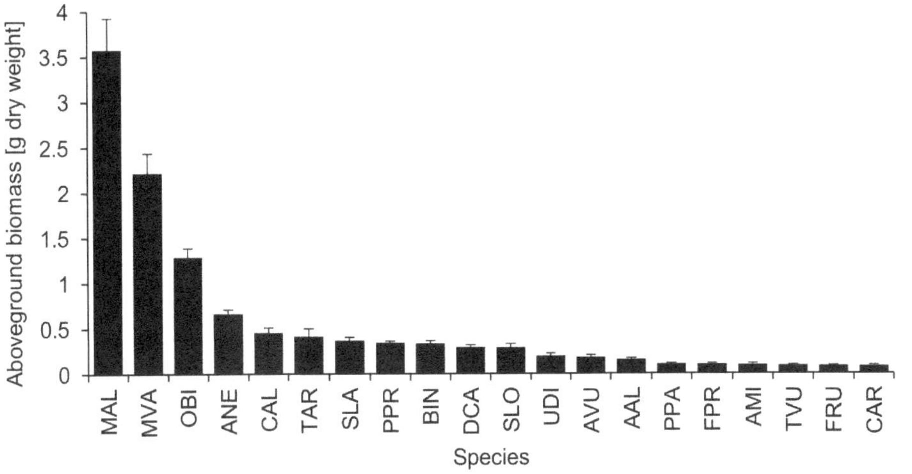

Figure 4. Species-specific plant productivities. Bars represent the mean dry weight of each species from the experiment. Error bars represent SE. Species abbreviations: ANE = *Acer negundo*, AMI = *Achillea millefolium*, AAL = *Ailanthus altissima*, AVU = *Artemisia vulgaris*, BIN = *Berteroa incana*, CAL = *Chenopodium album*, CAR = *Cirsium arvense*, DCA = *Daucus carota*, FPR = *Festuca pratensis*, FRU = *Festuca rubra*, MVA = *Medicago* x *varia*, MAL = *Melilotus albus*, OBI = *Oenothera biennis*, PPR = *Petrorhagia prolifera*, PPA = *Poa pratensis*, SLA = *Silene latifolia*, SLO = *Sisymbrium loeselii*, TVU = *Tanacetum vulgare*, TAR = *Trifolium arvense*, UDI = *Urtica dioica*.

ships. Besides N-fixation, it was the species' floristic status that explained variation in community productivity second most in our experiment. This finding underlines the fact that exotic plants pose significant impacts at the community and ecosystem level.

In addition, we hope that this article serves as a methodological reference for future ecological multi-species studies in two ways. First, manipulative studies on functional diversity may assume the use of a continuous trait-based approach, and second, the species' attribute of being exotic or native may be seen in a functional context and included in the set of functional traits that are commonly used. In this paper we avoided to label the floristic status as 'functional trait' because it breaks with the strict definition of a functional trait as being any morphological, physiological or phenological feature. It should be considered rather as an 'effect trait' that can determine effects of plants on ecosystem functions [51]. However, the difficulty in classifying this trait should not prevent community ecologists from its inclusion in studies using trait based approaches, since it can significantly explain variation in functions such as productivity and invasibility.

Acknowledgments

We thank Thomas Dürbye from the Botanic Garden and Botanical Museum Berlin-Dahlem for providing us with seeds of our experimental plant species. Further, we thank Ilja Sonnemann for being available for subject questions at any time and for giving comments on an early version of the manuscript.

Author Contributions

Conceived and designed the experiments: CS MH SW. Performed the experiments: CS MH. Analyzed the data: CS MH SW. Wrote the paper: CS.

References

1. Sala O, Chapin F, Armesto J, Berlow E, Bloomfield J, et al. (2000) Global biodiversity scenarios for the year 2100. Science 287: 1770–1774.
2. Hooper DU, Chapin FS, Ewel JJ, Hector A, Inchausti P, et al. (2005) Effects of biodiversity on ecosystem functioning: A consensus of current knowledge. Ecol Monogr 75: 3–35.
3. Myers N (1996) Environmental services of biodiversity. Proc Natl Acad Sci U S A 93: 2764–2769.
4. Tilman D, Knops J, Wedin D, Reich P, Ritchie M, et al. (1997) The Influence of Functional Diversity and Composition on Ecosystem Processes. Science 277: 1300–1302.
5. Díaz S, Cabido M (2001) Vive la difference: plant functional diversity matters to ecosystem processes. Trends Ecol Evol 16: 646–655.
6. Naeem S, Wright JP (2003) Disentangling biodiversity effects on ecosystem functioning: deriving solutions to a seemingly insurmountable problem. Ecol Lett 6: 567–579.
7. Tilman D (1997) The Influence of Functional Diversity and Composition on Ecosystem Processes. Science 277: 1300–1302.
8. Hooper D, Vitousek P (1997) The effects of plant composition and diversity on ecosystem processes. Science 277: 1302–1305.
9. Hector A, Schmid B, Beierkuhnlein C (1999) Plant diversity and productivity experiments in European grasslands. Science 286: 1123–1127.
10. Marquard E, Weigelt A, Temperton VM, Roscher C, Schumacher J, et al. (2009) Plant species richness and functional composition drive overyielding in a six-year grassland experiment. Ecology 90: 3290–3302.
11. Petchey OL, Gaston KJ (2006) Functional diversity: back to basics and looking forward. Ecol Lett 9: 741–758.
12. McGill BJ, Enquist BJ, Weiher E, Westoby M (2006) Rebuilding community ecology from functional traits. Trends Ecol Evol 21: 178–185.
13. Roscher C, Schumacher J, Gubsch M, Lipowsky A, Weigelt A, et al. (2012) Using plant functional traits to explain diversity-productivity relationships. PLoS One 7: e36760.
14. Díaz S, Lavorel S, de Bello F, Quétier F, Grigulis K, et al. (2007) Incorporating plant functional diversity effects in ecosystem service assessments. Proc Natl Acad Sci U S A 104: 20684–20689.

15. Lavorel S, Grigulis K, Mcintyre S, Williams NSG, Garden D, et al. (2008) Assessing functional diversity in the field – methodology matters! Funct Ecol 22: 134–147.
16. Grime JP (1998) Benefits of plant diversity to ecosystems: immediate, filter and founder effects. J Ecol 86: 902–910.
17. Mouchet MA, Villéger S, Mason NWH, Mouillot D (2010) Functional diversity measures: an overview of their redundancy and their ability to discriminate community assembly rules. Funct Ecol 24: 867–876.
18. Petchey OL, Gaston KJ (2002) Functional diversity (FD), species richness and community composition. Ecol Lett 5: 402–411.
19. Mokany K, Ash J, Roxburgh S (2008) Functional identity is more important than diversity in influencing ecosystem processes in a temperate native grassland. J Ecol 96: 884–893.
20. Schumacher J, Roscher C (2009) Differential effects of functional traits on aboveground biomass in semi-natural grasslands. Oikos 118: 1659–1668.
21. Petchey OL, Hector A, Gaston KJ (2004) How Do Different Measures of Functional Diversity Perform? Ecology 85: 847–857.
22. Celesti-Grapow L, Pyšek P, Jarošik V, Blasi C (2006) Determinants of native and alien species richness in the urban flora of Rome. Divers Distrib 12: 490–501.
23. Klotz S, Kühn I, Durka W (2002) BIOLFLOR - Eine Datenbank zu biologisch-ökologischen Merkmalen der Gefäβpflanzen in Deutschland. Schriftenreihe für Vegetationskunde 38. Bonn: Bundesamt für Naturschutz.
24. Pérez-Harguindeguy N, Díaz S, Garnier E, Lavorel S, Poorter H, et al. (2013) New handbook for standardised measurement of plant functional traits worldwide. Aust J Bot 61: 167.
25. McIntyre S, Lavorel S, Landsberg J, Forbes TDA (1999) Disturbance response in vegetation - towards a global perspective on functional traits. J Veg Sci 10: 621–630.
26. Violle C, Navas M-L, Vile D, Kazakou E, Fortunel C, et al. (2007) Let the concept of traits be functional! Oikos 116: 882–892.
27. Elton CS (1958) The Ecology of Invasions by Animals and Plants. University Of Chicago Press.
28. Pokorny ML, Sheley RL, Zabinski CA, Engel RE, Svejcar TJ, et al. (2005) Plant Functional Group Diversity as a Mechanism for Invasion Resistance. Restor Ecol 13: 448–459.
29. Byun C, Blois S, Brisson J (2013) Plant functional group identity and diversity determine biotic resistance to invasion by an exotic grass. J Ecol 101: 128–139.
30. Kowarik I (2003) Biologische Invasionen: Neophyten und Neozoen in Mitteleuropa. Stuttgart: Ulmer. 320 p.
31. Prasse R, Ristow M, Klemm G, Machatzi B, Raus t, et al. (2001) Liste der wildwachsenden Gefäβpflanzen des Landes Berlin mit Roter Liste. Berlin: Kulturbuch-Verlag.
32. Kleyer M, Bekker RM, Knevel IC, Bakker JP, Thompson K, et al. (2008) The LEDA Traitbase: a database of life-history traits of the Northwest European flora. J Ecol 96: 1266–1274.
33. Schroeder FG (1968) Zur Klassifizierung der Anthropochoren. Veg Acta Geobot 16: 225–238.
34. Casanoves F, Pla L, Di Rienzo JA, Díaz S (2011) FDiversity: a software package for the integrated analysis of functional diversity. Methods Ecol Evol 2: 233–237.
35. Morgan SP (1988) Logistic regression: Description, examples, and comparisons. J Marriage Fam 50: 929–936.
36. Zhao L, Chen Y, Schaffner DW (2001) Comparison of logistic regression and linear regression in modeling percentage data. Appl Environ Microbiol 67: 2129–2135.
37. Kieschnick R, McCullough BD (2003) Regression analysis of variates observed on (0, 1): percentages, proportions and fractions. Stat Modelling 3: 193–213.
38. R Core Team (2013) R: A language and environment for statistical computing. R Foundation for Statistical Computing, Vienna, Austria.
39. Tilman D, Reich P, Knops J, Wedin D (2001) Diversity and productivity in a long-term grassland experiment. Science 294: 843–845.
40. Spaekova I, Leps J (2001) Procedure for separating the selection effect from other effects in diversity-productivity relationship. Ecol Lett 4: 585–594.
41. Lepš J, Osbornová-Kosinová J, Rejmánek M (1982) Community stability, complexity and species life history strategies. Vegetatio 50: 53–63.
42. Loreau M, Hector A (2001) Partitioning selection and complementarity in biodiversity experiments. Nature 412: 72–76.
43. Flombaum P, Sala OE, Rastetter EB (2014) Interactions among resource partitioning, sampling effect, and facilitation on the biodiversity effect: a modeling approach. Oecologia 174: 559–566.
44. Vilà M, Espinar JL, Hejda M, Hulme PE, Jarošík V, et al. (2011) Ecological impacts of invasive alien plants: a meta-analysis of their effects on species, communities and ecosystems. Ecol Lett 14: 702–708.
45. Crawley M (2009) The R Book. West Sussex: John Wiley & Sons Ltd.
46. Tilman D, Wedin D, Knops J (1996) Productivity and sustainability influenced by biodiversity in grassland ecosystems. Nature 379: 718–720.
47. Tilman D (1999) Diversity and Production in European Grasslands. Science 286: 1099–1100.
48. Fargione JE, Tilman D (2005) Diversity decreases invasion via both sampling and complementarity effects. Ecol Lett 8: 604–611.
49. Prieur-Richard A-H, Lavorel S, Dos Santos A, Grigulis K (2002) Mechanisms of resistance of Mediterranean annual communities to invasion by *Conyza bonariensis*: effects of native functional composition. Oikos 99: 338–346.
50. Abdul-Rahman AA, Habib SA (1989) Allelopathic effect of alfalfa (*Medicago sativa*) on bladygrass (*Imperata cylindrica*). J Chem Ecol 15: 2289–2300.
51. Lavorel S, Garnier E (2002) Predicting changes in community composition and ecosystem functioning from plant traits: revisiting the Holy Grail. Funct Ecol 16: 545–556.

17

How Can We Identify and Communicate the Ecological Value of Deep-Sea Ecosystem Services?

Niels Jobstvogt[1,2]*, Michael Townsend[3], Ursula Witte[1], Nick Hanley[4]

1 Oceanlab, University of Aberdeen, Aberdeen, United Kingdom, **2** Aberdeen Centre for Environmental Sustainability (ACES), University of Aberdeen, Aberdeen, United Kingdom, **3** National Institute of Water and Atmospheric Research (NIWA), Hamilton, New Zealand, **4** Department of Geography and Sustainable Development, University of St Andrews, St Andrews, United Kingdom

Abstract

Submarine canyons are considered biodiversity hotspots which have been identified for their important roles in connecting the deep sea with shallower waters. To date, a huge gap exists between the high importance that scientists associate with deep-sea ecosystem services and the communication of this knowledge to decision makers and to the wider public, who remain largely ignorant of the importance of these services. The connectivity and complexity of marine ecosystems makes knowledge transfer very challenging, and new communication tools are necessary to increase understanding of ecological values beyond the science community. We show how the Ecosystem Principles Approach, a method that explains the importance of ocean processes via easily understandable ecological principles, might overcome this challenge for deep-sea ecosystem services. Scientists were asked to help develop a list of clear and concise ecosystem principles for the functioning of submarine canyons through a Delphi process to facilitate future transfers of ecological knowledge. These ecosystem principles describe ecosystem processes, link such processes to ecosystem services, and provide spatial and temporal information on the connectivity between deep and shallow waters. They also elucidate unique characteristics of submarine canyons. Our Ecosystem Principles Approach was successful in integrating ecological information into the ecosystem services assessment process. It therefore has a high potential to be the next step towards a wider implementation of ecological values in marine planning. We believe that successful communication of ecological knowledge is the key to a wider public support for ocean conservation, and that this endeavour has to be driven by scientists in their own interest as major deep-sea stakeholders.

Editor: James P. Meador, Northwest Fisheries Science Center, NOAA Fisheries, United States of America

Funding: This research project was funded by MASTS (Marine Alliance for Science and Technology for Scotland; URL: www.masts.ac.uk). MASTS is funded by the Scottish Funding Council (grant reference HR09011) and contributing institutions. Townsend's involvement was funded by NIWA (National Institute of Water and AtmosphericResearch; URL: www.niwa.co.nz) under the Coasts and OceansResearch Programme 3 (2013/14 SCI). The funders had no role in study design, data collection and analysis, decision to publish, or preparation of the manuscript.

Competing Interests: The authors have declared that no competing interests exist.

* Email: niels.jobstvogt@abdn.ac.uk

Introduction

The concept of ecosystem services (ES) has inspired a movement away from conservation for the sake of nature's inherent value to one that explicitly identifies, links and communicates the benefits of conservation to human wellbeing [1–3]. The endeavour of describing, quantifying and valuing the economic benefits that nature provides to society through ES has been identified as a powerful tool to make ecosystems count in cost-benefit analysis for environmental decision making [1,4]. Throughout this paper, however, the term 'value' is used in a broader sense, as a holistic concept which can include social, ecological and economic values. This broadening of the concept of value is needed because for the remotest places on earth like the deep sea, it is particularly challenging to make direct links between changes in system functioning and effects on the delivery of final ES (and thus on human well-being) [5,6].

The deep sea accounts for nearly 91% of the world's oceans with depths ranging from 200 m to almost 11,000 m. Despite its remoteness and size, its ecosystems are far from being unaffected by anthropogenic impacts such as fishing, climate change, and pollution [7–10]. To date many knowledge gaps remain around the functioning of deep-sea ecosystems. This is partially explained through the high costs, difficulties, and risks that are associated with deep-sea research. This lack of ecological knowledge means that we also know very little about the social and economic value of protecting the deep sea. By identifying and quantifying the ES benefits provided by the deep sea it is likely that appreciation for these benefits will change. This should lead to a larger emphasis on mitigating anthropogenic impacts in the oceans.

The major challenges of accounting for deep-sea ES stem from most people's lack of awareness about the deep-sea environment, and from the prevalence of intermediate services relative to easier-to-appreciate final services. Intermediate services in this paper refer to the indirect services that the ecosystem provides, such as habitat provision and nutrient cycling (the Millennium Ecosystem Assessment [1] refers to this category as supporting services). Intermediate services are the functional basis of the final services supplied by the system (Figure 1). The final services are considered as the ecosystem's contribution to human well-being [11] and include the ES categories of provisioning (e.g. commercial fish species), regulating (e.g. waste absorption and detoxification) and

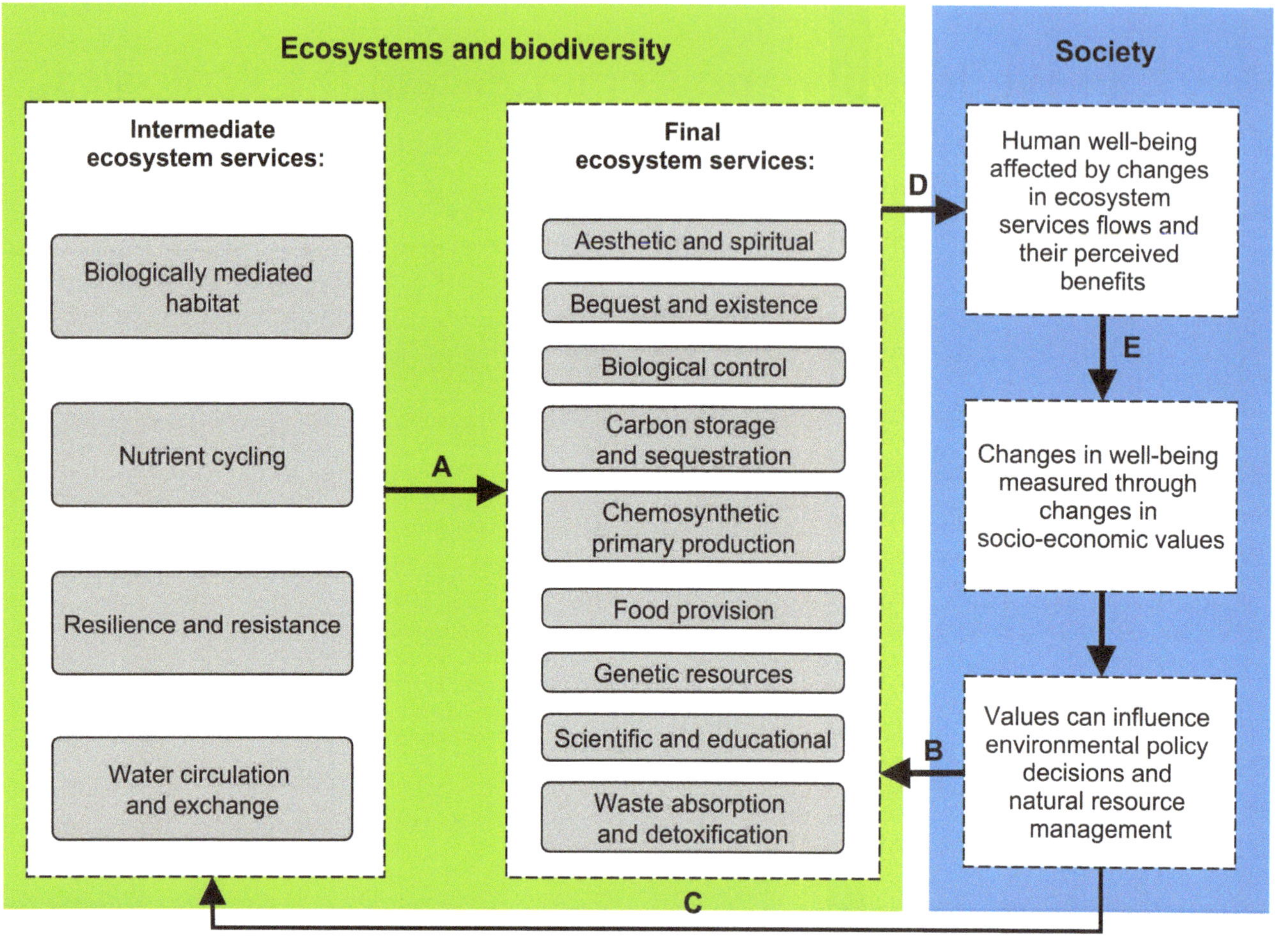

Figure 1. The ecosystem services framework for the example of deep-sea ecosystem services (ES). Environmental policies can either influence the management of final ES directly (arrow B) or indirectly via the intermediate ES (arrow C). The latter requires a sufficient understanding of the dependencies between intermediate and final ES (arrow A). Our understanding for the benefits provided by deep-sea ES (arrow D) and the values associated with them (arrow E) is currently very limited. The framework was simplified from [53] and adapted to the ES used for this expert consultation.

cultural (e.g. aesthetic values) [1]. There is a need to improve the integration of intermediate services and the processes that sustain them into the way in which we assess ES [12], since this ecological understanding is essential for demonstrating how human well-being ultimately depends on ecological processes and biodiversity [13,14].

In this paper we analyse the links between ES and the underlying processes that sustain them. This analysis was undertaken with the help of an expert panel of deep-sea scientists, focussing on submarine canyons. Within the deep sea, submarine canyons are an important ecosystem, which host unique reservoirs of biodiversity [15–18]. Along with much of the deep sea, they remain mostly underexplored and unprotected [9,19,20]. We selected this deep-sea ecosystem to test the Ecosystem Principles Approach (EPA) [21,22]. The EPA has recently been developed as a way of incorporating and translating ecological knowledge into easily understandable 'units' of information ('ecosystem principles') suitable for a wide range of audiences and thus for use in an ecosystem management context. The focus is on known and broadly-accepted information, with scientists from a wide field of expertise condensing this knowledge into principles that explain the linkages between ES, and their dependencies on underlying processes. The ecosystem principles also provide marine managers with qualitative information on temporal, spatial, and causal dependencies of ES flows. In the New Zealand case study by Townsend and colleagues [22], the EPA highlighted the importance of accounting for intermediate services in marine management, which were often provided by different geographical areas relative to the location at which final services were taken into account in the ES assessment. Economic theory suggests that intermediate ES can and should be valued through the final services that they support and the resultant direct benefits to people [40]. Indeed, ES are often perceived as a purely economic concept [23], but they also have social and ecological values, which when integrated with economic values produce a more holistic ecosystem assessment that can better inform natural resource management decisions [24]. This dominance of economic approaches and monetary valuation of ES stems from the often-felt pressure among the nature conservation sector to "speak the same language" as business and policy sectors in order to make conservation count [25–29].

In contrast, the EPA's advantage lies where economic valuation reaches its limits, by offering a more holistic picture which might help non-experts to better understand the high ecological value that scientists associate with the deep sea. Links between ecosystem processes and ES are less well known for the deep sea. For this reason it may be more helpful to highlight the links between deep-sea ES and processes themselves, rather than presenting decision makers with a set of economic values that are likely to underestimate the ecosystems' holistic value due to the omission of important ecological aspects of the deep sea [23]. In this paper we test the applicability of the EPA to little known and remote deep-sea ecosystems, such as submarine canyons, and demonstrate how the approach can provide decision makers with an accessible knowledge base for conservation decisions, despite some deficiencies in scientific data and associated uncertainty. We further discuss the approach's utility for expert consultation and cross-disciplinary knowledge transfer.

Methodology

2.1 Case study area: The Nazaré Canyon

The Nazaré Canyon on the Portuguese continental margin (Figure 2), also described as "Europe's Grand Canyon" [16], was chosen as the case study area of this paper to test the applicability of the EPA for deep-sea ecosystems. The Nazaré Canyon has attracted scientific interest due to its habitat heterogeneity and is considered to be a biodiversity hotspot [16]. Like other submarine canyon ecosystems it plays an important role in transportation processes at the continental margins [30,31]. The canyon is shallowest at 1 km off the Portuguese coast (50 m depth) and with a total length of 210 km it extends into the Iberian abyssal plain where it reaches depths of over 4,900 m [32].

2.2.1 Sampling and survey structure. Twenty-three researchers with knowledge of the Nazaré Canyon, covering a broad range of disciplines such as ecology, biology, microbiology, biogeochemistry, geography, geomorphology, geology, sedimentology and oceanography, were invited to participate in a HERMIONE (Hotspot Ecosystem Research and Man's Impact On European Seas project; URL: www.eu-hermione.net; last access July 2013) workshop in September 2012 and in an online pre-workshop survey (Figure 3). These opportunities were used to gather ideas and feedback for the two main surveys that followed. We used an email-based Delphi process to gather structured information by consensus from the invited expert panel in two consecutive rounds of surveying (post-workshop survey I and II; Figure 3). The Delphi process was originally developed as an interactive forecasting technique, where an expert panel goes through iterative survey rounds. The group results of each round are fed back to participants who are able to adapt their responses in the next survey round. The main idea of the Delphi process is to lead the group towards a consensus through the indirect exchange of information via a process coordinator. This process allowed us to subsequently include experts' requests for changes and additional information into the post-workshop surveys (Figure 3). The Delphi process makes it less likely that some researchers dominate the discussion and the outcomes, by maintaining anonymity throughout the communication process, thus avoiding the potential peer pressure of an expert workshop setting (further detail on the Delphi process in [33]). The survey questionnaires are available on request from the corresponding author.

2.2.2 Ethics statement. The nature of this research did not require ethical approval according to the University of Aberdeen Research Ethics Framework (*University of Aberdeen Research Ethical Review Checklist, Appendix A,* pp. 26–28; URL: www.abdn.ac.uk/documents/research-governance-framework-appendix4.pdf; last access March 2013). All study participants were recruited using an opt-in strategy and therefore consent was not explicitly recorded. Workshop and surveys did not include any sensitive personal questions. We asked participants to state their age, years of research experience, and field of studies; answering these questions was optional. Participants were provided with information on study objectives, sponsors, the participants' role, survey and workshop durations, potential benefits to the participants, summarised methodology, destination of gathered data and research results, the potential science impact and a contact address for further questions. Throughout the post-workshop surveys participants were identifiable via their email addresses. However, data on stated opinions and personal information was stored anonymously and kept confidential at all times. We chose email as the preferred communication method to facilitate the exchange of information during the Delphi phase of the expert consultation.

2.3 Submarine canyon ecosystem services

The experts helped to identify ES that were either perceived as less important or not relevant for the submarine canyon based on Table 1. Subsequently, the ES 'genetic resources', 'biological control', 'aesthetic and spiritual', 'scientific and educational', and 'chemosynthetic primary production' were excluded from Table 1 as less important relative to the other deep-sea ES. Accordingly eight ES (Table 1) were taken forward as a focus for the development of ecosystem principles.

2.4 Ecosystem Principles Approach

One of the main goals of the expert consultation was to develop a list of submarine canyon ecosystem principles, which could then be linked to ES. As an initial step, a review on the submarine canyon literature identified relatively well-explored ecosystem processes and relationships. The review findings were then discussed in the expert workshop and principles added or refined according to experts' suggestions. The following paragraph shows how the concept of ecosystem principles was described to canyon experts: "[An ecosystem principle] explicitly defines a key element

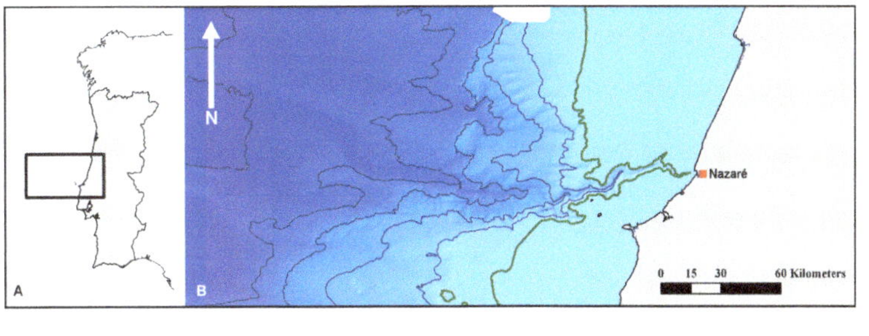

Figure 2. Nazaré Canyon. (**A**) Overview map of Portugal and the Nazaré Canyon area. (**B**) Nazaré Canyon bathymetry map with the Portuguese coastline to the east. Contour lines (blue) at 1000 m intervals; the 200 m depth contour, indicating the shelf edge, is marked in green. Data courtesy of Instituto Hidrografico, Lisbon and National Oceanography Centre, Southampton.

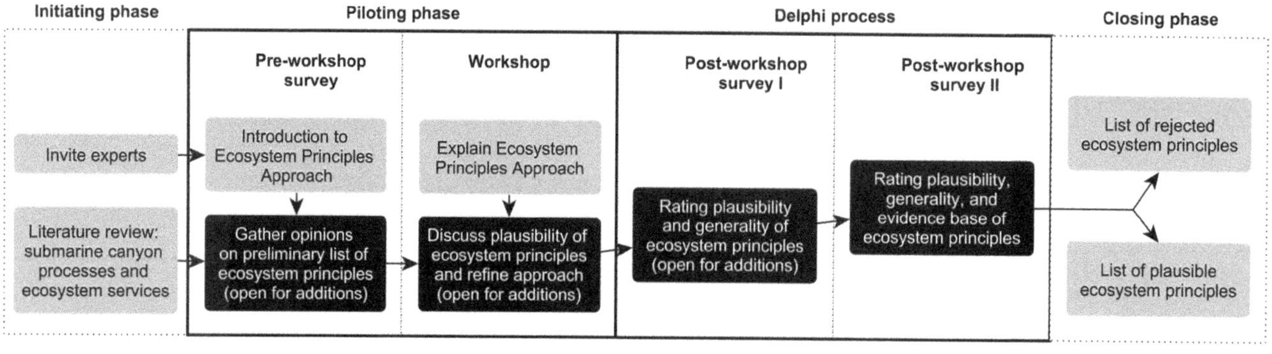

Figure 3. The survey phases of the submarine canyon expert consultation. Survey steps where experts were directly involved are highlighted as black boxes.

of how we expect the ecological system to operate" [22]. The workshop invitation summarised the EPA and included the methodology paper by Townsend and colleagues [22] as preparation for the workshop. During the workshop we provided further information in the form of a presentation that helped to increase participants' familiarity with the EPA. Principles were excluded from the initial list after each of the two consecutive post-workshop surveys (Delphi process; Figure 3) when fewer than 50% of the experts agreed with the plausibility of the principle. Experts were also able to propose new principles or suggest changes to the list of principles that was identified from the literature and subsequently refined throughout the piloting and Delphi phases (Figure 3).

We also asked the expert panel to categorise ecosystem principles according to their level of generality. The following categories were available: (i) general deep-sea principle, (ii) general canyon principle, (iii) shelf-incising canyon specific principles, to (iv) Nazaré Canyon-specific principle. The option with the highest frequency was then presented as the group vote in the subsequent survey.

Only in the second stage of the Delphi process were experts asked to distinguish their rating based on evidence on the one side and their expert view (as individuals) on the other side. The evidence base was rated on a five-point scale from 'very poor' to 'very good'. For the presentation of the group result, this was then divided into three categories of good, intermediate or poor evidence, according to the average group scores. These evidence scores had no influence on the decision to include or exclude any principle, but were introduced to separate personal opinion from the levels of evidence that existed in support of the principle.

During the workshop experts stressed the importance of the connectivity function of submarine canyons at the continental

Table 1. Submarine canyon ecosystem services.

Ecosystem services	Descriptions
Provisioning services:	
Carbon sequestration and storage	The uptake, storage, and burial of organic material within the canyon.
Food provision	The provision of marine organisms for human consumption.
Genetic resources and chemical compounds*	The use of canyon organisms in biotechnological, pharmaceutical, or industrial applications.
Regulating services:	
Biological control*	The control of diseases and invasive species.
Waste absorption and detoxification	The burial, decomposition and transformation of waste within the canyon ecosystem.
Cultural services:	
Aesthetic and spiritual*	The canyon ecosystem aesthetic and spiritual or inspirational source for religion, arts, movies, documentaries, books and folklore.
Bequest and existence	Safeguarding the canyon ecosystem for future generations and for the existence of marine species.
Scientific and educational*	The cognitive use of the canyon ecosystem for science and education.
Intermediate services:	
Biologically mediated habitat	Canyon habitats formed by marine organisms that provide nursery and refuge sites for other marine life.
Nutrient cycling	The storage and recycling of nutrients by canyon organisms.
Chemosynthetic primary production*	Primary productivity that is not dependent on energy from the sun.
Resilience and resistance	The amount of disturbance that the canyon ecosystem can cope with and its ability to regenerate after disturbance.
Water circulation and exchange	The currents, such as up-and down-welling, dense shelf water cascading, and mixing of water masses.

Services are grouped into four categories: provisioning, regulating, cultural and intermediate.
Listed items taken from [5,7,36,52] with alterations.
*Deep-sea ES that were not taken forward for the development of submarine canyon principles.

margin and we therefore chose 'water circulation and exchange' as an example to demonstrate how ecosystem principles link the supply of ES with their underlying processes. In the final results section, those ecosystem principles which in their description indicated a relationship between the ES 'water circulation and exchange' and other services were linked. The ecosystem principles relevant to this particular ES were assigned by the authors of this paper.

Results

3.1 Sample characteristics

The workshop was attended by 14 deep-sea researchers, 11 of whom had completed the pre-workshop pilot survey (Figure 3). All 14 workshop attendees were invited to participate in the two stages of the Delphi process (Figure 3), which 11 did for the first round and 10 for the second round. The average survey participant had 21 years of research experience and the survey covered academics from senior professors to PhD researchers, with male and female researchers equally represented.

3.2 Ecosystem principles

Over the course of the Delphi process, 21 ecosystem principles were identified from the literature and were then assessed and refined by the expert group (Tables 2 and 3). To highlight the nature of ecosystem principles, we present principle P1 as an example, which was rated to be plausible by all experts: 'canyons host a large number of different habitats and as a result increase species diversity at a regional scale' (for further principle descriptions we refer the reader to Tables 2 and 3). Four principles were discarded, whereas 17 principles were rated as plausible. Ten ecosystem principles fell into the category 'general submarine canyon principles', five into category 'general deep-sea principles', and two into 'shelf-incising canyon specific principles' (Tables 2 and 3). This indicated that the majority of ecosystem principles were at an appropriate level to describe processes and linkages between ES for submarine canyons in general and that they can be readily transferred to other canyons. The Delphi process had the expected effect of driving opinions closer towards consensus. Seven of the ten experts who participated in Delphi-rounds I and II (Figure 3) were closer to the group rating after the second Delphi-round.

Comparing the ratings of evidence and plausibility, we recognised that the existence of supporting evidence was not necessarily a requisite for an ecosystem principle to be plausible. Seven principles obtained intermediate evidence scores and for P16 and P17 evidence was rated as poor. However, this lack of evidence did not translate into a lack of plausibility. It was therefore an advantage to separate the two ratings from each other to distinguish between the experts' opinions and their evidence-based judgments. However, plausibility was clearly lower overall when evidence ratings were poor (Tables 2 and 3).

Developing principles to link food availability with biodiversity was challenging, and none were rated as plausible (P18 and P20). Experts had strong concerns of oversimplification when it came to the type, quality and amount of organic matter as a source of food and how changes of those parameters affected biodiversity. For biodiversity there was again a concern of over-simplification by omitting information on the spatial scale of biodiversity. In the same way, geographical scale mattered to experts, and lack of information on depth ranges and exact geographical position was criticised. The rating on generality provided a preliminary solution for implementing information on the geographical transferability of principles. Giving experts the chance to express uncertainty about the generality of principles as well as disentangling opinions about the rating of evidence allowed them to express their expectations for submarine canyons based on their research experience, and to transfer widely accepted knowledge from other ecosystems.

3.3 Linking principles and services

Many of the principles in Tables 2 and 3 have the capacity to provide information on *where* and *when* principles are likely to operate: for the principles included in Figure 4, particularly P8 and P7 reflect these spatio-temporal components. Other principles such as P3 and P12 explain *how* certain ES are provided and go into more detail on the processes involved. Principles like P17 and P16 that address effects of high biodiversity on ecosystem processes are capable of linking a broader set of ES such as 'carbon storage', 'food provision', 'bequest and existence' and 'waste absorption'. The ecosystem principles associated with 'biologically mediated habitat' were mainly thought to have an effect on biodiversity (e.g. P1 and P13) and to indirectly affect final ES such as 'food provision' and 'bequest and existence values' (P5).

For sustainable ecosystem management it can be equally important to understand the processes and principles that are involved in the provision of ES, as it is to understand the social and economic benefits of those services. We used 'water circulation and exchange' as an example to showcase how ecosystem principles explain links between ecological processes and ES (Figure 4; see also Table S1 for further details). 'Water circulation and exchange' has an important connectivity function in the submarine canyon (P11) and upwelling effects can lead to enhanced 'nutrient cycling' and as a result enhance productivity (P7 and P6; Figure 4). Further, 'nutrient cycling' might be important as an intermediate service for two different final ES, 'food provision' and 'bequest and existence values', because it can enhance fish abundance (P7). The 'bequest and existence' value can arise through the value that people tend to hold for iconic species (including fish), whereas 'food provision' relies on the abundance of commercially important fish as a consumptive resource. Trophic relationships, enhanced biomass, maintenance of deep-sea organisms (including non-iconic and non-commercial species) are important processes that sustain 'bequest and existence' as well as 'food provision' (P3, P4, P9; Figure 4) and should therefore be considered for management purposes. For 'carbon storage' and 'waste absorption' ecosystem management might be more concerned with other processes such as the transportation of organic and inorganic material, means of transportation, sedimentation rates, storage time, and burial processes that are important in parts of the submarine canyon (P3, P8, P12; Figure 4). How 'water circulation and exchange' is linked to 'resilience and resistance' as well as 'biologically mediated habitats' could not be resolved through the ecosystem principles developed in our workshop. This might be an indication that either too little evidence exists to support any ecosystem principles or that the links with processes that sustain these two ES are too complex to be described in the simplified form of ecosystem principles.

Discussion

The deep-sea case study for the Nazaré Canyon resulted in new insights on how to address the difficulties of assessing marine ES for ecosystem management purposes, especially when uncertainty is high due to lack of scientific data. In times where the demands on deep-sea resources are increasing, and scientific data on the potential impacts on marine biodiversity is scarce [5,34–36],

Table 2. Submarine canyon ecosystem principles with expert ratings on their plausibility and evidence base.

ID	Ecosystem principles	Plausibility	Evidence (mean score ± SE)
	General submarine canyon principles:		
P1	Canyons host a large number of different habitats and as a result increase species diversity at a regional scale.	100%	GOOD (3.8±0.2)
P2	The canyon topography tends to have a focusing or channelling effect for sediment and organic material.	100%	GOOD (3.8±0.2)
P3	The strength of large scale transportation events varies and occurrence ranges from a yearly to decadal pattern. They can be triggered by storms, high sediment load in the water column, cooling and increasing salinity of surface waters, or slope failures.	100%	GOOD (3.9±0.3)
P4	The transport of organic material from shallower waters to the deep seabed, which is mainly driven by large scale transportation events, is an important source of food for deep-sea organisms.	100%	MEDIUM (3.3±0.3)
P5	Canyons can serve as fish feeding ground, refuge and nursery area and therefore often show higher abundance of fish than their surroundings.	90%	MEDIUM (2.8±0.3)
P6	Canyons can enhance the mixing of water masses and as a result influence the exchange of nutrients, heat and salt between the shelf and the deep sea.	90%	MEDIUM (3.4±0.4)
P7	The canyon topography affects up- and down-welling of water masses at the continental margin. Upwelling events around the canyon head enhances productivity locally; as a result fish abundance can be higher.	90%	MEDIUM (3.3±0.3)
P8	By transporting large amounts of organic material from the shelf into deeper waters, canyons act as temporary stores of sediment and carbon. It can take decades or even centuries until the transported material reaches the abyssal plain, where it is then deposited on geological time scales.	80%	GOOD (3.8±0.2)
P9	Food quantity and quality tends to be higher within some canyon areas compared to the surrounding slope. This can enhance the biomass of the benthic and pelagic fauna.	80%	MEDIUM (3.3±0.4)
P10	Many species that are found in canyons are not found on the slope. They are therefore contributing to regional diversity.	80%	MEDIUM (3.1±0.4)

ID = principle identification number. The plausibility rating: ten experts participated in the full rating process (i.e. 100% = 10 experts). The evidence rating (1–5 from 'very good' to 'very poor'): poor (mean score <2.5), medium (2.5≤ mean score <3.5) and good (mean score ≥3.5); SE = standard error.

approaches such as the Ecosystem Principles Approach (EPA) are crucial to draw the link between the ecological and socio-economic dimensions of the ecosystem. Currently this linkage is poorly understood, contributing to the under-valuation of deep-sea ecosystems which is likely to undermine conservation efforts. We briefly outline the utility of expert consultation under these circumstances and reflect on the ability of the EPA to integrate more ecology into the assessment of ES and into the decision-making process. We first discuss how the EPA can help to communicate the overall importance of deep-sea ecosystems for the provision of ecosystem services. Further, we explain how the EPA can improve marine ecosystem-based management by promoting the inclusion of information on ecosystem processes into an ecosystem services assessment.

4.1 Communicating ecological values

The EPA has the ability to broaden access to ecological knowledge so that decision makers are not dependent on science advisors alone, but can take informed decisions on the basis of simplified ecological knowledge made available to them [37,38].

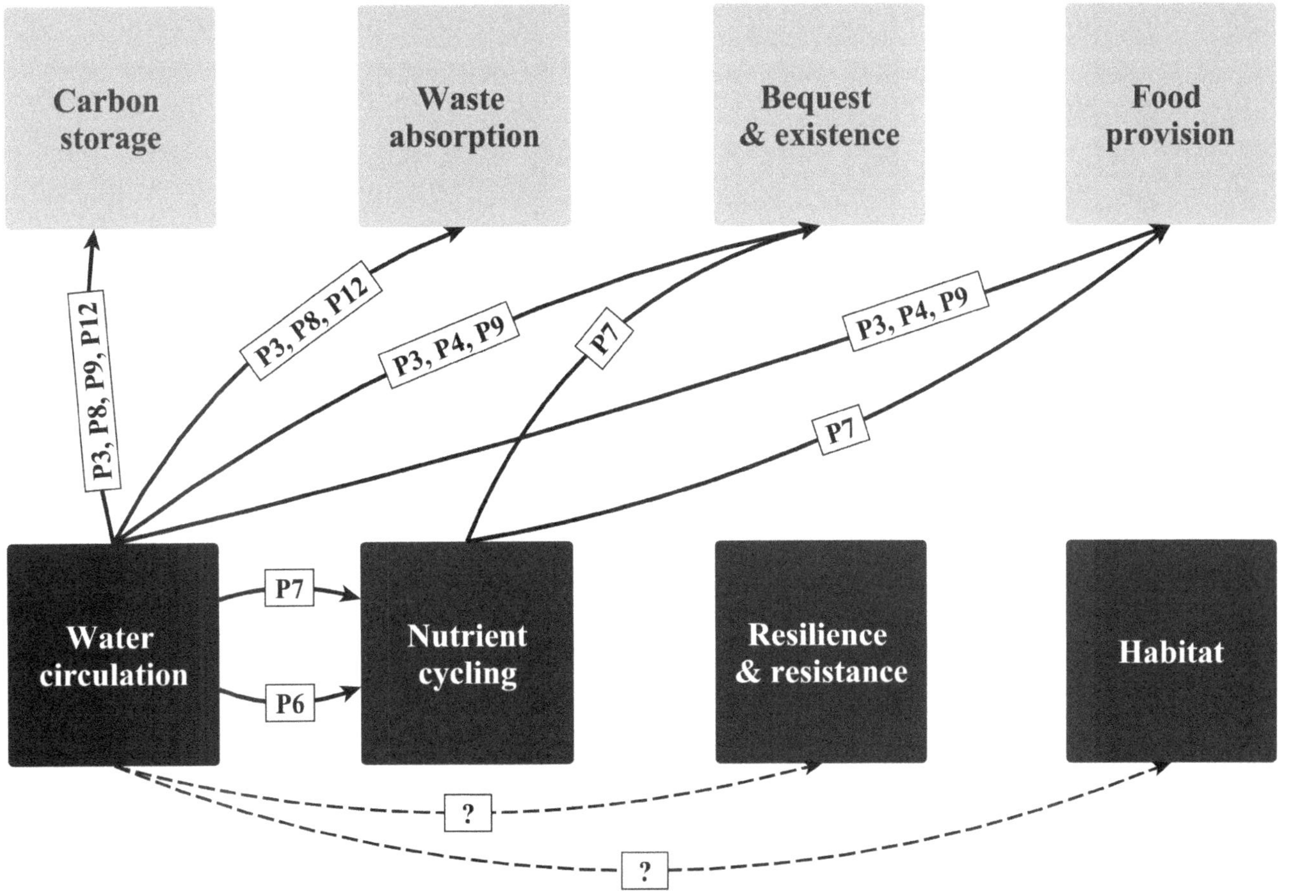

Figure 4. Links between 'water circulation' (black box) and other canyon ecosystem services explained through ecosystem principles. The intermediate services are in the lower half (dark grey and black boxes) and final services in the upper half (light grey boxes) of the diagram. Principles are indicated as arrows with their respective ID (cf. explanation in Table S1). Research gaps highlighted as question marks with dotted lines. Principles unrelated to 'water circulation' were omitted from this figure.

This broadening of access could also assist in increasing information about deep-sea ES amongst the general public. The process of developing ecosystem principles for the submarine canyon environment made clear that experts were able and willing to make predictions on connections between ES, generalisations on important canyon processes, and to link canyon characteristics to effects on ES. Also, while there remain research gaps and uncertainties, the list of ecosystem principles presented in Tables 2 and 3 provides the best available science knowledge to date, presented in easily understandable units of information. The list includes spatial and temporal information, as well as information on how the principles influence the supply of ES. We found that it was very important for the participating scientists to distinguish between their opinions (i.e. plausibility of the principles) and the (less subjective) rating of the existing scientific evidence base to inform principles. It was not imperative for principles to have a good evidence base, but rather to be generally accepted as plausible (cf. P5, P17 and P16). Herein lies a predictive strength of the EPA, in backing up the uncertainty associated with deep-sea science with a consensus-based approach, thereby decreasing uncertainty about ES linkages. The generality of principles was equally important, accounting for concerns that some ecosystem principles were valid on a larger scale than others i.e. 'general deep-sea principles' or 'shelf-incising canyon specific principles'. This additional type of information is crucial to highlight the ability of the EPA to transfer principles to other submarine canyons or even other deep-sea ecosystems. The majority of experts rated the principles as either very broadly applicable to the deep sea or to submarine canyons in general, irrespective of the type of canyon. The broad applicability of principles was thought to be an effect of reducing the complexity of ecological information.

We share the view of one workshop participant who stated that it will be difficult to determine when the list of ecosystem principles is complete. New evidence, the inclusion of researchers with different academic backgrounds, and assigning more time to the task might increase the number of principles on the list. Thus including a broad range of disciplines into the principle development process is crucial. Nonetheless, there exists an asymptotic relationship between effort expended and the number of principles, where spending more time on identifying and reviewing submarine canyon principles might increase the detail of principles, but not their utility for management decisions. The EPA's utility lies not in providing large amounts of detail, but in providing meaningful, concise information to better understand the overall functioning of the ecosystem in conjunction with the ES it provides. The EPA is based on what we know today and the ecosystem principles in this paper cover a broad range of topics, with further workshops or surveys being likely to provide diminishing returns of new principles to our established list.

There are many aspects of the deep sea that the science community remains uncertain about, but the EPA helped to focus and assimilate known information with the underlying ecosystem processes that are better understood and agreed on. However, the fact that some ecosystem principles were discarded (Table 3) indicated that there remain gaps in understanding on how canyon biodiversity is influenced by current regimes, and by different types and quality of organic material, and also the importance of recruitment processes between deeper and shallower waters. The high specificity and complexity of these and other processes might not allow us to develop these processes as ecosystem principles at the current time.

4.2 Utility of the Ecosystem Principles Approach

The deep sea is hard to sample and poorly understood, yet we were able to draw on experts' knowledge and condense what they know about submarine canyons, one of the deep-sea's biodiversity hotspots. We showed that the EPA, in combination with a Delphi process, can be a useful tool for working at the fringes of our current knowledge, using collective expert opinions to evaluate and arbitrate on the content of ecological understanding. Through this process we can also highlight where knowledge and hence research gaps lie, and where funding is needed. The EPA might be seen as a balancing act between the precautionary principle on the one side and economic reasoning on the other. The precautionary principle as framed in Rio in 1992 states that lack of scientific certainty shall not be used as an excuse to postpone actions that might prevent environmental degradation [39]. Lubchenco [37] lists guidance for decision-making under uncertainty as one of the roles that science should play in society. This might include reliance on more subjective approaches such as the EPA to support more holistic decision-making in marine resource management until we have a greater body of scientific evidence to prove or disprove what researchers have outlined as ecosystem principles.

The principles for 'water circulation and exchange' (Figure 4) demonstrate the EPA's ability to provide information on the ecological value of the ecosystem, and how these are linked to the kinds of final ES which economists are likely to value in monetary terms. The approach does not focus on final ES alone, but provides information at multiple levels without losing sight of the indirect impacts on ecosystems through intermediate services and underlying processes, and of the multiple connections between ecosystems. By linking processes with services through ecosystem principles, we draw the attention towards the network character of ecosystems. The highly interlinked nature of this network means that it is actually far from straightforward to categorise and separate services for ES valuation, especially given the multitude of connections within the marine environment. This is so irrespective of whether social, ecological, or economic definitions of 'value' are assessed. Presenting the information on final and intermediate services together with their underlying processes in a network style can better inform future management scenarios with more realistic ecological information than assessments that are limited to final ES alone.

Current ES valuation frameworks suggest that intermediate services should be valued only in terms of the final services that they support and not be included directly in a valuation of ES flows to avoid double-counting of their social or economic value [25,40–42]. The resulting requirement for effective management is that underlying processes and linkages are sufficiently well understood [23] and the ES they support are provided within the managed area [5]. However, the spatial and temporal distances between marine intermediate and final services can span millennia and act on a global scale, as is the case with the ocean nutrient cycle [35,43]. Marine ecosystems are highly connected systems with many processes being important for the provision of intermediate and final ES and crossing ecosystem boundaries [5]. Hence, if intermediate and final ES are spatially separated, the chances are that recommendations which focus only on final ES will be poor for marine resource management [22,44]. The field of ES valuation has its roots in terrestrial ecosystems where ecosystems with their services and processes are less open than in marine ecosystems [45,46]; valuation approaches might have to be adapted for the marine environment to account for its higher connectivity. Also, to capture the holistic value of intermediate services, we would have to successfully value all final ES (including cultural ES values and other non-marketed ES), which is still more of a research aspiration rather than a currently-achievable outcome. A failure to recognise the contribution of intermediate services for the ES valuation in ecosystems where they dominate, such as the deep sea, will likely lead to misguided policy decisions [47].

The EPA should be seen as an addition to baseline ecological research and economic ES valuation, rather than as a substitute for either. The EPA's advantage lies where monetary valuation reaches its limits, in highlighting links between ES and their underlying processes, and in linking intermediate services with final ES. While economic ES values can help set marine management priorities that are socially and economically desirable (Daily et al., 2009), the EPA focusses on the ecological 'value' of the ecosystem and can provide important information on how such management priorities can be achieved. Where economic values require empirical links to well-being and monetary quantification, ecological 'value' is more focussed on the importance of the ecosystem processes and characteristics that lead to such economically valued benefits being produced.

4.3 Future research opportunities and lessons learnt

The EPA might not only enhance the availability of ecological information and its uptake by decision makers, but can also improve how research results are shared across disciplines. Inter- as well as trans-disciplinary collaborations are complicated by the existence of language barriers. The use of different key terms or jargon restricts access to the pool of knowledge to only a small number of experts. The lack of frameworks that translate research findings into understandable and meaningful formats has been described as one of the major reasons why information might not reach beyond disciplinary boundaries [48,49]. Different methodologies, attitudes and perceptions between disciplines might further decrease the flow of scientific evidence [38,48]. Thus to allow economists, geologists, biologists, oceanographers and other disciplines involved in marine science to share information it would be beneficial to work on a global matrix of ecosystem principles similar to the ES valuation databases provided by the Marine Ecosystem Services Partnership (MESP; URL: www.marineecosystemservices.org; last access August 2013) which gathers studies on monetary ES values. In contrast to the MESP database, the EPA would be able to add to the evidence base not just on economic and social, but also on ecological values. Also, a more extensive dataset on ecosystem principles for marine ecosystems would increase the chances that more complex management scenarios could be developed, such as in Bayesian belief networks (BBNs), which depend heavily on the availability of baseline information on ecosystem processes, even though BBNs are able to deal with knowledge gaps when expert knowledge is available [50]. Other fields that are using approaches like habitat mapping as well as biological value mapping might benefit from

Table 3. Submarine canyon ecosystem principles continued from Table 2.

ID	Ecosystem principles	Plausibility	Evidence (mean score ± SE)
	Shelf-incising canyon specific principles:		
P11	Canyons function as major transport pathways between the shelf and the deep sea.	100%	GOOD (4.0±0.3)
P12	Sediment, organic material, and pollutants that are transported alongshore get trapped by the canyon and transported down the canyon slope.	90%	MEDIUM (3.4±0.3)
	General deep-sea principles:		
P13	Areas with reef forming or habitat creating organisms can support higher diversity than their surroundings. These habitats are most common on hard substrates, such as areas with steep slopes, rocks, boulders, vertical walls, or overhangs.	100%	GOOD (4.1±0.3)
P14	The biomass of invertebrates living in and on the seafloor can constitute an important food source for commercially important deep-sea species.	100%	GOOD (3.7±0.3)
P15	The organisms inhabiting soft substrates play a major role for the recycling of nutrients. The process is largely dominated by bacteria, but is to a smaller extent also attributed to the animals living in and on the sediment.	70%	GOOD (3.5±0.3)
P16	Higher biodiversity can support higher rates of ecosystem processes.	70%	POOR (2.4±0.5)
P17	Higher biodiversity increases the insurance value of an ecosystem by increasing the likelihood that the ecosystem is able to provide the same ecosystem functions after an ecosystem impact occurred.	60%	POOR (2.0±0.4)
	Rejected ecosystem principles*:		
P18	Diversity tends to be lower in areas with high food availability.	10%	Not assessed
P19	Space and resource occupancy by native species can decrease invasion risk.	30%	POOR (1.4±0.2)
P20	Where strong bottom currents are common, food availability and substrate characteristics become less important and current speed becomes the main driver for species abundance and diversity.	40%	Not assessed
P21	The disturbances caused by strong bottom currents keep species diversity and abundance at low levels.	40%	POOR (2.3±0.4)

ID = principle identification number. The plausibility rating: ten experts participated in the full rating process (i.e. 100% = 10 experts). The evidence rating (1–5 from 'very good' to 'very poor'): poor (mean score <2.5), medium (2.5≤mean score <3.5) and good (mean score ≥3.5); SE = standard error.
*Principles P18–P21 were rejected by the majority of experts, i.e. their plausibility was below 50%.

the EPA as well, given that spatial ecosystem principles were developed [51].

Using the EPA it should be possible to provide more precise temporal and spatial information on ecosystem principles, and to develop management strategies based on the list of principles and evidence on social and economic values and resulting management priorities. Showing the EPA's potential to improve people's understanding of ecosystem functioning was beyond the scope of this study, but future research involving the wider public and decision makers would be beneficial to test the effect of such simplified ecological knowledge on their decisions.

Three insights in particular emerge from our interdisciplinary workshop, which might improve future marine conservation initiatives and their acceptance:

(1) To further the field of marine ES valuation it would be beneficial to acknowledge that the traditional approaches to ES valuation, which have their roots in terrestrial research, might not be easily transferable to a highly linked marine

environment. Marine ecosystem boundaries are much less clearly defined than in terrestrial environments, and ES flows are less easily traceable. We might therefore need different approaches to valuing ES flows in the marine environment. The EPA is but one potential approach to improve integration of ecological values with social and economic values.

(2) The precautionary principle demands that we are cautious with our exploitation of the environment, but in the same time that management recommendations are made on a timely basis to the best of our knowledge, without postponing decisions for indefinite time until more certainty has been gained. The marine science community should more willingly embrace its important societal role in providing recommendations for nature conservation management with the support of social science approaches.

(3) We propose greater transparency in decisions on the conservation importance of marine areas. It should be possible to enhance understanding of the social, ecological as well as the economic values of certain areas, and to justify their protection, by providing easy understandable information on marine ES and how they relate to underlying ecosystem processes.

Acknowledgments

We are deeply thankful to the workshop and survey participants for their time and input, and to Dr Rene van der Wal, Dr Nicolas Krucien, and Prof Graham Pierce (University of Aberdeen) for constructive feedback on early drafts of this paper. We also like to thank Dr Veerle Huvenne (National Oceanography Centre) and the Instituto Hidrografico for the Nazaré Canyon bathymetry data, and the conference organizers for the workshop opportunity during the HERMIONE Annual Meeting 2012. We also thank two anonymous reviewers for their comments.

Author Contributions

Conceived and designed the experiments: NJ NH UW MT. Performed the experiments: NJ. Analyzed the data: NJ MT. Contributed reagents/materials/analysis tools: MT. Wrote the paper: NJ MT NH UW.

References

1. MEA (2005) Millenium Ecosystem Assessment: ecosystems and human well-being synthesis report. Washington DC: Island Press, 141.
2. Salles J-M (2011) Valuing biodiversity and ecosystem services: Why put economic values on Nature? Comptes rendus biologies 334: 469–482.
3. Lele S, Springate-baginski O, Lakerveld R, Deb D, Dash P (n.d.) Ecosystem services: origins, contributions, pitfalls, and alternatives. Conservation & Society: 1–46.
4. Hanley N, Barbier E (2009) Pricing Nature: Cost-Benefit Analysis and Environmental Policy. Cheltenham: Edward Elgar, 353.
5. Armstrong CW, Foley NS, Tinch R, Van den Hove S (2012) Services from the deep: Steps towards valuation of deep sea goods and services. Ecosystem Services 2: 2–13.
6. Jobstvogt N, Hanley N, Hynes S, Kenter J, Witte U (2014) Twenty thousand sterling under the sea: Estimating the value of protecting deep-sea biodiversity. Ecological Economics 97: 10–19.
7. Van den Hove S, Moreau V (2007) Deep-sea biodiversity and ecosystems: A scoping report on their socio-economy, management and governance. Cambridge: UNEP-WCMC, 84.
8. Benn AR, Weaver PP, Billet DSM, Hove S Van Den, Murdock AP, et al. (2010) Human activities on the deep seafloor in the North East Atlantic: an assessment of spatial extent. PloS one 5(9): 1–15.
9. Ramirez-Llodra E, Tyler PA, Baker MC, Bergstad OA, Clark MR, et al. (2011) Man and the Last Great Wilderness: Human Impact on the Deep Sea. PLoS ONE 6(7): 1–25.
10. Jones DOB, Yool A, Wei C-L, Henson SA, Ruhl HA, et al. (2013) Global reductions in seafloor biomass in response to climate change. Global change biology: 1–12.
11. Haines-Young R, Potschin M (2013) Common International Classification of Ecosystem Services (CICES): Consultation on Version 4, August–December 2012. Nottingham, 34.
12. Turner RK, Paavola J, Cooper P, Farber S, Jessamy V, et al. (2003) Valuing nature: lessons learned and future research directions. Ecological Economics 46: 493–510.
13. Cardinale BJ, Duffy JE, Gonzalez A, Hooper DU, Perrings C, et al. (2012) Biodiversity loss and its impact on humanity. Nature 486: 59–67.
14. Mace GM, Norris K, Fitter AH (2012) Biodiversity and ecosystem services: a multilayered relationship. Trends in ecology & evolution 27: 19–26.
15. Stiles ML, Ylitalo-Ward H, Faure P, Hirshfield MF (2007) There's no place like home: Deep seafloor ecosystems of New England and the mid-Atlantic. Oceana, 40.
16. Tyler P, Amaro T, Arzola R, Cunha M, De Stigter H, et al. (2009) Europe's Grand Canyon: Nazaré Submarine Canyon. Oceanography 22: 46–57.
17. Danovaro R, Company JB, Corinaldesi C, D'Onghia G, Galil B, et al. (2010) Deep-sea biodiversity in the Mediterranean Sea: the known, the unknown, and the unknowable. PloS one 5(8): 1–25.
18. Auster PJ, Moore J, Heinonen KB, Watling L (2005) A habitat classification scheme for seamount landscapes: assessing the functional role of deep-water corals as fish habitat. In: Freiwald A, Roberts JM, editors. Cold-Water corals and ecosystems. Berlin Heidelberg: Springer, 761–769.
19. Glover AG, Smith CR (2003) The deep-sea floor ecosystem: current status and prospects of anthropogenic change by the year 2025. Environmental Conservation 30: 219–241.
20. Würtz M (2012) Mediterranean submarine canyons: Ecology and governance. Gland, Switzerland and Malaga, Spain: IUCN, 216.
21. Townsend M, Thrush S (2010) Ecosystem functioning, goods and services in the coastal environment. Prepared by the National Institute of Water and Atmospheric Research for Auckland Regional Council. Auckland Regional Council Technical Report 2010/033.
22. Townsend M, Thrush S, Carbines M (2011) Simplifying the complex: an "Ecosystem Principles Approach" to goods and services management in marine coastal ecosystems. Marine Ecology Progress Series 434: 291–301.
23. Cornell S (2010) Valuing ecosystem benefits in a dynamic world. Climate Research 45: 261–272.
24. De Groot RS, Wilson M a, Boumans RM. (2002) A typology for the classification, description and valuation of ecosystem functions, goods and services. Ecological Economics 41: 393–408.
25. Fisher B, Turner K, Zylstra M, Brouwer R, De Groot R, et al. (2008) Ecosystem services and economic theory: integration for policy-relevant research. Ecological Applications 18: 2050–2067.
26. Brink P, Mazza L, Badura T, Kettunen M, Withana S (2012) Nature and its role in the transition to a green economy. The Economics of Ecosystems & Biodiversity (TEEB), 72.
27. Kallis G, Gómez-Baggethun E, Zografos C (2013) To value or not to value? That is not the question. Ecological Economics 94: 97–105.
28. Balmford A, Fisher B, Green RE, Naidoo R, Strassburg B, et al. (2010) Bringing Ecosystem Services into the Real World: An Operational Framework for Assessing the Economic Consequences of Losing Wild Nature. Environmental and Resource Economics 48: 161–175.
29. Peterson MJ, Hall DM, Feldpausch-Parker AM, Peterson TR (2010) Obscuring ecosystem function with application of the ecosystem services concept. Conservation biology: the journal of the Society for Conservation Biology 24: 113–119.
30. De Leo FC, Smith CR, Rowden AA, Bowden DA, Clark MR (2010) Submarine canyons: hotspots of benthic biomass and productivity in the deep sea. Proceedings Biological sciences/The Royal Society 277: 2783–2792.
31. Harris PT, Whiteway T (2011) Global distribution of large submarine canyons: Geomorphic differences between active and passive continental margins. Marine Geology 285: 69–86.
32. Lastras G, Arzola RG, Masson DG, Wynn RB, Huvenne V a I, et al. (2009) Geomorphology and sedimentary features in the Central Portuguese submarine canyons, Western Iberian margin. Geomorphology 103: 310–329.
33. Linstone HA, Turoff M (2002) The Delphi Method: Techniques and Applications. New Jersey: New Jersey Institute of Technology, 616.
34. Mengerink KJ, Van Dover CL, Ardron J, Baker M, Escobar-Briones E, et al. (2014) A Call for Deep-Ocean Stewardship. Science 344: 696–698.
35. Thurber AR, Sweetman AK, Narayanaswamy BE, Jones DOB, Ingels J, et al. (2013) Ecosystem function and services provided by the deep sea. Biogeosciences Discussions 10: 18193–18240.
36. Armstrong CW, Foley N, Tinch R, Van den Hove S (2010) Ecosystem goods and services of the deep sea. Deliverable D6.2 HERMIONE Project, 68.

37. Lubchenco J (1998) Entering the Century of the Environment: A New Social Contract for Science. Science 279: 491–497.
38. Rudd MA, Lawton RN (2013) Scientists' prioritization of global coastal research questions. Marine Policy 39: 101–111.
39. Myers N (2002) The Precautionary Principle Puts Values First. Bulletin of Science, Technology & Society 22: 210–219.
40. UK NEA (2011) UK NEA: Synthesis of the key findings. UK National Ecosystem Assessment, Cambridge, UNEP-WCMC, 87.
41. Boyd J, Banzhaf S (2007) What are ecosystem services? The need for standardized environmental accounting units. Ecological Economics 63: 616–626.
42. Wallace KJ (2007) Classification of ecosystem services: Problems and solutions. Biological Conservation 139: 235–246.
43. Roberts CM, Andelman S, Branch G, Bustamante RH, Carlos Castilla J, et al. (2003) Ecological criteria for evaluating candidate sites for marine reserves. Ecological Applications 13: 199–214.
44. McLeod K, Leslie H (2009) Ecosystem-based management for the oceans. Washington, DC: Island Press, 368.
45. Hawkins SJ (2004) Scaling up: the role of species and habitat patches in functioning of coastal ecosystems. Aquatic Conservation: Marine and Freshwater Ecosystems 14: 217–219.
46. Raffaelli D (2006) Biodiversity and ecosystem functioning: issues of scale and trophic complexity. Marine Ecology Progress Series 311: 285–294.
47. Johnston RJ, Russell M (2011) An operational structure for clarity in ecosystem service values. Ecological Economics 70: 2243–2249.
48. Davies AL, Colombo S, Hanley N (2013) Improving the application of long-term ecology in conservation and land management. Journal of Applied Ecology 51: 63–70.
49. Peters DPC (2010) Accessible ecology: synthesis of the long, deep, and broad. Trends in Ecology & Evolution 25: 592–601.
50. Campbell D, Chilton S, Clark S, Fitzsimmons C, Gazzola P, et al. (2012) Bayesian Belief Networks as an Interdisciplinary Marine Governance and Policy Tool. Valuing Nature Network Report, 27.
51. Townsend M, Thrush SF, Lohrer AM, Hewitt JE, Lundquist CJ, et al. (2014) Overcoming the challenges of data scarcity in mapping marine ecosystem service potential. Ecosystem Services 8: 44–55.
52. Beaumont N, Townsend M, Mangi S, Austen MC (2006) Marine biodiversity: an economic valuation. Building the evidence base for the Marine Bill. Department for Environment Food and Rural Affairs, 64.
53. Mace GM, Bateman I, Albon S, Balmford A, Brown C, et al. (2011) UK NEA Technical Report: Conceptual Framework and Methodology. UK National Ecosystem Assessment: Technical Report. Cambridge: UK National Ecosystem Assessment, UNEP-WCMC, 11–26.

18

Spatial and Temporal Variations of Ecosystem Service Values in Relation to Land Use Pattern in the Loess Plateau of China at Town Scale

Xuan Fang[1], Guoan Tang[1]*, Bicheng Li[2], Ruiming Han[3]

1 Key Laboratory of Virtual Geographic Environment, Ministry of Education, School of Geography Science, Nanjing Normal University, Nanjing, China, **2** Research Center of Soil and Water Conservation and Ecological Environment, Chinese Academy of Sciences, Yangling, Shaanxi, China, **3** School of Geography Science, Nanjing Normal University, Nanjing, China

Abstract

Understanding the relationship between land use change and ecosystem service values (ESVs) is the key for improving ecosystem health and sustainability. This study estimated the spatial and temporal variations of ESVs at town scale in relation to land use change in the Loess Plateau which is characterized by its environmental vulnerability, then analyzed and discussed the relationship between ESVs and land use pattern. The result showed that ESVs increased with land use change from 1982 to 2008. The total ESVs increased by 16.17% from US$ 6.315 million at 1982 to US$ 7.336 million at 2002 before the start of the Grain to Green project, while increased significantly thereafter by 67.61% to US$ 11.275 million at 2008 along with the project progressed. Areas with high ESVs appeared mainly in the center and the east where largely distributing orchard and forestland, while those with low ESVs occurred mainly in the north and the south where largely distributing cropland. Correlation and regression analysis showed that land use pattern was significantly positively related with ESVs. The proportion of forestland had a positive effect on ESVs, however, that of cropland had a negative effect. Diversification, fragmentation and interspersion of landscape positively affected ESVs, while land use intensity showed a negative effect. It is concluded that continuing the Grain to Green project and encouraging diversified agriculture benefit to improve the ecosystem service.

Editor: Ricardo Bomfim Machado, University of Brasilia, Brazil

Funding: This study was sponsored by the Jiangsu Planned Projects for Postdoctoral Research Funds (No. 1401033C), the National Natural Science Foundation of China (No. 41401441), and the Priority Academic Program Development of Jiangsu Higher Education Institutions (PAPD) (No. 164320H101). The funders had no role in study design, data collection and analysis, decision to publish, or preparation of the manuscript.

Competing Interests: The authors have declared that no competing interests exist.

* Email: tangguoan@njnu.edu.cn

Introduction

Ecosystem contributes to human welfare by providing goods and services directly and indirectly [1–2]. With widely spreading of environmental problems, ecosystem service received increasing attention. Many studies showed human factors, such as urban sprawl [3,4,5], socioeconomic changes [6], agricultural policies [7,8], could affect natural or artificial ecosystems. Land use, an original and foundational human activity and represents the most substantial human alteration to systems on the planet of earth for long-term study [9], plays an important role in providing ecosystem services, including biodiversity, water filtration, retention of soil, etc. [10] Inappropriate land use may lead to significant degradation of local and regional ecological services [11]. Moreover, there were studies showed that ecosystem service trade-offs could successful apply to land use planning [12,13]. Understanding the relationship between ecosystem services and land use change is essential for maintaining a healthy ecosystem and getting sustainable services.

The growing body of literatures focused on how ecosystem service changes in response to land use change of different regions [14,15,16,17,18]. However, these studied focus on the impact of land use type on ecosystem service, while the spatial pattern of land that reflects ecological processed and functions [19] get less attention. Monitoring the characteristic of landscape patterns including area, shape, diversity, etc., is helpful to deeply understand the relationship between ecosystem service and land use change and then to provide complete references for land use planning.

The Loess Plateau is the area suffered from the most severe soil erosion in the world, and it is also a major agricultural production region in China [20]. Long-term poor land use has resulted in vegetation destruction and accelerated soil erosion [21]. To control soil erosion and restore the ecosystem, the Grain for Green project converting slope cropland to grassland or forestland was implemented in 1999 by the Chinese Government [22]. The land use on the plateau under the project has changed significantly. Studying the ecosystem service in relation to land use change before and after the Grain to Green project was crucial for ecosystem protection and agricultural sustainability for the area. Researchers have analyzed ecosystem service at different scales within the Loess Plateau [17,18,23]. However, town is a basic administrative area in China. Exploring the characteristic of ecosystem services change at town scale is of practical significance to provide operable land use planning.

Ecosystem service values (ESVs) is monetary assessment of ecosystem services. This paper examined the characteristics of ESVs at Hechuan town, a typical town in the hilly and gully region of the Loess Plateau. The objectives of this study were: 1) to analyze the changes in land use pattern from 1982 to 2008; 2) to access the spatial and temporal variation in ESVs in response to land use during this period; 3) to quantitively analysis the relationship between ESVs and land use pattern; and 4) to discuss how land use management is favorable for ecosystem service supply and the ecological and economic sustainable development.

Data and Methods

2.1 Ethics statement

No specific permits were required for the described studies, and the work did not involve any endangered or protected species.

2.2 Study area

The study area, Hechuan town (106°18′43″~106°32′16″E, 35°54′59″~36°06′05″N), is located in Guyuan city of the Ningxia Hui Autonomous Region of northwest China (Fig. 1), consisting 12 villages with 16,524 people. The reasons that Hechuan Town was chosen as the study area were, on the one hand, Hechuan town has the typical characteristics of Loess Plateau including the

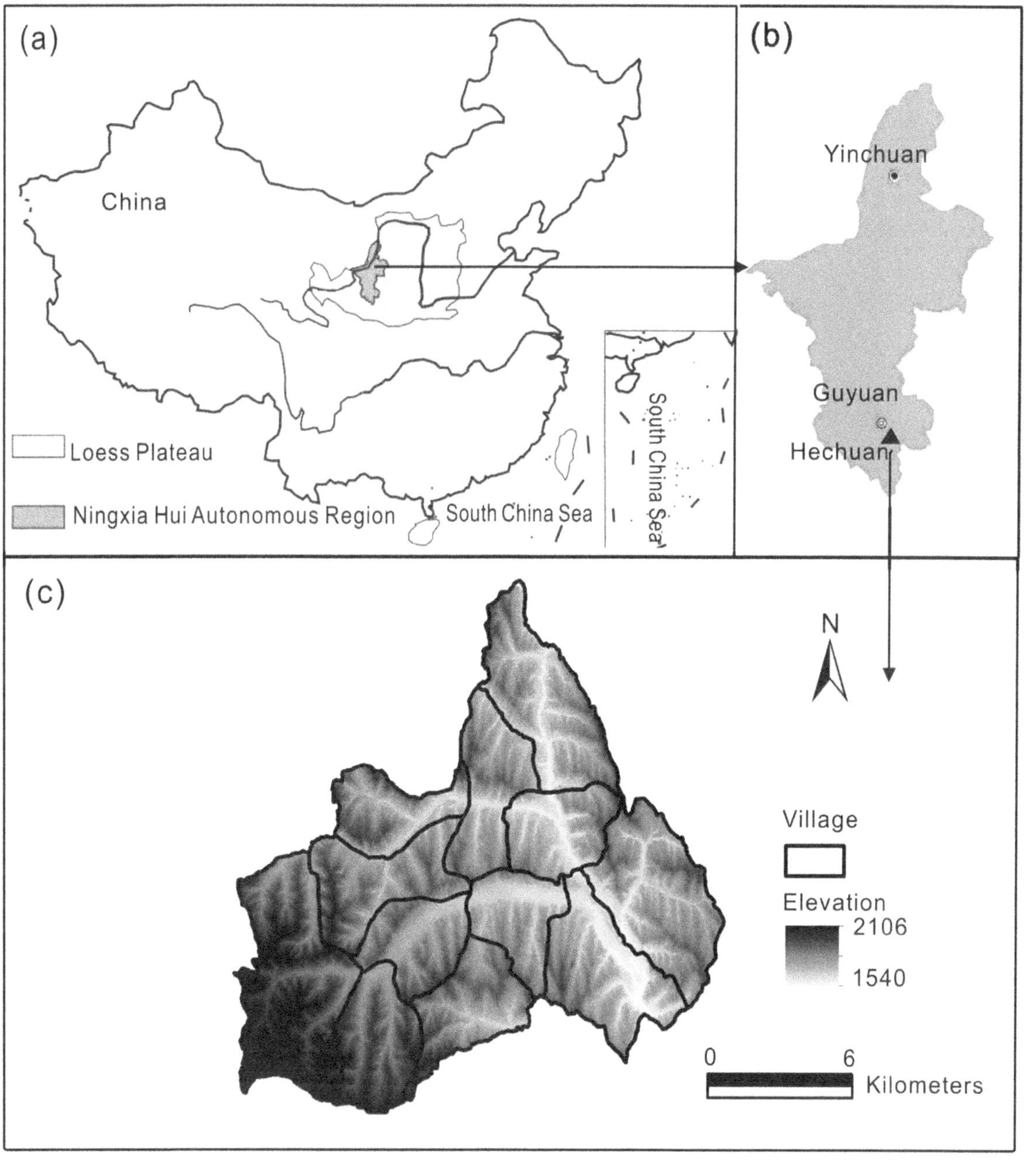

Figure 1. Location of the study area. Ningxia Province and the Loess Plateau, China (a), the location of Hechuan Town in the Loess area of Ningxia Province (b) and, the village boundary and the digital elevation model (DEM) map of Hechuan Town (c).

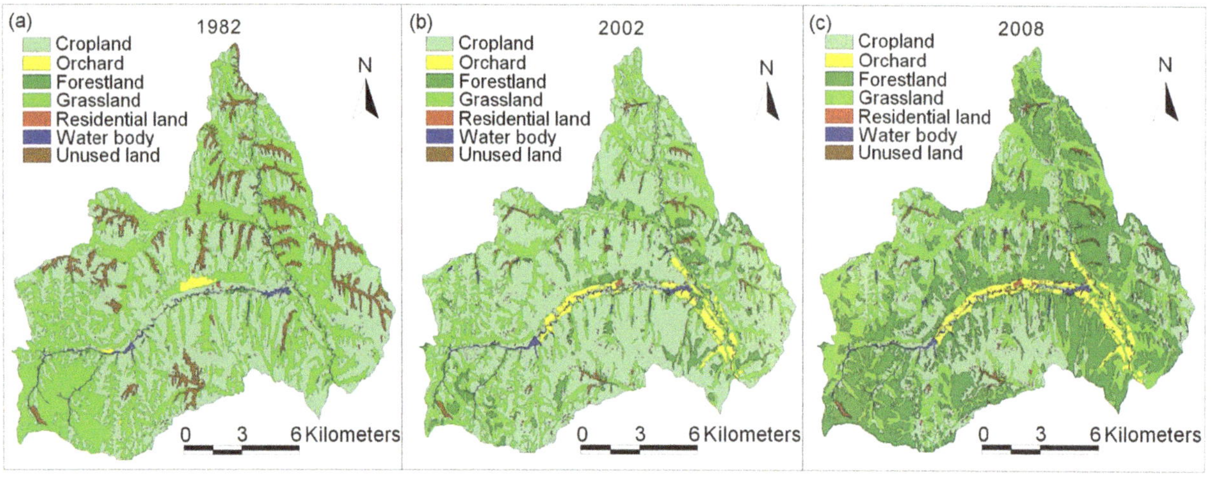

Figure 2. Land use maps of Hechuan town in 1982 (a), 2002 (b) and 2008 (c).

terrain of hill and gull, the fragile ecosystem and the backward economy; on the other hand, there was a long term ecological observation and experiment station in the study area, which facilitated the survey of land use and ecosystems. This town has an altitude ranging from 1540 to 2106 m, covering an area of 215.58 km^2. There exist the topographic differences in the town. The central area with river terrace stretches smoothly with a low elevation. The terrain in the northern area is fragmented while that of southern area is relatively simple. Hechuan town has a semi-arid continental temperate climate with the average annual temperature of 6.9°C and precipitation of 419 mm (1982–2002). Most of the annual precipitation is concentrated between June to September in the form of heavy storms that can cause severe soil erosion. The soil is composed of loessial soil and Dark loessial soils, which is erodible due to its weak cohesion and high infiltrability.

The ecosystem in Hechuan town is fragile with serious soil erosion and frequent natural disasters. Human disturbances of excessive land use, such as deforestation, overgrazing and over-reclamation further destructed the native natural grassland. Therefore, this area has long been in a vicious circle, endless cultivation and poverty. Since the early 1980s, a variety of comprehensive investigation of soil erosion was practiced by Chinese Academy of Sciences. Shanghuang watershed, located in the east of Hechuan town, was taken as a key test area. The ecological restoration covering the whole town was started from implementing the Grain for Green project after 2002 (launched in 1999 by China government). Since then, abandoned cropland, shrubland (*Caragana korshinskii*, *Hippophae rhamnoides*) and artificial grassland (*Medicago sativa*) was generated, which made a significant change on landscape pattern and ecosystem components providing a variety of ecosystem services. Meanwhile, farming and grazing, the traditional way of living, had to be changed, and raising livestock, orchards, and migrant working diversified their incomes.

2.3 Data acquisition and preprocessing

Land use data was the key data for evaluating landscape pattern and ecosystem service. The land use data of 1982 was obtained by digitizing the land use patches from the 1:10,000 scale topographic maps of 1982, in which the information of land use types and its boundary are clearly shown. The 10 m resolution of remote sensing image could be considered to be corresponding with the scale of 1:50000 [24,25]. The land use data of 1982 acquired from 1:10000 topographic maps was therefore generalized to be at 1:50000 scale [26]. The land use data of 2002 and 2008 were respectively extracted from the 10 m resolution multispectral Spot-5 image of 2002 and 2008 by updating the land use patches of

Table 1. Equivalent weight factor of ecosystem service values (ESVs) per hectare of terrestrial ecosystem in China [30].

	Cropland	Forestland	Grass land	Water body	Barren land
Gas regulation	0.72	4.32	1.5	0.51	0.06
Climate regulation	0.97	4.07	1.56	2.06	0.13
Water supply	0.77	4.09	1.52	18.77	0.07
Soil formation and retention	1.47	4.02	2.24	0.41	0.17
Waste treatment	1.39	1.72	1.32	14.85	0.26
Biodiversity protection	1.02	4.51	1.87	3.43	0.40
Food production	1.00	0.33	0.43	0.53	0.02
Raw material	0.39	2.98	0.36	0.35	0.04
Recreation and culture	0.17	2.08	0.87	4.44	0.24
Total	7.90	28.12	11.67	45.35	1.39

Table 2. The ecosystem service values (ESVs) per hectare of different land use types in Hechuan town (US$·ha-1·yr-1).

	Cropland	Orchard	Forestland	Grass land	Water body	Unused land
Gas regulation	22.570	91.222	135.422	47.022	15.987	1.881
Climate regulation	30.407	88.244	127.585	48.902	64.576	4.075
Water supply	24.138	87.930	128.212	47.649	588.397	2.194
Soil formation and retention	46.081	98.118	126.018	70.219	12.853	5.329
Waste treatment	43.573	47.649	53.918	41.379	465.514	8.150
Biodiversity protection	31.975	99.999	141.378	58.620	107.523	12.539
Food production	31.348	11.912	10.345	13.480	16.614	0.627
Raw material	12.226	52.351	93.416	11.285	10.972	1.254
Recreation and culture	5.329	46.238	65.203	27.272	139.184	7.523
Total	247.647	623.663	881.498	365.828	1421.619	43.573

1982 one by one in visual interpretation method. The interpretation sign was established by understanding the Spot image characteristics and carrying out field surveys in order to further determine the relationship between the true ground and the image. The kappa accuracy index [27] was used to assess the accuracy of the interpretation. The stratified random sampling method was used to generate the reference points on the classified image for the accuracy test. These reference points were located in the field with a GPS with 5-m precision for ground truth. The total kappa indexes are all higher than 0.85, which are higher than the minimum acceptable (0.7) [28]. Considering the characteristic of the land use in study area and the interpretation level of the data and to facilitate the calculation of ESVs, the land use was classified into seven types: cropland, orchard, forestland, grassland, residential area, water area, and unused land (Fig. 2).

To acquire accurate area data of the land use for ESVs estimation and facilitate analyzing the spatial distribution of ESVs, the topographic maps and Spot images were transformed to the same projection and coordinate system (the Albers-Conical-Equal-Area projection system and Krasovsky 1940 coordinate system) before the extraction of land use data, and all acquired land use data were transformed to Arc-grid formats with the same grid size (10 m×10 m). The above data processing was completed using ERDAS and ArcGIS software.

2.4 Analysis on land use pattern

The transfer matrix analysis of land use was produced to understand how land use changed. Landscape metrics analysis was used for spatial pattern analysis of land use. Landscape metrics has been adopted widely; meanwhile, its abilities to indicate ecological process gained increasing attention [29,30,31]. Conceptual flaws in landscape pattern analysis, limitations inherent in landscape metrics and the improper use of pattern analysis may lead to the misuse of landscape metrics [32]. For better explanations and predictions of ecological phenomena from ecological pattern, the landscape metrics in this study was therefore selected by two steps. Firstly, the diversity, the fragmentation and the dominance of landscape were all considered, and then 34 metrics was selected, by understanding the knowledge of the landscape pattern and the ecological services indication of landscape metrics [33,34] and referring to the previous studies on landscape pattern [4,31,35,36]. Secondly, a correlation analysis for the 34 metrics was employed to ensure the low redundancy among landscape metrics. If the coefficient between two metrics was significant at 0.05 level, only one metric of them could be eventually selected.

Landscape-level metrics providing general landscape information and class-level metrics providing more specific information about variations at the local level and spatial patterns of land use classes [37] were used to monitor the characteristics of landscape pattern. The selected landscape-level metrics were patch density (PD), area-weighted mean shape index (SHAPE_AM), Interspersion and Justaposition Index (IJI), and Shannon's diversity index (SHDI). The selected class-level metrics were PD, the percentage of landscape (PLAND), SHAPE_AM and IJI. PD and SHAPE_AM could show the fragmentation of landscape. SHDI and PLAND reflect the dominance of some land use type and the diversity of landscape, respectively. IJI reflects whether the patches or classes are contiguous. Landscape metrics analysis was conducted with above metrics by FRAGSTATS 3.3, in which the eight-neighbor rule was used to derive the patch number. Besides these metrics, the land use intensity index (LUII) was also used to describe the landscape pattern. It was calculated by the following equation [31]:

$$LUII = \sum_{i=1}^{n} A_i \times C_i \qquad (1)$$

where $LUII$ is the land use intensity index, A_i is the percentage of for a give land use type i, and C_i is the coefficient value of intensity for a give land use type i, that is assigned 4 for build-ups, 3 for farmland and 2 for forest, orchard, grassland and water bodies, and 1 for unused land.

2.5 Estimation of ESVs

Costanza et al.'s model of ESVs estimation was adopted in this study [1,2]. The model classified ecosystem service into 17 types of service functions and estimated the ESVs by placing an economic value on different biomes [34]. For the defects of this model, such as overestimating the agriculture ESVs and underestimating the wetland ESVs, Xie et al. proposed refined coefficients for ESVs assessment both solving the above problem and making it apply to China [33,34]. Based on this model, the total ESVs in the study area was calculated using the following formulas:

$$ESV_k = \sum_{f} A_k VC_{kf} \qquad (2)$$

Table 3. Land use transition matrix from 1982 to 2002 and from 2002 to 2008 (%).

	2002								
1982	**Cropland**	**Orchard**	**Forestland**	**Grassland**	**Residential land**	**Water body**	**Unused land**	**Total**	**Loss**
Cropland	44.42	2.33	1.18	2.50	0.35	0.04	0.01	50.83	6.41
Orchard	0.17	0.05	0.26	0.09	0.00	0.00	0.00	0.57	0.53
Forestland	0.10	0.01	0.15	0.02	0.00	0.00	0.00	0.28	0.13
Grass land	12.63	0.39	4.74	22.15	0.01	0.08	0.00	40.01	17.86
Residential land	0.05	0.01	0.00	0.01	0.14	0.00	0.00	0.22	0.07
Water body	0.01	0.00	0.01	0.00	0.00	0.80	0.00	0.81	0.02
Unused land	1.39	0.00	0.16	3.98	0.00	0.07	1.66	7.27	5.61
Total	58.76	2.78	6.51	28.77	0.51	0.99	1.68	100.00	
Gain	14.35	2.73	6.36	6.62	0.37	0.19	0.01		
	2008								
2002	**Cropland**	**Orchard**	**Forestland**	**Grass land**	**Residential land**	**Water body**	**Unused land**	**Total**	**Loss**
Cropland	27.09	1.00	20.75	9.84	0.07	0.00	0.00	58.76	31.68
Orchard	0.00	2.75	0.01	0.00	0.01	0.00	0.00	2.78	0.03
Forestland	0.05	0.04	5.84	0.57	0.01	0.01	0.00	6.51	0.67
Grass land	0.12	0.00	7.17	21.42	0.02	0.04	0.00	28.77	7.35
Residential land	0.00	0.00	0.00	0.00	0.51	0.00	0.00	0.51	0.00
Water body	0.02	0.02	0.00	0.01	0.00	0.94	0.00	0.99	0.05
Unused land	0.00	0.00	0.27	0.08	0.00	0.00	1.33	1.68	0.35
Total	27.27	3.82	34.05	31.91	0.62	1.00	1.33	100.00	
Gain	0.18	1.07	28.21	10.50	0.11	0.05	0.00		

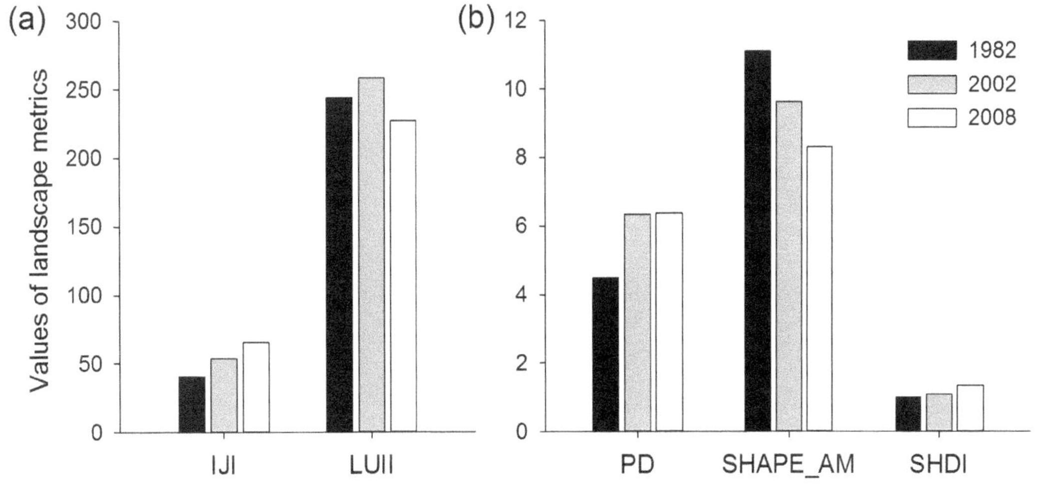

Figure 3. Landscape metrics at the landscape level in Hechuan Town in 1982, 2002 and 2008. IJI: Interspersion and Justaposition Index; LUII: land use intensity index; PD: patch density; SHAPE_AM: area-weighted mean shape index; SHDI: Shannon's diversity index.

$$ESV_f = \sum_k A_k VC_{kf} \tag{3}$$

$$ESV = \sum_k \sum_f A_k VC_{kf} \tag{4}$$

where ESV_k, ESV_f, ESV are the ESVs of land use type k, the ESVs of ecosystem service function type f, and the total ESVs respectively. A_k is the area (ha) for land use types. VC_{kf} is the value coefficient (US$·ha-1·yr-1) for land use type k and ecosystem service function type f, which is the key for ESVs estimating. Xie et al.'s model was used to determine VC_{kf}, which can be expressed as follows:

$$VC_{kf} = R_{kf} \times V_f \tag{5}$$

where R_{kf} is the equivalent weight factor of ecosystem service, V_f is food production values of agriculture land per area per year.

The equivalent weight factor was presented for customizing Chinese terrestrial ecosystem based on Costanza et al.'s model by surveying 500 Chinese ecologists (Table 1) [34]. It is the ratio of the ESVs to the economic value of average natural food production provided by agricultural land per hectare per year. The factors of land use types in our study were basically assigned based on the nearest ecosystems in Xie et al.'s model. However, minor adjustments were made. The equivalent weight factor of orchard which was not put forward clearly in Xie et al.'s model was determined by the mean of grassland and forestland by referring some researches [5,18]. The factor of unused land equates to that of barren land, and that of residential land was determined to zero.

The value of food production service of agriculture land per area per year was considered to be 1/7 of the actual price of food production in Xie et al.'s model. With the average actual food production of cropland in Hechuan town from 1982 to 2008 of 901.77 kg/ha which was get from *Statistic yearbook of the Yuanzhou District, Guyuan City, Ningxia Hui Autonomous Region* and the average grain price of US$ 0.243 per kilogram (i.e. an equivalent of RMB Yuan 1.69 according to the average exchange rate of 2008) in 2008, the value of food production service of cropland per area per year was calculated to be US$ 31.348 (i.e. an equivalent of RMB Yuan 217.713 according to the average exchange rate of 2008). ESVs of one unit area of each land use types were then assigned as shown in Table 2.

After the ESVs were calculated by above processing, a sensitivity analysis was conducted to test the land use type's representative for ecosystem types and the certainty of the coefficients value for ecosystem service. A coefficient of sensitivity (CS) was used to indicate the degree of sensitivity of ESVs to a coefficients value, calculated by the following formula [5]:

$$CS = \left| \frac{(ESV_j - ESV_i)/ESV_i}{(VC_{jk} - VC_{ik})/VC_{ik}} \right| \tag{6}$$

where ESV_j an ESV_i are the total ESVs of the initial status j and the adjusted status i, and VC_{jk} and VC_{ik} are the initial and adjusted coefficients. A 50% adjustment in the coefficients was made in the study. The greater the CS responded to the adjustment, the more critical is the use of an accurate coefficient [38]. A CS lower than 1 indicates the ESVs is inelastic to the coefficient and the estimation of ESVS is reliable. Otherwise, a CS greater than 1 indicates the estimation of ESVs is sensitive to the coefficient.

2.6 Correlation and regression analysis

The data of ESVs and landscape metrics was used to analysis the relationship between ecosystem service and land use pattern change. Because the spatial variation of landscape pattern exist among 12 villages in Hechuan town, the land use data of the three years (1982, 2002 and 2008) for the 12 villages can be considered as representing different landscape pattern on a time-for-space perspective [39]. Therefore, there were totally 36 sample data. Correlation and regression was employed for the relationship analysis, in which Multiple stepwise regression was specifically chosen considering the multicollinearity among landscape metrics. The dependents were the nine categories and total ESVs, while the corresponding independents were the landscape-level and class-level landscape metrics.

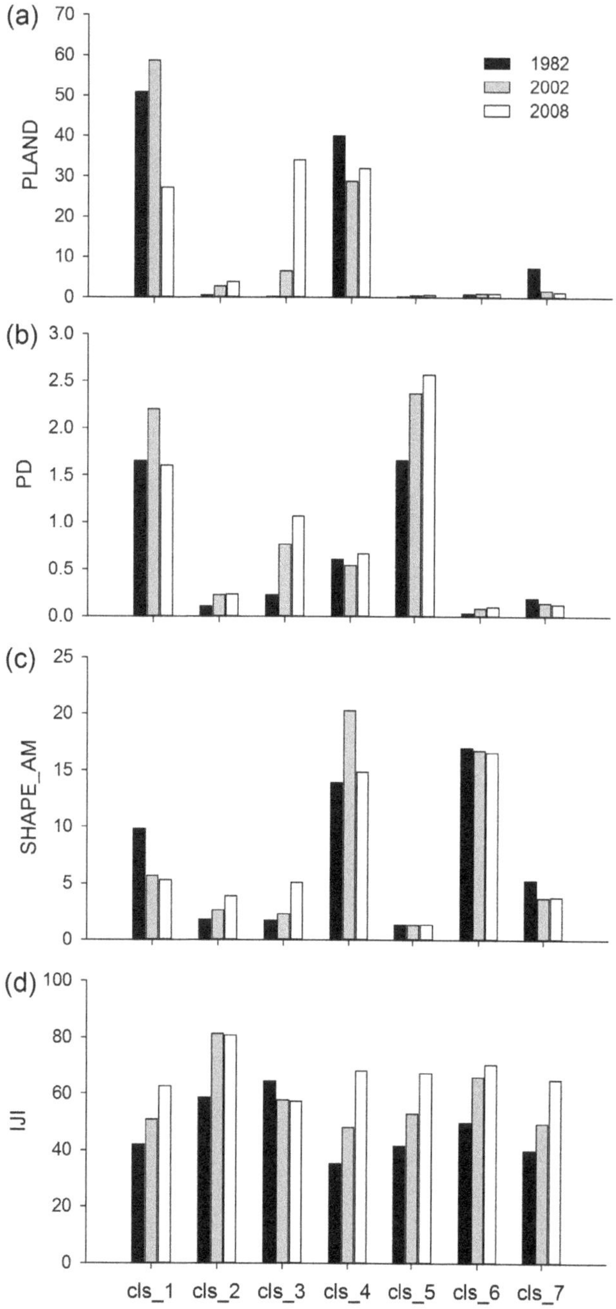

Figure 4. Landscape metrics at the class-level in Hechuan Town in 1982, 2002 and 2008. cls_1, cls_2, cls_3, cls_4, cls_5, cls_6, and cls_7 represent cropland, orchard, forestland, grassland, residential land, water body and unused land. PLAND: the percentage of landscape; PD: patch density; SHAPE_AM: area-weighted mean shape index; IJI: Interspersion and Justaposition Index.

Results

3.1 Changes of land use pattern

Table 3 showed the land use transition matrix. From 1982 to 2002, cropland as the dominant land use type increased from 50.83% to 58.76%. Grassland was the land use type with the largest change in area, decreasing from 40.01% to 28.77%. Orchard increased by 6.24% of total area, indicating the economic driving force of fruit trees on land use change. Forestland

Table 4. The change of ecosystem service values (ESVs) in Hechuan Town from 1982 to 2008.

		Cropland	**Orchard**	**Forestland**	**Grass land**	**Water body**	**Unused land**	**Total**
ESVs	1982	2.714	0.077	0.051	3.155	0.249	0.068	6.315
(10^6 US\$ yr^{-1})	2002	3.137	0.374	1.237	2.269	0.304	0.016	7.336
	2008	1.456	0.514	6.470	2.517	0.305	0.013	11.275
Change of ESVs	1982–2002	0.423	0.297	1.186	−0.887	0.054	−0.053	1.021
(10^6 US\$ yr^{-1})	2002–2008	−1.681	0.140	5.234	0.248	0.002	−0.003	3.939
	1982–2008	−1.258	0.437	6.419	−0.638	0.056	−0.056	4.960
Change of ESVS	1982–2002	2.248	55.387	333.431	−4.045	3.145	−11.081	2.328
(%)	2002–2008	−7.716	5.404	60.934	1.574	0.072	−2.992	7.731
	1982–2008	−6.674	81.578	1805.410	−2.913	3.232	−11.771	11.309
Average annual Change	1982–2002	0.112	2.769	16.672	−0.202	0.157	−0.554	0.117
(% yr^{-1})	2002–2008	−1.286	0.901	10.155	0.262	0.012	−0.498	1.289
	1982–2008	−0.256	3.137	69.439	−0.112	0.124	−0.452	0.435

Table 5. Values of different ecosystem service functions in 1982, 2002, and 2008.

	1982			2002			2008		
	ESVs (10^6 US$·$yr^{-1}$)	%	Rank	ESVs (10^6 US$·$yr^{-1}$)	%	Rank	ESVs (10^6 US$·$yr^{-1}$)	%	Rank
Gas regulation	0.678	10.73	6	0.826	11.26	6	1.529	13.56	5
Climate regulation	0.791	12.53	5	0.936	12.75	5	1.540	13.65	4
Water supply	0.800	12.67	4	0.960	13.09	4	1.610	14.28	3
Soil formation and retention	1.141	18.06	1	1.260	17.17	1	1.764	15.65	1
Waste treatment	0.938	14.85	2	1.015	13.84	3	1.078	9.56	6
Biodiversity protection	0.915	14.49	3	1.054	14.37	2	1.738	15.42	2
Food production	0.466	7.38	7	0.506	6.90	7	0.367	3.25	9
Raw material	0.247	3.91	9	0.390	5.32	8	0.881	7.81	7
Recreation and culture	0.339	5.37	8	0.388	5.29	9	0.768	6.81	8
Total	6.315	100.00		7.336	100.00		11.275	100.00	

increased from 0.57% to 2.78%, reflecting that ecological restoration began to gain attention. From 2002 to 2008, cropland and forestland changed significantly, decreasing from 58.76% to 27.27% and increasing from 6.51% to 34.05% respectively. Land use structure was transferred from cropland dominated (58.76%) to cultivated land (27.27%), forestland (34.05%) and grassland (31.91%) relatively balanced distributed.

The most notable change of land use from 1982 to 2002 was the conversion from grassland to cropland and forestland with 12.63% and 4.74% of the total area respectively. The conversions from cropland (2.50%) and unused land (3.98%) to grassland were not adequate to compensate for the grass loss. From 2002 to 2008, the notable changes of land use were cropland to forestland, cropland to grassland, and grassland to forestland, with the rates of 20.75%, 9.84%, and 7.17% respectively. It was found that the conversion among land use types was more outstanding and concentrated than that before 2002, reflecting that the Grain for Green project as an ecological policy had great influence on land use change.

The results of landscape-level metric analysis were exhibited in Fig. 3. The significant increased PD from 1982 to 2002 reflected the landscape fragmentation. It was relative to the increase of patches on the land use types with intense human disturbance, such as cropland, residential land and artificial reservoir. Oppositely, the slight change of PD from 2002 to 2008 reflected that human disturbance became stable. The change of human disturbance was also demonstrated by the change of LUII which increased before 2002 and decreased after 2002. SHAPE_AM decreased in the study period, showing the landscape became more regular in shape. The increase of IJI suggested that the landscape became more contiguous and the ecological connectivity among land use types increased. SHDI increase obviously from 2002 to 2008, which related to that the land use structure became even.

Fig. 4 showed the change of class-level metrics. The PLAND of land use types indicated that cropland, forestland, and grassland had significantly influence on land use pattern. PD in orchard, forestland, and residential land increased obviously, attributing to the increasing area of these land use types and the fragmental terrain. SHAPE_AM showed that cropland and unused land became more regular in shape, while orchard and forestland more complicated. IJI increased generally in land use types. Orchard was the most contiguous with high IJI, which was relative to its concentrated distribution across the river terrace.

3.2 ESVs from 1982 to 2008

The ESVs of each land use type and the total ESVs was shown in Table 4. The total ESVs of Hechuan town was US$ 6.315, US$ 7.336 and US$ 11.275 million in 1982, 2002 and 2008, respectively. From 1982 to 2002, the decline of ESVs caused by the decrease of grassland was offset by the increase of forestland, orchard and cropland, resulting that the total ESVs increased by US$ 1.021 million. From 2002 to 2008, the total ESVs increased by US$ 3.939 million, mainly due to the increase of forestland. The average annual change rate of total ESVs before and after 2002 was quite different, that is 0.81% and 8.95% respectively. It indicated the Grain to Green project implemented since 2002 had a significant effect on the ecosystem service. It was also shown from the value of ESVs produced by forestland occupying 57.39% of the total ESVs. Overall, the total ESVs increased US$ 4.960 million during the study period, mainly due to the increase of ESVs by the increase of forestland and orchard far beyond the decrease of ESVs by the decrease of cropland and grassland. It was essentially because of the higher coefficient value of forestland and orchard than that of cropland and grassland.

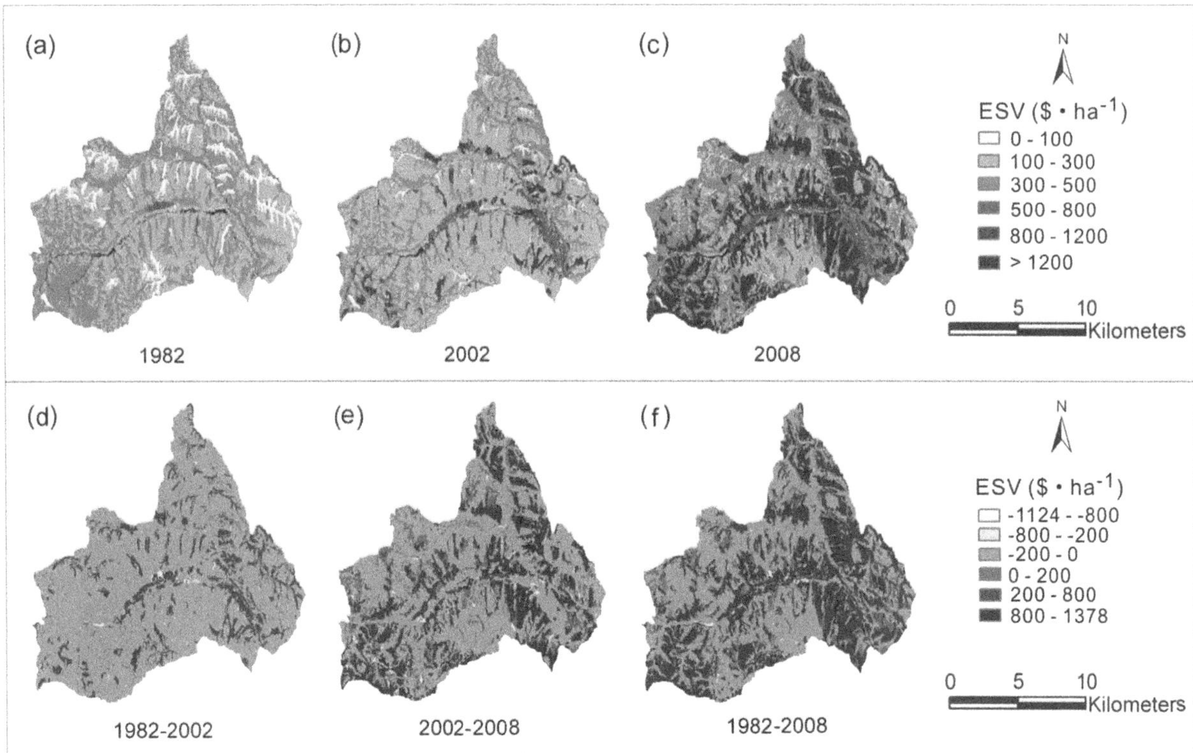

Figure 5. Spatial and temporal distribution of ecosystem service values (ESVs) in Hechuan Town from 1982 to 2008. The spatial distribution of ESVs in 1982 (a), 2002 (b) and 2008 (c), and the spatial-temporal changes of ESVs between time intervals from 1982 to 2002 (d), 2002 to 2008 (e) and 1982 to 2008 (f).

The ESVs of each ecosystem function type was shown in Table 5. Expect for food production, the values of ecosystem service functions increased especially after 2002. The decrease of food production was due to the great decline of cropland in the Grain to Green project. The ESVs proportion of each ecosystem function type to the total ESVs represented the contribution of each ecosystem function to the total ESVs. It was found that the functions of soil formation and retention, waste treatment, and food production were decline during 1982 to 2008, while other functions were improved. The rank of the contribution by each ecosystem service function was also estimated. It was basically stable except for relatively obvious decline in the rank of waste treatment and food production. In 2008, the rank order for each ecosystem service was as follows from high to low, soil formation and retention, biodiversity protection, water supply, climate regulation, gas regulation, waste treatment, raw material, recreation and culture, and food production. Soil formation and retention was the highest during the study period.

3.3 Spatial distribution of ESVs

Maps of ESVs in different periods (Fig. 5) showed the spatial distribution of ESVs of unit area in Hechuan town, directly reflecting the difference of ESVs among land use types. In 1982, the ESVs>4000 mostly appeared in the center of the town where river and river terrace located. It was because water body and orchard which intensely distributed in river terrace for its high water demand both had high ESVs. Therefore, due to the orchard increasing intensely and the forest increasing scatteredly, the increase of ESVs also mainly happened across the river terrace in 2002. Since 2008, the ESVs>4000 spread widely with the increase of forestland transformed from cropland. The lowest ESVs mostly occurred in the gully where unused land was distributed in 1982. With vegetation recovery in the gully, the low ESVs happened from gully to terraced hillside where cropland with low ESVs was distributed in 2008. Fig. 5d–f showed the temporal change of ESVs spatial distribution. The change characteristic of 2002 to 2008 was adjacent to that during the total study period, reflecting that the change of ESVs mainly occurred after 2002, just after the Grain to Green project.

3.4 Relationship between ESVs and land use pattern

From the above analysis on the change of land use and ESVs in quantity and spatial distribution, we could infer there was some relationship between land use change and ecosystem service. To quantitively understand the relationship, the correlation analysis and regression analysis between ESVs and landscape pattern metrics was conducted.

Table 6 showed there existed significant correlations between ESVs and many landscape metrics ($p<0.01$), which explained that landscape pattern affected ESVs significantly. For example, the correlation coefficients between total ESVs and landscape metrics showed that there existed significantly positive relationship between SHDI (0.433), PLAND_3 (0.677), SHAPE_AM_3 (0.744), IJI_4 (0.513) and ESVs, and negative relationship between LUII (−0.634), PLAND_1 (−0.752) and ESVs. It reflected that the diversity and intensity of land use had important effects on total ESVs. It also reflected that cropland, forestland and grassland were the land use types which had significant effects on total ESVs. On quantity,the less the cropland and the more the forestland, the higher the total ESVs were. As to the landscape shape, the more regular the cropland and the more complex the forestland, the higher the total ESVs were. The higher the IJI of grassland, the

Table 6. Correlation coefficients between ecosystem service values (ESVs) and landscape pattern metrics.

	TESVs	ESVs_1	ESVs_2	ESVs_3	ESVs_4	ESVs_5	ESVs_6	ESVs_7	ESVs_8	ESVs_9
PD	0.035	0.497*	0.509*	0.539*	0.477*	0.516*	0.499*	−0.221	0.547	0.478*
SHAPE_AM	0.326	−0.216	−0.220	−0.188	−0.177	−0.088	−0.205	0.026	−0.290	−0.166
IJI	0.292	0.624*	0.635*	0.639*	0.597*	0.534*	0.621*	−0.293	0.687	0.586*
SHDI	0.433*	0.763*	0.764*	0.766*	0.741*	0.507*	0.765*	−0.636*	0.775*	0.765*
LUII	−0.634*	−0.681*	−0.658*	−0.618*	−0.675*	−0.113	−0.684*	0.977*	−0.599*	−0.734*
PLAND_1	−0.752*	−0.810*	−0.795*	−0.772*	−0.811*	−0.334	−0.815*	0.952*	−0.742*	−0.853*
PD_1	0.045	−0.055	−0.063	−0.056	−0.054	−0.113	−0.051	−0.134	−0.091	−0.022
SHAPE_AM_1	−0.476*	−0.369	−0.358	−0.330	−0.368	−0.063	−0.369	0.495	−0.328	−0.390
IJI_1	0.189	0.542*	0.552*	0.530*	0.510*	0.405	0.533*	−0.199	0.619	0.485*
PLAND_2	0.323	0.527*	0.541*	0.590*	0.520*	0.595*	0.534*	−0.246	0.558	0.525*
PD_2	0.420	0.457*	0.471*	0.515*	0.449	0.529*	0.463*	−0.193	0.491	0.452
SHAPE_AM_2	0.159	0.450	0.468*	0.541*	0.439	0.629*	0.460*	−0.149	0.493	0.452
IJI_2	0.207	0.392	0.409	0.483*	0.376	0.583*	0.402	−0.115	0.438	0.396
PLAND_3	0.677*	0.984*	0.983*	0.941*	0.975	0.558*	0.980*	−0.770*	0.988*	0.961*
PD_3	0.276	0.631*	0.637*	0.629	0.625	0.477*	0.629*	−0.383	0.653	0.606*
SHAPE_AM_3	0.744*	0.828*	0.827*	0.780*	0.820	0.449*	0.821*	−0.623*	0.836*	0.799*
IJI_3	0.040	0.231	0.241	0.291	0.208	0.356	0.236	−0.069	0.276	0.231
PLAND_4	0.224	−0.192	−0.212	−0.203	−0.159	−0.298	−0.181	−0.264	−0.311	−0.110
PD_4	−0.294	0.199	0.201	0.190	0.160	0.101	0.192	−0.086	0.257	0.171
SHAPE_AM_4	0.455	−0.061	−0.065	−0.049	−0.028	−0.029	−0.053	−0.086	−0.125	−0.022
IJI_4	0.513*	0.717*	0.719*	0.705*	0.697*	0.457*	0.715*	−0.539*	0.739*	0.701*
PLAND_5	−0.035	0.290	0.313	0.381	0.279	0.578*	0.297	0.081	0.357	0.271
PD_5	−0.047	0.244	0.269	0.322	0.244	0.548*	0.248	0.188	0.314	0.208
SHAPE_AM_5	0.118	0.081	0.082	0.053	0.072	−0.009	0.075	−0.004	0.105	0.055
IJI_5	0.307	0.461*	0.470*	0.525*	0.446*	0.512*	0.471*	−0.312	0.482	0.477*
PLAND_6	0.047	0.139	0.167	0.360	0.148	0.852	0.170	0.064	0.153	0.207
PD_6	−0.160	0.122	0.137	0.198	0.105	0.371	0.128	0.080	0.172	0.118
SHAPE_AM_6	0.378	0.088	0.086	0.140	0.080	0.141	0.099	−0.218	0.067	0.137
IJI_6	0.020	0.201	0.217	0.276	0.197	0.438	0.208	0.028	0.239	0.197
PLAND_7	−0.385	−0.447	−0.474*	−0.525*	−0.507*	−0.769*	−0.456*	−0.041	−0.442	−0.422
PD_7	−0.270	−0.105	−0.129	−0.182	−0.154	−0.494*	−0.114	−0.245	−0.108	−0.089
SHAPE_AM_7	−0.236	−0.313	−0.329	−0.434	−0.348	−0.650*	−0.333	0.150	−0.279	−0.358
IJI_7	0.210	0.416	0.408	0.287	0.402	−0.083	0.394	−0.254	0.443	0.341

TESVs: the total ecosystem service values (ESVs); ESVs_1: the ESVs of gas regulation; ESVs_2 climate regulation; ESVs_3: the ESVs of water supply; ESVs_4: the ESVs of soil formation and retention; ESVs_5: the ESVs of waste treatment, ESVs_6: the ESVs of biodiversity protection; ESVs_7 the ESVs of food production; ESVs_8: the ESVs of raw material; ESVs_9: the ESVs of recreation and culture.
PD: patch density; SHAPE_AM: area-weighted mean shape index; IJI: Interspersion and Justaposition Index; SHDI: Shannon's diversity index; LUII: land use intensity index; PLAND: percentage of landscape. The 1, 2, 3, 4, 5, 6, 7 after the above landscape metrics respects different landscape, that is cropland, orchard, forestland, grassland, residential land, water body and unused land, respectively.
*significant at 0.01 level.

Table 7. Regression analysis between ecosystem service values (ESVs) and landscape patterns (n = 36).

Dependent	Standardized coefficients regression	R^2	Sig.
Gas regulation	0.878×PLAND_3+0.166×PLAND_2-0.099×PLAND_1-0.068×IJI_1	0.990	*
Climate regulation	0.790×PLAND_3-0.197×PLAND_7-0.190×LUII+0.081×PLAND_2	0.998	*
Water supply	0.665×PLAND_3-0.317×PLAND_7-0.254×LUII+0.106×PLAND_2	0.955	*
Soil formation and retention	0.684×PLAND_3-0.301×PLAND_7-0.284×LUII+0.066×PLAND_2	0.998	*
Waste treatment	0.672×PLAND_6+0.365×PLAND_3-0.352×PLAND_7+0.051×PLAND_2+0.049×PLAND_5	0.993	*
Biodiversity protection	0.059×SHDI +0.861×PLAND_3-0.033×SHAPE_AM_3+0.133×PLAND_1	0.967	*
Food production	0.742×LUII-0.173×PLAND_3-0.052×SHDI+0.106×PLAND_1	0.991	*
Raw material	0.964×PLAND_3+0.091×SHDI+0.068×LUII	0.981	*
Recreation and culture	−0.747×PLAND_1+0.380×IJI_1	0.853	*
Total	-0.588×PLAND_1+0.569× SHAPE_AM_3-0.303×SHDI	0.709	*

*significant at 0.01 level.
PLAND_1: the percentage of cropland; PLAND_2: the percentage of orchard; PLAND_3: the percentage of forestland; PLAND_5: the percentage of residential land; PLAND_6: the percentage of water body; PLAND_7: the percentage of unused land; SHAPE_AM_3: the area-weighted mean shape index of forestland; IJI_1: the Interspersion and Justaposition Index of cropland; LUII: land use intensity index; SHDI: Shannon's diversity index.

higher the total ESVs were. This indicated that the connectivity of grassland was important for ecosystem service.

Correlation also occurred between ESVs of all the functions and landscape metrics (Table 6). However, the relationships between ESVs of different functions and landscape pattern were different. For example, the correlation between food production and landscape pattern was almost opposite from that between other ecosystem functions and landscape pattern. For example, PLAND_1 had a positive effect on food production; SHDI, PLAND_3, SHAPE_3, and IJI_4 had a negative effect on food production. It could infer that there were contradictions between food production and other ecosystem functions.

As shown in Table 7, the result of regression analysis further explained that the ESVs was correlated significantly with landscape pattern. The total ESVs could be predicted by PLAND on cropland, SHAPE_AM on grassland, and SHDI. ESVs of all kinds of ecosystem functions also could be explained by landscape metrics. These regression equations indicated that landscape-level metrics (such as SHDI and LUII) and class_level metrics (such as PLAND of forestland, orchard, and cropland, unused land, SHAPE of forestland, IJI of cropland) acted as predictors for categories of ecosystem services. Specifically, the proportion of forest (PLAND_3) accounted for almost all of the categories of ecosystem services.

Discussion

4.1 Reliability of ESVs

This study estimated ESVs by multiplying the area for each land use types by the corresponding value coefficients. As discussed in the previous researches, estimations using this method was coarse with high variation and uncertainty for the following reasons, limitations on the economic evaluation [1], problems of double counting and scales [40,41,42], the complex, dynamic and nonlinear ecosystems [43], the imperfect matches of land use categories as proxies [38] and the accuracy of the ecosystem value coefficients [5]. This study also existed such uncertainty on ESVs estimation. For example, the value coefficient of orchard, determined by the average of forest and grassland, was an approximate estimation and need a further exploration. However, the estimation of temporal variation on ESVs was considered to be more reliable than that of cross-sectional analysis [5]. In addition, the sensitivity analysis of the estimated ESVs with 50% adjustment in the value coefficients was conducted. The result showed that the sensitivity coefficients of all land use categories were lower than 1 (Table 8), which suggested that despite of the above limitations, the estimated ESVs are reliable and useful for subsequent study.

4.2 Relationship between ESVs and landscape pattern

It is usually assumed that land use can affect the ecosystem service. Moreover, a few studies showed that there was a correlation between landscape pattern and ESVs [41,44]. This study signified this statement at town scale on the Loess Plateau. Land use configuration, land use intensity, landscape diversity, fragmentation and connectivity all affected ecosystem service.

The correlation analysis between ESVs and PLAND implied land use structure had significant impact on ecosystem service. Especially, the increase of forestland and the decrease of cropland played an important part in improving the ESVs in the past twenty years. It is closely related to the Grain to Green project comprehensively started in study area since 2002. In the project,

Table 8. The coefficient of sensitivity (CS) resulting from adjustment of ecosystem valuation coefficients.

	Cropland	Orchard	Forestland	Grass land	Water body	Unused land
1982	−0.430	−0.012	−0.008	−0.500	−0.039	−0.011
2002	−0.428	−0.052	−0.169	−0.309	−0.041	−0.002
2008	−0.129	−0.048	−0.574	−0.223	−0.027	−0.001

measures for optimizing land use structure were implemented, including restoring slope cropland into forest and grassland, banning grazing, transforming slopes into terraces, and building reservoirs, etc. Forestland and grassland increased by 423.19% (27.54% of the study area) and 10.93% (3.15% of the study area), and cropland decreased by 53.59%(31.49% of the study area) (Table 3). The increase of ESVs due to the increase of forestland occupied 46.28% of the total ESVs in 2008 (Table 4). The result of the correlation analysis between ESVs and PLAND reflected that vegetation recovery could strongly enhance ecosystem service, and it was coincident with many other studies on the Loess Plateau [17,18,23,45]. LUII, which also related to the proportion of land use types, implied the intensity of human activities. This study showed land use intensity had a negative effect on ecosystem service with negative correlation coefficients (−0.634) (Table 6). It was coincident with some studies on ESVs change under urbanization [5,31]. These studies showed that urbanization which means the increase of land use intensity led to considerable declines in ESVs.

Landscape diversity always presents high positive relevance with biodiversity [46]. Our results were coincident to previous statements given the positive relationships between SHDI and biodiversity conservation. However, there were studies reporting the negative relationships between them, in which the increase of SHDI was the result of rapid urban sprawl [31]. In our study, the increase of SHDI was because land use structure became more balanced, which was the result of the increase of forestland. In addition, landscape diversity could also promote agricultural production [47]. Our study disagreed with this statement, and showed that food production was weakened with landscape diversification. It was because that the increase of SHDI was the result of a larger number of conversion from cropland to forestland. Therefore, the relationship between landscape diversity and biodiversity conservation as well as food production should not be treat as the same but be understood considering the driving force of SHDI change.

Fragmentation could lead to declining habitat quality, lower wildlife survival, and limited movement of soil microorganisms [48], and subsequently cause the decrease of ecosystem service [30]. Our study disagreed with this statement. For example, PD of the total landscape, PD_Forest, PD_orchard and shape_ Forest revealed significantly positive impacts on most categories of ESVs (Table 6–7). The increase of PD and the decrease of connectivity of landscape were usually simultaneous, which is disagreed in our study (Fig. 3 and Fig. 4). The landscape became more contiguous as IJI shown. Table 6 showed the IJI had significantly positive impacts on ESVs. Especially, the increase of IJI of grassland promoted the total ESVs and all categories of ESVs. This maybe because the connectivity of landscape has contribution to habitat corridors [49] and forest production [50].

Based on the relationship between ESVs and landscape pattern, we could improve the ecosystem service by the adjustment of land use policy. On the one hand, continuing to implement the Grain to Green project is helpful for improving ESVS, because it could increase the vegetation coverage, decline the intensity of land use, and make cropland become regular by canceling the slope cropland. On the other hand, diversified agriculture gathering planing fruit trees, planting crops and breeding, which could promote the diversification of land use, should be encouraged to increase both ESVs and farmer's incomes.

Conclusion

ESVs at town scale in the Loess Plateau were estimated in Hechuan town of Ningxia Hui Autonomous Region from 1982 to 2008. It was concluded that ESVs varied with land use change. ESVs in 1982, 2002, and 2008 were US$ 6.315, US$ 7.336 and US$ 11.275 million respectively. Among all the land use types, forestland, grassland and cropland had important contribution (> 90%) on ESVs. The total ESVs increased slowly by 16.17% due to the decrease of grassland from 1982 to 2002, while the total ESVS increased significantly by 67.61% due to the increase of forestland from 2002 to 2008. Areas with high services level were mainly located in the center due to orchard and east due to forestland, while areas with low services level mainly located in the north and south sides due to cropland.

Land use pattern had a significant effect on ecosystem service in our study by analyzing and discussing the relationship between landscape pattern and ESVs. The proportion of forestland had a positive effect on ecosystem service while that of cropland had a negative effect on ESVs. The diversity and interspersion of landscape both had a positive effect on ESVs. Land use intensity which reflects the intensity of human activities had a negative effect on ESVs. Fragmentation had positive effect on ESVs, which was disagreed with the previous studies because the fragmentation in study area was related to the increased patch of such land use types as forestland, water body, orchard.

Based on the results of this study, it was conclude that land use pattern was important for ecosystem service. Therefore, we could improve the ecosystem service by the adjustment of land use policy. Continuing the Grain to Green project is reasonable and significant because it could increase the vegetation coverage and decline land use intensity. Diversified agriculture collecting planing fruit trees, growing food and breeding should be encouraged, because it could not only promote ecosystem service by increasing landscape diversification but also improve people's incomes.

Author Contributions

Conceived and designed the experiments: XF. Analyzed the data: XF. Contributed reagents/materials/analysis tools: GAT BCL. Contributed to the writing of the manuscript: XF GAT RMH.

References

1. Costanza R, Arge DR, Groot DR, Farber S, Grasso M, et al. (1997) The value of the world's ecosystem services and natural capital. Nature 387: 253−260.
2. Costanza R, Cumberland J, Daly H, Goodland R, Norgaard R (1997) An Introduction to ecological economics. Delray Beach Fla USA: St Lucie Press.
3. Kreuter UP, Harris HG, Matlock MD, Lacey RE (2001) Change in ESVs in the San Antonio area, Texas. Ecological Economics 39: 333−346.
4. Ronald CE, Yuji M (2013) Landscape pattern and ESV changes: Implications for environmental sustainability planning for the rapidly urbanizing summer capital of the Philippines. Landscape and Urban Planning 116: 60−72.
5. Li TH, Li WK, Qian ZH (2010) Variations in ESV in response to land use changes in Shenzhen. Ecological Economics 69: 1427−1435.
6. Cai YB, Zhang H, Pan WB, Chen YH, Wang XR (2013) Land use pattern, socio-economic development, and assessment of their impacts on ESV: study on natural wetlands distribution area (NWDA) in Fuzhou city, southeastern China. Environ Monit Assess 185: 5111−5123.
7. Zaehle S, Bondeau A, Carter RT, Cramer W, Erhard M, et al. (2007) Projected changes in terrestrial carbon storage in europe under climate and land-use change, 1990–2100. Ecosystems 10: 380−401.
8. Eliska L, Jana F, Edward N, David V (2013) Past and future impacts of land use and climate change on agricultural ecosystem services in the Czech Republic. Land Use Policy, 33: 183−194.
9. Vitousek PM, Mooney HA, Lubchenco J, Melillo JM (1997) Human domination of earth's ecosystems. Science 277: 494–499.
10. Nasiri F, Huang GH (2007) Ecological viability assessment: A fuzzy multi-pleattribute analysis with respect to three classes of ordering techniques. Ecol Inform 2: 128–137.

11. Collin ML, Melloul AJ (2001) Combined land-use and environmental factors for sustainable groundwater management. Urban Water 3: 229–237.
12. Schmidta JP, Mooreb R, Alber M (2014) Integrating ecosystem services and local government finances into land use planning: A case study from coastal Georgia. Landscape and Urban Planning 122: 56–67.
13. Ernesto FV, Federico CF (2006) Land-use options for Del Plata Basin in South America: Tradeoffs analysis based on ecosystem service provision. Ecological Economics 57: 140–151.
14. Christine F, Susanne F, Anke W, Lars K, Franz M (2013) Assessment of the effects of forestland use strategies on the provision of ecosystem services at regional scale. Journal of Environmental Management 127: 96–116.
15. Ignacio P, Berta M, Pedro Z, David GDA, Carlos M (2014) Deliberative mapping of ecosystem services within and around Donana National Park (SW Spain) in relation to land use change. Reg Environ Change14: 237–251.
16. Mendoza-Gonzalez G, Martinez ML, Lithgow D, Perez-Maqueo O, Simonin P (2012) Land use change and its effects on the value of ecosystem services along the coast of the Gulf of Mexico. Ecological Economics 82: 23–32.
17. Su CH, Fu BJ (2013) Evolution of ecosystem services in the Chinese Loess Plateau under climatic and land use changes. Global and Planetary Change 101: 119–128.
18. Si J, Nasiri FZ, Han P, Li TH (2014) Variation in ESVs in response to land use changes in Zhifanggou watershed of Loess plateau: a comparative study. Environmental Systems Research 3: 2.
19. Turner MG, Gardner RH, O'Neill RV (2001) Landscape Ecology in theory and practice. New York: Springer-Verlag.
20. Ritsema CJ (2003) Introduction: soil erosion and participatory land use planning on the Loess Plateau in China. Catena 54: 1–5.
21. Fu BJ, Wang YF, Lu YH, He CS, Chen LD, et al. (2009) The effects of land-use combinations on soil erosion: a case study in the Loess Plateau of China. Progress in Physical Geography 33: 793–804.
22. Fu BJ, Chen DX, Qiu Y, Wang J, Meng QH (2002) Land Use Structure and Ecological Processes in the LoessHilly Area, China. Beijing: Commercial Press. (in Chinese).
23. Jing L, Zhiyuan R (2011) Variations in ESV in Response to Land use Changes in the Loess Plateau in Northern Shaanxi Province, China. Int. J. Environ. Res 5: 109–118.
24. Zhang TB, Tang JX, Liu DZ (2006) Feasibility of Satellite Remote Sensing Image About Spatial Resolution. Journal of Earth Sciences and Environment 28: 79–83.
25. Chu YF, Li ES, Lu J, Zhang KK (2007) The Adaptability Analysis to the Satellite Image Spatial Resolution and Mapping Scale. Hydrographic Surveying and Charting 27: 47–50.
26. Li Q, Liu C, Xi CY, Liu ML (2002) Cartographic Generalization of Digital Land Use Current Situation Map. Bulletin of Surveying and Mapping 9: 59–63.
27. Congalton RG (1991) A review of assessing the accuracy of classifications of remotely sensed data. Remote Sensing of Environment 37: 35–46.
28. Wang Y, Gao JX, Wang JS, Qiu J (2014) Value Assessment of Ecosystem Services in Nature Reserves in Ningxia, China: A Response to Ecological Restoration. PloS One 9: e89174. doi:10.1371/journal.Pone.0089174.
29. Ribeiro SC, Lovett A (2009) Associations between forest characteristics and socio-economic development: a case study from Portugal. Journal of Environmental Management 90: 2873–2881.
30. Su S, Jiang Z, Zhang Q, Zhang Y (2011) Transformation of agricultural landscapes under rapid urbanization: a threat to sustainability in Hang-Jia-Hu region, China. Applied Geography 31: 439–449.
31. Su SL, Xiao R, Jiang ZL, Zhang Y (2012) Characterizing landscape pattern and ESV changes for urbanization impacts at an eco-regional scale. Applied Geography, 34: 295–305.
32. Li H, Wu J (2004) Use and misuse of landscape indices. Landscape Ecology 19: 389–399.
33. Xie GD, Lu CX, Xiao Y, Zheng D (2003) The Economic Evaluation of Grassland Ecosystem Services in Qinghai Tibet Plateau, Journal of Mountain Science 21: 50–55. (in Chinese).
34. Xie GD, LU CX, Leng YF, Zheng D, Li SC (2008) Ecological assets valuation of Tibetan Plateau. Journal of Natural Resources 18: 190–196. (in Chinese).
35. Liu DL, Li BC, Liu Xianzhao Z, Warrington DN (2011) Monitoring land use change at a small watershed scale on the Loess Plateau, China: applications of landscape metrics, remote sensing and GIS. Environmental Earth Sciences 64: 2229–2239.
36. Pan WKY, Walsh SJ, Bilsborrow RE, Frizzelle BG, Erlien CM, et al. (2004) Farm-level models of spatial patterns of land use and land cover dynamics in the Ecuadorian Amazon. Agriculture, Ecosystems and Environment 101: 117–134.
37. de Groot RS, Wilson MA, Boumans RMJ (2002) A typology for the classification, description and valuation of ecosystem functions, goods and services. Ecological Economics 41: 393–408.
38. Kreuter UP, Harris HG, Matlock MD, Lacey RE (2001) Change in ESVs in the San Antonio area, Texas. Ecological Economics 39: 333–346.
39. Wu J, Jenerette GD, Buyantuyev A, Redman CL (2011) Quantifying spatiotemporal patterns of urbanization: the case of the two fastest growing metropolitan regions in the United States. Ecological Complexity 8: 1–8.
40. Turner RK, Paavola J, Coopera P, Farber S, Jessamya V, et al. (2003) Valuing nature: lessons learned and future research directions. Ecological Economics 46: 493–510.
41. Hein L, Koppen VK, de Groot RS, van Ierland EC (2006) Spatial scales, stakeholders and the valuation of ecosystem services. Ecological Economics 57: 209–228.
42. Konarska KM, Sutton PC, Castellon M (2002) Evaluating scale dependence of ecosystem service valuation: a comparison of NOAA-AVHRR and Landsat TM datasets. Ecological Economics 41: 491–507.
43. Limburg KE, O' Neill RV, Costanza R, Farber S (2002) Complex systems and valuation. Ecological Economics 41: 409–420.
44. Zhang MY, Wang KL, Liu HY, Zhang CH (2011) Responses of Spatial-temporal Variation of Karst Ecosystem Service Values to Landscape Pattern in Northwest of Guangxi, China. Chin. Geogra. Sci. 21: 446–453.
45. Deng L, Shangguan ZP, Li R (2012) Effects of the grain-for-green program on soil erosion in China. International Journal of Sediment Research 27: 120–127.
46. Nagendra H (2002) Opposite trends in response for the Shannon and Simpson indices of landscape diversity. Applied Geography, 22: 175–186.
47. Shrestha RP, Schmidt-Vogt D, Gnanavelrajah N (2010) Relating plant diversity to biomass and soil erosion in a cultivated landscape of the eastern seaboard region of Thailand. Applied Geography 30: 606–617.
48. Sherrouse BC, Clement JM, Semmens DJ (2011) A GIS application for assessing, mapping, and quantifying the social values of ecosystem services. Applied Geography 31: 748–760.
49. Li M, Zhu Z, Vogelmann JE, Xu D, Wen W, et al. (2011) Characterizing fragmentation of the collective forests in southern China from multitemporal Landsat imagery: a case study from Kecheng district of Zhejiang province.Applied Geography 31: 1026–1035.
50. Long JA, Nelson TA, Wulder MA (2010) Characterizing forest fragmentation: distinguishing change in composition from configuration. Applied Geography 30: 426–435.

19

Early Spring, Severe Frost Events, and Drought Induce Rapid Carbon Loss in High Elevation Meadows

Chelsea Arnold, Teamrat A. Ghezzehei, Asmeret Asefaw Berhe*

School of Natural Sciences, University of California Merced, Atwater, California, United States of America

Abstract

By the end of the 20th century, the onset of spring in the Sierra Nevada mountain range of California has been occurring on average three weeks earlier than historic records. Superimposed on this trend is an increase in the presence of highly anomalous "extreme" years, where spring arrives either significantly late or early. The timing of the onset of continuous snowpack coupled to the date at which the snowmelt season is initiated play an important role in the development and sustainability of mountain ecosystems. In this study, we assess the impact of extreme winter precipitation variation on aboveground net primary productivity and soil respiration over three years (2011 to 2013). We found that the duration of snow cover, particularly the timing of the onset of a continuous snowpack and presence of early spring frost events contributed to a dramatic change in ecosystem processes. We found an average 100% increase in soil respiration in 2012 and 2103, compared to 2011, and an average 39% decline in aboveground net primary productivity observed over the same time period. The overall growing season length increased by 57 days in 2012 and 61 days in 2013. These results demonstrate the dependency of these keystone ecosystems on a stable climate and indicate that even small changes in climate can potentially alter their resiliency.

Editor: Benjamin Poulter, Montana State University, United States of America

Funding: Funding for this study was provided by University of California Merced startup funds and Graduate Research Council grants to AAB and TAG. The funders had no role in study design, data collection and analysis, decision to publish, or preparation of the manuscript.

Competing Interests: The authors have declared that no competing interests exist.

* Email: AABerhe@UCMerced.edu

Introduction

Magnitude and timing of extreme weather events have recently gained attention for their potential to alter ecosystem processes [1–4]. The presence of more extreme weather events has increased concerns over the ability of natural ecosystems to respond to such rapid changes [5]. Extreme interannual change in weather (for example, from a very "wet" to a very "dry" year and vice versa) may trigger rapid carbon loss from an ecosystem [6,7]. For high elevation mountain ecosystems in particular, the seasonal timing of the accumulation and melting of the snowpack is crucial for supplying abundant water to low-lying communities and high-elevation forests [8]. It is also essential for promoting meadow productivity [9] and soil carbon storage [6,10]. It is expected that earlier snowmelt will result in drying of meadow soils over the course of the growing season. This drying may lead to increased carbon storage through an increase in the net primary productivity of the system, but it can also lead to a loss of carbon through increased rates of decomposition. Whether the ecosystem remains a sink for or shifts to a source of atmospheric carbon dioxide (CO_2) will have large implications on the ability of the meadow to filter, store, and release water to the river systems. Prolonged conditions that result in a significant loss of carbon can eventually trigger a tipping point to an ecological regime shift in the meadow.

The coupled hydrological and biogeochemical cycles in high elevation meadows are influenced by the depth and duration of the annual winter snowpack that acts as an insulating blanket during the winter [11]. Not only does the winter snowpack protect the meadow soils from large temperature fluctuations and winter desiccation, it also functions to recharge the meadow soils during the spring snowmelt [12]. There is a vital two-way relationship between hydrology and soil organic matter (SOM) dynamics in such high elevation systems. Hydrology exerts a strong control on storage, stability, and composition of SOM [13] in the meadow soils, while SOM dynamics controls the ability of the meadows to provide ecosystem services such as filtering, storing and releasing water to the river systems. Without these wetland systems to slow the passage of water from the snowpack to the streams, the watersheds become less resilient to flood pulses [14]. The essential nature of those ecosystem services warrants a "keystone" status of mountain meadows in terms of mountain hydrology. A keystone species is one that has a disproportionately large impact to an ecosystem in comparison to its abundance. Mountain meadows, though small in aerial extent in the Sierra Nevada, are an essential component of the mountain water cycle. Watersheds that have lost meadow functioning due to degradation have limited water storage capacity and ability to attenuate floods [15]. Degradation of meadows results in a flashy system, where the surface and shallow subsurface flows in the watershed respond rapidly to precipitation events. Furthermore, high-elevation meadows, which are hotspots of biodiversity [16] and function as breeding grounds for many organisms in the central Sierra Nevada Mountains and

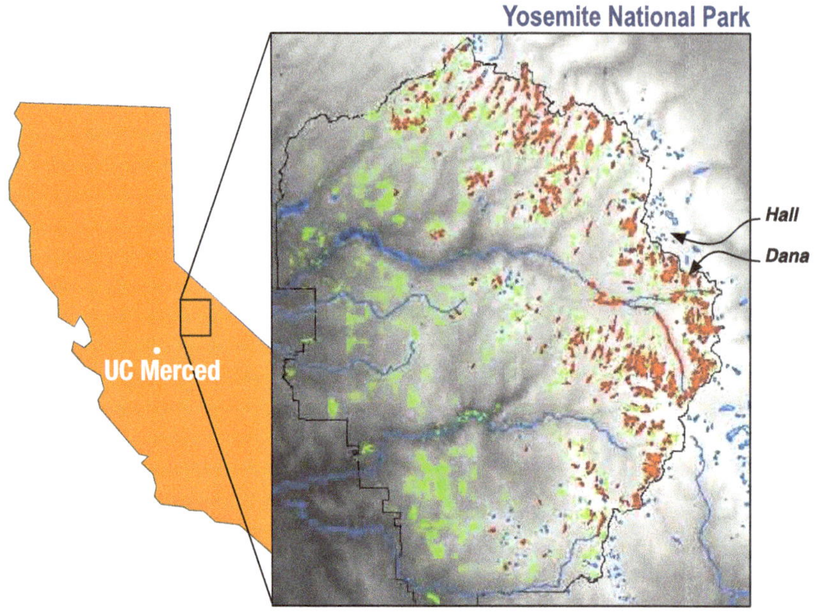

Figure 1. Map of study sites along the boundary of Yosemite National Park, California. Polygons represent the extent of meadow area in Yosemite National Park, with subalpine meadows (>3 hectares and between 2600–3200 m) highlighted in red.

other similar ecosystems, are likely to be a key indicator of the overall health of the watersheds.

While interannual variations in snowpack depth and duration are normal in the Sierra Nevada [17], consecutive years with extreme water conditions can significantly increase or decrease the overall length of the summer growing season and duration of snow free days, which will directly affect soil carbon storage. Previously, hydrologic modeling research in the Sierra Nevada has highlighted the sensitivity of that region's watersheds to earlier onset of spring and increased duration of low flows [18]. It was shown that some watersheds that are highly vulnerable to an increase in duration of low flows with climate warming, also occupy the largest mountain meadow area. An increase in the duration of low flows can cause meadows systems to dry down significantly, causing feedbacks to ecosystem processes such as primary productivity and soil respiration. If the trend in the onset of spring [19] continues, and meadows dry down earlier in the growing season, we can expect an increase in the decomposition of soil organic matter as the normally saturated soils become aerobic. This could also potentially impact river systems through a reduction in the ability of meadows to contribute to baseflow as they dry down. In addition, the timing of the onset of snowcover in the early winter can impact meadow soils and biota due to the widely fluctuating soil and air temperatures. The meadow soils in theses systems remain at 0°C as soon as snow accumulates in a continuous snowpack. This insulating layer protects overwintering biota and prevents drying of the meadow soil. Colder winter temperatures coupled to lack of continuous snowpack renders meadow soils and biota susceptible to severe desiccation which will impact ecosystem processes such as soil respiration and productivity in the following summer growing season.

An increase in interannual variation in the onset of spring has the potential to dramatically affect the balance between carbon storage and loss. This would occur mainly through changes to the input of carbon from above and belowground biomass [20], and loss through soil respiration and leaching [21,22] in meadow soils. A period of rapid carbon loss for the organic-rich high-elevation meadow soils can trigger a positive feedback loop that contributes to declining soil moisture [23], further organic matter decomposition and reduced plant productivity through changes in soil structure [24].

In mountain meadow ecosystems, mean changes in the timing of spring snowmelt have already been shown to influence plant phenology [25,26], interactions between plants and pollinators [27], and longer term changes in meadow vegetation community structure [28]. In order to examine how the timing and duration of snow cover and presence of early season frost events can influence ecosystem processes, (net primary productivity and soil respiration), we monitored changes in surface carbon dioxide flux and above ground net primary productivity over three consecutive summers (2011 to 2013) in two high elevation meadows in the Central Sierra Nevada mountain range of California.

Methods

Methodology

The objective of this study was to track ecological responses of high elevation meadows to extreme seasonality. We combine field-based measurements of soil respiration and aboveground net primary productivity with remote sensing techniques to gauge how the amount and timing of precipitation, and seasonality impact meadow systems.

Site description

Our study was conducted in two subalpine meadows with different hydrologic regimes located at the crest of the Sierra Nevada mountain range along the boundary of Yosemite National Park (YNP) (Figure 1). Both meadows were formed as a direct result of past glaciation. Their resulting geomorphic position in the landscape remains conducive to high water tables throughout much of the growing season. One meadow is located in the Harvey Monroe Hall Research Natural Area (Hall RNA) at 3200-m elevation on a large medial moraine on the eastern side of the central Sierra Nevada. The mean daily temperatures range from

Table 1. Historic ranking of SWE in Dana Meadow.

The top 12 driest years on record (1927–2012)				The top 12 wettest years on record (1927–2012)			
Rank	**Driest**	**SWE (cm)**	**% of normal**	**Rank**	**Wettest**	**SWE (cm)**	**% of normal**
1	**2012**	**19.2**	**25**	1	1969	147.32	192.0
2	1977	19.81	25.8	2	1995	133.35	173.8
3	1976	30.23	39.4	3	2006	122.43	159.6
4	1931	33.27	43.4	4	1986	121.16	157.9
5	**2012**	**37.59**	**49.0**	5	1978	120.40	156.9
6	1990	38.61	50.3	6	1998	119.89	156.3
7	1964	39.88	52.0	**7**	**2011**	**119.38**	**155.6**
8	1929	40.64	53.0	8	1952	118.36	154.3
9	1994	44.70	58.3	9	1956	115.57	150.7
10	1992	45.21	58.9	10	1980	113.54	148.0
11	1939	45.47	59.3	11	2005	106.93	139.4
12	1930	46.74	60.9	12	1993	105.41	137.4

Recent years that were in the top 10 wettest or driest years are in bold.

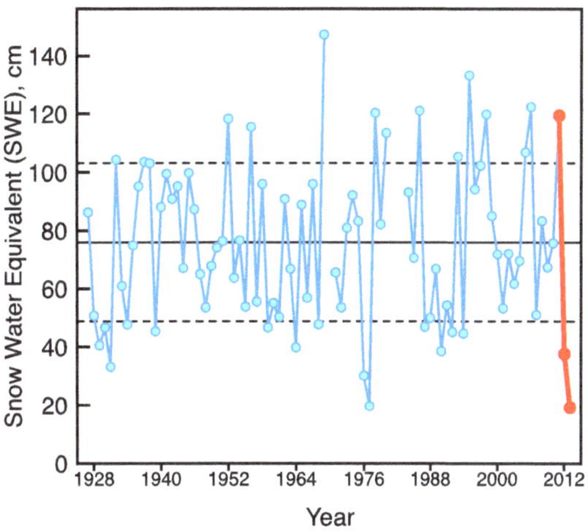

Figure 2. Historic SWE record for Dana meadows (Yosemite National Park) with 2011–2013 highlighted in red.

−4.9°C to 12.9°C [29]. The soils in the Hall RNA are characterized as Inceptisols with the suborders Andic Cryumbrepts and Lithic Cryumbrepts [30]. For a contrasting type of meadow, we chose Dana Meadows, which is located at a 3000-m elevation along YNP's Tioga Pass Road in a U-shaped glacial valley with hummocky ablation till. Dana Meadows exhibits mean temperatures similar to those of the Hall RNA and an average precipitation of 1000 mm/year. The soils in Dana Meadows are classified as Inceptisols with the suborders Xeric Dystrocryepts and Vitrandic Eutrocryepts [31]. The research permits for these two study sites were granted by the United States Department of Interior–National Parks, Yosemite National Park for study site Dana Meadows (37.893100N, −119.256900W), and the United States Department of Agriculture–Forest Service, Pacific Southwest Region for study site Hall RNA (37.958056N, −119.296111W). The field studies did not involve endangered or protected species.

Field methods

In July 2010, transects were established along a hydrologic gradient at two locations in the Hall RNA and two locations in Dana Meadows. The hydrologic gradient in the meadow was established using vegetation associations as a proxy for water table depth [32]. In all four transects, the same vegetation type was utilized to identify each meadow region: *Carex filifolia* in the xeric sites, *Ptilagrostis kingii* in the mesic sites and *Carex scopulorum/Carex subnigricans* in the hydric sites. Three replicate soil collars were inserted in the soil at depths of approximately 3–5 cm (the variation was due to differences in soil characteristics) at three hydrologically distinct regions of the transect (designated as: dry, intermediate, and wet). The collars are located approximately 2 meters apart. The soil carbon dioxide (CO_2) efflux was measured using a LI-COR 8100A portable infrared gas analyzer (LI-COR Biosciences, Lincoln, Nebraska USA), fitted with a portable 10-cm soil respiration chamber. After a 45-second pre-purge, one-minute measurements were recorded and were followed by a 30-second post-purge. Weekly measurements were recorded during the first half of the growing season, followed by biweekly measurements through September. All measurements were taken at mid-day from collars with vegetation left intact. In each of the 4 transects, there were 6 collars in 2011, and 18 collars in 2012/2013. Above ground productivity was estimated by harvesting the total biomass

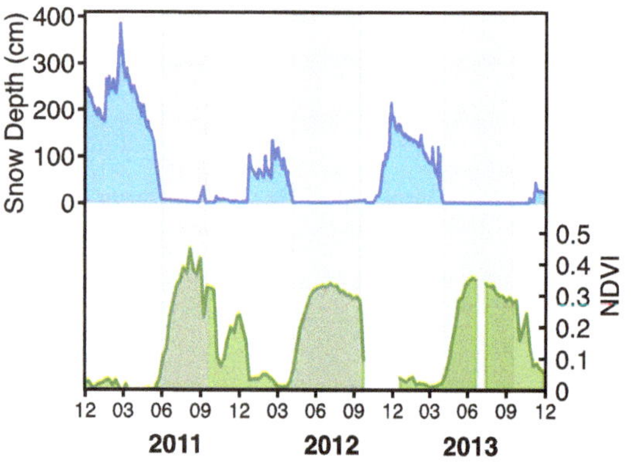

Figure 3. Snow depth (top panel) from Dana Meadows and NDVI (weighted mean average of all subalpine meadows in Yosemite National Park). Gray panels denote the growing season in the meadows as defined by the first day the meadow is snow free and the date where the NDVI crosses a threshold of 0.3.

in six 20 cm square quadrats in each region (dry, intermediate and wet) of the transect at peak production each year. Vegetation samples were oven dried at 50°C, and weighed to determine biomass. Historical and current meteorological data were obtained from the California Department of Water Resources station for Dana Meadows (ID: DAN). Meteorological data utilized for this study include, maximum air temperature, minimum air temperature, snow depth and snow water equivalent (SWE). SWE is the amount of water contained in a unit of snowpack. The April 1 SWE is an important metric for water resource managers in California. It represents the time where historically there has already been the maximum snowpack accumulation for the year, and thus represents the amount of available water to downstream users. Soil temperature and water content were measured at one site in the Hall RNA. Decagon 5TM sensors (Decagon Devices, Inc.) were inserted at 5, 15 and 25 cm below the soil surface. They were continuously monitored using a Decagon EM-50 datalogger.

Satellite-based remote sensing imagery

The Terra/MODIS surface reflectance (MOD09Q1.5) 8-day L3 global 250-m product was downloaded directly from the Land Processes Distributed Active Archive Center of the United States Geological Survey (https://lpdaac.usgs.gov). This level 3 surface reflectance product, which had been radiometrically corrected and georeferenced, provided a measure of the surface reflectance at the ground level in the absence of atmospheric scattering or absorption. The data were projected in a custom sinusoidal projection specific to the MODIS imagery. The eight-day composite images represented the maximum surface reflectance value for that time period and minimized the impacts of clouds and aerosols.

Processing MODIS imagery to NDVI

In the first stage of processing, the MODIS product MOD09Q1.5 was reprojected from a custom sinusoidal projection to the California Albers projection. The latter is a version of the Albers Equal Area projection optimized for statewide calculations. Bands 1 (620–670 nm) and 2 (841–876 nm) were utilized to calculate the NDVI over the entire MODIS image. The following equation was used: NDVI = (band 2 − band 1)/(band 2 + band 1). The resulting NDVI product was resampled down to 30 m and was used to produce an average NDVI for the entire meadow

Table 2. Mean maximum air temperature (°C) in Dana Meadow.

	2013		2012		2011		2010	Historic Mean (2000–2010)
JAN	3.12		4.78		4.93			2.21
FEB	3.21	**1.82**	2.82	**3.38**	0.89	**2.53**		2.44
MAR	6.81		3.64		2.47			5.35
APR	9.11		8.06		5.78			6.62
MAY	10.75	**8.89**	11.86	**7.85**	7.67	**5.31**		11.24
JUN	15.97		16.28		13.48			15.44
JUL	20.05		19.87		18.39			19.62
AUG	18.57	**18.20**	21.33	**19.16**	19.48	**17.12**		18.72
SEPT	15.11				17.17			14.59
OCT	10.43		13.24		11.94			7.06
NOV	7.00	**10.85**	6.17	**9.70**	3.68	**10.93**		5.04
DEC	4.68		−0.88		2.53		1.76	1.83

Mean seasonal temperatures (DJF, MAM, JJA, SON) are highlighted in gray. Historic mean maximum air temperatures from 2000–2010 are referenced in far right column.

Table 3. Mean minimum air temperature (°C) in Dana Meadow.

	2013		2012		2011		2010	Historic Mean (2000–2010)
JAN	−11.20		−7.85		−9.48			−10.92
FEB	−12.64	**−12.12**	−11.07	**−9.69**	−12.57	**−10.45**		−11.40
MAR	−8.01		−9.57		−11.00			−10.11
APR	−6.33		−7.18		−9.54			−8.56
MAY	−2.03	**−5.46**	−3.39	**−6.71**	−7.35	**−9.30**		−3.75
JUN	1.53		0.44		−2.20			0.34
JUL	4.87		3.46		2.29			4.25
AUG	2.85	**3.08**	5.89	**3.26**	2.76	**0.95**		2.84
SEPT	0.17				1.56			−0.93
OCT	−4.89		−3.06		−2.62			−4.59
NOV	−5.78	**−3.50**	−6.36	**−4.71**	−8.43	**−3.16**		−7.51
DEC	−10.09		−12.51		−10.16		−9.28	−10.37

Mean seasonal temperatures ((DJF, MAM, JJA, SON) are highlighted in gray. Historic mean minimum air temperatures from 2000–2010 are referenced in far right column.

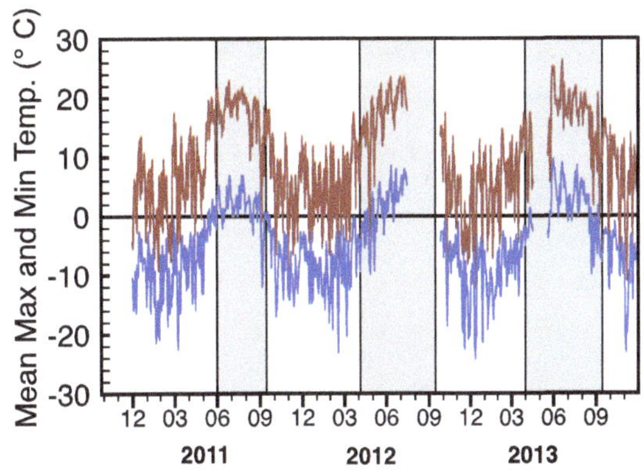

Figure 4. Maximum (red) and minimum (blue) daily air temperatures for Dana Meadow in 2011, 2012 and 2013. Gray panels denote the growing season in the meadow as defined by the first day the meadow is snow free and the date where the NDVI crosses a threshold of 0.3.

polygon region. The meadow polygons, resulting derived data layer and the associated metadata are currently being prepared as a spatial data product by the U.S. Geological Survey's Western Ecological Research Center at the Yosemite Field Station [33].

Statistical Analysis

Soil respiration rates for each collar were integrated over time to determine the cumulative CO_2 efflux for the growing season. Missing data was filled in via linear interpolation between the prior sampling date and the next date sampled. Repeated measures analysis of variance (RM-ANOVA) was used to determine significant differences between the effects of moisture class across the three years for both ANPP and cumulative CO_2 flux data. The different sites were utilized as replicates. If the RM-ANOVA model was significant, a Tukey's post hoc test ($p<0.05$) was used to assess differences between means. In addition, in order to determine the effect of year within a moisture region of the meadow, a subset of data was created for dry, intermediate and wet sites and a one-way RM-ANOVA model was utilized to determine significance within moisture classes across years. If the model was significant, a Tukey's post hoc ($p<0.05$) was used to determine differences between means. Data was tested for normality prior to analysis using the Shapiro-Wilks test. All statistical analyses were conducted using R statistical software (r-project.org).

Results and Discussion

Meteorological Data

The last several years in California have been marked by extreme seasonal weather on either ends of the spectrum. The 2011 water year (October 2010 through September 2011) was the seventh-wettest year on record (1929–2012) in YNP, with the April 1 SWE in Dana meadows reaching 156% of the 50 year mean (1951–2000) (Table 1). The 2012 water year was the fifth-driest year on record with only 49% of the mean SWE and the 2013 water year ranked the driest year on record with 25% of the mean SWE (Table 1). Looking at the entire historic record of Dana Meadows SWE, there is an increase after 1969 in the number of years with SWE values greater or less than one standard deviation from the mean (Figure 2). This apparent increase in the SWE variability corresponds to trends found in increase in the variability

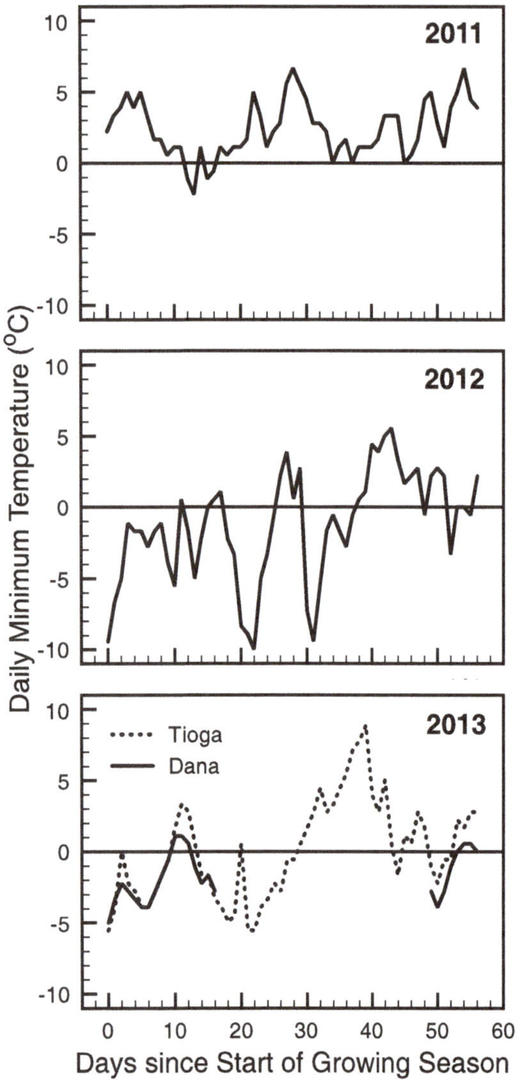

Figure 5. Daily minimum temperatures were used to compare frost events in the first sixty days since the start of the growing season (2011–2013) in Dana Meadow. Growing season was determined by the first day that the meadow was snow free each year. Dotted line represents data from a nearby meteorological station (Station id:TES) at north end of meadow was used for missing data in 2013.

of streamflow in Central California around the same time period [34].

In ecosystems dependent upon the enduring winter snowpack to insulate them from freezing events, the timing of the first day of continual snow cover for the winter can be critical to biological communities [35]. Likewise, the duration of that snow cover and timing of subsequent spring melt plays an essential role in microbial turnover [36,37], plant phenology [38], and meadow hydrology [39,40]. In addition, recent research has shown that winter warming in arctic ecosystems is contributing to a decline in plant productivity during the subsequent summer growing season [41]. Not only was the depth of snowpack distinctly different in the three consecutive years, but also the duration of snow cover differed greatly in all three years (Figure 3). The water year 2012 was especially anomalous with no continual snow cover until mid January. A significant ice storm occurred over the bare soils on

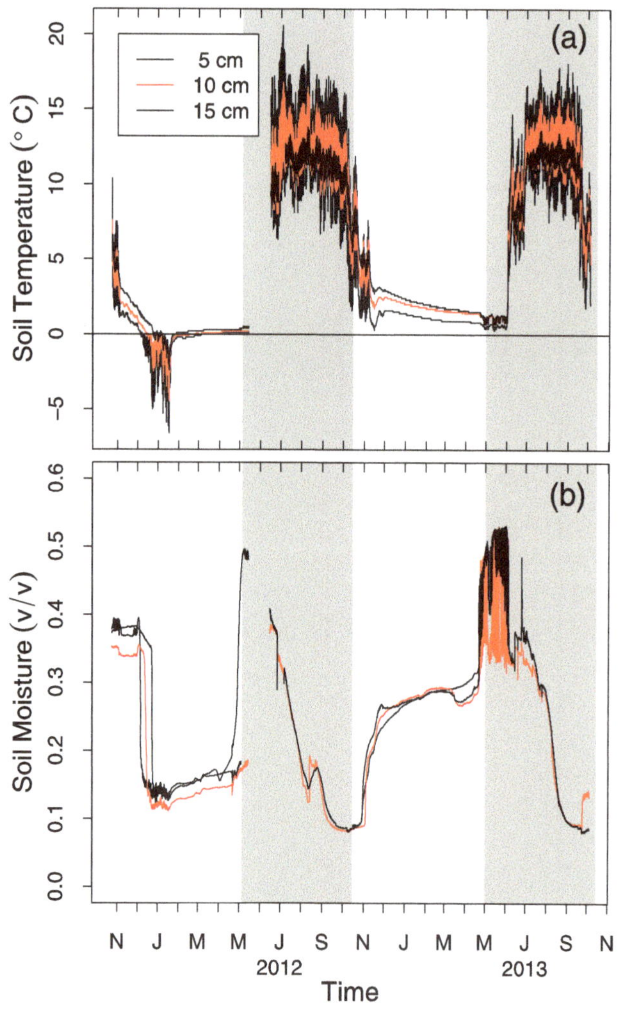

Figure 6. Time series of soil temperatures and volumetric water content for a dry meadow site from November 2011 to November 2013. The shaded panels indicate the growing season in 2012 and 2013.

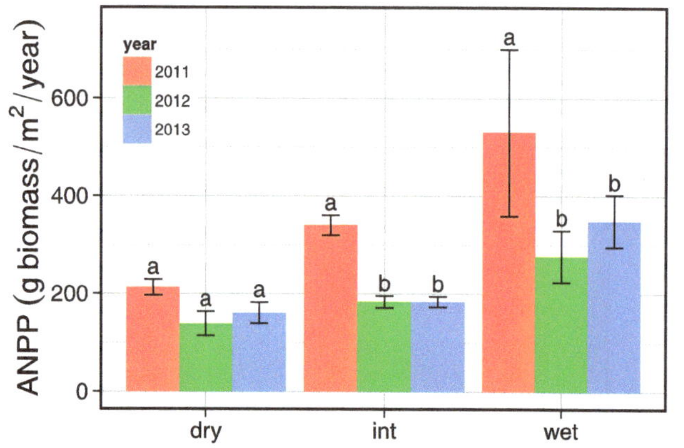

Figure 7. Mean annual aboveground net primary productivity averaged across all four sites for each moisture region in 2011, 2012, and 2013. Error bars represent standard error among sites. Letters denote significant differences in homogenous groups across years as determined by a Tukey post hoc test ($p<0.05$).

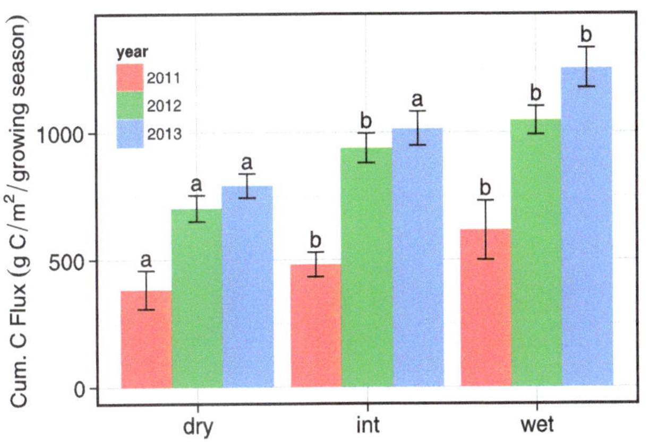

Figure 8. Mean cumulative carbon flux averaged across all four sites for each moisture region in 2011, 2012, and 2013. Error bars represent standard error among sites. Letters denote significant differences in homogenous groups across years as determined by a Tukey post hoc test ($p<0.05$).

December 4, 2012, with widespread needle damage to conifers noted at the study site after snowmelt.

With the exception of January, the average maximum air temperature in Dana Meadows was warmer in 2013 and 2012 through July of each year, as compared to 2011, with the spring (Mar–May) mean temperature increasing by 2–3°C (Table 2). This early warming is a contributing factor to the onset of an early spring in those years. The mean monthly minimum temperatures show warmer spring and summer temperatures in 2012 and 2013 as compared to 2011 (Table 3).

Ecosystem Response

The extreme seasonal changes from 2011 and 2012/2013 caused dramatic shifts in the onset of spring and in the number of snow-free days in the meadows of YNP. In Dana Meadows, the first snow-free day in 2012 and 2013 occurred 57–61 days earlier than in 2011, and the growing season increased by 35–37%, from approximately 106 days in 2011 to 163/167 days in 2012/2013. The documented shift to an earlier onset of spring in the Sierra Nevada [19] appears to have altered the response of the meadow ecosystems; rather than increasing their productivity, the earlier spring has rendered the meadows more sensitive to late winter/early spring frost events [11,28,42,43].

The maximum and minimum temperatures in the meadow show a clear seasonal trend, with very few frost events occurring within a normal growing season (Figure 4). The seasonal trends are similar between years, but the time point when the growing season is initiated is critical for assessing potential frost impacts on newly sprouting vegetation. As the snow melts, it causes saturation of the meadow soil and plants respond rapidly to this moisture and available nutrients by sending up green shoots. This leaves them susceptible to freezing temperatures. Tranquillini (1964) has shown that high elevation plants are very frost resistant, however notes that plants dependent on an insulating snowpack are susceptible to frost damage even in minor frost events [44]. If spring arrives earlier, as in 2012 and 2013, there is an increased likelihood of a severe frost event to damage newly sprouting vegetation. This pattern was evident during two frost events that occurred in 2012 after the snow had cleared from the meadow (Figures 4 and 5). The first event occurred over a four-day period that peaked on May 27, when the temperature dropped to −10°C. This event occurred approximately 20 days into the growing season. The second event occurred over three days beginning on June 5 that included a low temperature of −9.4°C. Because meadows undergo a rapid greening within days of a snowmelt, frost can damage sensitive meadow species and reduce overall productivity [43,45]. In 2013, there were several frost events that occurred on May 19 and 20 and on May 22 and 23 with a low at −5.5°C.

Frost damage to vegetation was apparent on a larger scale using satellite-derived Normalized Difference Vegetation Index (NDVI) mapping. The NDVI time series for the three years is shown in Figure 3. A rapid greening was apparent at the beginning of the growing season for both years; however, instead of reaching a peak in 2012, the NDVI plateaued before the vegetation senesced in mid-summer, which could indicate that the meadow vegetation was stressed and never reached maximum greenness in 2012. Normal meadow NDVI ranges from 0 during snowcover to a maximum of 0.45–0.65 at peak production and then falls to around 0.3 during senescence. In 2011, the peak NDVI occurred around 0.45, but in 2012 and 2013, the peak NDVI was just above the senescence value. Since the meadow soils begin the season saturated due to snowmelt and dry down over the growing season, it is unlikely that the plateau was caused by the meadows drying earlier in the growing season. If this were the case, we would expect to see a peak NDVI soon after the rapid increase at the beginning of the season. The pattern in 2013 is similar, though there is a small peak in early season NDVI, but the overall pattern is much lower than in 2011, indicating that although aboveground productivity did recover slightly from 2012, the meadow was still in a stressed state. There is an anomalous peak of NDVI that occurs between November 2011 and January 2012. This value was reflective of the senesced vegetation and bare ground that lacked a continuous snowpack until mid January 2012.

Another potential explanation for the decreased peak NDVI in 2012 is the winter desiccation of the meadow soils that may have damaged overwintering roots. Figure 6 shows a time series of soil temperature and volumetric water content at three depths in the soil from a drier region of the meadow. The soil temperatures dropped well below freezing for an extended period of time in early 2012 before the snowpack began to accumulate in mid-January. This drop in soil temperature triggered a desiccating event in the soil, as seen in the lower panel. Volumetric soil

Table 4. One-Way RM-ANOVA results from homogeneous moisture regions of the meadow.

	Dry	Intermediate	Wet
2011–2012	$P<0.003$	$P<0.0007$	$P<0.006$
2011–2013	$P<0.001$	$P<0.0004$	$P<0.002$
2012–2013	$P=0.798$	$P=0.888$	$P=0.667$

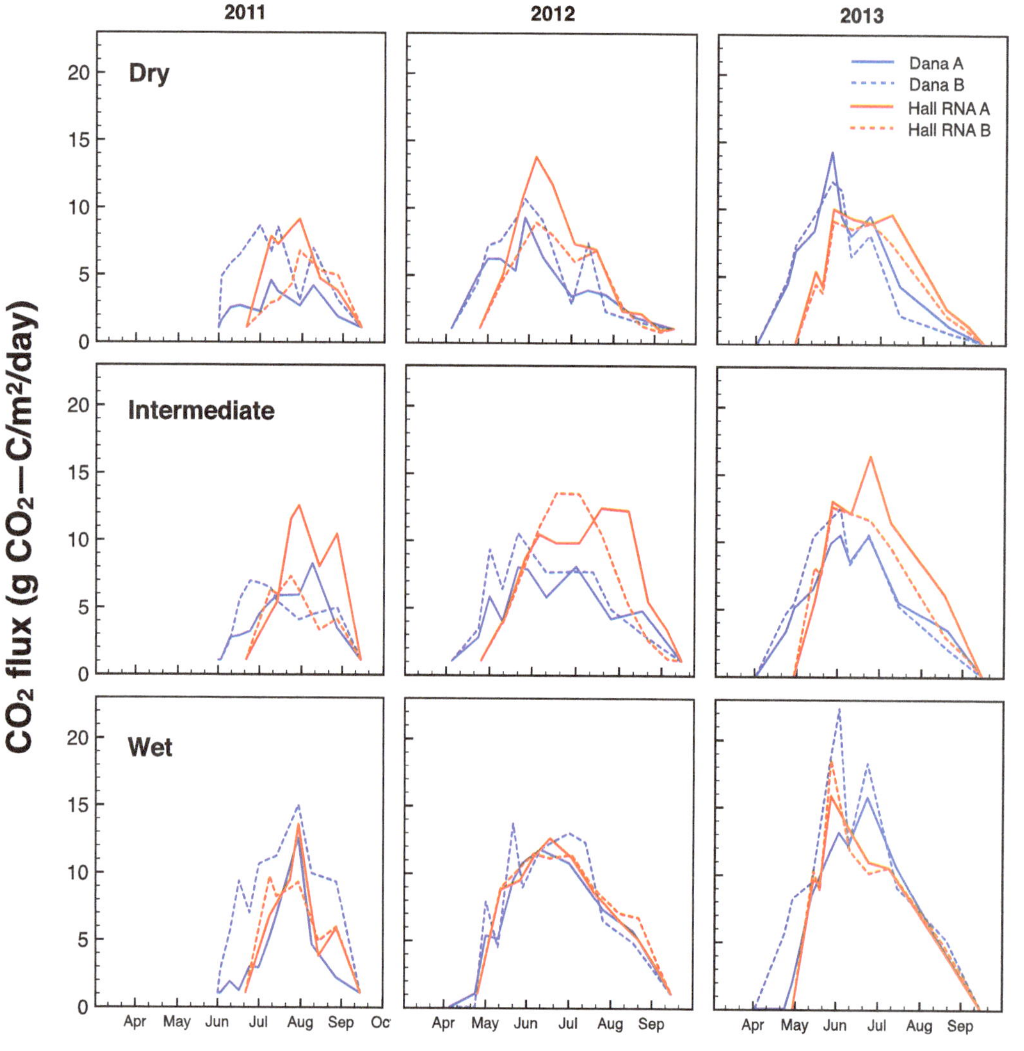

Figure 9. LI-COR surface CO_2 flux data for 2011, 2012, and 2013 in Dana Meadows and the Hall RNA.

moisture content was subsequently extremely low before snowpack accumulated. In a similar situation, Bokhorst et al have found that winter warming in the arctic, where snow melts off of the surface renders plants less productive the following spring [46].

The prolonged growing season in 2012 and 2013 should have led to increased plant productivity if water and temperature were not limiting, but the productivity actually declined by an average of 39% in all the regions of the meadow (Figure 7). Averaged across all sites, the drop in productivity from 2011 to 2012 was the most significant of all three years ($p<0.000001$). The aboveground productivity in 2013 was still significantly lower than 2011 ($p<0.0001$) and didn't significantly increase from 2012, indicating that the system is still at a stressed state and not recovering rapidly. However, since belowground biomass was not quantified in this study, it is unknown if the meadow systems are adapting to the change in seasonality by putting more energy into belowground biomass. However, due the fact that meadows already allocate a greater proportion of their carbon inputs (60–80%) to roots, it is unlikely that a shift to more belowground production could offset the carbon losses via soil respiration in 2012 and 2013. Studies have also shown that 90% of the fixed carbon is re-respired in the same season in peats and fens [47]. There is a delicate balance between productivity and carbon loss via respiration and unless the plants drastically increased the belowground biomass in response to the early spring, the losses via soil respiration will override the carbon inputs from the additional belowground biomass.

Different hydrologic regions of the meadows (dry – wet) responded differently to the change in duration and amount of snow cover (Figure 7). There was no significant difference in the productivity of the dry regions of the meadow across the three years. However, both the intermediate and wet regions showed significant reduction in biomass from 2011–2012 (intermediate, $p<.0001$; wet, $p<0.0001$). Both regions of the meadow still had significantly lower productivity in 2013 than 2011 (intermediate, $p<0.0001$; wet, $p<0.0001$), but no significant change from 2012 to 2013.

The mean cumulative carbon flux shows an increasing trend over time (Figure 8). Averaged across all moisture regions of the meadows, the mean cumulative carbon flux from the meadows was significant with respect to year (RM ANOVA model; year,$p<0.00001$, moisture, $p<0.00001$, moisture:year, $p=0.447$). However, this was driven mainly by the significance between 2011–2012 ($p<0.00001$) and 2011–2013 ($p<0.00001$). While 2013 continued to have larger cumulative fluxes in all regions of the meadow, they were not significantly different than 2012 ($p = 0.481$). There was no interaction between moisture region and year. Comparing the individual moisture regions of the meadow, there was significant change in the cumulative carbon flux of the dry, intermediate and wet regions of the meadow from 2011 to 2012 and 2011 to 2013, but no significant change from 2012 to 2013 (Table 4).

This result suggests that easily decomposable soil organic matter rapidly decomposed with the shift in environmental conditions in 2012 and 2013 [48–51]. The short growing season in 2011 and wet conditions throughout the growing season effectively reduced the overall soil carbon efflux in all regions of the meadow. Although the 2011 water year was extremely wet, and the meadows experienced little drying, this had no apparent effect on the available moisture in 2012. One explanation for this finding may be the dryness caused by a lack of snowpack in December 2011 through mid-January 2012; when the snow finally accumulated in January, the meadows were extremely dry underneath the snowpack (Figure 6). These dry soil conditions at the start of the 2012 growing season and below average snowpack coupled to above average surface temperatures throughout the 2012 growing season led to the extreme drying of the meadow soils and subsequently large carbon fluxes. The 2013 water year was the driest on record for Dana Meadows and there was a continuing trend of high mean cumulative carbon flux from the meadows. Although the cumulative flux was similar in both 2012 and 2013, there was a difference in the timing of peak soil carbon efflux. In 2013, the peak soil carbon efflux occurred early in the growing season and rapidly declined in all regions of the meadow, whereas in 2012, the peak occurred in the middle of the growing season (Figure 9). This shift indicates how responsive these ecosystems are to seasonal variation. The sustained magnitude of the cumulative carbon flux from the system over two summers has resulted in a loss of over 6% of the total carbon stock in the meadows we studied.

With climate extremes occurring at an increasing frequency around the world, our data demonstrate that sensitive ecosystems respond rapidly to the changes in seasonality and may reach a tipping point sooner rather than later. Multiple years of ecosystem stresses such as frost or drought can potentially cause a regime shift in vegetation with ramifications to the cycling of carbon in these systems. The magnitude of loss was significant given the small areal extent of these meadows, which is not proportional to their importance to overall ecosystem functioning and keystone position on the landscape. If the frequency of extreme events continues in this region, coupled to a decline in meadow aboveground productivity, we can expect carbon stocks in the meadows to rapidly decline, leading to meadow degradation and a reduction in ecosystem services in these watersheds.

Acknowledgments

The authors wish to thank J.R. Matchett, Eric L. Berlow, Samuel J. Traina, and John Harte for their comments on earlier versions of this manuscript, and Jesseca Burkhart for help in the field.

Author Contributions

Conceived and designed the experiments: CA TAG AAB. Performed the experiments: CA TAG AAB. Analyzed the data: CA TAG AAB. Contributed reagents/materials/analysis tools: CA TAG AAB. Contributed to the writing of the manuscript: CA TAG AAB.

References

1. Walther G-R, Post E, Convey P, Menzel A, Parmesan C, et al. (2002) Ecological responses to recent climate change. Nature 416(6879): 389–395.
2. Holmgren M, Stapp P, Dickman CR, Gracia C, Graham S, et al. (2006) Extreme climatic events shape arid and semiarid ecosystems. Frontiers in Ecology and the Environment 4: 87–95.
3. Jentsch A, Beierkuhnlein C (2008) Research frontiers in climate change: effects of extreme meteorological events on ecosystems. Comptes Rendus Geoscience 340: 621–628.
4. Jentsch A, Kreyling J, Elmer M, Gellesch E, Glaser B, et al. (2011) Climate extremes initiate ecosystem-regulating functions while maintaining productivity. Journal of Ecology 99: 689–702.
5. Craine JM, Nippert JB, Elmore AJ, Skibbe AM, Hutchinson SL, et al. (2012) Timing of climate variability and grassland productivity. Proceedings of the National Academy of Sciences 109: 3401–3405.
6. Galvagno M, Wohlfahrt G, Cremonese E, Rossini M, Colombo R, et al. (2013) Phenology and carbon dioxide source/sink strength of a subalpine grassland in response to an exceptionally short snow season. Environmental Research Letters 8: 025008.
7. Reichstein M, Bahn M, Ciais P, Frank D, Mahecha MD, et al. (2013) Climate extremes and the carbon cycle. Nature 500: 287–295.
8. Trujillo E, Molotch NP, Goulden ML, Kelly AE, Bales RC (2012) Elevation-dependent influence of snow accumulation on forest greening. Nature Geoscience 5: 705–709.
9. Baptist F, Flahaut C, Streb P, Choler P (2010) No increase in alpine snowbed productivity in response to experimental lengthening of the growing season. Plant Biology 12: 755–764.
10. Aurela M, Laurila T, Tuovinen J-P (2004) The timing of snow melt controls the annual CO2 balance in a subarctic fen. Geophysical Research Letters 31: L16119.
11. Inouye DW (2008) Effects of climate change on phenology, frost damage, and floral abundance of montane wildflowers. Ecology 89: 353–362.
12. Loheide II SP, Lundquist JD (2009) Snowmelt-induced diel fluxes through the hyporheic zone. Water Resour Res 45: W07404.
13. Heimann M, Reichstein M (2008) Terrestrial ecosystem carbon dynamics and climate feedbacks. Nature 451: 289–292.
14. Hammersmark CT, Rains MC, Mount JF (2008) Quantifying the hydrological effects of stream restoration in a montane meadow, northern California, USA. River Research and applications 24: 735–753.
15. Brown KA (2013) Groundwater storage in a mountain meadow northern Sierra Nevada California [Ph.D.]. Stanislaus: California State University.
16. Myers N, Mittermeier RA, Mittermeier CG, Da Fonseca GA, Kent J (2000) Biodiversity hotspots for conservation priorities. Nature 403: 853–858.
17. Kapnick S, Hall A (2009) Observed changes in the Sierra Nevada snowpack: potential causes and concerns. Prepared for the CEC and Cal/EPA ECE-500-2009-016-D: 09-016.
18. Null SE, Viers JH, Mount JF (2010) Hydrologic response and watershed sensitivity to climate warming in California's Sierra Nevada. PLoS One 5: e9932.
19. Cayan DR, Kammerdiener SA, Dettinger MD, Caprio JM, Peterson DH (2001) Changes in the onset of spring in the western United States. Bulletin-American Meteorological Society 82: 399–416.
20. Chivers M, Turetsky M, Waddington J, Harden J, McGuire A (2009) Effects of experimental water table and temperature manipulations on ecosystem CO2 fluxes in an Alaskan rich fen. Ecosystems 12: 1329–1342.
21. Xiang S-R, Doyle A, Holden PA, Schimel JP (2008) Drying and rewetting effects on C and N mineralization and microbial activity in surface and subsurface California grassland soils. Soil Biology and Biochemistry 40: 2281–2289.
22. Alm J, Schulman L, Walden J, Nykänen H, Martikainen PJ, et al. (1999) Carbon balance of a boreal bog during a year with an exceptionally dry summer. Ecology 80: 161–174.
23. Orchard VA, Cook F (1983) Relationship between soil respiration and soil moisture. Soil Biology and Biochemistry 15: 447–453.
24. Stephens JC, Allen Jr LH, Chen E (1984) Organic soil subsidence. Man-induced land subsidence Reviews in Engineering Geology VI Geological Society of America: 107–122.
25. Wipf S, Stoeckli V, Bebi P (2009) Winter climate change in alpine tundra: plant responses to changes in snow depth and snowmelt timing. Climatic Change 94: 105–121.
26. Price MV, Waser NM (1998) Effects of experimental warming on plant reproductive phenology in a subalpine meadow. Ecology 79: 1261–1271.
27. Thomson JD (2010) Flowering phenology, fruiting success and progressive deterioration of pollination in an early-flowering geophyte. Philosophical Transactions of the Royal Society B: Biological Sciences 365: 3187–3199.
28. Forrest J, Inouye DW, Thomson JD (2010) Flowering phenology in subalpine meadows: Does climate variation influence community co-flowering patterns? Ecology 91: 431–440.

29. Taylor DW (1984) Vegetation of the Harvey Monroe Hall Research Natural Area, Inyo National Forest, California: Unpub. report for the US Forest Service, Pacific Southwest Forest and Range Experiment Station.
30. United States Department of Agriculture (n.d.) Staff SS United States General Soil Map (STATSGO2). Natural Resource Conservation Service, United States Department of Agriculture.
31. United States Department of Agriculture (n.d.) Staff SS Soil Survey Geographic (SSURGO) Database. Natura Resource Convservation Service, United States Department of Agriculture.
32. Allen-Diaz BH (1991) Water table and plant species relationships in Sierra Nevada meadows. American Midland Naturalist: 30–43.
33. Berlow EL, Knapp RA, Ostoja SM, Williams RJ, McKenny H, et al. (2013) A network extension of species occupancy models in a patchy environment applied to the Yosemite toad (Anaxyrus canorus). PLoS One 8: e72200.
34. Pagano T, Garen D (2005) A recent increase in western US streamflow variability and persistence. Journal of Hydrometeorology 6: 173–179.
35. Schimel JP, Bilbrough C, Welker JM (2004) Increased snow depth affects microbial activity and nitrogen mineralization in two Arctic tundra communities. Soil Biology and Biochemistry 36: 217–227.
36. Lipson D, Schadt C, Schmidt S (2002) Changes in soil microbial community structure and function in an alpine dry meadow following spring snow melt. Microbial ecology 43: 307–314.
37. Nemergut DR, Costello EK, Meyer AF, Pescador MY, Weintraub MN, et al. (2005) Structure and function of alpine and arctic soil microbial communities. Research in Microbiology 156: 775–784.
38. Walker MD, Ingersoll RC, Webber PJ (1995) Effects of interannual climate variation on phenology and growth of two alpine forbs. Ecology: 1067–1083.
39. Bales RC, Molotch NP, Painter TH, Dettinger MD, Rice R, et al. (2006) Mountain hydrology of the western United States. Water Resources Research 42: 8432.
40. Jordan RP (1978) The snowmelt hydrology of a small alpine watershed. [Ph.D.]. University of British Columbia: University of British Columbia.
41. Bokhorst S, Bjerke J, Street L, Callaghan T, Phoenix G (2011) Impacts of multiple extreme winter warming events on sub-Arctic heathland: phenology, reproduction, growth, and CO2 flux responses. Global Change Biology 17: 2817–2830.
42. Inouye D (2001) The ecological and evolutionary significance of frost in the context of climate change. Ecology Letters 3: 457–463.
43. Inouye DW, Morales MA, Dodge GJ (2002) Variation in timing and abundance of flowering by Delphinium barbeyi Huth (Ranunculaceae): the roles of snowpack, frost, and La Niña, in the context of climate change. Oecologia 130: 543–550.
44. Tranquillini W (1964) The physiology of plants at high altitudes. Annual Review of Plant Physiology 15: 345–362.
45. Neuner G, Hacker J (2012) Ice formation and propagation in alpine plants. Plants in Alpine Regions: 163–174.
46. Bokhorst SF, Bjerke JW, Tømmervik H, Callaghan TV, Phoenix GK (2009) Winter warming events damage sub-Arctic vegetation: consistent evidence from an experimental manipulation and a natural event. Journal of Ecology 97: 1408–1415.
47. Clymo RS (1983) Peat. In: Gore, A.J.P. (Ed.), Ecosystems of the World, 4A. Mires: swamp, bog, fen and moor. Amsterdam: Generall Studies, Elsevier.
48. Laiho R (2006) Decomposition in peatlands: reconciling seemingly contrasting results on the impacts of lowered water levels. Soil Biology and Biochemistry 38: 2011–2024.
49. Davidson EA, Janssens IA (2006) Temperature sensitivity of soil carbon decomposition and feedbacks to climate change. Nature 440: 165–173.
50. Berhe AA, Suttle KB, Burton SD, Banfield JF (2012) Contingency in the Direction and Mechanics of Soil Organic Matter Responses to Increased Rainfall. Plant and Soil 358: 371–383.
51. Berhe AA, Kleber M (2013) Erosion, deposition, and the persistence of soil organic matter: mechanistic considerations and problems with terminology. Earth Surface Processes and Landforms 38: 908–912.

20

The Effects of Body Mass on Dung Removal Efficiency in Dung Beetles

Beatrice Nervo*, Claudia Tocco, Enrico Caprio, Claudia Palestrini, Antonio Rolando

Department of Life Sciences and Systems Biology, University of Torino, Torino, Italy

Abstract

Understanding of the role of body mass in structural-functional relationships is pressing, particularly because species losses often occur non-randomly with respect to body size. Our study examined the effects of dung beetle body mass on dung removal at two levels. First, we used the lab experiment to evaluate the efficiency of eight dung beetle species belonging to two functional groups (tunnelers, dwellers) on dung removal. Second, the same species employed in the lab were used in field mesocosms to examine the effects of the two functional groups on dung removal maintaining realistic differences in the total body mass between tunneler and dweller assemblages. Furthermore, the experimental assemblages contained one and four species within each functional group, so the effect of body mass heterogeneity was examined. We used a statistical approach (offset method) which took into account *a priori* constraints due to the study design allowing us to analyse the effect of larger species in mesocosm style experiments. Body size played a crucial role in dung removal: large beetles were more efficient than small ones and the percentage of removed dung increased with higher body mass heterogeneity. Tunnelers were more efficient than dwellers over both short and long time periods (one month and one year). Significant effects of dwellers were found only after one year. Moreover, our study showed that not including the body mass as an offset in the model resulted in sometimes different results, as the offset expresses dung removal independently of the body mass. This approach confirmed that body size is likely a pivotal factor controlling dung removal efficiency at multiple levels, from single species to overall dung beetle assemblages. Even though other specific traits should be examined, this study has begun to address the consequences of losing individuals with specific traits that are especially sensitive to perturbations.

Editor: Norman W. H. Mason, Landcare Research, New Zealand

Funding: These authors have no support or funding to report.

Competing Interests: The authors have declared that no competing interests exist.

* Email: beatrice.nervo@unito.it

Introduction

The last two decades of intensive research have provided compelling evidence for a link between biodiversity and ecosystem functioning (BD-EF). In particular, BD-EF research has concentrated on species richness and on the impact of anthropogenic environmental changes on ecosystem functions and services [1,2,3]. Species richness, one of the key components of biological diversity, may indeed be important for maintaining functional processes due to interspecific differences in how species process resources, affect the physical environment, and interact with one another [4]. However, the perceived artificiality of the random community assemblages used to create various experimental gradients of diversity is a constant criticism of experimental BD-EF research [5]. Several authors have emphasized that natural and anthropogenic diversity gradients show non-random patterns in the order and characteristics of species lost [6,7,8,9]. Such extinction bias raises questions about how useful inferences from random-assembly designs will be for informing conservation efforts. Since species and their associated functions are currently being lost at an unprecedented rate, understanding of the role of body mass in structural-functional relationships is pressing, particularly because species losses occur non-randomly with respect to body size [10]. Interspecific differences in body mass can have potentially profound effects across multiple scales of biological organization, since many life-history traits are body mass correlated. Body mass indeed may represent an indicator of the niche of each species and by extension of the entire ecological network [10]. For this reason, in BD-EF studies on bacteria, protists and animals, we need to take into account the body mass of the individuals within the biodiversity entity (e.g. species or functional groups) because ecosystem processes may depend not only on ecological traits but also on the metabolic rate of the individuals [3]. According to this scenario, there may be different patterns in the consumption of resources within the same functional group dependent on whether is composed of small or large species, since the metabolic rate per unit body mass is thought to decrease with increasing body size [11,12].

Dung beetles (Coleoptera: Scarabaeoidea) play a key role in dung removal and soil processes, and as such they are considered important in terms of providing ecosystem functions and services. Largely coprophagous, they use three broad nesting strategies driving a series of ecological processes such as dung removal, nutrient cycling, bioturbation, plant growth enhancement, secondary seed dispersal, trophic regulation, parasite suppression and fly control [13]. Dweller species lay eggs and brood their larvae inside the dung mass itself, or just at the soil-dung interface. Tunneler species bury brood balls in vertical chambers in close proximity to the original deposition site. Roller species transport dung balls some distance away, before burial beneath the soil surface. These varied patterns of consumption and relocation of dung are included within functional diversity, another relevant component of biological diversity characterizing species assemblages. Here, we focused on dung beetle assemblages of alpine pastures, where tunnelers are represented by several medium-large species, while dwellers are usually much smaller. There is great concern about the decline of dung beetles in Western Europe [14,15], especially with regard to large body sized species. For example, in Finland there has been a significant loss of large, tunneling Geotrupidae beetles (one out of three species now regionally extinct) and tunnelling *Onthophagus* (two out of three species lost) [16], in the Mediterranean region with a decline of large roller species populations [17], and the Padana Plain (northern Italy) has experienced the virtual extinction of at least one species of *Scarabaeus* and three species of *Gymnopleurus* [18]. Moreover, some studies have shown the ecological role of dung beetle assemblages in Finnish and UK pastures [16,19,20], but there is limited research on the ecological role of dung beetles in the Italian Alps, where land-use changes are affecting biodiversity and threatening local ecosystem services [21].

Our study examines the effects of dung beetle body mass on dung removal at two levels (individual species and the functional group). First, we used the lab experiment to evaluate the efficiency of eight dung beetle species belonging to two functional groups (tunnelers and dwellers) on dung removal. Second, the same species employed in the lab were used in field mesocosms to examine the effect of the two functional groups on dung removal. Since the experimental assemblages contained one and four species within each functional group, the influence of body mass heterogeneity on dung removal was analysed too. Finally, we developed an original statistical approach which took into account *a priori* constraints due to the study design allowing us to analyse the effect of larger species in mesocosm style experiments and compare assemblages with different body mass.

Material and Methods

Study area and species collection

The study area was a pasture dominated by graminaceous plants, which is located in a private protected area (Oasi Zegna), included in the Sessera Valley (45°40′16″N; 8°05′07″E, N-W Italy, 1400 m a.s.l.). Oasi Zegna (www.oasizegna.com), gave permission to conduct this research and collect dung beetles. No endangered or protected species were involved in the field studies. The climate is temperate sub-alpine with a mean annual temperature of 10°C and a mean annual rainfall of 1800 mm. A preliminary study in this area showed that the dung beetle assemblage is characterized by 27 species belonging to two families and to two functional groups (Scarabaeidae and Geotrupidae as tunnelers and Aphodiinae, a subfamily of Scarabaeidae, as dwellers) [22]. Field and lab experiments were carried out in June 2012 and 2013. In both years, four tunneler species (*Anoplotrupes stercorosus* (Scriba, 1796), *Trypocopris pyrenaeus* (Charpentier, 1825), *Onthophagus fracticornis* (Preyssler, 1790), *Geotrupes stercorarius* (Linnaeus, 1768)) and four dweller species (*Teuchestes fossor* (Linnaeus, 1758), *Acrossus depressus* (Kugelann, 1792), *Parammoecius corvinus* (Erichson 1848), *Acrossus rufipes* (Linnaeus, 1758)) were collected by hand or using dung-baited pitfall traps. Collected beetles were transferred to terraria until the start of the experiments. A minimum of ten dried individuals of each species were weighed to estimate body mass using an analytical balance (0.01 mg). Measured body mass was used as a proxy for body size. Fresh cattle dung was collected for the experiments from a closed barn in a livestock farm and was mixed to make it of uniform consistency; no insects nor larva were found in the dung samples. The cattle-grazed pasture had not been treated with anthelmintics such as ivermectin.

Lab experiments

Lab experiments were carried out to test the efficiency of dung removal for each of the eight species. Two individuals of each species were added to a 25 cm diameter terrarium with 45 g dung for dwellers and 90 g dung for tunnelers (5×5 cm dung size). When possible, beetles were sexed to ensure the presence of a male and a female in the pair. The amount of dung provided to dwellers was halved compared to tunnelers because of their smaller size. There were eight replicates for each species (four with *A. rufipes* because too few individuals were collected), and four controls only with dung. Each terrarium was covered by nylon mesh cloth to prevent beetles from escaping and others from entering. The residual dung was collected, oven-dried and weighed after 80 hours from the beginning of the experiment. This interval was set because it was the time taken by the most efficient species in monoculture (*A. stercorosus*) to remove most of the dung (Pers. Obs). This experimental protocol allowed the evaluation of the dung removal rate by adults only, because no larvae could hatch from eggs in such a short time.

Field experiments

The same eight dung beetle species were used to set the experimental field treatments in which we compared assemblages with one and four species of tunnelers (T1, T4) and dwellers (D1, D4) in order to evaluate the role of the functional groups and body mass heterogeneity in dung removal. Controls (C) were set up by using dung without beetles. The number of individuals was varied in each treatment in order to maintain the same total body mass within the tunneler treatments, and within the dweller treatments. The species used for the monocolture (10 individuals of *A. stercorosous* for tunneler treatments- T1, and 24 individuals of *T. fossor* for dweller treatments- D1) were chosen based on two criteria: (i) the abundance of the species in the study area; (ii) the dung removal efficiency showed in the lab experiment. In tunnelers, *A. stercorosus* was the most efficient and the most abundant species; in dwellers, *A.rufipes* showed the highest efficiency in dung removal, but unfortunately it was the less abundant. For this reason we decided to use *T. fossor* in monoculture which was the second most efficient species in terms of dung removal. The total body mass of tunnelers in the treatments (1.634 g) was doubled compared to dwellers (0.864 g) which reflected approximately the proportion of individuals between the two functional groups under natural field conditions in the Alps [22] (Table 1). We also took into account body mass heterogeneity, BH, calculated as σ/μ, where σ is the standard deviation and μ is the mean of the total body mass in each treatment. Maintaining differences in body mass between tunnelers and dwellers, instead of holding total dung beetle body mass constant (as per [19]), and using small and large species in our mesocosms allowed a more realistic representation of the natural composition of dung beetle assemblages in the study area. Fifty frames (PVC pipes, 40 cm diameter, 15 cm height) were fixed in the ground to a depth of 10 cm and randomly assigned to the experimental treatments and control. Two experimental units were considered: twenty-five frames were used to analyze dung removal efficiency after one month and twenty-five to analyze it after one year. Five replicates were considered in each experimental unit (4 treatments and 1 control×5 = 25). The control has served to detect for the presence of earthworms or other dung

Table 1. Dung beetle species used in the experiments.

Functional groups and species	Total number of beetles	Mean body mass (mg)	N of individuals for each frame			
			T1	T4	D1	D4
Tunnelers						
Anoplotrupes stercorosus	140	163.4	10	4		
Geotrupes stercorarius	20	220.31		2		
Trypocopris pyrenaeus	20	165.13		2		
Onthophagus fracticornis	140	15.54		14		
Dwellers						
Teuchestes fossor	420	36			24	18
Acrossus depressus	60	7.8				6
Acrossus rufipes	20	34.19				2
Parammoecius corvinus	400	2.06				40
Body mass heterogeneity (BH)			18.3	111.1	14.9	118.2

Total number of individuals used in the field experiment, mean dry body mass, body mass heterogeneity and number of individuals used in each treatment frame per species (10 replicates for each frame, five for short-term effects and five for long-term effects). Key to treatment codes: T1 = one tunneler species; T4 = four tunneler species; D1 = one dweller species; D4 = four species of dwellers.

consumers. In according to other studies such as [19,23], 600 g of dung and the dung beetles were added to the treatments and each pipe was covered by nylon mesh cloth. Dung residuals were collected, oven-dried, and weighed respectively after one month and one year in each experimental plot.

Statistical analysis

The effects of each tunneler and dweller species on the proportion of dung removed in lab experiments were tested using generalized linear models (GLM). Grams of residual dung were expressed as a percentage of removed dung as follows: [1-(grams of residual dry dung/grams of starting dry dung)]*100. Species was modeled as an independent categorical variable. The same analysis was used to assess the effects of different functional groups and body mass heterogeneity on the proportion of dung removed in the field. The experimental treatments (and the control) and body mass heterogeneity within the treatments were modeled as independent categorical variables (low and high body mass heterogeneity). Body mass heterogeneity, expressed as the coefficient of variation of the body mass within each treatment, was tested as a numerical variable too (BH). The body mass of each species used in the lab experiments was added to the models as an offset in order to statistically adjust for body size, as there was a significant correlation between dung removal rates and body mass (Table S1 in Appendix S1). Body mass *per se* is likely to be a key determinant of dung removal in such systems (e.g. [16,24]), hence it was statistically adjusted for in order to better compare the actual effects of different species on dung removal. In the field experiments, differences in the total body mass between tunneler and dweller treatments were therefore considered using the total body mass within the treatments as an offset. The offset term adjusts the model's dependent variable taking into account possible bias linked to the data collection or sample design [25,26]. Visual inspection of frequency distributions and Shapiro-Wilk tests confirmed the normality of errors in most cases. The gamma distribution was specified otherwise. *Post-hoc* Interaction Analysis calculated pairwise differences between the species (lab experiment) and the treatments (field experiment) using the Chi-square test (Package Phia in R [27]) and correcting the p-values according to the Holm method. All analyses were carried out using R 3.0.3. [28].

Results

Lab experiments

Dung removal efficiency increased with increasing body mass both in tunnelers ($p<0.01$) and dwellers ($p<0.01$) (Table S1 in Appendix S1). Larger species removed more dung per unit of weight, i.e. they were more efficient even when accounting for body size. Within tunneler species, *A. stercorosus* removed a significantly higher proportion of dung compared to the other three tunneler species when adjusting for body size (Table 2, Figure 1). All the tunneler species removed a significant part of the dung mass compared to controls (removed dung varied between 5% and 33%), with the exception of *O. fracticornis* (9%–10%). Post-Hoc Interaction tests showed that *A. stercorosus* was the most efficient species compared to the other three tunnelers and that *G. stercorarius* removed more dung than *O. fracticornis*. *A. rufipes* showed a high efficiency in dung removal (12–30%), while none of the other dweller species removed a significant part of the dung mass compared to controls after 80 hours (0–7%) (Table 2, Figure 1). The Interaction test between dweller species was significant only between *A. rufipes* and *P. corvinus* (Table 2).

Field experiments

The percentage of dung removed in tunneler treatments compared to control was significantly higher after one month and in the treatments of both functional groups after one year (Figure 2). The Interaction test between treatments showed that T4 removed significantly more dung than T1, D1 and D4 after one month, while differences were no more significant after one year. Comparisons between dung removal efficiency of the two functional groups showed that tunnelers removed a significantly higher proportion of dung both after one month and one year (Table 3). Tunneler treatments with high body mass heterogeneity (four species, T4) removed significantly more dung compared to those with low body mass heterogeneity (one species, T1) after one month, but these differences were no longer significant after one

Table 2. Effects of tunneler and dweller species on dung removal in the lab.

TUNNELER SPECIES				**DWELLER SPECIES**			
Reference level: Control				**Reference level: Control**			
Distribution: Gaussian	**Estimate**	**t value**	*p*	**Distribution: Gamma**	**Estimate**	**t value**	*p*
Intercept	67.2	14.83	***	Intercept	64.55	17.82	***
Anoplotrupes stercorosus	27.16	5.36	***	*Parammoecius corvinus*	4.23	1.03	NS
Geotrupes stercorarius	17.38	3.43	**	*Acrossus depressus*	6.21	1.50	NS
Onthophagus fracticornis	9.31	1.84	NS	*Teuchestes fossor*	7.9	1.90	NS
Trypocopris pyreneaus	16.83	3.28	**	*Acrossus rufipes*	16.1	3.33	**
Interaction Test	**Value**	**Chisq**	*p*	***Interaction Test***	**Value**	**Chisq**	*p*
A. stercorosus - G. stercorarius	9.78	9.32	*	*A. rufipes - P. corvinus*	11.87	10.08	*
A. stercorosus - O. fracticornis	17.85	31.04	***	*A. rufipes - A. depressus*	9.89	6.89	NS
A. stercorosus - T. pyrenaeus	10.33	9.69	*	*A. rufipes - T. fossor*	8.19	4.67	NS
G. stercorarius - O. fracticornis	8.07	6.34	*	*P. corvinus - A. depressus*	−1.98	0.51	NS
G. stercorarius - T. pyrenaeus	0.54	0.03	NS	*P. corvinus - T. fossor*	−3.67	1.71	NS
O. fracticornis - T. pyrenaeus	−7.52	5.15	NS	*A. depressus - T. fossor*	−1.69	0.35	NS

Factor estimates and statistical significance (GLM) for the dung removed in relation to the tunneler (sample size = 33) and dweller (sample size = 30) species. This table shows the results of the statistical models with the "body mass" of each species as an offset. Significance codes:
'***' $p<0.001$;
'**' $p<0.01$.
'*' $p<0.05$.

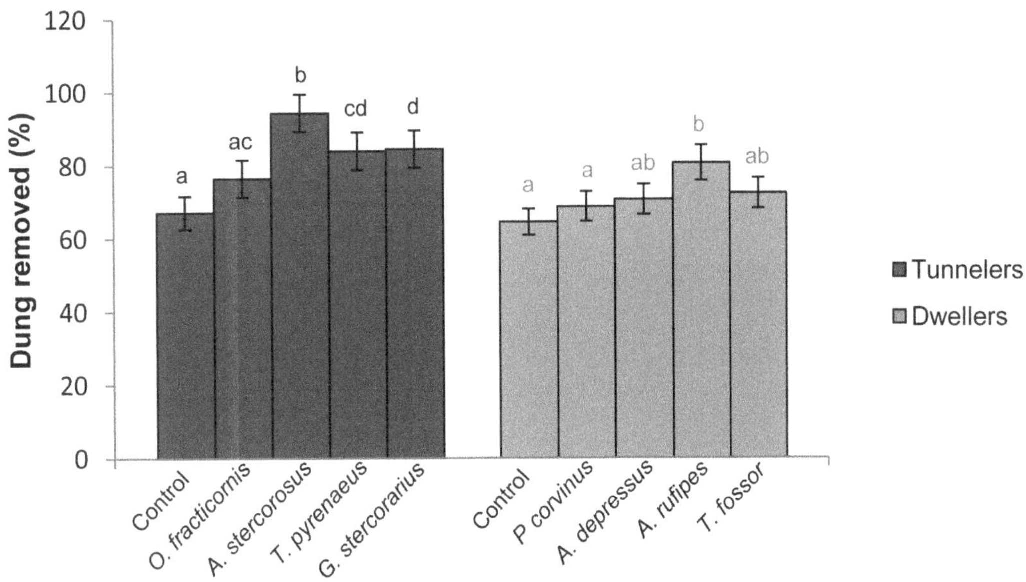

Figure 1. Lab experiment. Parameter estimates of dry dung removed (%) after 80 hours in tunneler (left) and dweller (right) species, derived from GLMs. Species are ordered in terms of their body mass (ascending order). Letters over the error bars indicate the differences among species and control.

year. Even though percentage of removed dung tended to increase in dweller treatments with four species compared to controls after one month, differences between low and high body mass hetrogeneity (D1, D4) were not significant either one month or one year later. The percentage of dung removed increased significantly with higher body mass heterogeneity (BH) after one month but no longer after one year (Table S1 in Appendix S2). Both the models with offset (Tables 2–3) and without offset (Table S2 in Appendix S1 and Table S2 in Appendix S2) differed for the value of the estimates and levels of significance. The most evident difference regarded the effect of the treatment D4 after one month, which removed significantly higher proportion of dung in the models without offset (Table S2 in Appendix S2).

Discussion

Functional group identity

Both our lab and field experiments clearly demonstrated that tunnelers were more efficient than dwellers in dung removal. Even though both groups in the field showed a significant removal effect after one year, dwellers appeared unable to provide a significant contribution to this ecosystem function compared to tunnelers in the short term, after one month. Dwellers take more time to remove dung due to their nesting behavior: larvae bred inside the dung consume the more fibrous material, while adults feed on the liquid component [13]. Only first and second instars were found in the dung collected after one month, meaning that they would have continued to consume dung, with an apparent effect over a longer period of time. Indeed, [29] showed that *A. rufipes* larvae are particularly active during the third instar when larvae are known to dig tunnels and display kleptoparasitic behavior [30]. As a result, the dung attains a much more loose and friable texture than material of a similar age which has not been worked through by larvae. For the same reason, with the notable exception of *A. rufipes*, no significant effects of dwellers were found in the lab experiments: an 80 hour interval was not long enough to highlight the delayed effects of dwellers. In contrast, tunnelers were more efficient in providing dung removal in the short term because adults not only feed on dung, but they also remove dung to provision nests in tunnels excavated in the ground beneath the dung pad. The long term effect of dung beetles on dung removal allows an evaluation of pasture quality after one year, before the beginning of the new season of pastoral activity. It is known that an area up to 12 times larger than the dung pad itself will remain ungrazed by livestock for several months, and even up to a year, if it remains in situ. This is due to its strong odour and to plant lignification [31]. However, if the dung pad is quickly buried or consumed, such areas are not avoided, hence there will be a larger area of pasture available for grazing [32].

Body mass

This study confirmed that body size is likely a pivotal factor controlling dung removal efficiency at multiple levels, from single species to overall dung beetle assemblages. Our lab experiment, as elsewhere [23,33,34,35], showed that body mass is an ecologically relevant characteristic because smaller beetles had relatively little effect on dung removal. For example, the effect of *O. fracticornis* on dung removal was certainly lower compared to the other tunnelers used in the experiment and it was not significant after 80 hours (i.e. its body mass was 1/14 than that of the largest *G. stercorarius*). Even though body size is considered an universally important trait influencing biodiversity-ecosystem function relationships within the detritivore trophic level [36,37], we expect that other species-specific traits in addition to body mass may influence dung utilization when differences between species' size are relatively small. The lab experiment, indeed, showed that *A. stercorosus* was the most efficient species in dung removal even if it was 30% smaller than *G. stercorarius*. Body mass may also explain differences between functional groups because local tunnelers were usually much larger than dwellers (i.e. *G. stercorarius* was 106 times larger than *P. corvinus*). Field experiments confirmed that large beetles, represented mainly by tunnelers in our alpine study area, have a dominant effect on dung removal. The reduced efficiency in dung removal due to a smaller body size could be

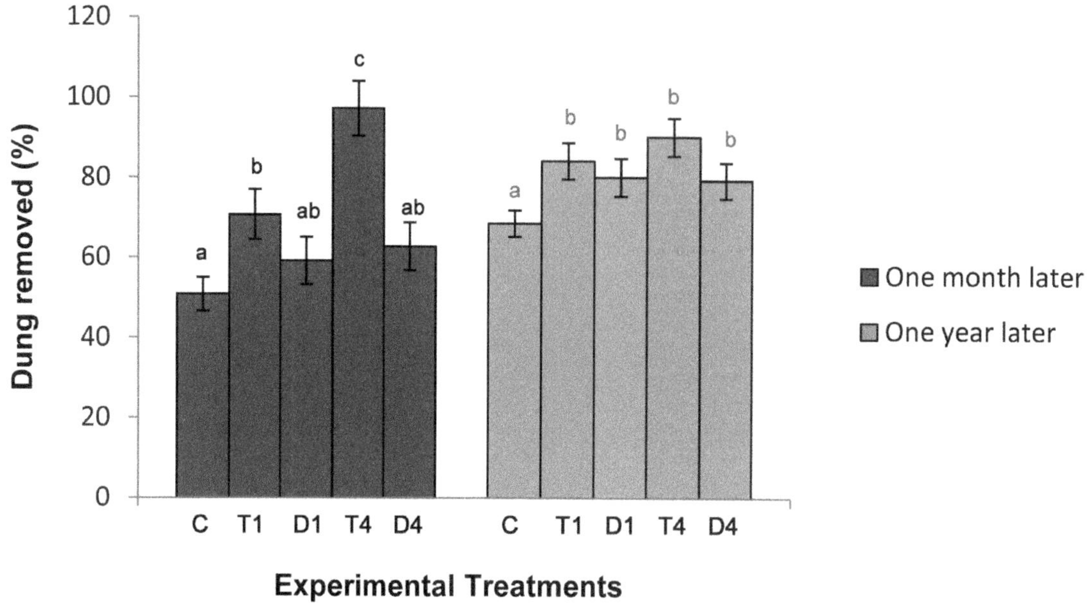

Figure 2. Field experiment. Parameter estimates of dry dung removed (%) after one month and one year in the experimental field treatments, derived from GLMs. Letters over the error bars indicate the differences among treatments and control.

compensated by a higher density of individuals [38]. However, even if dwellers are the dominant component of the Sessera Valley, large tunnelers showed high abundances [22] and consequently higher total body mass compared to smaller dwellers in the pastures of our study area. This suggests that the loss of larger species, which are often more prone to extinction [16] may have important consequences on the ecosystem functioning. Furthermore, by enlarging field mesocosms to also include geotrupid species, our study differed to that of [19], who considered only relatively small *Aphodius* and *Onthophagus* species in order to maintain the total body mass constant. The inclusion in the field treatments of larger species such as *G. stercorarius* and *A. stercorosus*, which are among the most abundant species in the pastures of the Sessera Valley [22], allowed a more realistic representation of the composition of the dung beetle community in the study area. Our results, which showed an enhancement of dung removal with higher body mass heterogeneity (BH), indicated that this may be an important driver in promoting dung burial and removal. Focusing on the treatments with one and four species within the same functional group (low and high body mass heterogeneity), we found that *A. stercorosus*, the most efficient species in dung removal as demonstrated by our lab experiments, increased its efficiency in the field mesocosms when it was in association with the other three tunneler species (T4). Since the most efficient species (and not the less efficient) was used as a term of comparison, this result suggests that body mass heterogeneity and probably the number of species may have increased ecosystem functioning in the short-term. This result was confirmed also by the increase in the percentage of dung removed after one month in relation to higher body mass heterogeneity (BH). Body mass heterogeneity may promote resource partitioning and facilitation, leading to a more efficient acquisition of resources. The form of resource partitioning and facilitation that we suppose consists of different patterns of food selection for different species due to their different ecological traits (i.e. body size) [39] and to the mutual benefit from one another through biophysical interactions of the resource [40]. Moreover species with different body size will also have different metabolism and consumption rates of resources [11] which may promote a more efficient dung removal in heterogeneous assemblages.

However, differences between dweller treatments with one and four species were not significant both one month and one year later. The slight tendency in treatment D4 (four dweller species) to remove a greater percentage of dung compared to the control after one month may have been attributed to the presence of *A. rufipes* which was the only dweller able to remove a significant part of the dung mass in the lab experiments.

Offset feature

Finally, using the offset term in the models allowed us to include the larger species in our study and compare treatments with different total body mass. The use of an offset allowed us to adjust the parameters of our models taking into account complicating effects of variations in body mass on the relationship between treatments (or species in the lab experiment) and dung removal. This approach represents one of the most novel aspect of our study because it presents a new way of considering the effects of larger species in mesocosm-style experiments. To our knowledge, the GLM offset feature has been over looked in other studies which compared assemblages with different total body mass. In our case, even though the results with offset (Tables 2–3) and without offset (Table S2 in Appendix S1, Table S2 in Appendix S2) were similar, the omission of total body mass from our models resulted in more notable differences in parameter estimates and significance levels of the field experiment than those of the lab experiment. In particular, the dweller treatment with high heterogeneity (D4) was significant compared to control when offset was omitted, but was not significant when taking into account variations in total body mass of the dung beetles among the treatments. This shows that results may differ depending on whether total dung removal or dung removal efficiency are being considered since different assemblages perform not only according to their mean specific traits, but also to the total body mass of the assemblage. In the lab experiment, models with or without body mass as an offset showed

Table 3. Effects of dung beetles on dung removal after one month and after one year in the field experiment.

AFTER ONE MONTH				**AFTER ONE YEAR**			
FIELD TREATMENTS				**FIELD TREATMENTS**			
Reference level: Control				*Reference level: Control*			
Distribution: Gaussian	**Estimate**	**t value**	***p***	*Distribution: Gaussian*	**Estimate**	**t value**	***p***
Intercept	50.85	12.16	***	Intercept	68.54	20.56	***
T1	19.89	3.17	**	T1	15.63	3.49	**
D1	8.35	1.41	NS	D1	11.6	2.46	*
T4	46.39	6.79	***	T4	21.64	4.59	***
D4	11.99	2.03	NS	D4	10.79	2.41	*
Interaction Test	**Value**	**Chisq**	***p***	*Interaction Test*	**Value**	**Chisq**	***p***
T1-D1	11.54	3.38	NS	T1-D1	4.02	0.81	NS
T1-T4	−26.49	13.77	**	T1-T4	−6.00	1.8	NS
T1-D4	7.9	1.59	NS	T1-D4	4.84	1.32	NS
D1-T4	−38.08	31.04	***	D1-T4	−10.04	4.53	NS
D1-D4	−3.64	0.38	NS	D1-D4	0.81	0.03	NS
T4-D4	34.39	25.39	***	T4-D4	10.85	5.89	NS
FUNCTIONAL GROUP				**FUNCTIONAL GROUP**			
Reference level: Dwellers				*Reference level: Dwellers*			
Distribution: Gaussian	**Estimate**	**t value**	***p***	*Distribution: Gaussian*	**Estimate**	**t value**	***p***
Intercept	61.02	17.5	***	Intercept	79.69	33.42	***
Tunnelers	21.08	3.88	**	Tunnelers	7.15	2.12	*
BODY MASS HETEROGENEITY				**BODY MASS HETEROGENEITY**			
Tunnelers (T1 vs T4)				**Tunnelers (T1 vs T4)**			
Reference level: T1				*Reference level: T1*			
Distribution: Gaussian	**Estimate**	**t value**	***p***	*Distribution: Gaussian*	**Estimate**	**t value**	***p***
Intercept	70.73	20.36	***	Intercept	85.8	19.95	*
T4	26.49	4.99	**	T4	6.02	0.95	NS
BODY MASS HETEROGENEITY				**BODY MASS HETEROGENEITY**			
Dwellers (D1 vs D4)				**Dwellers (D1 vs D4)**			
Reference level: D1				*Reference level: D1*			
Distribution: Gaussian	**Estimate**	**t value**	***p***	*Distribution: Gaussian*	**Estimate**	**t value**	***p***

Table 3. Cont.

BODY MASS HETEROGENEITY				BODY MASS HETEROGENEITY			
Dwellers (D1 vs D4)				**Dwellers (D1 vs D4)**			
Reference level: D1				*Reference level: D1*			
Distribution: Gaussian	**Estimate**	**t value**	***p***	***Distribution: Gaussian***	**Estimate**	**t value**	***p***
Intercept	60.06	20.43	***	Intercept	80.14	39.71	***
D4	3.62	0.87	NS	D4	−0.83	−0.3	NS

Treatment factor estimates and statistical significance (GLM) for the dung removed in relation to the experimental treatments (sample size = 22), functional groups (sample size = 18) and body mass heterogeneity (sample size = 9). This table shows the results of the statistical models with the "total body mass" within the treatments as an offset. Key to treatment codes: T1 = one tunneler species; T4 = four tunneler species; D1 = one dweller species; D4 = four species of dwellers.
Significance codes:
'***'p<0.001;
'**' p<0.01.
'*' p<0.05.

slight differences, highlighting that larger species removed more dung even when accounting for body size. Consequently, the performance of our species did not depend exclusively on the trait considered (i.e. body size), but probably also on the metabolic rates of the individuals and other species-specific traits which need to be investigated further.

As a caveat, we recognise the limitations of our study which considered a single study area in the Alps, and which was conducted on a limited number of species which were all tested in monoculture, but only in the lab. Accordingly, while we have not found differences in dung removal within dweller treatments under present local conditions, they may emerge under others, as shown by [19]. Admittedly, this design left many species combinations unstudied mainly for lack of sufficient number of individuals. Nonetheless, the trait-based approach used in our study could start to address the consequences of losing individuals with specific traits that are especially sensitive to particular perturbations. For instance, habitat loss usually affects large-bodied organisms disproportionately [3,41]. Future studies should be aimed at considering other specific traits beyond body mass, and at analyzing multiple functions. In this framework, only appropriate management aimed at preserving heterogeneous dung beetle assemblages in body size (especially medium and large tunneler species) can warrant the maintenance of ecosystem services in the Alps, threatened by the land use changes driven mainly by the overall decline of traditional agro-pastoral systems [42], climate change [43], and human leisure activities [44,45].

Supporting Information

Appendix S1 Details of GLMs used for the lab experiment without the offset term. Factor estimates and statistical significance (GLM) for the dung removed in relation to the tunneler (sample size = 33) and dweller (sample size = 30) species. Also, models with body mass of each species in relation to the percentage of removed dung are presented. Significance codes: '***'p<0.001; '**' p<0.01 '*' p<0.05.

Appendix S2 Details of GLMs used for the field experiment without the offset term. Factor estimates and statistical significance (GLM) for the percentage of removed dung in relation to the experimental treatments (sample size = 22) after one month and one year. Also, models with body mass heterogeneity in relation to the percentage of removed dung are presented. Significance codes: '***'p<0.001; '**' p<0.01 '*' p<0.05.

Acknowledgments

We thank Alberto Berlen and Gabriele Iussig for helping in the field and Michele Lonati, Giampiero Lombardi, and Luisella Celi for their contribution in defining the field experimental protocol. We are grateful for the important suggestions of Dan Chamberlain who greatly improved the manuscript and checked the English. We are also indebted to Shaun Forgie, Liz Nichols and four other anonymous referees for their useful and stimulating comments on a previous draft of the paper.

Author Contributions

Conceived and designed the experiments: BN CT EC CP AR. Performed the experiments: BN CT. Analyzed the data: BN EC. Contributed reagents/materials/analysis tools: AR CP. Wrote the paper: BN EC AR.

References

1. Naeem S, Duffy JE, Zavaleta E (2012) The Functions of Biological Diversity in an Age of Extinction. Science 336: 1401–1406.
2. Cardinale BJ, Duffy JE, Gonzalez A, Hooper DU, Perrings C, et al. (2012) Biodiversity loss and its impact on humanity. Nature 486: 59–67.
3. Reiss J, Bridle JR, Montoya JM, Woodward G (2009) Emerging horizons in biodiversity and ecosystem functioning research. Trends Ecol Evol 24(9): 505–514.
4. Larsen TH, Williams NM, Kremen C (2005) Extinction order and altered community structure rapidly disrupt ecosystem functioning. Ecol Lett 8: 538–547.
5. Duffy JE (2003) Biodiversity loss, trophic skew and ecosystem functioning. Ecol Lett 6(8): 680–687.
6. Wardle DA (1999) Is "sampling effect" a problem for experiments investigating biodiversity-ecosystem function relationships? Oikos 87: 403–407.
7. Huston MA, Aarsen LW, Austin MP, Cade BS, Fridley JD, et al. (2000) No consistent effect of plant diversity on productivity. Science 289: 1255a.
8. Srivastava DS (2002) The role of conservation in expanding biodiversity research. Oikos 98: 351–360.
9. Díaz S, Symstad AJ, Chapin FS III, Wardle DA, Huenneke LF (2003) Functional diversity revealed by removal experiments. Trends Ecol Evol 18: 140–146.
10. Woodward G, Ebenman B, Emmerson M, Montoya JM, Olesen JM, et al. (2005) Body size in ecological networks. Trends Ecol Evol 20(7): 402–409.
11. Reichle D (1968) Relation of Body Size to Food Intake, Oxygen Consumption, and Trace Element Metabolism in Forest Floor Arthropods. Ecology 49:538–542.
12. Peters RH (1983) The Ecological Implications of Body Size. Cambridge University Press
13. Nichols E, Spector S, Louzada J, Larsen T, Amezquita S, et al. (2008) Ecological functions and ecosystem services provided Scarabaeinae dung beetles. Biol Conserv 141: 1461–1474.
14. Carpaneto GM, Mazziotta A, Valerio L (2007) Inferring species decline from collection records: roller dung beetles in Italy (Coleoptera, Scarabaeidae). Divers Distrib 13(6): 903–919.
15. Dortel E, Thuiller W, Lobo JM, Bohbot H, Lumaret JP, et al. (2013) Potential effects of climate change on the distributions of Scarabaeidae dung beetles in Western Europe. J Insect Conserv 17 (5): 1059–1070.
16. Rosenlew H, Roslin T (2008) Habitat fragmentation and the functional efficiency of temperate dung beetles. Oikos 117: 1659–1666.
17. Lobo JM, Lumaret JP, Jay-Robert P (2001) Diversity, distinctiveness and conservation status of the Mediterranean coastal dung beetle assemblage in the Regional Natural Park of the Camargue (France). Divers Distrib 7(6): 257–270.
18. Barbero E, Palestrini C, Rolando A (1999) Dung beetle conservation: effects of habitat and resource selection (Coleoptera: Scarabaeoidea). J Insect Conserv 3(2): 75–84.
19. Beynon S, Mann DJ, Slade EM, Lewis OT (2012) Species rich dung beetle communities buffer ecosystem services in perturbed agro-ecosystems. J Appl Ecol 49: 1365–1372.
20. Penttilä A, Slade EM, Simojoki A, Riutta T, Minkkinen K, et al. (2013) Quantifying Beetle-Mediated Effects on Gas Fluxes from Dung Pats. PLOS ONE 8(8): e71454.
21. Falcucci A, Maiorano L, Boitani L (2007) Change in land use/land cover patterns in Italy and their implications for biodiversity conservation. Landscape Ecol 22: 617–631.
22. Tocco C, Negro M, Rolando A, Palestrini C (2013) Does natural reforestation represent a potential threat to dung beetle diversity in the Alps?. J Insect Conserv 17: 207–217.
23. Slade EM, Mann DJ, Villanueva JF, Lewis OT (2007) Experimental evidence for the effects of dung beetle functional group richness and composition on ecosystem function in a tropical forest. J Anim Ecol 76: 1094–1104
24. O'Hea NM, Kirwan L, Finn JA (2010) Experimental mixtures of dung fauna affect dung decomposition through complex effects of species interactions. Oikos 119: 1081–1088.
25. Shi SG (2010) Direct analysis of pre-adjusted loss cost, frequency or severity in tweedie models. Casualty Actuarial Society E-Forum, 1–13.
26. Yan G, Guszcza J, Flynn M, Wu CSP (2009) Applications of the offset in property-casualty predictive modeling. Casualty Actuarial Society E-Forum, 366.
27. De Rosario-Martinez H (2013) Phia: Post-Hoc Interaction Analysis. Package version 0.1-5. Available: http://CRAN.R-project.org/package=phia.
28. R Development Core Team (2013) R: A language and environment for statistical computing. R Foundation for Statistical Computing, Vienna, Austria. . ISBN 3–900051–07–0. Crawley 2002.
29. Holter P (1977) An experiment on dung removal by Aphodius larvae (Scarabaeidae) and earthworms. Oikos 28: 130–136.
30. Klemperer HG (1980) Kleptoparasitic behavior of *Aphodius rufipes* (L.) larvae in nests of *Geotrupes spiniger* Marsh. (Coleoptera, Scarabaeidae). Ecol Entomol 5: 143–151.
31. Haynes RJ, Williams PH (1993) Nutrient cycling and soil fertility in the grazed pasture ecosystem. Adv Agron (USA).
32. Miranda CB, Santos JD, Bianchin I (2000) The role of *Digitonthophagus gazella* in pasture cleaning and production as a result of burial of cattle dung. Pasturas Tropicales 22(1): 14–18.
33. Andresen E (2002) Dung beetles in a Central Amazonian rainforest and their ecological role as secondary seed dispersers. Ecol Entomol 27: 257–270.
34. Amézquita S, Favila ME (2010) Removal rates of native and exotic dung by dung beetles (Scarabaeidae: Scarabaeinae) in a fragmented tropical rain forest. Environ Entomol 39: 328–336.
35. Kaartinen R, Hardwick B, Roslin T (2013) Using citizen scientists to measure an ecosystem service nationwide. Ecology 94(11): 2645–2652.
36. Nichols E (2013) Fear begets function in the 'brown' world of detrital food webs. J Anim Ecol 82(4): 717–720.
37. Wardle DA, Bardgett RD (2004) Human-induced changes in large herbivorous mammal density: the consequences for decomposers. Front Ecol Environ 2(3): 145–153.
38. Andresen E (2003) Effect of forest fragmentation on dung beetle communities and functional consequences for plant regeneration. Ecography 26: 87–97.
39. Holter P, Scholtz CH, Wardhaugh KG (2002) Dung feeding in adult scarabaeines (tunnellers and endocoprids): even large dung beetles eat small particles. Ecol Entomol 27(2): 169–176.
40. Cardinale BJ, Palmer MA, Collins SL (2002) Species diversity enhances ecosystem functioning through interspecific facilitation. Nature 415(6870): 426–429.
41. Raffaelli D (2004) How extinction patterns affect ecosystems. Science 306(5699): 1141–1142.
42. Tocco C, Probo M, Lonati M, Lombardi G, Negro M, et al. (2013) Pastoral Practices to Reverse Shrub Encroachment of Sub-Alpine Grasslands: Dung Beetles (Coleoptera, Scarabaeoidea) Respond More Quickly Than Vegetation. PLoS ONE 8(12): e83344.
43. Jacqueline GF, Guisan A, Zimmermann NE (2007) Tree line shifts in the Swiss Alps: Climate change or land abandonment? J Veg Sci 18: 571–582.
44. Rolando A, Caprio E, Rinaldi E, Ellena I (2007) The impact of high-altitude ski-runs on alpine grassland bird communities. J Appl Ecol 44: 210–219.
45. Caprio E, Chamberlain DE, Isaia M, Rolando A (2011) Landscape changes caused by high altitude ski-pistes affect bird species richness and distribution in the Alps. Biol Conserv 144: 2958–2967.

Ecosystem Service Valuations of Mangrove Ecosystems to Inform Decision Making and Future Valuation Exercises

Nibedita Mukherjee[1,2]*, William J. Sutherland[3], Lynn Dicks[3], Jean Hugé[1,4], Nico Koedam[2¶], Farid Dahdouh-Guebas[1,2¶]

1 Laboratory of Systems Ecology and Resource Management, Université Libre de Bruxelles, Brussels, Belgium, **2** Laboratory of Plant Biology and Nature Management, Vrije Universiteit Brussel, Brussels, Belgium, **3** Conservation Science Group, Department of Zoology, University of Cambridge, Cambridge, England, United Kingdom, **4** Centre for Sustainable Development, Ghent University, Gent, Belgium

Abstract

The valuation of ecosystem services is a complex process as it includes several dimensions (ecological, socio-cultural and economic) and not all of these can be quantified in monetary units. The aim of this paper is to conduct an ecosystem services valuation study for mangroves ecosystems, the results of which can be used to inform governance and management of mangroves. We used an expert-based participatory approach (the Delphi technique) to identify, categorize and rank the various ecosystem services provided by mangrove ecosystems at a global scale. Subsequently we looked for evidence in the existing ecosystem services literature for monetary valuations of these ecosystem service categories throughout the biogeographic distribution of mangroves. We then compared the relative ranking of ecosystem service categories between the monetary valuations and the expert based analysis. The experts identified 16 ecosystem service categories, six of which are not adequately represented in the literature. There was no significant correlation between the expert based valuation (the Delphi technique) and the economic valuation, indicating that the scope of valuation of ecosystem services needs to be broadened. Acknowledging this diversity in different valuation approaches, and developing methodological frameworks that foster the pluralism of values in ecosystem services research, are crucial for maintaining the credibility of ecosystem services valuation. To conclude, we use the findings of our dual approach to valuation to make recommendations on how to assess and manage the ecosystem services provided by mangrove ecosystems.

Editor: Anna R. Armitage, Texas A&M University at Galveston, United States of America

Funding: NM was funded by the 'Mandat d'Impulsion Scientifique' (MIS ID 1765914) of the National Science Foundation (FNRS), Belgium and "CREC" (EU IRSES # 247514). NM would like to thank the British Ecological Society for the Parkyn bursary that made this research possible. WJS is funded by Arcadia. LD is funded by the Natural Environment Research Council (Grant codes NE/K015419/1, NE/J500665/1). The funders had no role in study design, data collection and analysis, decision to publish, or preparation of the manuscript.

Competing Interests: The authors received funding from a commercial source "Arcadia". This does not alter the authors' adherence to PLOS ONE policies on sharing data and materials.

* Email: nibedita.41282@gmail.com

¶ NK and FDG are joint last authors on this work.

Introduction

The sustainable provision of the goods and services that we derive from nature (i.e. ecosystem services) is essential to human well-being and survival [1–3]. The consequences of the widespread decline of these ecosystem services (ES) have been amply demonstrated by research in the past decade [4,5]. Over half (approx. 60%) of the major global ES have either been degraded or used unsustainably [2]. These ES range from provisioning services such as freshwater and fisheries, to regulating services such as air and water purification and climate regulation, to cultural and aesthetic services. For example, two-thirds of the world population is projected to be under water stress by 2025 [6] while one third of the world's major fisheries had already collapsed by 2003 and many continue to decline [7,8].

One of the steps in addressing this situation may be through valuation of the critical ES particularly in monetary terms. There is a widespread notion that valuation exercises might help decision makers appreciate the value of ES to society and the anticipated cost of their imminent loss [9,10]. Economic valuation in particular is often expected to be a useful tool to support conservation policy decisions and governance [10,11]. Following the seminal work by Costanza et al. [3] and the Millennium Ecosystem Assessment [2] which involved 1300 scientists, valuation of ES has received unprecedented attention in the last decade [11–13]. A range of different valuation methods have been designed and new networks have been formed for better exchange of information in this rapidly evolving field [14,15]. In a recent study, Costanza et al. [16] state that valuing ecosystems and their services is inevitable (even if it is implicit) in any decision involving trade-offs concerning them. Economic valuations of ES improve the transparency of the valuation process and may thereby usher in better decision making about ES [16].

Until now, valuation methods, data and classification systems for ecosystems were developed predominantly for terrestrial ecosystems while coastal ecosystems have received scant attention [17,18]. Peer-reviewed literature on global economic valuations of coastal forests like mangrove ecosystems is rather limited [19,20]. Mangrove ecosystems are tidally influenced wetland forests present in 123 countries [21]. Some worldwide assessments have considered mangroves as a subset of other coastal ecosystems in the economic evaluations of ES. However, the contribution of mangrove ecosystems to the aggregate economic value is often hard to disentangle. The possible pitfall in such large-scale studies is that there is considerable overlap with several other ecosystem types, possibly leading to double counting. For instance, mangroves are either combined with tidal marshes (wetlands) in Costanza et al. [16] or divided into 'tropical forests', 'coastal systems' and 'coastal wetlands' in de Groot et al. [15].

Mangrove ecosystems merit further attention in their own right. The number of people living within 10 km of significant mangrove areas might rise to 120 million by 2015 [22]. The bulk of this population resides in developing countries in Asia and West and Central Africa and is significantly dependent on mangrove resources for daily sustenance and livelihood. In coastal regions dominated by sandy beaches where timber species are scarce, mangrove plants are often the only available source of fuelwood and timber for construction of houses in tropical developing countries [23]. The linkages between mangroves and fisheries have been documented in ecological literature [23–26]. Even though the absolute economic value (monetary value) of the resources may not be high [27] (e.g. some species of snails and crabs have no market value but they are consumed when no other food or protein source is available), the relevance of these biological resources may be paramount for the communities dependent on them. Mangroves in such cases may be considered human life-support systems. Mangrove forests are also important for their role in providing coastal protection against recurrent storms and other natural hazards [28]. The dense network of roots bind the soil and trap the sediment and suspended particulate matter in deltaic settings [29]. Mangroves are also known to be the most carbon rich forests in the tropics, reported to have 1023 Mg C per hectare of forest including soil carbon [30]. Per unit area, this is higher than any other marine ecosystem, such as seagrass beds and salt marshes. In spite of their socioeconomic importance, mangrove area has declined by 30–50% in the past 50 years, a rate higher than most other biomes [31]. Remnant mangroves are severely threatened, with up to 40% of the mangrove plant species being susceptible to extinction in some regions [32]. This loss and degradation may seriously undermine the ability of mangroves to provide valuable ES for present and future generations [33]. Stemming this loss is urgent and requires better management, and restoration of, intact and damaged mangrove ecosystems. It also calls for systematic assessments of current 'stocks' and 'flows' of ES to ensure the sustainable use of these resources [11]. Since mangroves have not received their due share of conservation attention and the rate of decline has not been curbed, the issue of giving incentives after establishing monetary value of ES becomes more important.

To date, there are only three large-scale economic assessments specifically targeted towards mangrove ecosystems [19,20,34]. While Vo et al. [34] provided a review of the methods used for valuation of mangrove ES, Brander et al. [20] focused only on mangroves in South East Asia. Salem and Mercer [19] on the other hand, looked at economic valuation of mangroves globally. None of these studies had attempted to bridge over to management implications of the ES values or identify the gaps in the current valuations of mangrove ES. All the above mentioned studies were conducted by economists, and were based on aggregation of case studies of economic valuations of mangrove ES in different geographic areas. They are thereby limited to only those estimates available in the existing literature and do not go beyond them. Since a large proportion of the ES derived from mangroves may lie in the 'grey market' (i.e. may not have a direct market price), it may be difficult for economists to identify and value the full range of nature-human interlinkages that involve mangroves especially in the case of developing countries. Martín-López et al. [35] emphasize the importance of a multidimensional valuation of ecosystem services, including their non-monetary values.

Combining monetary assessments with valuation by expert based knowledge is one such approach, which may provide interesting insights that are otherwise not possible to attain through conventional monetary valuations alone. The expert based Delphi technique is particularly useful in this context and has been found to be useful in other ecosystems [36,37]. Since most ES are heavily reliant on the proper functioning of the ecosystems [38], integrating ecological knowledge from experts in the relevant field can be considered as an important first step in the mangrove valuation exercise. Though some economists recognize that functionality of ecosystems per hectare has been declining in several cases [16] and this in turn affects the supply of ES, rarely has this aspect been given its due attention in economic assessments except in a few cases [39]. The current study proposes a slightly different framework that turns the valuation process on its head by beginning primarily with ecologists rather than economists.

The aim of this paper is to identify the gaps in current economic valuations of mangrove ES and to suggest ways to inform decision making for better management of mangrove resources. We used an inclusive approach to value ES of mangroves that could then be used to inform governance and decision-making. We used an expert-based participatory approach to identify, categorize and rank the various mangrove ES categories. Hence this study presents new information beyond traditional meta-analysis [19] or systematic reviews of existing economic valuations, and thereby captures the ES that have not yet been economically valued. Further on, we searched for evidence in the existing ES literature for monetary valuations of these ranked ES categories. We compared the relative ranking of ES categories by the monetary valuations and by the expert based knowledge. In order to make the expert based ES categories comparable to other ES valuations, they were collated according to the Common International Classification of Ecosystem Services (CICES) v4.3 (http://cices.eu/). This allowed us to make practical suggestions at the level of management. Lastly, we discuss the possible future trajectories of the valuation and management options for the different ES categories.

Methods

1. The Delphi technique

1.1. Brief description of the method. The Delphi technique is defined as 'a method for structuring a group communication process so that the process is effective in allowing a group of individuals as a whole to deal with a complex problem' [40]. In this technique, expert judgement is elicited in an iterative, anonymous survey with feedback to the participants between each round. The Delphi technique allows all the participants to evaluate the information produced by the group and weigh dissenting views and the consensus is expected to increase from round to round. Individual participants may reconsider or explain their suggestions based upon their evaluation of new information provided. Essentially, the Delphi technique transforms diverse individual knowledge to create a collective wisdom without the domination of individual views [41–43].

1.2. Mangrove Delphi technique. We invited 106 mangrove experts (scientists, reserve managers and field-based conservationists) to participate in the survey. The current work is part of a larger global survey on biodiversity and ecosystem functioning of mangroves [44]. The criteria used for selection and invitation of experts have been explained in detail in [44] and have also been included in the material S1. Briefly, the experts consisted of established mangrove ecologists, mangrove managers and on-ground restoration biologists who were/are involved in mangrove research and management for at least 8 years. Care was taken to select experts outside our research group.

In the first round of the survey, thirty-five experts participated (34% of those invited), while nineteen experts participated in the second round (54% of the first round participants). Respondents of the first round (n = thirty-five) had carried out field research on mangroves in fifty-five countries (Fig. 1). The respondents who completed the entire survey (both rounds) had published 691 peer-reviewed co-authored publications on mangroves and had been cited over 10,829 times (without self-citations). The respondents had a median of 20 years of experience in mangroves and their cumulative expertise covers all mangrove species and environments based on spatial outline in [21].

1.3 Procedure of the survey. The Delphi technique survey consisted of two rounds, conducted within a time-frame of four months (5th November, 2011 to 5th March, 2012). The entire survey was conducted online and a website was designed specifically for this survey [45]. The online survey questionnaires in both rounds were designed using Google Forms.

The survey consisted of five steps. (i) The first round of questions of the Delphi technique was prepared and an online invitation for participation was sent to the selected 106 experts. Typical of the Delphi technique process, the first round questions were open-ended and the experts were asked to suggest the various ES provided by mangrove ecosystems based on their field experience. In order to avoid the confusion between ecosystem functions (defined by Reiss et al. [46] as the changes in energy and matter over time and space through biological activity) and ecosystem services (defined as products of ecosystem functioning that are of (usually socioeconomic) value to humans by [46], only the latter has been dealt with in this study. The question on ecosystem functions has already been addressed in [44]. (ii) The respondents completed the survey and sent it back to us. (iii) The responses were analyzed and collated into a list of 16 ES categories based on their similarity. A feedback report was prepared and uploaded on the website. (iv) In the second round, the experts scored these 16 categories of ES that they had suggested in the previous round. The experts were asked to score options on a Likert scale of 1–5, where "1" indicated low value and "5" indicated high value [47]. Later, to rank the categories, the Likert scale scores for each ES category were given corresponding weights (e.g.Likert score 1 = weight 1, Likert score 5 = weight 5) and multiplied by the number of votes for that option to generate a total weighted score for that ES. Further on, these weighted scores were converted to a percentage scale to generate a ranking of the ES categories. Since the experts who participated in the first round were requested to participate in the second round, the participants were self-selected in the second round, contained within the first selection. (v) Thereafter, the second round responses were analysed, compiled into a feedback report and uploaded on the website. Based on the categorization of von der Gracht [48] for consensus measurement, we followed a 'subjective analysis' approach. It was felt that a third round would not add to the understanding provided by the first two rounds. Thus, the Delphi technique was terminated after the second round. Unless otherwise stated, the results from the second round of the Delphi technique are presented here.

We did not obtain ethics approval for this exercise as only those experts who were willing to give their views took part in the survey.

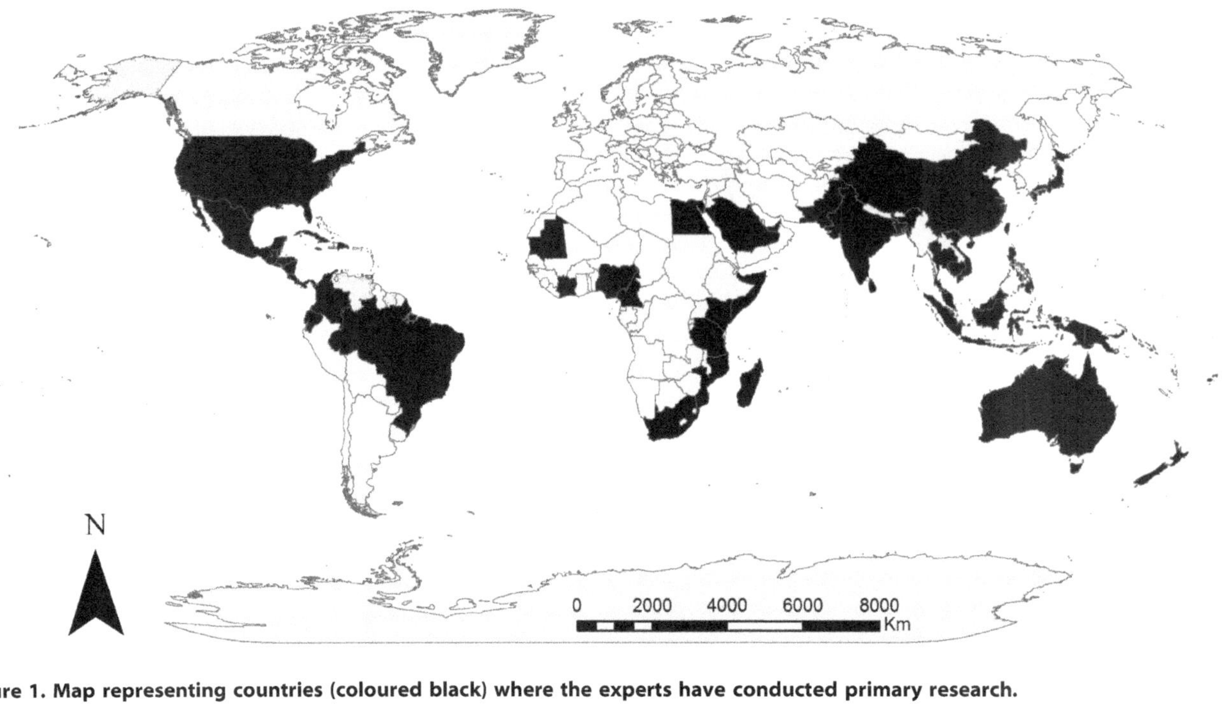

Figure 1. Map representing countries (coloured black) where the experts have conducted primary research.

The respondents were given the choice of being acknowledged or remaining anonymous at the end of the second round of the survey. Names of those mangrove experts who wished to be acknowledged can be found in the acknowledgements section. In addition, the survey was anonymous while it was being conducted, similar to other studies in the literature [49].

2. Estimates of economic value

Existing global databases like the Ecosystem Services Valuation Database [50] formed a valuable starting point for the data collation. The Economics of Ecosystems and Biodiversity (TEEB) database covers more than 1310 values of ES for a range of different ecosystems from 267 references. We selected only those values from these case studies that specifically belonged to mangroves. The bibliometric search was performed using the keywords ("mangrove" or "mangroves") and ("ecosystem service" or "economic valuation" or "value") in a search query within the ISI Web of Science database (http://apps.webofknowledge.com), Google and Google Scholar from 1955–2013 (as of 14th January, 2014). After an initial screening of over 6000+ records, only those studies in peer reviewed literature were retained that specifically mentioned (a) a monetary value for the ecosystem service in USD, (b) the valuation method, (c) the location (country) where the study was conducted. Datasets from published reviews (peer-reviewed) were also taken into consideration [15,19]. We do acknowledge that economic valuations of mangrove ES published in 'grey literature' may exist, but it was beyond the scope of this paper to include them. In addition, documents published in any language other than English have not been covered in this study.

All values which were published after [15] were standardized to 2007 estimates based on the procedure described in detail in the TEEB database [50] to maintain parity. Briefly, all economic estimates in the original case studies were converted into the official local currency. Then these values were adjusted to 2007 values and finally they were converted to international dollars using the purchase power parity (PPP) conversion factor ('local currency per international $' series). The official exchange rates, GDP deflators and PPP conversion factors from the World Bank World Development Indicators 2009 were used to standardize values estimated in different years and different currencies.

Mathematically,

$$\text{1 local currency unit (LCU) on date D1} =$$

$$x * y \text{ International USD on Date D2}$$

(where, $x =$ deflation of LCU between D1 and D2, $y =$ PPP forex rate in USD per LCU on D2).

Since estimates of the sample size and variance of the original studies were absent in most of the original estimates it was difficult to conduct a meta-analysis following the guidelines set by Vetter et al. [51].

Values obtained from the various sources (n = thirty four) were collated according to the Common International Classification of Ecosystem Services (CICES) classification system (http://cices.eu/) and the response categories of the Delphi technique (see previous section on the Delphi technique). Average values for each category were calculated. These average economic values were sorted to produce a ranking of the ES categories. The correlation between the ranking produced by the economic valuations and by the experts in the Delphi technique was tested using the Spearman's Rank correlation calculated in the statistical programme 'R'. In case a category suggested by the Delphi technique fell across two or more categories in the CICES framework, the economic values were aggregated to the Delphi ES category for comparison.

Results

1. Expert based valuation

There was a high level of consensus amongst the experts (as indicated by the stability of responses after only two rounds of the Delphi technique) even though they worked in a range of different field sites, biogeographical regions and socio-economic settings across the globe. The 16 ES categories identified by the expert panel are shown in Table 1 based on the CICES framework. The role of mangroves in fisheries, coastal protection, protection from sedimentation and provisioning for wood and timber were identified to be the top three ES of mangrove ecosystems (Fig 2). Three of these ES fall under the category of regulation and maintenance services according to CICES, with "fisheries" being spread over both provisioning (nutrition) and regulation and maintenance (nursery function). Mangrove ecosystems were also identified to be important environmental risk indicators and carbon sequesters. In the context of climate change, the emphasis on coastal protection and protection from sedimentation are particularly important, given the location of mangroves close to the coast and the rapid decline of mangrove area [52] in the past few decades.

2. Economic valuation

The economic values of the mangrove ES standardized for the 2007 international dollar is given in Table 2. According to the existing peer-reviewed literature, ecotourism and fisheries generated the highest economic value (including subsistence) based on the 2007 estimates (Fig 3). It should be noted here that the role of mangroves in fisheries has been split into three categories based on the CICES classification but their combined value is presented in Table 2. Coastal protection also ranked highly (third) according to the average estimates of economic values.

Six of the ES categories identified and valued by the experts in the Delphi technique were not represented in the economic valuation literature in our database (*viz*: fodder, water bio-remediation, protection from salt intrusion, aesthetic value, pharmaceuticals and environmental risk indicator). While aesthetic value maybe evaluated to a certain degree based on the eco-tourism and recreation potential (e.g. [53]), the others may need appropriate indicators or new methods to be incorporated in the valuation framework because no other indicator covers them currently. The underlying ecological functions which lead the provisioning of these services may also need further attention. There was no significant correlation between ES ranks according to the two ranking approaches (Spearman's rank correlation coefficient $r_s = 0.42$, $p = 0.23$).

Discussion

In this section, we focus on the implications for management and conservation of mangroves and future ES research. The recommendations broadly follow the CICES framework for ES classification and the current methods used to value ES.

Provisioning services

The key provisioning services of mangroves identified were fisheries (important for subsistence, livelihood and commercial fisheries), wood and timber, honey, energy sources, fodder and pharmaceuticals. In developing countries where a large proportion of the rural poor population live on less than $1/day [54], these ES

Table 1. Ecosystem service categories identified by the mangrove Delphi, grouped according to the Common International Classification of Ecosystem Services (CICES) v4.3 (http://cices.eu/).

CICES for ecosystem accounting

Section	Division	Group	Class	Class type	Delphi technique categories
Provisioning	Nutrition	Biomass	Wild animals and their outputs	Animals by amount, type	Fisheries (food)
					Honey
			Animals from *in-situ* aquaculture	Animals by amount, type	Fisheries (aquaculture)
	Materials	Biomass	Fibres and other materials from plants, algae and animals for direct use or processing	Material by amount, type, use, media (land, soil, freshwater, marine)	Wood and timber
					Pharmaceuticals
			Materials from plants, algae and animals for agricultural use		Fodder
	Energy	Biomass-based energy sources	Plant-based resources	By amount, type, source	Energy resources
Regulation & Maintenance	Mediation of waste, toxics and other nuisances	Mediation by biota	Bio-remediation by micro-organisms, algae, plants, and animals	By amount, type, use, media (land, soil, freshwater, marine)	Water bio-remediation
		Mediation by ecosystems	Filtration/sequestration/storage/ accumulation by ecosystems	By amount, type, use, media (land, soil, freshwater, marine)	Pollution abatement, Environmental risk Indicator
	Mediation of flows	Mass flows	Mass stabilization and control of erosion rates	By reduction in risk, area protected	Protection from sedimentation
			Buffering and attenuation of mass flows		Protection from salt intrusion
		Gaseous/air flows	Storm protection	By reduction in risk, area protected	Coastal protection
	Maintenance of physical, chemical, biological conditions	Lifecycle maintenance, habitat and gene pool protection	Maintaining nursery populations and habitats	By amount and source	Fisheries (nursery)
		Soil formation and composition	Decomposition and f ixing processes		Carbon sequestration
		Atmospheric composition and climate regulation	Global climate regulation by reduction of greenhouse gas concentrations	By amount, concentration or climatic parameter	Carbon sequestration
Cultural	Physical and intellectual interactions with biota, ecosystems, and land-/seascapes (environmental settings)	Intellectual and representative interactions	Entertainment		Ecotourism and recreation
			Aesthetic		Aesthetic value

The first five columns belong to the CICES framework and the results of the Delphi technique are included in the last column.

are crucial life support systems even though some of them might have a low monetary value (e.g. wood and timber) [27,55]. Since all of these services are exhaustible in nature and the consumption or harvest by one individual reduces the available stock for the next individual, it is imperative to direct adequate management resources to safeguard these ES in the long term. As a first step, we recommend creating a baseline of the underlying ecological potential (stocks) of the mangrove forests in each site, based on ecological knowledge (scale dependent on the extent of the forest and the available management resources). As a second step, we recommend assessing the level of sustainable yield (flows) that can be supported by the chosen mangrove forest under consideration. Finally, we recommend the creation of conditions for recovery or effective restoration of mangrove forests in the areas where they existed in the past and creation of alternative resources (e.g. alternative energy and timber resources from managed plantations) for the consumption by local communities. In addition, investing resources in strengthening the cultural capital of the local communities may be useful for conservation. In developing countries, where non-monetized mechanisms have been in place

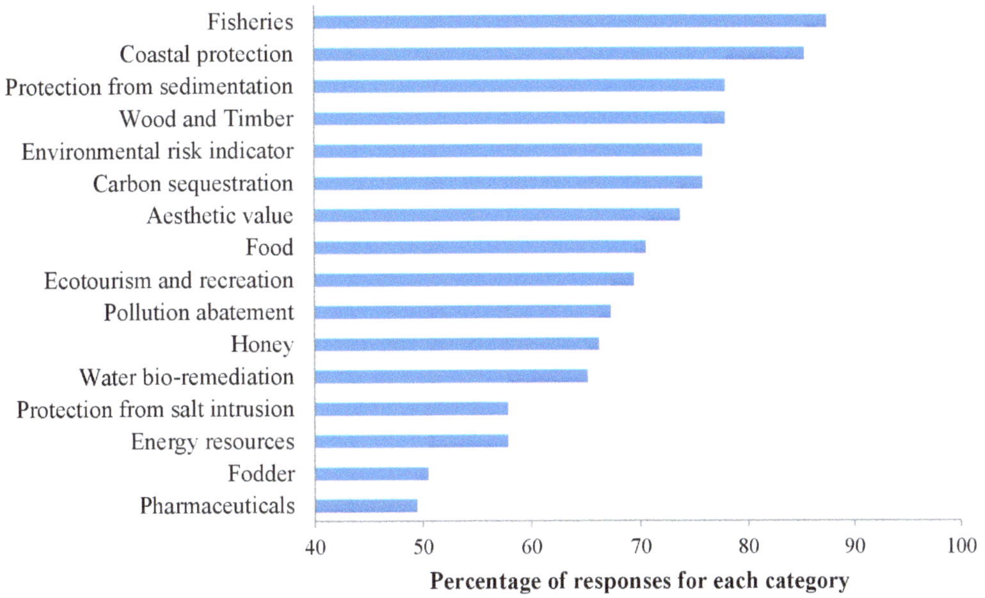

Figure 2. Ranking of the ecosystem service categories of mangroves based on the scores given by the experts in the Delphi technique.

for decades in managing common property resources, creating incentives for community based management of mangroves maybe more beneficial than payment for ES schemes as suggested by Gómez-Baggethun et al. [56] and Ostrom and Nagendra [57].

Regulation and maintenance services

According to the CICES framework, half (eight out of sixteen) of the ES categories mentioned by the experts fall in the regulation and maintenance ES section, four in the top five ranking ES alone (considering the nursery element of the fisheries service). Out of these, only coastal protection has permeated adequately in economic valuation literature (nine estimates). These regulating ES do not have a direct market price in most cases and are estimated by contingent valuation methods (based on stated willingness to pay or willingness to accept a change). Therefore, it is harder to generate economic incentives for conservation of mangrove ecosystems in the short term. For example, unsustainable harvest of mangroves for provisioning services like timber or conversion of mangrove area to aquaculture ponds for short-term economic gain may seriously jeopardize their capability to provide regulating services like coastal protection in the long term. Recent studies indicate that the ecological bases for many of these ES are rapidly changing due to climate change and other impacts [58]. One of the many examples of severe consequences of mangrove destruction is the devastation caused by the recent typhoon Haiyan in November, 2013 in coastal areas of the Philippines where over half (approx. 51%) of mangroves have been destroyed in the last century alone [59]. Conservation and restoration of the ecological status is thereby urgently needed for continued existence of these ES. Strong cross-country policy measures and creation of mangrove protected areas may be useful in this regard particularly in countries where mangroves areas are shared between several nations such as Sunderbans. In addition, adequate valuation mechanisms (monetary and non-monetary) are needed for those ES categories mentioned by the experts but not represented in the literature. It is worth noting here that in several mangrove areas, the power to make decisions leading to either conservation/restoration of mangrove ecosystems may not lie on the local communities who depend on mangrove ES for their livelihood and subsistence. A recent study, [44] showed that degradation due to large scale development (e.g. building of highways, ports and harbours) is the biggest threat to mangroves globally. Often these large infrastructure development projects are initiated by bodies for whom the actual values (monetary and otherwise) of mangrove ES are non-existent.

Cultural services

The two categories of cultural ES identified by the Delphi technique were ecotourism and recreation, and aesthetic services. While the economic valuations revealed that ecotourism has high economic potential, most of the estimates originated from the Caribbean islands and the Atlantic East Pacific (AEP) distribution of mangroves (also see [60,61]). There could be two possible explanations; either ecotourism is infrequent in mangroves in the Indo West Pacific or there are fewer valuation exercises. If the latter is true then adequate economic valuations of the revenue generated by tourism, need to be done in the Indo West Pacific distribution of mangroves where the bulk of mangrove species and forests are located. One such example is the study conducted by Uddin et al. [53]. If the former is true and assuming that there is potential for sustainable tourism, then investing in the development of infrastructure for facilitating ecotourism in collaboration with local communities would be useful for the local economy. However, careful attention needs to be paid in choosing the appropriate incentives used for promoting tourism, i.e. the same principle of stock and flow applies here too for sustainable use of resources [13] or else there may be a collapse of the supply of ES. It should also be noted that there could be economic valuations in the grey literature or in other languages that were not covered in this study.

Table 2. Ecosystem services provided by mangroves sorted according to their mean economic values (2007 Int$/ha*yr).

Delphi technique categories	Mean economic value (2007 Int$/ha*yr)	No. of estimates	Economic rank	Delphi technique rank
Fisheries (nursery and aquaculture)	17090.1	25	1	1
Ecotourism and recreation	14072.14	10	2	7
Coastal protection	8459.12	9	3	2
Pollution abatement	7859.92	2	4	8
Food	1535.21	16	6	6
Protection from sedimentation	579.28	1	7	3
Energy resources	306.92	8	8	12
Wood and timber	247.34	3	9	3
Carbon sequestration	195.23	3	10	4
Honey	4.23	2	11	9
Fodder	0	0	0	13
Water bio-remediation	0	0	0	10
Protection from salt intrusion	0	0	0	11
Aesthetic value	0	0	0	5
Pharmaceuticals	0	0	0	14
Environmental risk indicator	0	0	0	4

Six of the 16 ecosystem services (highlighted in bold and italics) identified by experts do not have adequate valuations in the peer-reviewed ecological economic literature.

Conclusions

This study complements the conventional monetary valuation of ES with non-economic valuation of ES by a range of mangrove experts who have been working in the field for over 20 years across the entire biogeographic range of mangroves. The deliberate primary focus on ecologists is both a bias and a scope of this study since ecosystem functioning is at the root of all the ES generation. Integrating the knowledge of ecologists who have prior experience in investigating ecosystem functioning of mangroves was our main aim. It should also be noted that some of the experts who participated and the authors who designed the questionnaire also

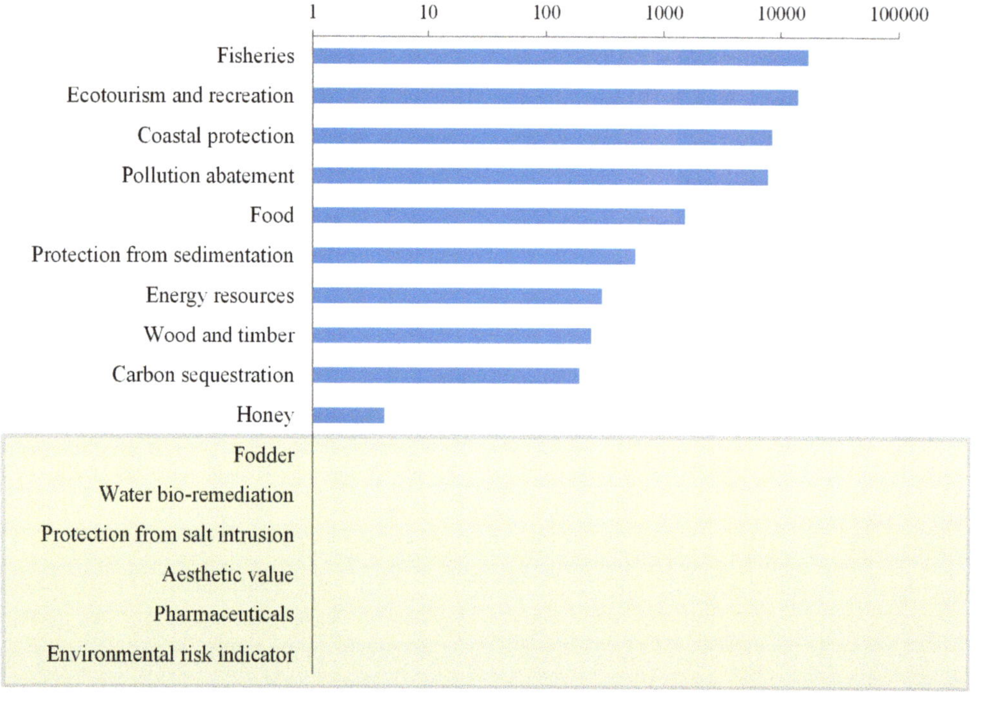

Figure 3. Average economic values (log scale) of the different ecosystem services noted in literature standardized for 2007 Int$/ha*yr. The five ecosystem services for which no economic values were found, are highlighted in the shaded box.

had research experience on socio-economic assessments in mangroves [61–64]. This research adds to the existing literature on mangrove ES in an unconventional way and this approach can be easily replicated for other ecosystems (or later periods for the same questions). While expert consultation has been recommended by economists for estimating economic value transfer [16], this study approaches the issue of valuation by experts from an epistemological perspective rather than that of a merely technical nature for adjusting values [65] or identifying indicators[66]. It is thereby more in the lines of harnessing expert knowledge for progressing thought and improving the *status quo* as demonstrated by Wallington and Moore [67].

The lack of correlation between mangrove ES ranks based on expert based valuation (Delphi technique) and economic valuation, indicates that economic valuations may have missed out crucial ES and the scope of valuation of ES needs to be broadened. Even though both approaches show that mangrove ecosystems are particularly important for the provision of a range of ecosystem services, the relative importance of each of the ES categories differs markedly. Acknowledging this diversity, and developing methodological frameworks fostering value pluralism in ecosystem valuation research is key to maintaining the credibility of the ES valuation approach. Different valuation results might indeed lead to different trade-offs among ecosystem services [35] or other land uses. When decision-makers use ES valuation in guiding their plans for mangrove management and conservation [68], they need to be supported by critical analysis and a plurality of ES valuation methods to prevent mismanagement of these key tropical ecosystems. This certainly does not imply commodification of ES as claimed by [69] but rather the much needed acknowledgement of the ES which are often implicitly recognized in governance and policy issues [70].

Although the Delphi technique is a very useful method for expert knowledge elicitation, the use of the method might tend towards subjectivity, especially when dealing with complex systems as indicated in the study by Benitez-Capistros et al. [71]. This situation is linked to the inherent difficulty in determining who is a 'knowledgeable respondent, i.e. an expert' – a shortcoming we addressed by a transparent definition of the expert selection criteria (see material S1). Related to this problem are the difficulties of recruiting participants and avoiding dropouts in each round. Nevertheless, when conducted transparently, the Delphi technique is a rigorous method as it brings more objectivity and accuracy in the overall outcome [72]. In addition, in order to have stakeholders who cannot be easily reached through the Delphi technique as applied here, complementary methods like integral valuation may be necessary.

This paper goes further than the valuation alone, as it addresses the issues related to management options and future ES assessments for the continued sustainable use of mangrove ES in the long term. Further research on ES valuation in selected mangrove sites is needed to complement this global scale explorative research. Looking forward, an open question that remains to be answered is how the newly emphasized ES categories (lacking economic valuations) influence decision making in conservation, management and restoration of mangroves in the future.

Acknowledgments

Katrien Quisthoudt, Pinaki Bhattacharya and Maria Voigt are thanked for their helpful suggestions in improving the manuscript and for the map. Geoff Hyde is thanked for his critical inputs in editing an earlier version of the manuscript. The authors would also like to thank the experts who took part in both rounds of the survey and gave their consent for acknowledgements: Colin Field, Cyril Piou, Fiona Wilmot, Francesca Porri, Gerard van der Velde, Gonasageran Naidoo, Jorge López-Portillo, Kandasamy Kathiresan, Mami Kainuma, Marcus Sheaves, Narayanasamy Rajendran, Peter Saenger, Robin Lewis, Satyanarayana Behara, Tarik Meziane and Wim Giesen.

Author Contributions

Conceived and designed the experiments: NM NK FDG. Analyzed the data: NM LD. Wrote the paper: NM. Performed the survey: NM. Edited and revised the document: LD WS NK JH FDG.

References

1. Hooper DU, Adair EC, Cardinale BJ, Byrnes JE, Hungate BA, et al. (2012) A global synthesis reveals biodiversity loss as a major driver of ecosystem change. Nature 486: 105–108.
2. Millennium Ecosystem Assessment (2005) Ecosystems and Human Well-being: Synthesis. Washington, DC.: World Resources Institute.
3. Costanza R, d'Arge R, de Groot R, Farber S, Grasso M, et al. (1997) The value of the world's ecosystem services and natural capital. Nature 387: 8.
4. Balvanera P, Pfisterer AB, Buchmann N, He JS, Nakashizuka T, et al. (2006) Quantifying the evidence for biodiversity effects on ecosystem functioning and services. Ecol Lett 9: 1146–1156.
5. Cardinale BJ, Duffy JE, Gonzalez A, Hooper DU, Perrings C, et al. (2012) Biodiversity loss and its impact on humanity. Nature 486: 59–67.
6. World Water Assessment Programme (2012) The United Nations World Water Development Report 4: Managing Water under Uncertainty and Risk. Paris: UNESCO.
7. Worm B, Barbier EB, Beaumont N, Duffy JE, Folke C, et al. (2006) Impacts of biodiversity loss on ocean ecosystem services. Science 314: 787–790.
8. Clavel J, Julliard R, Devictor V (2011) Worldwide decline of specialist species: toward a global functional homogenization? Frontiers in Ecology and the Environment 9: 222–228.
9. Ruckelshaus M, McKenzie E, Tallis H, Guerry A, Daily G, et al. (2013) Notes from the field: Lessons learned from using ecosystem service approaches to inform real-world decisions. Ecological Economics.
10. Laurans Y, Rankovic A, Bille R, Pirard R, Mermet L (2013) Use of ecosystem services economic valuation for decision making: questioning a literature blindspot. J Environ Manage 119: 208–219.
11. Bateman IJ, Harwood AR, Mace GM, Watson RT, Abson DJ, et al. (2013) Bringing ecosystem services into economic decision-making: land use in the United Kingdom. Science 341: 45–50.
12. Kareiva P, Chang A, Marvier M (2008) Environmental economics- Development and Conservation Goals in World Bank Projects. Science 321: 1638–1639.
13. Bateman IJ, Mace GM, Fezzi C, Atkinson G, Turner K (2010) Economic Analysis for Ecosystem Service Assessments. Environmental and Resource Economics 48: 177–218.
14. Brouwer R, Brander L, Kuik O, Papyrakis E, Bateman I (2013) A synthesis of approaches to assess and value ecosystem services in the EU in the context of TEEB. VU University Amsterdam.
15. de Groot R, Brander L, van der Ploeg S, Costanza R, Bernard F, et al. (2012) Global estimates of the value of ecosystems and their services in monetary units. Ecosystem Services 1: 50–61.
16. Costanza R, de Groot R, Sutton P, van der Ploeg S, Anderson SJ, et al. (2014) Changes in the global value of ecosystem services. Global Environmental Change 26: 152–158.
17. Liquete C, Piroddi C, Drakou EG, Gurney L, Katsanevakis S, et al. (2013) Current status and future prospects for the assessment of marine and coastal ecosystem services: a systematic review. PLoS One 8: e67737.
18. Barbier EB (2011) Progress and Challenges in Valuing Coastal and Marine Ecosystem Services. Review of Environmental Economics and Policy 6: 1–19.
19. Salem ME, Mercer DE (2012) The Economic Value of Mangroves: A Meta-Analysis. Sustainability 4: 359–383.
20. Brander LM, J. Wagtendonk A, S. Hussain S, McVittie A, Verburg PH, et al. (2012) Ecosystem service values for mangroves in Southeast Asia: A meta-analysis and value transfer application. Ecosystem Services 1: 62–69.

21. Spalding M, Kainuma M, Collins L (2010) World atlas of mangroves. London, UK;: Washington, DC: Earthscan.
22. van Bochove J, Sullivan E, Nakamura T (2014) The Importance of Mangroves to People: A Call to Action. United Nations Environment Programme.
23. Walters BB, Rönnbäck P, Kovacs JM, Crona B, Hussain SA, et al. (2008) Ethnobiology, socio-economics and management of mangrove forests: A review. Aquatic Botany 89: 220–236.
24. McNally CG, Uchida E, Gold AJ (2011) The effect of a protected area on the tradeoffs between short-run and long-run benefits from mangrove ecosystems. Proc Natl Acad Sci U S A 108: 13945–13950.
25. Jones DL, Walter JF, Brooks EN, Serafy JE (2010) Connectivity through ontogeny: fish population linkages among mangrove and coral reef habitats. Marine Ecology Progress Series 401: 245–258.
26. Unsworth RKF, De León PS, Garrard SL, Jompa J, Smith DJ, et al. (2008) High connectivity of Indo-Pacific seagrass fish assemblages with mangrove and coral reef habitats. Marine Ecology Progress Series 353: 213–224.
27. Hussain SA, Badola R (2010) Valuing mangrove benefits: contribution of mangrove forests to local livelihoods in Bhitarkanika Conservation Area, East Coast of India. Wetlands Ecology and Management 18: 321–331.
28. Das S, Crépin A-S (2013) Mangroves can provide protection against wind damage during storms. Estuarine, Coastal and Shelf Science 134: 98–107.
29. Lee SY, Primavera JH, Dahdouh-Guebas F, McKee K, Bosire JO, et al. (2014) Ecological role and services of tropical mangrove ecosystems: a reassessment. Global Ecology and Biogeography 23: 726–743.
30. Donato DC, Kauffman JB, Murdiyarso D, Kurnianto S, Stidham M, et al. (2011) Mangroves among the most carbon-rich forests in the tropics. Nature Geoscience 4: 5.
31. Balmford A, Bruner A, Cooper P, Costanza R, Farber S, et al. (2002) Economic reasons for conserving wild nature. Science 297: 950–953.
32. Polidoro BA, Carpenter KE, Collins L, Duke NC, Ellison AM, et al. (2010) The loss of species: mangrove extinction risk and geographic areas of global concern. PLoS One 5: e10095.
33. Feller IC, Lovelock CE, Berger U, McKee KL, Joye SB, et al. (2010) Biocomplexity in Mangrove Ecosystems. Annual Review of Marine Science 2: 395–417.
34. Vo QT, Kuenzer C, Vo QM, Moder F, Oppelt N (2012) Review of valuation methods for mangrove ecosystem services. Ecological Indicators 23: 431–446.
35. Martín-López B, Gómez-Baggethun E, García-Llorente M, Montes C (2014) Trade-offs across value-domains in ecosystem services assessment. Ecological Indicators 37: 220–228.
36. Scolozzi R, Morri E, Santolini R (2012) Delphi-based change assessment in ecosystem service values to support strategic spatial planning in Italian landscapes. Ecological Indicators 21: 134–144.
37. James P, Tzoulas K, Adams MD, Barber A, Box J, et al. (2009) Towards an integrated understanding of green space in the European built environment. Urban Forestry & Urban Greening 8: 65–75.
38. Spangenberg JH, von Haaren C, Settele J (2014) The ecosystem service cascade: Further developing the metaphor. Integrating societal processes to accommodate social processes and planning, and the case of bioenergy. Ecological Economics 104: 22–32.
39. Liquete C, Zulian G, Delgado I, Stips A, Maes J (2013) Assessment of coastal protection as an ecosystem service in Europe. Ecological Indicators 30: 205–217.
40. Turoff M (1970) The Design of a Policy Delphi. Technological Forecasting & Social Change 2: 149–171.
41. Dalkey N, Helmer O (1963) An Experimental Application of the Delphi Method to the Use of Experts. Management Science 9: 458–467.
42. De Villiers MR, De Villiers PJT, Kent AP (2005) The Delphi technique in health sciences education research. Medical Teacher 27: 639–643.
43. Mukherjee N, Hugé J, Sutherland WJ, McNeill J, Van Opstal M, et al. (in prep) The role of the Delphi technique: an assessment of strengths, weaknesses and applications. Methods in Ecology and Evolution.
44. Mukherjee N, Sutherland WJ, Khan MNI, Berger U, Schmitz N, et al. (2014) Using expert knowledge and modeling to define mangrove composition, functioning, and threats and estimate time frame for recovery. Ecology and Evolution 4: 2247–2262.
45. Mukherjee N (2011) Welcome to the Delphi Survey on Mangroves. Brussels: Google Sites.
46. Reiss J, Bridle JR, Montoya JM, Woodward G (2009) Emerging horizons in biodiversity and ecosystem functioning research. Trends Ecol Evol 24: 505–514.
47. Likert R (1932) A Technique for the Measurement of Attitudes. Archives of Psychology 140: 1–55.
48. von der Gracht HA (2012) Consensus measurement in Delphi studies. Technological Forecasting and Social Change 79: 1525–1536.
49. Sutherland WJ, Bellingan L, Bellingham JR, Blackstock JJ, Bloomfield RM, et al. (2012) A collaboratively-derived science-policy research agenda. PLoS One 7: e31824.
50. Van der Ploeg S, De Groot R, Wang Y (2010) The TEEB Valuation Database:overview of structure, data and results. Wageningen, the Netherlands.: Foundation for Sustainable Development.
51. Vetter D, Rücker G, Storch I (2013) Meta-analysis: A need for well-defined usage in ecology and conservation biology. Ecosphere 4: art74.
52. Giri C, Ochieng E, Tieszen LL, Zhu Z, Singh A, et al. (2011) Status and distribution of mangrove forests of the world using earth observation satellite data. Global Ecology and Biogeography 20: 154–159.
53. Uddin MS, de Ruyter van Steveninck E, Stuip M, Shah MAR (2013) Economic valuation of provisioning and cultural services of a protected mangrove ecosystem: A case study on Sundarbans Reserve Forest, Bangladesh. Ecosystem Services 5: 88–93.
54. Barbier EB (2012) Natural Capital, Ecological Scarcity and Rural Poverty. World Bank. pp.38.
55. Badola R, Hussain SA (2005) Valuing ecosystem functions: an empirical study on the storm protection function of Bhitarkanika mangrove ecosystem, India. Environmental Conservation 32: 85–92.
56. Gómez-Baggethun E, de Groot R, Lomas PL, Montes C (2010) The history of ecosystem services in economic theory and practice: From early notions to markets and payment schemes. Ecological Economics 69: 1209–1218.
57. Ostrom E, Nagendra H (2006) Insights on linking forests, trees, and people from the air, on the ground, and in the laboratory. Proc Natl Acad Sci U S A 103: 19224–19231.
58. Lee SY, Primavera JH, Dahdouh-Guebas F, McKee K, Bosire JO, et al. (2014) Ecological role and services of tropical mangrove ecosystems: a reassessment. Global Ecology and Biogeography: n/a-n/a.
59. Long J, Napton D, Giri C, Graesser J (2014) A Mapping and Monitoring Assessment of the Philippines' Mangrove Forests from 1990 to 2010. Journal of Coastal Research 294: 260–271.
60. Avau J, Cunha-Lignon M, De Myttenaere B, Godart MF, Dahdouh-Guebas F (2011) The commercial images promoting Caribbean mangroves to tourists: Case studies in Jamaica, Guadeloupe and Martinique. Journal of Coastal Research Special Issue 64: 1277–1281.
61. Satyanarayana B, Bhanderi P, Debry M, Maniatis D, Fore F, et al. (2012) A socio-ecological assessment aiming at improved forest resource management and sustainable ecotourism development in the mangroves of Tanbi Wetland National Park, The Gambia, West Africa. Ambio 41: 513–526.
62. Satyanarayana B, Mulder S, Jayatissa LP, Dahdouh-Guebas F (2013) Are the mangroves in the Galle-Unawatuna area (Sri Lanka) at risk? A social-ecological approach involving local stakeholders for a better conservation policy. Ocean & Coastal Management 71: 225–237.
63. Dahdouh-Guebas F, Collin S, Lo Seen D, Ronnback P, Depommier D, et al. (2006) Analysing ethnobotanical and fishery-related importance of mangroves of the East-Godavari Delta (Andhra Pradesh, India) for conservation and management purposes. J Ethnobiol Ethnomed 2: 24.
64. Dahdouh-Guebas F. Les biens et services écosystémiques: l'exemple des mangroves. In: Meerts P, editor; 2013; Brussels, Belgium. Centre Paul Duvigneaud de Documentation Ecologique. pp.182–193.
65. Curtis IA (2004) Valuing ecosystem goods and services: a new approach using a surrogate market and the combination of a multiple criteria analysis and a Delphi panel to assign weights to the attributes. Ecological Economics 50: 163–194.
66. Orsi F, Geneletti D, Newton AC (2011) Towards a common set of criteria and indicators to identify forest restoration priorities: An expert panel-based approach. Ecological Indicators 11: 337–347.
67. Wallington TJ, Moore SA (2005) Ecology, Values, and Objectivity: Advancing the Debate. BioScience 55: 873–878.
68. Laurans Y, Mermet L (2014) Ecosystem services economic valuation, decision-support system or advocacy? Ecosystem Services 7: 98–105.
69. Gomez-Baggethun E, Ruiz-Perez M (2011) Economic valuation and the commodification of ecosystem services. Progress in Physical Geography 35: 613–628.
70. Hauck J, Görg C, Varjopuro R, Ratamäki O, Jax K (2013) Benefits and limitations of the ecosystem services concept in environmental policy and decision making: Some stakeholder perspectives. Environmental Science & Policy 25: 13–21.
71. Benitez-Capistros F, Hugé J, Koedam N (2014) Environmental impacts on the Galapagos Islands: Identification of interactions, perceptions and steps ahead. Ecological Indicators 38: 113–123.
72. Hasson F, Keeney S (2011) Enhancing rigour in the Delphi technique research. Technological Forecasting and Social Change 78: 1695–1704.

Does Land-Use Intensification Decrease Plant Phylogenetic Diversity in Local Grasslands?

Eugen Egorov[1]*, Daniel Prati[2], Walter Durka[3], Stefan Michalski[3], Markus Fischer[2], Barbara Schmitt[2], Stefan Blaser[2], Martin Brändle[1]

1 Department of Ecology, Faculty of Biology, Philipps Universität, Marburg, Germany, **2** Institute of Plant Sciences, University of Bern, Bern, Switzerland, **3** Department Community Ecology, Helmholtz Centre for Environmental Research, Halle, Germany

Abstract

Phylogenetic diversity (PD) has been successfully used as a complement to classical measures of biological diversity such as species richness or functional diversity. By considering the phylogenetic history of species, PD broadly summarizes the trait space within a community. This covers amongst others complex physiological or biochemical traits that are often not considered in estimates of functional diversity, but may be important for the understanding of community assembly and the relationship between diversity and ecosystem functions. In this study we analyzed the relationship between PD of plant communities and land-use intensification in 150 local grassland plots in three regions in Germany. Specifically we asked whether PD decreases with land-use intensification and if so, whether the relationship is robust across different regions. Overall, we found that species richness decreased along land-use gradients the results however differed for common and rare species assemblages. PD only weakly decreased with increasing land-use intensity. The strength of the relationship thereby varied among regions and PD metrics used. From our results we suggest that there is no general relationship between PD and land-use intensification probably due to lack of phylogenetic conservatism in land-use sensitive traits. Nevertheless, we suggest that depending on specific regional idiosyncrasies the consideration of PD as a complement to other measures of diversity can be useful.

Editor: Maarja Öpik, University of Tartu, Estonia

Funding: The DFG (German Research Foundation; http://www.dfg.de/en/index.jsp) funded the study in the framework of the Biodiversity Exploratories SSP 1374 "Infrastructure-Biodiversity-Exploratories"(BR 1967/9-1). The funders had no role in study design, data collection and analysis, decision to publish, or preparation of the manuscript.

Competing Interests: The authors have declared that no competing interests exist.

* Email: eugen.egorov@staff.uni-marburg.de

Introduction

Land-use change is one of the primary drivers of biodiversity loss [1,2]. Despite a large amount of studies dealing with the effects of land use on biodiversity, there are still gaps in the understanding of land use – biodiversity relationships. For example, the negative effects of different land-use types on biodiversity can differ in strength or vary in their effects. In addition, regional idiosyncrasies might interact with land use and affect biodiversity responses to land-use intensification, thus impeding general predictions [3]. Recent studies have advocated the consideration of phylogenetic diversity (PD) in ecological analyzes [4–7]. In brief, PD is defined as the total amount of phylogenetic space covered by species in a community. It therefore encapsulates the entire trait space of a community [8] and thus, may serve as a complement to trait diversity if the traits cannot be measured or trait data are not available [4]. Moreover, PD is an important factor for ecosystem function itself. It has been shown that PD can explain more variance in productivity in grasslands than species richness or functional diversity [9]. Plant productivity increased with mycorrhizal PD, which may be caused by niche differentiation, as increasing number of mycorrhizal families provide different advantages to their host plants [10]. Higher Plant PD also increases diversity of higher trophic levels and affects several ecosystem functions and processes [11–13]. That is, higher plant PD reinforces the positive effects of plant species richness on higher trophic levels when species richness is held constant [13]. Finally it has been found that PD promotes ecosystem stability and resilience [12] as well as interacts with plant species richness and alters its effect on herbivory [11]. Despite a consensus that PD is an important factor in understanding biodiversity – ecosystem functions relationships [7] or community assembly rules [14], little effort has been done in analyzing the effects of land-use intensity on PD [15].

In Central Europe managed grasslands are one of the most abundant and species-rich ecosystems [16]. In Germany, about 12% of area is covered by grasslands [17]. Most of these grasslands were established during a long period of low-intensity land-use and a large number of species have adapted to those conditions causing high levels of biodiversity. Land-use intensification in particular during the 20th century posed considerable threats to biodiversity in grasslands, e.g. due to dramatic habitat loss and extinction of less competitive species [18–20]. It is also likely that land-use intensification will be the major driver of biodiversity loss in grasslands during the next decades [1,21]. To attain a compromise between high land-use intensity and biodiversity conservation [1] and to assess the consequences of biodiversity loss a deeper understanding of the relationship between land-use intensification, biodiversity and ecosystem functioning is mandatory.

In general previous studies of plant biodiversity-ecosystem functioning relationships have shown that species richness enhances ecosystem functions [22–24]. Simply counting the number of species, however, is often not sufficient for analyzing the effects of biodiversity on ecosystem functions [25]. More comprehensive approaches consider functional diversity, defined as diversity of traits important for ecosystem level processes [26]. Functional diversity is thought to be the component of biodiversity with the largest effect on ecosystem processes [27–29]. However, implementation of trait data is subject to several limitations. For example, assessment of trait data is time-consuming and the *a priori* choice of specific traits is not always straightforward [26]. To overcome these shortcomings, PD has been proposed as a proxy for functional diversity [5,30]. Recent studies, however, question PD as a proxy and propose it rather as a complement to functional diversity [31]. Despite the current discussion on the use of community phylogenetics in analyzes of assembly processes under several biotic and abiotic conditions [32] the importance of PD to ecosystem processes calls for its implementation into ecological analyzes [4]. While the negative effect of land-use intensification on species richness and functional diversity has been subject to many studies [2,33,34], a relatively small number of studies investigated how increasing land-use intensity affects PD of plant communities, particularly in grasslands. Studies that compared observed phylogenetic community structure of plants with expected patterns [30] revealed shifts in phylogenetic community structure with increasing disturbance and stress [15,35–38]. Similar patterns were also shown within animal communities [39–41]. Changes in phylogenetic community structure may include shifts from overdispersion, where co-occurring species are less phylogenetically related than expected by chance, to clustering, where co-occurring species are phylogenetically more related than expected by chance. Such a shift from overdispersion to clustering is thought to be caused by environmental filtering that selects species with similar ecological traits that are likely to be closely related [15,36; but see 32]. Increasing land use intensity should therefore favor plant species with traits adapted to cope with effects of land-use intensification like fertilization, cattle grazing and frequent mowing. If such traits are phylogenetically conserved and play a major role in the phylogenetic community assembly, communities are likely to become phylogenetically more clustered with increasing land-use intensity. If traits are convergent or show a low phylogenetic signal, plant communities should not exhibit phylogenetic clustering with increasing land-use intensities or even lead to an increase in PD [38].

For conservational purposes the response of rare species to land-use intensification is of great interest. Rare species are in general more vulnerable to land-use intensification than common species [18–20]. Assuming that common species might be better adapted to high land-use intensities, phylogenetic diversity of common species should be less sensitive to land-use intensification than that of rare species. However, to our knowledge there are no studies exploring the response of PD of rare and common species to land-use intensification separately.

Socher et al. [3] showed that strength and direction of the effects of land use on biodiversity can differ between regions. Regional idiosyncrasies can also alter the effect of land use on phylogenetic diversity due to different regional species pools, environmental and geographical variables. It is therefore necessary to compare the effects of land-use intensification on PD among regions. Other limitations of previous research on plant PD are that the majority of studies are either experimental or describe phylogenetic patterns along natural or environmental gradients and are restricted to certain, often narrow, taxonomic scales [23,42]. Descriptive studies of PD – land-use intensity relationships in human-disturbed systems are still scarce. When analyzing plant PD with respect to man-made disturbance, studies often focus on urban regions [37] or do not encompass the most common agricultural land-use categories such as fertilization, mowing and grazing. Including most common land-use types in descriptive studies of PD – land use relationships in agricultural systems could give new insights on these relationships under "real world" conditions. Previous studies may also suffer from the lack of considering species abundance data. Presence/absence data are highly sensitive to the chance and possible temporary occurrence of a single individual in unusual or unsuitable habitat. Interspecific relationships and interactions between species and ecosystems are based on interactions between individuals, which are cumulative in their effects. Neglecting abundance data may impede to discover important ecological relationships [6].

In this study we use species abundance data to analyze the PD of plant communities in local grasslands (150 sites) across land-use intensification gradients in three regions in Germany. In particular we aimed to answer the following questions:

1) Are there regional differences in the response of phylogenetic diversity to land use?
2) Does land-use intensification decrease phylogenetic diversity of plant communities in grasslands?
3) Does phylogenetic diversity of common and rare species assemblages show different relationships with respect to land-use intensification?

For a better understanding and interpretation of the relationship between PD and land-use intensification, information on the phylogenetic signals in traits relevant for landuse are of interest (i.e. related to a certain ecosystem function or environmental gradient). Thus, we used a set of traits that are likely to be sensitive toland use and tested for phylogenetic signal in those traits.

Materials and Methods

Study area

Our study is part of the Biodiversity Exploratories project, a large German research project to investigate the relationships between land-use, biodiversity and ecosystem functioning (www.biodiversity-exploratories.de). The Biodiversity Exploratories represent three typical regions in Germany covering a south-west – north-east gradient and each region comprises grasslands and forests under a range of land-use types and intensities [43]. The exploratory Schwäbische Alb (hereafter named Alb) is situated in the SW Germany and is part of the UNESCO Biosphere Reserve Schwäbische Alb. The exploratory Hainich-Dün (hereafter named Hainich) is situated in western Thuringia, central Germany. The exploratory Schorfheide-Chorin (hereafter named Schorfheide) is situated in NE Germany and is part of the UNESCO Biosphere Reserve Schorfheide-Chorin. In each region 50 experimental grassland plots representing gradients from semi-natural to intensive land-use were established (overall 150 plots). For more details see [43].

Field work permits were issued by the responsible state environmental offices of Baden-Württemberg, Thüringen, and Brandenburg (according to § 72 BbgNatSchG). The study did not involve protected or endangered species.

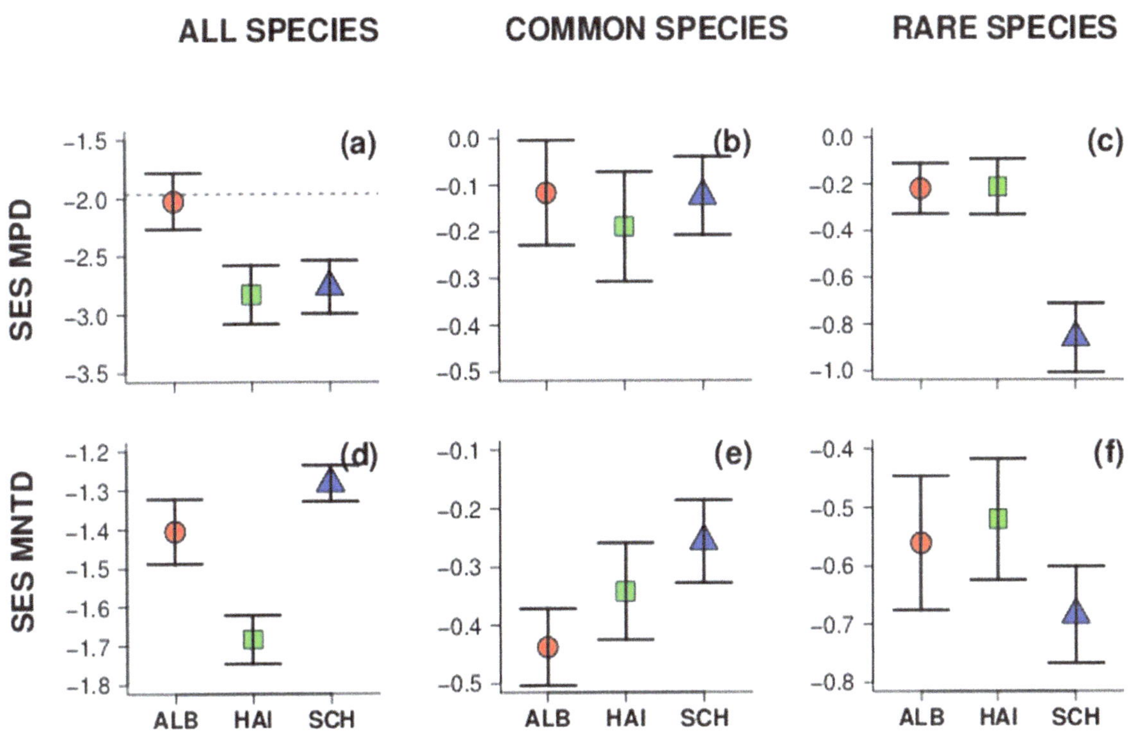

Figure 1. Mean (±SE) values of MPD and MNTD effect sizes for total, common and rare species assemblages in three regions in Germany. (a)–(c) Mean MPD and (d)–(f) mean MNTD for all, common and rare species assemblages in the three regions. Region abbreviations: ALB = SchwäbischeAlb (red circle); HAI = Hainich-Dün (green square); SCH = Schorfheide-Chorin (blue triangle). Error bars indicate ± SE. Points below the dashed line (<−1.96) are significantly clustered. Note different scales of y-axes.

Land-use

Land-use information for each of the 150 grassland plots was obtained by yearly interviews with farmers and land-owners between 2006 and 2010. The acquired information included fertilization level (kg nitrogen ha^{-1} $year^{-1}$), mowing frequency (number of cuts $year^{-1}$) and grazing intensity (livestock units × days of grazing ha^{-1} $year^{-1}$) [43]. The three land-use components were standardized by the respective mean intensity within each region to yield the fertilization, mowing and grazing intensity [44]. For each year the individual components were summed up to a combined quantitative land-use intensity index (LUI). The yearly LUI-values (2006–2010) were averaged for each plot and the obtained means were then used in all our analyses [44].

Vegetation releves and phylogeny

Between 2009 and 2011 we recorded the vegetation on a 4×4 m plot in each of the 150 grasslands three times (2009, 2010 and 2011). For each plot, vascular species richness and their relative abundance in percent cover was estimated. The species were further grouped into common and rare species based on their abundance for each year and region separately, taking into account local (plot) abundance and distribution (number of plots occupied) of each species. Common species were defined as the top 10% in terms of total abundance across plots occupied by a species, while the bottom 90% of the species were defined as rare. Based on these data we calculated the species richness of all, common and rare species as the average richness per plot across the three years. Note that the analyses of plant species richness from our study sites have been already published elsewhere [3,45]. We included these results here only for comparative purposes. Therefore our discussion focuses only on the effects of land-use on PD. A low number of gymnosperms and ferns with low site incidence were omitted from all analyzes.

Phylogenetic relatedness of species was obtained from a well resolved and dated phylogeny of the Central European flora [46]. In brief, this phylogeny was assembled by manually grafting subtrees on a backbone topology, dating of nodes based on fossil records using the bladj algorithm in PHYLOCOM [47] and calculating an ultrametric tree (for details see [46]). We pruned the overall phylogeny to match the species pool of each of the three regions. As a result we obtained three trees, one for each region, representing the phylogenetic relationships of the respective species pool.

According to the data sharing regulations of the Biodiversity Exploratory Project and in accordance with the rules of the German Science Foundation DFG, the data will be made publicly available no later than five years after collection.

Traits and phylogenetic signal

We compiled functional trait data from different data bases. As traits related to productivity we included the maximal plant height (cm) and specific leaf area (SLA; in cm^2/g). As traits related to reproduction we used data start of flowering (month of the year). Data on the SLA were taken from the LEDA trait data base [48], data on start of flowering and plant height were gathered from BiolFlor data base [49] and from floras [50,51]. Means were calculated when entries differed among the sources, but generally the values were highly consistent across sources. We further compiled performance and persistence traits relevant for agricultural grasslands: (1) soil nutrient indicator value (N, [52], (2) mowing tolerance (M), (3) grazing tolerance (G) and (4) trampling tolerance (T, all according to [53] from [54] and Briemle pers.

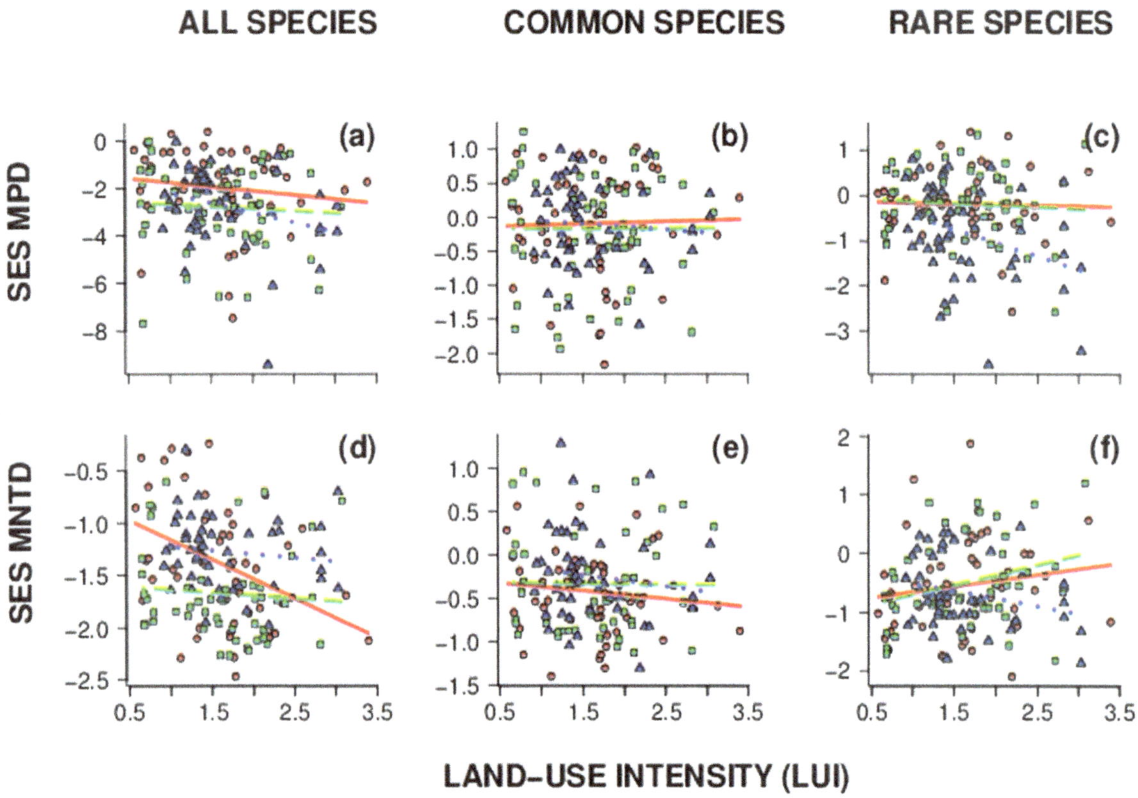

Figure 2. Relationships between mean pairwise distance (effect size MPD), mean nearest taxon distance (effect size MNTD) and land-use intensity (LUI) in three regions in Germany. Linear regression plots showing regression slopes for relationships between (a–c) mean pairwise distance and (d–f) mean nearest taxon distance for total, common and rare species assemblages and land-use intensity (LUI). Color and type code: red solid line/circle = Schwäbische Alb (Alb); green dashed line/square = Hainich-Dün (Hai); blue dotted line/triangle = Schorfheide-Chorin (Sch). Note different scales of y-axes. For significance of regression slopes see Appendix S4.

comm.). For all traits we hypothesized that different agricultural use, in particular fertilization, mowing and grazing selects for species with different traits values. All indicators have numeric values ranging from 1 (low) to 9 (high). Available trait data ranged from 77% (SLA, height and flowering onset) to 86% (G) of the species.

We tested for the strength and significance of phylogenetic signals in traits using Pagel's λ and Blomberg's K implemented in the phytools package [55] in R. We log transformed values for the maximum height to achieve normality. It has been proposed that Pagel's λ is an overall more robust metric than e.g. Blomberg's K [56], however, in general both metrics revealed similar results.

Phylogenetic diversity

Phylogenetic diversity estimates of plots were calculated with the "picante" package in R [57]. We calculated for each year and region separately the mean pairwise distance (MPD) and mean nearest taxon distance (MNTD) [30] weighted by species abundance (estimated % cover) as well as using presence/absence data. Considering % cover as a surrogate for species abundance may only approximate the "true" species abundance distribution within a community. However because of the large number of plots in our study individual counts of species would be very time-consuming and are thus not feasible. Estimates of % cover are at least rough approaches to estimate abundance and we suggest that analyses based on such approaches are more meaningful than considering only presence/absence data, especially in the context of the relative contribution of abundant, subordinate and transient species [59]. We used a slightly modified calculation of MPD based on abundance data as proposed by Gerhold et al. [60] to reduce effects of species richness. Abundance weighted and presence/absence versions of indices showed moderate correlations (MPD: $r = 0.41$; MNTD: $r = 0.58$). However, results based on the two indices did not differ considerably and therefore we present here only the results of abundance weighted indices (see Appendix S5 and S6 for presence/absence PD results). MPD measures the mean phylogenetic distance between two taxa in a sample and MNTD the mean phylogenetic distance to the nearest taxon in a sample. Hence MPD summarizes all phylogenetic distances including those of very distantly related species (e.g. between species of different orders) while MNTD considers only those between the most closely related species (e.g. between species within a genus). Thus, a stronger relationship of MNTD with land-use intensity compared to MPD would indicate that land-use has a stronger effect on the terminal than on the basal phylogenetic composition of a community. Both metrics depended on species richness and we therefore calculated standardized effect sizes ((observed metric - expected metric)/standard deviation of expected metric). We used a null model that shuffles the tip labels of the phylogeny maintaining all other properties of the sample matrix (i.e. species richness in plots and species prevalence). This null model was chosen since it tests for the null hypothesis, that phylogeny is not an important factor for structuring plants communities. Note that effect sizes of both metrics were calculated for each year and region separately. For each plot we then

Table 1. Effects of region, land-use intensity (LUI) and its interaction with region on effect size of (A) mean pairwise distance (MPD) and (B) mean nearest taxon distance (MNTD) for total, common and rare species assemblages.

A		MPD(total)		MPD(common)		MPD(rare)	
	df	**F**	**p**	**F**	**p**	**F**	**p**
Region	2	3.44	**0.035**	0.14	0.87	9.04	**0.0002**
LUI	1	3.65	0.06	0.01	0.91	3.71	0.06
Region × LUI	2	0.68	0.51	0.12	0.89	2.15	0.12
Residuals	144						

B		MNTD(total)		MNTD(common)		MNTD(rare)	
	df	**F**	**p**	**F**	**p**	**F**	**P**
Region	2	10.43	**<0.001**	1.51	0.22	0.73	0.48
LUI	1	7.60	**0.0066**	1.13	0.29	1.33	0.25
Region × LUI	2	3.02	0.052	0.25	0.78	3.51	**0.032**
Residuals	144						

ANOVA table with F and *p* values of the full models. Significant values in bold.

Table 2. Values for t-statistics and corresponding p values of the linear models with (A) MPD and (B) MNTD as dependent variables and LUI, rarity (two-level factor: common and rare) and their interaction as independent variables.

A	**MPD**							
	ALL		**ALB**		**HAI**		**SCH**	
	t	**p**	**t**	**p**	**t**	**p**	**t**	**p**
Intercept	−0.65	0.52	−0.55	0.58	−0.61	0.54	0.04	0.97
LUI	−0.10	0.92	0.19	0.85	0.03	0.98	−0.42	0.67
Rarity	0.25	0.8	0.04	0.97	0.29	0.77	0.19	0.85
LUI×Rarity	−1.39	0.17	−0.30	0.77	−0.36	0.72	−1.80	0.08
B	**MNTD**							
	ALL		**ALB**		**HAI**		**SCH**	
	t	**p**	**t**	**p**	**t**	**p**	**t**	**p**
Intercept	−1.58	0.12	−1.09	0.28	−1.30	0.20	−0.21	0.83
LUI	−0.87	0.38	−0.61	0.55	−0.08	0.94	−0.98	0.33
Rarity	−2.61	**0.009**	−1.57	0.12	−2.08	**0.041**	−0.66	0.51
LUI×Rarity	1.52	0.13	1.33	0.19	1.69	0.09	−0.72	0.47

Interaction term determines whether rare species PD response differs from that of common species PD. Significant values in bold.

calculated averages across the three years which were further used in all subsequent analyses (see above).

We used simple linear regressions and ANOVAs to analyze the relationships between plant PD and land-use intensification. We considered region (exploratory) as a factor to analyze whether PD differs among regions and whether the relationships between PD and LUI differ among regions (region×LUI interaction). To assess whether rare species assemblages respond more strongly to increasing land use than common species, we compared the slope of the regression lines with an ANCOVA by testing the significance of the LUI×"rarity" interaction. All statistical analyses were conducted in R [58].

Results

A total of 282 vascular plant species were recorded in the three regions from 2009 to 2010 (Appendix S1). We found depending on the considered species pool and the specific traits analyzed varying levels for Pagel's λ and Blomberg's K (Appendix S2). Based on Blomberg's K we found no strong phylogenetic conservatism in analyzed traits (Appendix S2). This suggests that PD cannot be seen as an overall proxy for functional diversity along land-use gradients.

Average total, rare and common species richness differed among regions (Appendix S3). Total and rare species richness decreased with increasing LUI with regional effects modulating the response of. In two regions (Alb, Hainich) total and rare species richness decreased with increasing LUI while in Schorfheide no effect was observed. The relationship between common species and LUI showed very contrasting patterns between regions but there was no overall decrease in species richness (Appendix S3).

Overall, average PD strongly varied among regions. But note that the differences depended on the PD-metric used and whether rare/common species were considered (Fig. 1a–f). When all species were considered, effect size of MPD showed strong significant clustering of communities in two regions (Hainich and Schorfheide) while MNTD estimates showed random patterns in all three regions. Mean phylogenetic community structure was random in respect to phylogeny for common and rare species assemblages in all three regions. After accounting for regional differences, total species MNTD decreased with increasing land-use intensity while MPD showed only a marginally significant decrease with similar relationships in all three exploratories (Table 1). Furthermore, land-use had slightly different effects on MNTD depending on region indicated by a marginally significant region×LUI interaction (Table 1), with a stronger decline of MNTD in one region (Alb: $r = -0.39$, $p<0.01$, Appendix S4), in particular. The other two regions showed a non-significant negative trend (Fig. 2). For MPD, only one region (Schorfheide) showed a significant decline with increasing land-use intensity ($r = -0.3$, $p<0.05$; Fig. 1, Appendix S4).

In general we found that for both common and rare species PD was not or only weakly affected by increasing land-use intensity. The relationships did not vary among regions except for rare species MPD (Table 1, Fig. 2). Overall, the strength of phylodiversity – land-use intensity relationships did not differ between common and rare species assemblages over three regions as indicated by non-significant LUI×rarity interaction terms in our models (Table 2).

Discussion

Land-use intensification is one of the major threats to global biodiversity in grasslands [21]. However, only a few studies have analyzed the effects of anthropogenic influence on PD of grassland plant communities. Several studies showed that anthropogenic influence can cause a decline in PD of species communities [15,37,61] which possibly may also decrease trait diversity and associated ecological functions [7]. In particular, PD can be important for ecosystem functioning when the ultimate processes, which depend on plant traits and trophic interactions, show a phylogenetic signal [7]. It has been shown that in grasslands PD can act as a better predictor of productivity than species richness or functional diversity [9,62]. Moreover, herbivory was stronger related to phylogenetic relatedness than to plant functional traits [63]. An experimental study by Pellissier et al. [38] revealed an increase in PD after strong fertilization and herbicide application while functional traits showed contrasting relationships presumably by selecting for convergent traits [38]. We found no evidence for strong phylogenetic signal in selected land-use sensitive traits (Appendix S2). Thus, phylogenetic diversity may not capture the relevant functional information leading to a relatively weak response to land-use intensification [31]. On the other side, the significant decrease of PD depending on region and metric used (see below), shows that PD might capture additional information beside the measured traits.

Dinnage [15] showed that the phylogenetic structure of plant communities in disturbed plots of old field sites is more clustered than expected, whereas phylogenetic structure in undisturbed plots does not differ from random expectations. This indicates, that land-use might act similarly to environmental filters and select for (presumably closely related) species with similar traits, which enable species to cope with disturbance. However, Dinnage analyzed the vegetation of an old field system with plowing being the disturbance that affected the phylogenetic diversity. This kind of disturbance mediates phylogenetic succession which can lead to increasing phylogenetic clustering of plant communities [64]. Our study sites are exposed to land-use types completely different to the former study and our results differ in the strength of the PD response to land-use intensification. Although land-use intensification slightly decreased phylogenetic diversity, considering the mean nearest taxon distance (MNTD) in particular, it did not lead to a shift form random to clustered community structures (Table 1, Fig. 2). In general, plant communities exhibited clustered and random phylogenetic structures on plots with both, low as well as high land-use intensities (points <1.96 on y-axis; Fig. 2). There are factors causing clustering of communities, especially when considering the tree-wide patterns (MPD, Fig. 1a) as was shown in several studies [e.g. 5,35]. Whether these factors refer to environmental filters [65,66] or exclusion of weak competitors [32] we cannot distinguish in our study. Land-use intensity, however, seems to play a minor role as determinant of phylogenetic community structure of plants in grasslands. This is contrary to the results of Dinnage [15] but such differences might be caused by different land-use types, with plowing causing a strong disturbance within habitats compared to our land-use types. Note also that in Dinnages study [11] no gradient of land-use intensity was analyzed and the definition of regional species pools was different from our study. Nevertheless, the slight decline of PD in our study may indicate that the influence of factors causing phylogenetic clustering of communities is mediated through or caused by increasing land-use intensity.

Many studies dealing with phylogenetic community structure use only one phylogenetic diversity index like NRI or NTI (equivalent to (−1 * effect size MPD) and (−1 * effect size MNTD), respectively) [e.g. 35,56]. Since the two metrics measure PD at different depths of phylogeny, with MPD (NRI) capturing tree-wide patterns and MNTD (NTI) being more sensitive to the tips of a phylogeny [30], depending on the distribution of traits, results of analyses might differ. However, when both metrics were used, similar results were reported [e.g. 57]. In our study, although the two metrics showed similar relationships with land use, MNTD was more sensitive to increasing land-use intensity. This emphasizes the importance of including different indices into analyzes of PD, as land-use sensitive traits might be conserved within a few relatively young clades (e.g. within families) and thus might be masked when using metrics considering a broader phylogenetic scale (e.g. MPD). Because MNTD shows a stronger response to land-use intensification it is possible that those traits are conserved in the younger nodes of phylogeny. Thus, using MPD might not capture relevant trait information when analyzing the effects of land use on phylogenetic diversity. In fact, as Blomberg's K can be thought of as the partitioning of variance with low values ($K<1$) indicating variance within clades, this might be the reason for MNTD being more sensitive to land use.

Although common and rare species might differ in several traits [67] or their sensitivity to soil biogeochemical parameters [68] and respond differently to land use and competition [69], we found no significant differences in their response to increasing land-use using analysis of covariance (Fig. 2, Tab 2). This suggests that traits that probably affect the abundance of species are randomly distributed across our plant phylogeny or/and are not affected by land-use. The only trait that was relatively strong conserved in both, common and rare species, was maximum height. Despite a relatively high phylogenetic signal in this trait, it seems that height is not a strong determinant of phylogenetic community structure in both, common and rare species assemblages. Another explanation might be that PD of common and rare species might respond differently to the single LUI components due to different traits not accounted for in our study and combining those to one index might neglect the differences in strength and direction of responses. Likewise, as the effects of land-use on PD did not differ in general between common and rare species communities, but rather showed slightly different patterns on a smaller scale, they should be examined separately if conservation efforts attempt to increase diversity for endangered taxa.

It is well known that regional peculiarities and species pools influence regional phylogenetic diversity [70,71]. For our study regions we found that considering all species Alb had overall high and Hainich overall low PD. Schorfheide showed contrasting patterns depending on the PD-metric used. Low MPD values suggest, that species in communities are closely related when accounting for the whole phylogeny, but high MNTD values indicate, that on lower phylogenetic scales (e.g. within families) species are distantly related. This might be explained by the fact that Schorfheide was more strongly affected by the Pleistocene glaciations than the other regions. One may argue that the plant communities of Schorfheide are still dominated by ecologically similar species belonging to closely related higher clades. Environmental filtering is then likely to cause strong phylogenetic clustering of communities considering the MPD (Fig. 1a). By contrast, within these clades PD might have increased due to limiting similarity [72] causing random community structure (Fig. 1d).

Differences in PD among regions may, to some extent, be also due historical land use rather than current [55] as suggested for species richness or functional diversity [70,73]. Such regional differences call for a careful consideration of regional particularities when providing management strategies to maintain or increase phylogenetic diversity of grassland plant communities under "real world" conditions.

The theory behind phylogenetic patterns along disturbance gradients relies on several hypotheses about distribution of ecological traits across phylogenetic trees [9,30,32,62,74]. We showed that although potentially land-use relevant traits show some levels of phylogenetic conservatism, PD still can provide additional information. The consideration of PD is therefore in particular importantin situations when functional traits of species are not available. Phylogenetic methods can complement ecological analyzes, but it must be pointed out that PD cannot be seen as a surrogate for other biodiversity metrics, functional diversity in particular.

Supporting Information

Appendix S1 Phylogenetic tree of plants in three regions in Germany used in this study.

Appendix S2 Phylogenetic signal in 7 traits considered as sensitive to land use for all, common and rare species in the three regions (ALB: Schwäbische Alb, HAI: Hainich-Dün and SCH: Schorfheide-Chorin) and in all regions combined. Significant values are in bold.

Appendix S3 Mean (±SE) values and regression slopes of species richness for total, common and rare species assemblages in three regions in Germany.

Appendix S4 Correlation coefficients and significance of the respective regression slopes from Fig. 1 and Appendix S3 and for all regions combined.

Appendix S5 Relationships between presence/absence based mean pairwise distance (effect size MPD), mean nearest taxon distance (effect size MNTD) and land-use intensity (LUI) in three regions in Germany.

Appendix S6 Mean (±SE) values of presence/absence based MPD and MNTD effect sizes for total, common and rare species assemblages in three regions in Germany.

Acknowledgments

We thank the managers of the three Exploratories, Swen Renner, Sonja Gockel, Kerstin Wiesner, and Martin Gorke for their work in maintaining the plot and project infrastructure; Simone Pfeiffer and Christiane Fischer giving support through the central office, Michael Owonibi for managing the central data base, and Eduard Linsenmair, Dominik Hessenmöller, Jens Nieschulze, Ingo Schöning, François Buscot, Ernst-Detlef Schulze, Wolfgang W. Weisser and the late Elisabeth Kalko for their role in setting up the Biodiversity Exploratories project. We also want to thank Steffen Boch, Stephanie Socher and Jörg Müller for providing vegetation data for our analyses.

Author Contributions

Conceived and designed the experiments: MB MF. Performed the experiments: EE. Analyzed the data: EE. Contributed reagents/materials/analysis tools: WD SM BS SB DP. Wrote the paper: EE. Provided editorial advice: DP WD MB.

References

1. Foley JA, DeFries R, Asner GP (2005) Global Consequences of Land Use. Science 309: 570–574. doi:10.1126/science.1111772.
2. Flynn DFB, Gogol-Prokurat M, Nogeire T, Molinari N, Richers BT, et al. (2009) Loss of functional diversity under land use intensification across multiple taxa. Ecol Lett 12: 22–33. doi:10.1111/j.1461-0248.2008.01255.x.
3. Socher SA, Prati D, Boch S, Müller J, Klaus VH, et al. (2012) Direct and productivity-mediated indirect effects of fertilization, mowing and grazing on grassland species richness. J Ecol 100: 1391–1399. doi:10.1111/j.1365-2745.2012.02020.x.
4. Cadotte MW, Cardinale BJ, Oakley TH (2008) Evolutionary history and the effect of biodiversity on plant productivity. Proc Natl Acad Sci 105: 17012–17017. doi:10.1073/pnas.0805962105.
5. Cavender-Bares J, Kozak KH, Fine PVA, Kembel SW (2009) The merging of community ecology and phylogenetic biology. Ecol Lett 12: 693–715.
6. Vamosi SM, Heard SB, Vamosi JC, Webb CO (2009) Emerging patterns in the comparative analysis of phylogenetic community structure. Mol Ecol 18: 572–592.
7. Srivastava DS, Cadotte MW, MacDonald AAM, Marushia RG, Mirotchnick N (2012) Phylogenetic diversity and the functioning of ecosystems. Ecol Lett: n/a–n/a. doi:10.1111/j.1461-0248.2012.01795.x.
8. Wiens JJ, Ackerly DD, Allen AP, Anacker BL, Buckley LB, et al. (2010) Niche conservatism as an emerging principle in ecology and conservation biology. Ecol Lett 13: 1310–1324. doi:10.1111/j.1461-0248.2010.01515.x.
9. Flynn DFB, Mirotchnick N, Jain M, Palmer MI, Naeem S (2011) Functional and phylogenetic diversity as predictors of biodiversity–ecosystem-function relationships. Ecology 92: 1573–1581. doi:10.1890/10-1245.1.
10. Maherali H, Klironomos JN (2007) Influence of Phylogeny on Fungal Community Assembly and Ecosystem Functioning. Science 316: 1746–1748. doi:10.1126/science.1143082.
11. Dinnage R (2013) Phylogenetic diversity of plants alters the effect of species richness on invertebrate herbivory. PeerJ 1: e93. Available: https://peerj.com/articles/93/. Accessed 14 January 2014.
12. Cadotte MW, Dinnage R, Tilman D (2012) Phylogenetic diversity promotes ecosystem stability. Ecology 93: S223–S233. doi:10.1890/11-0426.1.
13. Dinnage R, Cadotte MW, Haddad NM, Crutsinger GM, Tilman D (2012) Diversity of plant evolutionary lineages promotes arthropod diversity - Ecology Letters - Wiley Online Library. Available: http://onlinelibrary.wiley.com/doi/10.1111/j.1461-0248.2012.01854.x/full. Accessed 14 January 2014.
14. Mouquet N, Devictor V, Meynard CN, Munoz F, Bersier L-F, et al. (2012) Ecophylogenetics: advances and perspectives. Biol Rev 87: 769–785. doi:10.1111/j.1469-185X.2012.00224.x.
15. Dinnage R (2009) Disturbance Alters the Phylogenetic Composition and Structure of Plant Communities in an Old Field System. PLoS ONE 4: e7071. doi:10.1371/journal.pone.0007071.
16. Pärtel M, Bruun HH, Sammul M (2005) Biodiversity in temperate European grasslands: origin and conservation European Grassland Federation; Distributed by British Grassland Society.
17. Statistisches Bundesamt (2012) Statistisches Jahrbuch 2012. Wiesbaden.
18. Suding KN, Collins SL, Gough L, Clark C, Cleland EE, et al. (2005) Functional- and abundance-based mechanisms explain diversity loss due to N fertilization. Proc Natl Acad Sci 102: 4387–4392. doi:10.1073/pnas.0408648102.
19. Kleijn D, Kohler F, Báldi A, Batáry P, Concepción ED, et al. (2009) On the relationship between farmland biodiversity and land-use intensity in Europe. Proc R Soc B Biol Sci 276: 903–909. doi:10.1098/rspb.2008.1509.
20. Storkey J, Meyer S, Still KS, Leuschner C (2011) The impact of agricultural intensification and land-use change on the European arable flora. Proc R - Soc B Biol Sci 279: 1421–1429. doi:10.1098/rspb.2011.1686.
21. Sala OE, Chaplin FS, Armesto JJ, Berlow E, Bloomfield J, et al. (2000) Global Biodiversity Scenarios for the Year 2100 Science 287: 1770–1774. doi:10.1126/science.287.5459.1770.
22. Tilman D, Wedin D, Knops J (1996) Productivity and sustainability influenced by biodiversity in grassland ecosystems. Nature 379: 718–720. doi:10.1038/379718a0.
23. Hector A, Schmid B, Beierkuhnlein C, Caldeira M, Diemer M, et al. (1999) Plant diversity and productivity experiments in European grasslands. Science 286: 1123.
24. Cardinale BJ, Wright JP, Cadotte MW, Carroll IT, Hector A, et al. (2007) Impacts of plant diversity on biomass production increase through time because of species complementarity. Proc Natl Acad Sci 104: 18123–18128. doi:10.1073/pnas.0709069104.
25. Cardinale BJ, Srivastava DS, Emmett Duffy J, Wright JP, Downing AL, et al. (2006) Effects of biodiversity on the functioning of trophic groups and ecosystems. Nature 443: 989–992. doi:10.1038/nature05202.
26. Petchey OL, Gaston KJ (2006) Functional diversity: back to basics and looking forward. Ecol Lett 9: 741–758. doi:10.1111/j.1461-0248.2006.00924.x.
27. Tilman D, Knops J, Wedin D, Reich P, Ritchie M, et al. (1997) The Influence of Functional Diversity and Composition on Ecosystem Processes. Science 277: 1300–1302. doi:10.1126/science.277.5330.1300.
28. Chapin III FS, Zavaleta ES, Eviner VT, Naylor RL, Vitousek PM, et al. (2000) Consequences of changing biodiversity. Nature 405: 234–242. doi:10.1038/35012241.
29. Loreau M (2000) Biodiversity and ecosystem functioning: recent theoretical advances. Oikos 91: 3–17. doi:10.1034/j.1600-0706.2000.910101.x.
30. Webb CO, Ackerly DD, McPeek MA, Donoghue MJ (2002) Phylogenies and Community Ecology. Annu Rev Ecol Syst 33: 475–505. doi:10.1146/annurev.ecolsys.33.010802.150448.
31. Bernard-Verdier M, Flores O, Navas M-L, Garnier E (2013) Partitioning phylogenetic and functional diversity into alpha and beta components along an environmental gradient in a Mediterranean rangeland. J Veg Sci 24: 877–889. doi:10.1111/jvs.12048.
32. Mayfield MM, Levine JM (2010) Opposing effects of competitive exclusion on the phylogenetic structure of communities: Phylogeny and coexistence. Ecol Lett 13: 1085–1093. doi:10.1111/j.1461-0248.2010.01509.x.
33. Díaz S, Cabido M, Zak M, Martínez Carretero E, Araníbar J (1999) Plant functional traits, ecosystem structure and land-use history along a climatic gradient in central-western Argentina. J Veg Sci 10: 651–660. doi:10.2307/3237080.
34. Stevens CJ, Dise NB, Mountford JO, Gowing DJ (2004) Impact of Nitrogen Deposition on the Species Richness of Grasslands. Science 303: 1876–1879. doi:10.1126/science.1094678.
35. Kluge J, Kessler M (2011) Phylogenetic diversity, trait diversity and niches: species assembly of ferns along a tropical elevational gradient. J Biogeogr 38: 394–405. doi:10.1111/j.1365-2699.2010.02433.x.
36. Brunbjerg AK, Borchsenius F, Eiserhardt WL, Ejrnaes R, Svenning J-C (2012) Disturbance drives phylogenetic community structure in coastal dune vegetation. J Veg Sci 23: 1082–1094. doi:10.1111/j.1654-1103.2012.01433.x.
37. Knapp S, Kühn I, Schweiger O, Klotz S (2008) Challenging urban species diversity: contrasting phylogenetic patterns across plant functional groups in Germany. Ecol Lett 11: 1054–1064.
38. Pellissier L, Wisz MS, Strandberg B, Damgaard C (2014) Herbicide and fertilizers promote analogous phylogenetic responses but opposite functional responses in plant communities. Environ Res Lett 9: 024016. doi:10.1088/1748-9326/9/2/024016.
39. Graham CH, Parra JL, Rahbek C, McGuire JA (2009) Colloquium Papers: Phylogenetic structure in tropical hummingbird communities. Proc Natl Acad Sci 106: 19673–19678. doi:10.1073/pnas.0901649106.
40. Machac A, Janda M, Dunn RR, Sanders NJ (2011) Elevational gradients in phylogenetic structure of ant communities reveal the interplay of biotic and abiotic constraints on diversity. Ecography 34: 364–371. doi:10.1111/j.1600-0587.2010.06629.x.
41. Hoiss B, Krauss J, Potts SG, Roberts S, Steffan-Dewenter I (2012) Altitude acts as an environmental filter on phylogenetic composition, traits and diversity in bee communities. Proc R Soc B Biol Sci. Available: http://rspb.royalsocietypublishing.org/cgi/doi/10.1098/rspb.2012.1581. Accessed 26 September 2012.
42. Cavender-Bares J, Ackerly DD, Baum DA, Bazzaz FA (2004) Phylogenetic Overdispersion in Floridian Oak Communities. Am Nat 163: 823–843. doi:10.1086/386375.
43. Fischer M, Bossdorf O, Gockel S, Hänsel F, Hemp A, et al. (2010) Implementing large-scale and long-term functional biodiversity research: The Biodiversity Exploratories. Basic Appl Ecol 11: 473–485. doi:10.1016/j.baae.2010.07.009.
44. Blüthgen N, Dormann CF, Prati D, Klaus VH, Kleinebecker T, et al. (2012) A quantitative index of land-use intensity in grasslands: Integrating mowing, grazing and fertilization. Basic Appl Ecol 13: 207–220. doi:10.1016/j.baae.2012.04.001.
45. Allan E, Bossdorf O, Dormann CF, Prati D, Gossner MM, et al. (in press) Interannual variation in land-use intensity enhances grassland multidiversity. PNAS.
46. Durka W, Michalski SG (2012) Daphne: a dated phylogeny of a large European flora for phylogenetically informed ecological analyses. Ecology 93: 2297–2297. doi:10.1890/12–0743.1.
47. Webb CO, Ackerly DD, Kembel SW (2008) Phylocom: software for the analysis of phylogenetic community structure and trait evolution. Bioinformatics 24: 2098–2100. doi:10.1093/bioinformatics/btn358.
48. Kleyer M, Bekker RM, Knevel IC, Bakker JP, Thompson K, et al. (2008) The LEDA Traitbase: a database of life-history traits of the Northwest European flora. J Ecol 96: 1266–1274. doi:10.1111/j.1365-2745.2008.01430.x.
49. Klotz S, Kühn I, Durka W (2002) BIOLFLOR - Eine Datenbank zu biologisch-ökologischen Merkmalen der Gefäβpflanzen in Deutschland. Bonn: Bundesamt für Naturschutz.
50. Binz A, Heitz C (1990) Schul- und Exkursionsflora fur die Schweiz: mit Berucksichtigung der Grenzgebiete: Bestimmungsbuch fur die wildwachsenden Gefasspflanzen. 19. Aufl. Basel: Schwabe. 659 p.
51. Jäger EJ, Werner K (2005) Exkursionsflora von Deutschland. 10th ed. Heidelberg: Spektrum Akademischer Verlag.
52. Ellenberg H, Weber HE, Düll R, Wirth V, Werner W, et al. (1992) Zeigerwerte von Pflanzen in Mitteleuropa. Göttingen: Verlag Erich Goltze. 1–248.
53. Briemle G, Ellenberg H (1994) Zur Mahdverträglichkeit von Grünlandpflanzen. Nat Landschaft 69: 139–147.
54. Briemle G, Nitsche S, Nitsche L (2002) Nutzungswertzahlen für Gefäβpflanzen des Grünlandes.

55. Revell LJ (2012) phytools: an R package for phylogenetic comparative biology (and other things): phytools: R package. Methods Ecol Evol 3: 217–223. doi:10.1111/j.2041-210X.2011.00169.x.
56. Münkemüller T, Lavergne S, Bzeznik B, Dray S, Jombart T, et al. (2012) How to measure and test phylogenetic signal: How to measure and test phylogenetic signal. Methods Ecol Evol 3: 743–756. doi:10.1111/j.2041-210X.2012.00196.x.
57. Kembel SW, Cowan PD, Helmus MR, Cornwell WK, Morlon H, et al. (2010) Picante: R tools for integrating phylogenies and ecology. Bioinformatics 26: 1463–1464. doi:10.1093/bioinformatics/btq166.
58. R Core Team (2012) R: A language and environment for statistical computing. Vienna, Austria: R Foundation for Statistical Computing.
59. Grime JP (1998) Benefits of plant diversity to ecosystems: immediate, filter and founder effects. J Ecol 86: 902–910. doi:10.1046/j.1365-2745.1998.00306.x.
60. Gerhold P, Price JN, Püssa K, Kalamees R, Aher K, et al. (2013) Functional and phylogenetic community assembly linked to changes in species diversity in a long-term resource manipulation experiment. J Veg Sci 24: 843–852. doi:10.1111/jvs.12052.
61. Helmus MR, Keller W (Bill), Paterson MJ, Yan ND, Cannon CH, et al. (2010) Communities contain closely related species during ecosystem disturbance. Ecol Lett 13: 162–174. doi:10.1111/j.1461-0248.2009.01411.x.
62. Cadotte MW, Cavender-Bares J, Tilman D, Oakley TH (2009) Using Phylogenetic, Functional and Trait Diversity to Understand Patterns of Plant Community Productivity. PLoS ONE 4: e5695. doi:10.1371/journal.pone.0005695.
63. Paine CET, Norden N, Chave J, Forget P-M, Fortunel C, et al. (2012) Phylogenetic density dependence and environmental filtering predict seedling mortality in a tropical forest: Neighbourhood similarity predicts seedling mortality. Ecol Lett 15: 34–41. doi:10.1111/j.1461-0248.2011.01705.x.
64. Valiente-Banuet A, Verdú M (2007) Facilitation can increase the phylogenetic diversity of plant communities. Ecol Lett 10: 1029–1036. doi:10.1111/j.1461-0248.2007.01100.x.
65. Butterfield BJ, Cavieres LA, Callaway RM, Cook BJ, Kikvidze Z, et al. (2013) Alpine cushion plants inhibit the loss of phylogenetic diversity in severe environments. Ecol Lett 16: 478–486. doi:10.1111/ele.12070.
66. Culmsee H, Leuschner C (2013) Consistent patterns of elevational change in tree taxonomic and phylogenetic diversity across Malesian mountain forests. J Biogeogr: n/a–n/a. doi:10.1111/jbi.12138.
67. Farnsworth EJ (2007) Plant life history traits of rare versus frequent plant taxa of sandplains: Implications for research and management trials. Biol Conserv 136: 44–52. doi:10.1016/j.biocon.2006.10.045.
68. Kleijn D, Bekker RM, Bobbink R, De Graaf MCC, Roelofs JGM (2008) In search for key biogeochemical factors affecting plant species persistence in heathland and acidic grasslands: a comparison of common and rare species. J Appl Ecol 45: 680–687. doi:10.1111/j.1365–2664.2007.01444.x.
69. Dawson W, Fischer M, van Kleunen M (2012) Common and rare plant species respond differently to fertilisation and competition, whether they are alien or native. Ecol Lett 15: 873–880. doi:10.1111/j.1461-0248.2012.01811.x.
70. Anacker BL, Harrison SP (2012) Historical and Ecological Controls on Phylogenetic Diversity in Californian Plant Communities. Am Nat 180: 257–269. doi:10.1086/666650.
71. Blanchet S, Helmus MR, Brosse S, Grenouillet G (2013) Regional *vs* local drivers of phylogenetic and species diversity in stream fish communities. Freshw Biol: n/a–n/a. doi:10.1111/fwb.12277.
72. MacArthur R, Levins R (1967) The Limiting Similarity, Convergence and Divergence of Coexisting Species. Am Nat 101: 377–385.
73. Klaus VH, Hölzel N, Boch S, Müller J, Socher SA, et al. (n.d.) Direct and indirect associations between plant species richness and productivity in grasslands: regional differences preclude simple generalization of productivity-biodiversity relationships. Available: http://www.preslia.cz/P132Klaus.pdf. Accessed 19 December 2013.
74. Losos JB (2008) Phylogenetic niche conservatism, phylogenetic signal and the relationship between phylogenetic relatedness and ecological similarity among species. Ecol Lett 11: 995–1003. doi:10.1111/j.1461-0248.2008.01229.x.

Permissions

All chapters in this book were first published in PLOS ONE, by The Public Library of Science; hereby published with permission under the Creative Commons Attribution License or equivalent. Every chapter published in this book has been scrutinized by our experts. Their significance has been extensively debated. The topics covered herein carry significant findings which will fuel the growth of the discipline. They may even be implemented as practical applications or may be referred to as a beginning point for another development.

The contributors of this book come from diverse backgrounds, making this book a truly international effort. This book will bring forth new frontiers with its revolutionizing research information and detailed analysis of the nascent developments around the world.

We would like to thank all the contributing authors for lending their expertise to make the book truly unique. They have played a crucial role in the development of this book. Without their invaluable contributions this book wouldn't have been possible. They have made vital efforts to compile up to date information on the varied aspects of this subject to make this book a valuable addition to the collection of many professionals and students.

This book was conceptualized with the vision of imparting up-to-date information and advanced data in this field. To ensure the same, a matchless editorial board was set up. Every individual on the board went through rigorous rounds of assessment to prove their worth. After which they invested a large part of their time researching and compiling the most relevant data for our readers.

The editorial board has been involved in producing this book since its inception. They have spent rigorous hours researching and exploring the diverse topics which have resulted in the successful publishing of this book. They have passed on their knowledge of decades through this book. To expedite this challenging task, the publisher supported the team at every step. A small team of assistant editors was also appointed to further simplify the editing procedure and attain best results for the readers.

Apart from the editorial board, the designing team has also invested a significant amount of their time in understanding the subject and creating the most relevant covers. They scrutinized every image to scout for the most suitable representation of the subject and create an appropriate cover for the book.

The publishing team has been an ardent support to the editorial, designing and production team. Their endless efforts to recruit the best for this project, has resulted in the accomplishment of this book. They are a veteran in the field of academics and their pool of knowledge is as vast as their experience in printing. Their expertise and guidance has proved useful at every step. Their uncompromising quality standards have made this book an exceptional effort. Their encouragement from time to time has been an inspiration for everyone.

The publisher and the editorial board hope that this book will prove to be a valuable piece of knowledge for researchers, students, practitioners and scholars across the globe.

List of Contributors

Sébastien Personnic, Charles F. Boudouresque, Delphine Thibault-Botha, Mireille Harmelin-Vivien, Thierry Thibaut, Sandrine Ruitton and Jean-Christophe Poggiale
Aix-Marseille University, Mediterranean Institute of Oceanography (MIO), Université de Toulon, CNRS/INSU, IRD, UM 110, Marseille, France

Patrick Astruch and Patrick Bonhomme
GIS Posidonie, Pytheas Institute, Aix-Marseille University, Marseille, France

Enric Ballesteros
Centre d'Estudis Avanc¸ats de Blanes - CSIC, Blanes, Spain

Sylvain Blouet
Aire marine protégée de la côte Agathoise, site natura 2000, Agde, France

Denise Bellan-Santini and Florent Renaud
Aix-Marseille University, Institut Méditerranéen de Biodiversitéet d'Ecologie (IMBE), UMR 7263, Station Marine d'Endoume, Marseille, France

Eric Feunteun
Museum National d'Histoire Naturelle, UMR 7208, Station Marine de Dinard, France

Gérard Pergent and Christine Pergent-Martini
Equipe Ecosystèmes Littoraux, FRES 3041, University of Corsica, Corte, France

Jérémy Pastor
Universitéde Perpignan, Via Domitia, Centre de Formation et de Recherche sur les Environnements Me´diterrane´ens, UMR 5110, Perpignan, France

Yuguang Zhang, Hui Lu, and Diqiang Li
Institute of Forestry Ecology, Environment and Protection, and the Key Laboratory of Forest Ecology and Environment of State Forestry Administration, Chinese Academy of Forestry, Beijing, China

Jing Cong
Institute of Forestry Ecology, Environment and Protection, and the Key Laboratory of Forest Ecology and Environment of State Forestry Administration, Chinese Academy of Forestry, Beijing, China
School of Mineral Processing and Bioengineering, Central South University, Changsha, China

Caiyun Yang
Institute for Environmental Genomics and Department of Microbiology and Plant Biology, University of Oklahoma, Norman, Oklahoma, United States of America

Yunfeng Yang
State Key Joint Laboratory of Environment Simulation and Pollution Control, School of Environment, Tsinghua University, Beijing, China

Jizhong Zhou
Institute for Environmental Genomics and Department of Microbiology and Plant Biology, University of Oklahoma, Norman, Oklahoma, United States of America
State Key Joint Laboratory of Environment Simulation and Pollution Control, School of Environment, Tsinghua University, Beijing, China
Earth Sciences Division, Lawrence Berkeley National Laboratory, Berkeley, California, United States of America

Jie Zhao
Key Laboratory of Vegetation Restoration and Management of Degraded Ecosystems, South China Botanical Garden, Chinese Academy of Sciences, Guangzhou, China
Key Laboratory of Agro- Ecological Processes in Subtropical Region, Institute of Subtropical Agriculture, Chinese Academy of Sciences, Changsha, China

Songze Wan, Zhanfeng Liu, Lixia Zhou and Shenglei Fu
Key Laboratory of Vegetation Restoration and Management of Degraded Ecosystems, South China Botanical Garden, Chinese Academy of Sciences, Guangzhou, China

Chenlu Zhang
State Key Laboratory of Cotton Biology, Key Laboratory of Plant Stress Biology, College of Life Sciences, Henan University, Kaifeng, China

KathiJo Jankowski, Daniel E. Schindler and M. Claire Horner-Devine
University of Washington, School of Aquatic and Fisheries Sciences, Seattle, Washington, United States of America

Teresa Dias, Cristina Cruz and Maria Amélia Martins-Loução
Centro de Biologia Ambiental, Faculdade de Ciências da Universidade de Lisboa, Lisboa, Portugal

Adelaide Clemente
Centro de Biologia Ambiental, Faculdade de Ciências da Universidade de Lisboa, Lisboa, Portugal
Museu Nacional de História Natural e da Ciência, Jardim Botânico, Lisboa, Portugal

Lucy Sheppard
Centre for Ecology and Hydrology - Edinburgh, Bush Estate, United Kingdom

Roland Bobbink
B-Ware Research Centre, Radboud University Nijmegen, Nijmegen, The Netherlands

Jeremy Lundholm, Amy Heim and Stephanie Tran
Departments of Biology and Environmental Science, Saint Mary's University, Halifax, Nova Scotia, Canada

Tyler Smith
Eastern Cereal and Oilseen Resarch Centre, Agriculture and Agri-Food Canada, Ottawa, Ontario, Canada

Hanna Oskarsson and Linda Kumblad
Department of Ecology, Environment and Plant Sciences, Stockholm University, Stockholm, Sweden

Ann-Kristin Eriksson Wiklund
Department of Applied Environmental Science, Stockholm University, Stockholm, Sweden

Gabriela Danielsson
Department of Biochemistry and Biophysics, Stockholm University, Stockholm, Sweden

Gunnar Thorsén
Department of Analytical Chemistry, Stockholm University, Stockholm, Sweden

Antonio Pusceddu, Cristina Gambi and Cinzia Corinaldesi
Dipartimento di Scienze della Vita e dell'Ambiente, Università Politecnica delle Marche, Ancona, Italy

Mariaspina Scopa
Dipartimento di Scienze della Vita e dell'Ambiente, Università Politecnica delle Marche, Ancona, Italy
Istituto Zooprofilattico Sperimentale dell'Abruzzo e del Molise 'G. Caporale', Experimental Regional Centre for Fishery and Acquacolture, Termoli, Italy

Roberto Danovaro
Dipartimento di Scienze della Vita e dell'Ambiente, Università Politecnica delle Marche, Ancona, Italy
Stazione Zoologica Anton Dohrn, Napoli, Italy

Johanna Jacomina Heymans
Scottish Association for Marine Science, Scottish Marine Institute, Oban, Argyll, United Kingdom

Marta Coll
Institute of Marine Science (ICM-CSIC), Passeig Marítim de la Barceloneta, Barcelona, Spain
Ecopath International Initiative Research Association, Barcelona, Spain

Simone Libralato
Department of Oceanography, OGS (Istituto Nazionale di Oceanografia e di Geofisica Sperimentale), Borgo Grotta Gigante, Sgonico, Zgonik (TS), Italy

Lyne Morissette
M - Expertise Marine, Sainte-Luce, Quebec, Canada

Villy Christensen
Fisheries Centre, University of British Columbia, Vancouver, Canada

Yidong Wang
Tianjin Key Laboratory of Water Resources and Environment, Tianjin Normal University, Tianjin, China

Mingjie Xu
Qianyanzhou Ecological Station, Key Laboratory of Ecosystem Network Observation and Modeling, Institute of Geographic Sciences and Natural Resources Research, Chinese Academy of Sciences, Beijing, China

University of Chinese Academy of Sciences, Beijing, China

Xuefa Wen, Huimin Wang, Xiaoqin Dai, Xiaoli Fu, Yunfen Liu, Xiaomin Sun and Guirui Yu
Qianyanzhou Ecological Station, Key Laboratory of Ecosystem Network Observation and Modeling, Institute of Geographic Sciences and Natural Resources Research, Chinese Academy of Sciences, Beijing, China

Wenjiang Zhang
State Key Laboratory of Hydraulics and Mountain River Engineering, Sichuan University, Chengdu, China

Jie Song
Department of Geography, Northern Illinois University, DeKalb, Illinois, United States of America

Alexander K. Fremier
School of the Environment, Washington State University, Pullman, Washington, United States of America

Evan H. Girvetz
International Center for Tropical Agriculture, Nairobi, Kenya
School of Environmental and Forest Sciences, University of Washington, Seattle, Washington, United States of America

Steven E. Greco and Eric W. Larsen
Department of Human Ecology, University of California Davis, Davis, California, United States of America

Tianjiang Chu, Qiang Sheng, Sikai Wang and Jihua Wu
Ministry of Education Key Laboratory for Biodiversity Science and Ecological Engineering, and Institute of Biodiversity Science, Fudan University, Shanghai, China

Daniel J. Hocking and Kimberly J. Babbitt
Department of Natural Resources and the Environment, University of New Hampshire, Durham, New Hampshire, United States of America

Miao Liu and Li Gong
State Key Laboratory of Urban and Regional Ecology, Research Center for Eco-environmental Sciences, Chinese Academy of Sciences, Beijing, China
University of Chinese Academy of Sciences, Beijing, China

Guohua Liu
State Key Laboratory of Urban and Regional Ecology, Research Center for Eco-environmental Sciences, Chinese Academy of Sciences, Beijing, China

Dongbo Wang
Monitoring Station for Eco-environment of Hulunbuir, Hulunbuir, China

Jian Sun
Key Laboratory of Ecosystem Network Observation and Modeling, Institute of Geographic Sciences and Natural Resources Research, Chinese Academy of Sciences, Beijing, China

Jiang Wang, Tong Chen, Yi Bai, Bao Ying Qian, Chong Bang Zhang
School of Life Science, Taizhou University, Linhai, China

Yuan Ge
Bren School of Environmental Science and Management/Earth Research Institute, University of California Santa Barbara, Santa Barbara, California, United States of America

Conrad Schittko, Mahmoud Hawa and Susanne Wurst
Freie Universität Berlin, Functional Biodiversity, Dahlem Center of Plant Sciences, Berlin, Germany

Niels Jobstvogt
Oceanlab, University of Aberdeen, Aberdeen, United Kingdom
Aberdeen Centre for Environmental Sustainability (ACES), University of Aberdeen, Aberdeen, United Kingdom

Michael Townsend
National Institute of Water and Atmospheric Research (NIWA), Hamilton, New Zealand

Ursula Witte
Oceanlab, University of Aberdeen, Aberdeen, United Kingdom

Nick Hanley
Department of Geography and Sustainable Development, University of St Andrews, St Andrews, United Kingdom

Xuan Fang and Guoan Tang
Key Laboratory of Virtual Geographic Environment, Ministry of Education, School of Geography Science, Nanjing Normal University, Nanjing, China

Bicheng Li
Research Center of Soil and Water Conservation and Ecological Environment, Chinese Academy of Sciences, Yangling, Shaanxi, China

Ruiming Han
School of Geography Science, Nanjing Normal University, Nanjing, China

Chelsea Arnold, Teamrat A. Ghezzehei and Asmeret Asefaw Berhe
School of Natural Sciences, University of California Merced, Atwater, California, United States of America

Beatrice Nervo, Claudia Tocco, Enrico Caprio, Claudia Palestrini and Antonio Rolando
Department of Life Sciences and Systems Biology, University of Torino, Torino, Italy

Nibedita Mukherjee and Farid Dahdouh-Guebas
Laboratory of Systems Ecology and Resource Management, Université Libre de Bruxelles, Brussels, Belgium
Laboratory of Plant Biology and Nature Management, Vrije Universiteit Brussel, Brussels, Belgium

William J. Sutherland and Lynn Dicks
Conservation Science Group, Department of Zoology, University of Cambridge, Cambridge, England, United Kingdom

Nico Koedam
Laboratory of Plant Biology and Nature Management, Vrije Universiteit Brussel, Brussels, Belgium

Jean Hugé
Laboratory of Systems Ecology and Resource Management, Université Libre de Bruxelles, Brussels, Belgium
Centre for Sustainable Development, Ghent University, Gent, Belgium

Eugen Egorov and Martin Brändle
Department of Ecology, Faculty of Biology, Philipps Universität, Marburg, Germany

Walter Durka and Stefan Michalski
Department Community Ecology, Helmholtz Centre for Environmental Research, Halle, Germany

Markus Fischer, Barbara Schmitt, Daniel Prati and Stefan Blaser
Institute of Plant Sciences, University of Bern, Bern, Switzerland

Index